Emergi
Nanotechnologie
in Dentistr

Emerging Nanotechnologies in Dentistry

Second Edition

Edited by

Karthikeyan Subramani
Roseman University of Health Sciences, Henderson, NV, United States

Waqar Ahmed
University of Lincoln, Lincoln, United Kingdom

ELSEVIER

Elsevier
Radarweg 29, PO Box 211, 1000 AE Amsterdam, Netherlands
The Boulevard, Langford Lane, Kidlington, Oxford OX5 1GB, United Kingdom
50 Hampshire Street, 5th Floor, Cambridge, MA 02139, United States

British Library Cataloguing-in-Publication Data
A catalogue record for this book is available from the British Library

Library of Congress Cataloging-in-Publication Data
A catalog record for this book is available from the Library of Congress

ISBN: 978-0-12-812291-4

For Information on all Elsevier publications
visit our website at https://www.elsevier.com/books-and-journals

Publisher: Matthew Deans
Acquisition Editor: Simon Holt
Editorial Project Manager: Sabrina Webber
Production Project Manager: Sruthi Satheesh
Cover Designer: Greg Harris

Typeset by MPS Limited, Chennai, India

Dedication

Every good and perfect gift is from above, coming down from the Almighty. Thanks be to God for his blessings and this wonderful life!

I would like to dedicate this book to my wife, kids, and my parents who have been of immense moral support and encouragement toward editing the second edition of this book. This is a special moment to remember and thank all my teachers, research mentors, and professors who have been the guiding light throughout my life.

Karthikeyan Subramani

I would like to dedicate this book to my father and mother for their unconditional love, encouragement, and inspiration. Special thanks goes to my teachers and Professor Michael Hitchman, my PhD supervisor, for setting such high standards for me to aspire to and my children Adam, Usman, Omar, and Aisha who have made my life a joy. Lastly and most importantly I would like to thank Rihana for her constant love and support.

Waqar Ahmed

Contents

List of Contributors xix
Foreword xxiii
Acknowledgments xxv

CHAPTER 1 Nanotechnology and its applications in dentistry—An introduction 1
Karthikeyan Subramani and Waqar Ahmed
1.1 Introduction 1
1.2 Nanotechnology Approaches 2
1.3 Nanotechnology to Nanomanufacturing 3
1.3.1 Top-Down Approach 4
1.3.2 Bottom-Up Approach 6
1.4 Nanodentistry 11
1.5 Future Directions and Conclusions 14
References 14

CHAPTER 2 Nanoparticles for dental materials: Synthesis, analysis, and applications 17
Sumita B. Mitra
2.1 Introduction: Why Use Nanoparticles? 18
2.2 Synthesis of Nanoparticles 19
2.2.1 Synthesis by Mechanical Attrition 19
2.2.2 Synthesis Through Sol–Gel Process 20
2.2.3 Synthesis of Silsesquioxane Nanoparticles 22
2.2.4 Synthesis of Polymer-Templated Nanoparticles 23
2.3 Examples of Dental Materials Using Nanoparticles 23
2.3.1 Nanocomposites Containing Oxide Nanoparticles 23
2.3.2 Silsequioxane-Based Composites 27
2.3.3 Calcium Phosphate and Calcium Fluoride Nanoparticles-Based Composites 28
2.3.4 Nanoparticles in Glass Ionomer Systems 28
2.3.5 Nanotechnology in Dental Adhesives 30
2.4 Selected Properties of Dental Materials Containing Nanoparticles 30
2.4.1 Optical Properties 30
2.4.2 Wear Properties 32

2.4.3 From B.D. Craig, S.B. Mitra, G.A. Kobussen, M.C. Doruff, H.L. Lechuga, M.R. Atkinson, Polish Retention Comparison of Experimental and Commercial Restorative Composite Materials, J. Dent. Res. 88, (2009) (Spec Issue A Abstract 1506). Mechanical Properties 33
2.5 Clinical Experience With Dental Materials Containing Nanoparticles 35
2.6 Conclusions 35
References 36

CHAPTER 3 Antimicrobial nanoparticles in restorative composites 41
Nurit Beyth, Ira Yudovin-Farber, Arijit Basu, Ervin I. Weiss and Abraham J. Domb
3.1 Introduction 41
3.2 Antibacterial Restorative Composites 42
3.2.1 Filler Phase Modification 43
3.2.2 Matrix Phase Modification 44
3.3 Antimicrobial Macromolecules 45
3.3.1 Polycationic Disinfectants 46
3.3.2 Polyethyleneimine 46
3.4 Nanoparticles 47
3.4.1 Polyethylenimine Nanoparticles 48
3.5 Conclusions 53
References 54

CHAPTER 4 Nanotechnology in operative dentistry: A perspective approach of history, mechanical behavior, and clinical application 59
Eduardo G. Mota and Karthikeyan Subramani
4.1 Introduction 60
4.2 Historical Review: Nanotechnology Applications in Operative Dentistry 60
4.3 Biomimetics 61
4.4 Nanotechnology in CAD/CAM 63
4.5 Fillers in Composite Resins 65
4.6 SEM and EDS Evaluations 66
4.7 Filler Weight Content (wt%) 67
4.8 Water Sorption 67

4.9 Mechanical Behavior....69
4.9.1 Compressive Strength....70
4.9.2 Diametral Tensile Strength....71
4.9.3 Flexural Strength and Flexural Modulus....72
4.9.4 Microhardness....73
4.9.5 Nanohardness....74
4.9.6 Wear Resistance....75
4.10 Clinical Application....79
4.11 Conclusions....81
Acknowledgments....81
References....81

CHAPTER 5 Impact of nanotechnology on dental implants....83
Karthikeyan Subramani, Sandrine Lavenus, Julie Rozé, Guy Louarn and Pierre Layrolle
5.1 Introduction....83
5.2 Nanoscale Surface Modifications....86
5.3 Interactions of Surface Dental Implants With Blood....87
5.4 Interactions Between Surfaces and MSCs....89
5.4.1 Origin of MSCs....89
5.4.2 Migration, Adhesion, and Proliferation....89
5.4.3 Differentiation....90
5.5 Tissue Integration....91
5.6 Conclusion....93
Acknowledgments....93
References....93

CHAPTER 6 Titanium surface modification techniques for dental implants—From microscale to nanoscale....99
Karthikeyan Subramani, Reji T. Mathew and Preeti Pachauri
6.1 Introduction....99
6.2 Titanium Surface Modification Methods....100
6.2.1 Mechanical Modification of Titanium Surface....101
6.2.2 Physicochemical Modification of Titanium Surface....101
6.2.3 Biochemical Modification of Titanium Surface....104
6.2.4 Physical Modification of Titanium Surface....114
6.3 Recent Techniques....115
6.3.1 Discrete Crystalline Deposition (DCD)....115
6.3.2 Laser Ablation....115

6.3.3 Titanium Oxide Blasted and Acid-Etched Implants 115
6.3.4 Photofunctionalization 115
6.4 Limitations & Conclusion 116
Acknowledgments 116
References 117

CHAPTER 7 Titanium nanotubes as carriers of osteogenic growth factors and antibacterial drugs for applications in dental implantology 125
Karthikeyan Subramani, Waqar Ahmed and Preeti Pachauri
7.1 Introduction 125
7.2 Titanium Nanotubes 126
7.3 TiO_2 Nanotubes for Implant Fabrication 127
7.4 Functionalization of TiO_2 Nanotubes with Growth Factors and Antibacterial/Antiinflammatory Drugs 129
7.5 Recent Advancements 132
7.6 Conclusions 132
References 133

CHAPTER 8 Cellular responses to nanoscale surface modifications of titanium implants for dentistry and bone tissue engineering applications 137
Kaushik Subramanian, Daniel Tran and Kytai T. Nguyen
8.1 Introduction 138
8.2 Nanotopography Generated from Surface Modification of Ti Implants 138
8.2.1 Surface Modification of Ti Implants With Inorganic Materials/Nanoparticles 139
8.2.2 Surface Modifications of Ti Implants With Polymers 141
8.3 Nanotopography and Protein Absorption 142
8.4 Nanotopography Alters Osteoblast Responses 143
8.4.1 Cell Morphology 149
8.4.2 Cell Adhesion 149
8.4.3 Cell Proliferation 149
8.4.4 Bioactive Molecules 150
8.4.5 Osseointegration 150

8.5 Nanotopography and Stem Cell Responses 151
8.5.1 Effects of Nanotopography on Endothelial Progenitor Cells 151
8.5.2 Effects of Nanotopography on Bone Marrow Stem Cells 152
8.6 Conclusions 153
References 153

CHAPTER 9 Corrosion resistance of Ti–6Al–4V with nanostructured TiO_2 coatings **165**
Qian Chen, Gerald D. McEwen, Nikita Zaveri, Ramji Karpagavalli and Anhong Zhou
9.1 Introduction 165
9.1.1 SiO_2–CaO Coatings on Ti–6Al–4V Alloys 166
9.1.2 SiO_2 and SiO_2–TiO_2 Intermediate Coatings on Titanium and Ti–6Al–4V Alloy 167
9.1.3 Coated Hydroxyapatite on Ti–6Al–4V by Electrophoretic Deposition 167
9.1.4 Double-Layer Glass–Ceramic Coatings on Ti–6Al–4V 168
9.2 Nanostructured TiO_2 Deposited on Ti–6Al–4V 168
9.2.1 Preparation of the Ti–6Al–4V Electrode 168
9.2.2 TiO_2 Nanoparticles Coating 169
9.3 Characterization Techniques 169
9.3.1 Scanning Electron Microscopy 169
9.3.2 Raman Microscopy 169
9.4 Corrosion Test With Electrochemical Techniques 173
9.4.1 Open-Circuit Voltage (OCV) and Tafel Analysis 174
9.4.2 Electrochemical Impedance Spectroscopy 175
9.5 Conclusion 176
References 177

CHAPTER 10 Multiwalled Carbon nanotubes/hydroxyapatite nanoparticles incorporated GTR membranes **181**
Qing Cai and Xiaoping Yang
10.1 Introduction 182
10.2 Periodontal Defects and GTR 183
10.2.1 Studies Using Nonresorbable Membranes 183
10.2.2 Studies Using Bioresorbable Membranes 184

10.2.3 Layer-Designed Membranes for GTR 185
10.2.4 Cell-Sheet-Based Technology for GTR 186
10.3 Use of Electrospinning for Preparation of Nanocomposites 187
10.3.1 Electrospinning 187
10.3.2 Carbon Nanotubes Incorporated Into Nanofibers 188
10.3.3 Organic–Inorganic Composite Nanofibers 190
10.4 GTR Membranes Based on Electrospun CNT/HA Nanoparticles Incorporated Composite Nanofibers 192
10.4.1 Fabrication of MWCNTs/HA Hybrids 192
10.4.2 Electrospun Nanofibers With Different Fiber Arrangements 194
10.4.3 Fabrication of PLLA/MWCNTs/HA Composite Nanofibers 195
10.4.4 Characterization of PLLA/MWCNTs/HA Composite Nanofibers 195
10.4.5 Cell Culture on PLLA/MWCNTs/HA Composite Nanofibers Membranes 197
10.4.6 In Vivo Implantation of PLLA/MWCNTs/HA Membranes 199
10.5 Conclusions 200
References 200

CHAPTER 11 Nanoapatitic composite scaffolds for stem cell delivery and bone tissue engineering 211
Hockin H.K. Xu, Mike D. Weir, Liang Zhao, Jennifer L. Moreau, Dwayne D. Arola and Carl G. Simon
11.1 Introduction 211
11.2 Development of Nanoapatitic and Macroporous Scaffolds 213
11.3 Cell Infiltration into Scaffold 215
11.4 Biomimetic Nanoapatite–Collagen Fiber Scaffold 215
11.5 Fast Fracture of Nanoapatite Scaffold 217
11.6 Fatigue of Nanoapatite Scaffold 219
11.7 Nanoapatite Scaffold–Human Umbilical Cord Stem Cell Interactions 220
11.8 Seeding Bone Marrow Stem Cells on Nanoapatite Scaffolds 223
11.9 Conclusions 226
Acknowledgments 226
References 227

CHAPTER 12 Self-assembly of proteins and peptides and their applications in bionanotechnology and dentistry ... 231
Karthikeyan Subramani and Waqar Ahmed
12.1 Introduction ... 231
12.2 Mechanism of Molecular Self-Assembly ... 232
12.3 Classification of Self-Assembly ... 232
12.4 Self-Assembly of Proteins and Peptides ... 233
12.5 Bionanotechnology Applications ... 236
12.6 Peptide Nanofibers, Nanotubes, and Nanowires ... 236
12.7 Three-Dimensional Peptide Matrix Scaffolds ... 242
12.8 Advantages and Limitations of Self-Assembling Peptide Matrix Scaffolds ... 244
12.9 Self-Assembly in Regenerative Biology and Dentistry ... 244
12.10 Conclusions ... 246
References ... 246

CHAPTER 13 Surface engineering of dental tools with diamond for enhanced life and performance ... 251
Waqar Ahmed, Htet Sein, Mark Jackson, Christopher Rego, Israr-Ul Hassan and Karthikeyan Subramani
13.1 Tooth Materials ... 251
13.2 Dental Burs ... 253
13.3 Chemical Vapor Deposition of Diamond Films Onto Dental Burs ... 256
13.3.1 Plasma-Enhanced CVD ... 257
13.3.2 Hot Filament CVD ... 259
13.3.3 Controlling Structure and Morphology ... 263
13.4 Bur Performance Investigations ... 273
13.4.1 Tool Preparation ... 273
13.4.2 CVD Diamond Deposition on the Dental Burs ... 273
13.4.3 Dental Bur Machining: Drilling Experiments ... 273
13.4.4 Dental Bur Machining: Machining Experiments on Human Teeth ... 274
13.4.5 Performance Testing ... 276
13.4.6 Drilling Experiments ... 277
13.4.7 Performance Results ... 278
13.5 Conclusions ... 284
References ... 284
Further Reading ... 288

CHAPTER 14 Nanomechanical characterization of mineralized tissues in the oral cavity **289**
Karthikeyan Subramani, Yee L. Chan, Alfonso H.W. Ngan and Nigel M. King
14.1 Introduction 289
14.2 Basic Data Analysis Protocol for Nanoindentation 290
14.3 Nanoindentation of Oral Mineralized Tissues 292
14.3.1 Sample Preparation 292
14.3.2 Hydration 292
14.3.3 Indenter Tips 294
14.3.4 Load Function and Data Analysis 295
14.3.5 Microstructural Influence 297
14.4 Summary 302
References 303
Further Reading 306

CHAPTER 15 Nanoindentation techniques for the determination of mechanical properties of materials in dentistry **307**
Jaraslov Menčík
15.1 Introduction 307
15.2 Basic Information From the Load–Displacement Curves 309
15.2.1 Hardness and Elastic Modulus 309
15.2.2 Harmonic Contact Stiffness 310
15.2.3 Work of Indentation and Other Information From P–h Curves 311
15.2.4 Indenter Calibration 312
15.3 Characterization of Inelastic Properties 313
15.3.1 Stress–Strain Diagram 313
15.3.2 Yield Stress 315
15.4 Determination of Properties in Nonhomogeneous Bodies 315
15.4.1 Surface Layers and Coatings 315
15.4.2 Multiphase Microstructure 317
15.5 Characterization of Time-Dependent Load Response 317
15.6 Resistance Against Crack Propagation 320
15.7 Scratch Tests for the Evaluation of Friction and Wear Resistance 322
15.8 Devices for Nanoindentation 322
Acknowledgments 323
References 324

CHAPTER 16 Nanocharacterization techniques for dental implant development 327

Dermot Brabazon

16.1 Measurement of the Topology of Nanostructures 327

16.1.1 Field Emission Scanning Electron Microscope 327

16.1.2 Scanning Probe Microscopy 330

16.1.3 Confocal Microscopy and Interferometry 333

16.2 Measurement of Nanostructure Internal Geometries 336

16.2.1 Transmission Electron Microscope 336

16.2.2 Focused Ion Beam 338

16.2.3 X-Ray Diffraction 338

16.2.4 Mercury Porosimetry 340

16.3 Measurement of Composition of Nanostructures 342

16.3.1 Energy Dispersive X-Ray Spectroscopy 342

16.3.2 X-Ray Photoelectron Spectroscopy 343

16.3.3 Secondary Ion Mass Spectroscopy 345

16.3.4 Auger Electron Spectroscopy 346

16.4 Measurement of the Mechanical Properties of Nanostructures 347

16.4.1 Nanoscratch Testing 347

16.4.2 Nanohardness Test 348

16.5 Conclusions 349

References 350

CHAPTER 17 Nanoparticulate drug-delivery systems for oral cancer treatment 355

Karthikeyan Subramani and Waqar Ahmed

17.1 Introduction 355

17.2 Cancer-Treatment Techniques 356

17.3 Mechanism of Action of Chemotherapeutic Agents 357

17.3.1 Prevention of Synthesis of Pre-DNA Molecule Building Blocks 357

17.3.2 Chemical Damage of DNA in the Cell Nuclei 357

17.4 Oral Cancer 358

17.5 TNM Classification of Tumors 359

17.6 Management of Oral Cancer 359

17.7 Nanoparticulate-Based Drug Delivery in Cancer Treatment 360

17.7.1 Gold Nanoparticles for Anticarcinogenic Drug Delivery 361

17.7.2 Liposomes in Oral Cancer Treatment 363

17.7.3 Magnetic Nanoparticles in Oral Cancer Treatment 365
17.7.4 Polymeric micelles as Drug-Delivery Systems 366
17.8 Limitations of Nanoparticles and Conclusion 366
References 367

CHAPTER 18 Carbon nanotubes: Applications in cancer therapy and drug delivery research **371**
Waqar Ahmed, Abdelbary Elhissi, Vinod Dhanak and Karthikeyan Subramani
18.1 Introduction 371
18.2 Cellular Uptake of CNTs 373
18.3 CNTs as Carriers for Small and Large Drug Molecules...........373
18.3.1 CNTs as Carriers of Small Anticancer Molecules 374
18.3.2 CNTs as Carriers of Immunoactive Molecules, Proteins, and Genetic Materials 378
18.3.3 CNTs as Carriers for Antimicrobial Molecules 381
18.3.4 Photothermal Therapy of Cancer Using CNTs 382
18.4 Carbon Nanotubes for Oral Cancer Therapy............................383
18.5 Conclusion 384
References 385

CHAPTER 19 Nanodiagnostics in microbiology and dentistry **391**
Karthikeyan Subramani and Manjula Mehta
19.1 Introduction 391
19.2 Nanomaterials...392
19.2.1 Applications of Nanomaterials 393
19.3 Biomedical Applications of Nanotechnology and Its Limitations...393
19.4 Nanotechnology Applications in Drug-Delivery Systems, Nanodiagnostics, and Various Other Fields 394
19.4.1 Drug-Delivery System 394
19.4.2 Nanodiagnostics and Disease Prevention 395
19.4.3 Disease Prevention 397
19.4.4 Other Applications 399
19.5 Contribution of Microbiology to Nanotechnology...................402
19.6 AFM Imaging of Microorganisms 403
19.6.1 Yeast 403
19.6.2 Bacteria 404

19.6.3 AFM Study of the Structure–Function Relationship of the Biofilm-Forming Bacterium *Streptococcus mutans* ... 407
19.6.4 Viruses ... 407
19.7 Nanoplasmonic Sensors Detecting Live Viruses ... 407
19.8 Nanodentistry ... 410
19.8.1 The Impact of Nanotechnology ... 411
19.8.2 Nanotechnology in Periodontics ... 414
19.9 Conclusions ... 415
References ... 416

CHAPTER 20 Neurotoxicity of nanomaterials ... 421
Feng Xiaoli and Shao Longquan
20.1 Introduction ... 421
20.2 Possible Dental Nanomaterials ... 423
20.2.1 Composite Resins and Bonding Systems ... 423
20.2.2 Root-Filling Materials ... 423
20.2.3 Bioceramics and Associated Dental Prosthesis ... 426
20.2.4 Surface Modifications for Dental Implants ... 426
20.2.5 Target-Delivery and Imaging in Tumor Chemotherapy ... 426
20.3 Neurotoxicity of Dental Nanomaterials ... 427
20.3.1 Possible Pathways into the CNS ... 427
20.3.2 In Vivo Toxicity Studies ... 429
20.3.3 In Vitro Toxicity Studies ... 430
20.3.4 Major Toxicity Mechanisms ... 435
20.4 Future Research Prospects ... 437
20.5 Conclusion ... 438
Acknowledgments ... 438
References ... 438

Index ... 445

List of Contributors

Waqar Ahmed
University of Lincoln, Lincoln, United Kingdom

Dwayne D. Arola
University of Maryland, Baltimore, MD, United States

Arijit Basu
The Hebrew University of Jerusalem, Jerusalem, Israel; Koch Institute for Integrative Cancer Research, Cambridge, MA, United States; Massachusetts Institute of Technology, Cambridge, MA, United States

Nurit Beyth
The Hebrew University of Jerusalem, Jerusalem, Israel

Dermot Brabazon
Dublin City University, Dublin, Ireland

Qing Cai
Beijing University of Chemical Technology, Beijing, China

Yee L. Chan
The University of Hong Kong, Pokfulam, Hong Kong

Qian Chen
Utah State University, Logan, UT, United States

Vinod Dhanak
University of Liverpool, Liverpool, United Kingdom

Abraham J. Domb
The Hebrew University of Jerusalem, Jerusalem, Israel

Abdelbary Elhissi
Qatar University, Doha, Qatar

Israr-Ul Hassan
Dhofar University, Salalah, Oman

Mark Jackson
Purdue University, West Lafayette, IN, United States; Kansas State University, Salina, KS, United States

Ramji Karpagavalli
Utah State University, Logan, UT, United States

Nigel M. King
The University of Hong Kong, Pokfulam, Hong Kong

Sandrine Lavenus
University of Nantes, Nantes, France

Pierre Layrolle
University of Nantes, Nantes, France

Shao Longquan
Southern Medical University, Guangzhou, China

Guy Louarn
University of Nantes, Nantes, France

Reji T. Mathew
Midwestern University, Downers Grove, IL, United States

Gerald D. McEwen
Utah State University, Logan, UT, United States

Manjula Mehta
Panjab University, Chandigarh, India

Jaraslov Menčík
University of Pardubice, Pardubice, Czech Republic

Sumita B. Mitra
Mitra Chemical Consulting LLC, St. Pete Beach, FL, United States

Jennifer L. Moreau
University of Maryland, Baltimore, MD, United States

Eduardo G. Mota
Pontifical Catholic University of Rio Grande do Sul (PUCRS), Porto Alegre, RS, Brazil

Alfonso H.W. Ngan
The University of Hong Kong, Pokfulam, Hong Kong

Kytai T. Nguyen
University of Texas at Arlington, Arlington, TX, United States; University of Texas, Dallas, TX, United States

Preeti Pachauri
Rama Dental College, Hospital & Research Center, Kanpur, India

Christopher Rego
Manchester Metropolitan University, Manchester, United Kingdom

Julie Rozé
University of Nantes, Nantes, France

Htet Sein
Manchester Metropolitan University, Manchester, United Kingdom

Carl G. Simon
National Institute of Standards and Technology, Gaithersburg, MD, United States

Karthikeyan Subramani
Roseman University of Health Sciences, Henderson, NV, United States

Kaushik Subramanian
University of Texas at Arlington, Arlington, TX, United States; University of Texas, Dallas, TX, United States

Daniel Tran
University of Texas at Arlington, Arlington, TX, United States; University of Texas, Dallas, TX, United States

Mike D. Weir
University of Maryland, Baltimore, MD, United States

Ervin I. Weiss
The Hebrew University of Jerusalem, Jerusalem, Israel

Feng Xiaoli
Southern Medical University, Guangzhou, China

Hockin H.K. Xu
University of Maryland, Baltimore, MD, United States

Xiaoping Yang
Beijing University of Chemical Technology, Beijing, China

Ira Yudovin-Farber
The Hebrew University of Jerusalem, Jerusalem, Israel

Nikita Zaveri
Utah State University, Logan, UT, United States

Liang Zhao
University of Maryland, Baltimore, MD, United States

Anhong Zhou
Utah State University, Logan, UT, United States

Foreword

Paradigm-shifting applications are an exciting prospect if the novel properties, particularly involving quantum mechanical behavior, that nanoscale structures exhibit can be controlled and harnessed. New functional molecule-based devices are already being perfected and in the last few years there has been significant progress in the control of the fabrication methods for nanoscale architectures leading to operational quantum-based devices. The next decade should see an explosion in the number of new products based on intelligent nanosystems with new physicochemical properties at reasonable cost and with wide availability. Synergistic collaboration in marrying fundamental breakthroughs with strategic applications will only be possible as long as governments recognize that we must nurture carefully the fragile academic environments necessary for basic research to flourish and devise nonperturbing but effective conduits for the development of strategic applications. Maintaining this clear perspective on basic experimental/theoretical research and applications development will be vital for fundamental breakthroughs and ensuring the birth of the groundbreaking technologies that result in new products.

Nanoscience and Nanotechnology (N&N) has several aspects but the crucial one can be simply stated as "Atom by Atom, Molecule by Molecule Assembly of a Complex and/or Functional System." By this definition we recognize that N&N is essentially the art of chemical synthesis of advanced systems. This is the so-called "Bottom-up" approach to the creation of devices—indeed it is the "Chemistry of the 21st Century." Progress will require enhanced ability to work at atomic, molecular, supramolecular levels (scale 1–100 nm) in order to understand, create, and use material structures, devices, and systems with fundamentally new and desirable properties that only occur in tiny structures. The traditional "Top-down" approach has required the development of brilliant techniques to carve ever smaller elements with micron and near-nanoscale dimensions from macroscopic materials utilizing ever more accurate machining and/or etching techniques. "Top-down" approaches, such as photolithography, have been spectacularly successful in the semiconductor industry resulting in, for instance, the fabrication of the minute components in integrated circuits.

"Bottom-up" chemical (N&N) approaches possess the intrinsic ability to oust the top-down approach, but this will only happen if we can improve considerably our control of atomic and molecular self-organization and self-assembly. There is no obvious fundamental reason why this cannot be achieved—though it may not be easy even in the short term. N&N has much more to offer than just simple miniaturization down to molecular and atomic scales. By investigating and understanding the functionality of materials at the nanoscale dimensions, it should be possible to optimize functional efficiency, minimize energy consumption, and reduce considerably the use of resources. Extensive research is being carried out worldwide to understand the advantages and limitations of N&N and its applications in a wide range of disciplines from material science, biomedical research to space research. In the field of medicine, N&N advances are beginning to be used extensively, for instance in nanoparticle-based drug delivery, as well as diagnostics, tissue engineering, and biosensing. In dentistry, ongoing research is exploring the applications of N&N in dental biomaterials, dental implantology, dental instruments, and nanoparticles/scaffolds for bone regeneration around dental implants and maxillofacial region. This book is the first of its kind bridging the fundamental research and applications of N&N in the field of dentistry and is thus a valuable and timely contribution of great benefit to experts and novices alike.

The authors of the various chapters come from a number of countries including the United States, the United Kingdom, Israel, India, Brazil, France, China, Taiwan, Oman, Hong Kong, and Czech Republic and represent many laboratories in both academia and industry which are in the forefront of biomedical/dental N&N research. It brings together experts who are working in this cross-disciplinary field, bridging nanotechnology with biomedical/dental applications such as tissue engineering, dental and orthopedic implantology, dental materials, biotechnology and cell biology, and clinicians in dentistry. Recent advances in the applications of N&N in dental biomaterials, dental implantology, bone tissue engineering and regeneration, operative dental tools manufacturing, characterization of mineralized tissues in the oral cavity at the nanoscale level, carbon nanotubes and nanoparticle-based drug delivery for the treatment of oral cancer are all addressed in this book. It also focuses, in depth, on the manufacturing processes involved with specific emphasis on preclinical and clinical applications of nanomaterials for applications in dentistry.

This book is an extremely valuable and timely source of information for researchers and academicians working in the fields of dentistry, nanotechnology, oral biology, biomaterials, oral microbiology, biomedical nanotechnology, and tissue engineering.

Professor Sir Harold W. Kroto

FRS Nobel Laureate, Chemistry 1996

Acknowledgments

Nanotechnology is a broad multidisciplinary subject and to write a comprehensive book, focusing on its applications in dentistry is an enormous challenge. It required valuable contributions from a group of highly talented and dedicated experts. We were very fortunate to be able to call on such specialists to update and write comprehensive chapters for the second edition. During the 3-year quest of editing the first edition of this book, we were inspired by the achievements and support of Prof. Sir Harold Kroto FRS who was kind enough to write the Foreword for this book.

We are grateful to the chapter authors and this book is a sum and collective effort of all their expertise. Therefore, we would like to thank the following contributors for their valuable time and hard work: Dr. S.B. Mitra, Dr. N. Beyth, Dr. I. Yudovin-Farber, Dr. E.I. Weiss, Prof. A.J. Domb, Dr. E.G. Mota, Dr. S. Lavenus, Dr. J. Rozé, Dr. G. Louarn, Prof. P. Layrolle, Dr. R.T. Mathew, Dr. P. Pachauri, Dr. K. Subramanian, Dr. D. Tran, Prof. K.T. Nguyen, Dr. Q. Chen, Dr. G.D. McEwen, Dr. N. Zaveri, Dr. R. Karpagavalli, Prof. A. Zhou, Dr. Q. Cai, Dr. X. Yang, Prof. H.H.K. Xu, Dr. M.D. Weir, Dr. L. Zhao, Dr. J.L. Moreau, Dr. D.D. Arola, Dr. C.G. Simon, Dr. H. Sein, Dr. M. Jackson, Dr. C. Rego, Dr. I.U. Hassan, Dr. J. Yazdani, Dr. Y.L. Chan, Dr. A.H.W. Ngan, Prof. N.M. King, Prof. J. Mencík, Prof. D. Brabazon, Dr. A. Elhissi, Dr. V.R. Dhanak, Prof. M. Mehta, Dr. F. Xiaoli, and Dr. S. Longquan.

We are thankful to the production team from Elsevier to bring this book in its finest form and quality.

We hope that this book inspires our readers to explore the interdisciplinary applications of Nanotechnology in Dentistry and that they find this work useful.

Karthikeyan Subramani and Waqar Ahmed

CHAPTER

1 Nanotechnology and its applications in dentistry—An introduction

Karthikeyan Subramani[1] and Waqar Ahmed[2]

[1]*Roseman University of Health Sciences, Henderson, NV, United States*

[2]*University of Lincoln, Lincoln, United Kingdom*

CHAPTER OUTLINE

1.1 Introduction 1
1.2 Nanotechnology Approaches 2
1.3 Nanotechnology to Nanomanufacturing 3
1.3.1 Top-Down Approach 4
1.3.2 Bottom-Up Approach 6
1.4 Nanodentistry 11
1.5 Future Directions and Conclusions 14
References 14

1.1 INTRODUCTION

Although nanotechnology has been around since the beginning of time, the discovery of nanotechnology has been widely attributed to the American Physicist and Nobel Laureate Dr. Richard Phillips Feynman [1] who presented a paper called *There's Plenty of Room at the Bottom* in December 29, 1959 at the annual meeting of the American Physical Society meeting at California Institute of Technology. Feynman talked about the storage of information on a very small scale, writing and reading in atoms, about miniaturization of the computer, building tiny machines, tiny factories, and electronic circuits with atoms. He stated, "In the year 2000, when they look back at this age, they will wonder why it was not until the year 1960 that anybody began seriously to move in this direction." However, he did not specifically use the term "nanotechnology." The first use of the word "nanotechnology" has been attributed to Taniguchi [2] in a paper published in 1974 "On the Basic Concept of NanoTechnology." Dr. K. Eric Drexler, an MIT graduate, later took Feynman's concept of a billion tiny factories and added the idea that they could make more copies of themselves, via computer control instead of control by a human operator, in his 1986 book *Engines of*

Emerging Nanotechnologies in Dentistry. DOI: http://dx.doi.org/10.1016/B978-0-12-812291-4.00001-7

Creation: The Coming Era of Nanotechnology, to popularize the potential of nanotechnology.

Since then several definitions of nanotechnology have evolved. For example, the dictionary (Merriam Webster dictionary 2010) definition states that nanotechnology is "the art of manipulating materials on an atomic or molecular scale especially to build microscopic devices." Other definitions include the US government (US Government www.nano.gov) which state, "Nanotechnology is research and technology development at the atomic, molecular, or macromolecular level in the length scale of approximately 1–100 nm range, to provide a fundamental understanding of phenomena and materials at the nanoscale and to create and use structures, devices, and systems that have novel properties and functions because of their small and/or intermediate size." The Japanese [K. Shimizu, INC 2, USA (2006)] have come up with a more focused and succinct definition. "True Nano": as nanotechnology which is expected to cause scientific or technological quantum jumps, or to provide great industrial applications by using phenomena and characteristics peculiar in nanolevel.

Regardless of the definition that is used, it is evident that the properties of matter are controlled at a scale between 1 and 100 nm. For example, chemical properties take advantage of large surface-to-volume ratio for catalysis, interfacial and surface chemistry is important in many applications. Mechanical properties involve improved strength hardness in light-weight nanocomposites and nanomaterials, altered bending, compression properties, nanomechanics of molecular structures. Optical properties involve absorption and fluorescence of nanocrystals, single photon phenomena, and photonic band-gap engineering. Fluidic properties give rise to enhanced flow using nanoparticles and nanoscale adsorbed films are also important.

1.2 NANOTECHNOLOGY APPROACHES

Numerous approaches have been utilized successfully in nanotechnology, and as the technology develops further, approaches may emerge. The approaches employed thus far have generally been dictated by the technology available and the background experience of the researchers involved. Nanotechnology is a truly multidisciplinary field involving chemistry, physics, biology, engineering, electronics, social sciences, etc., which need to be integrated together in order to generate the next level of development in nanotechnology (Fig. 1.1). Fuel cells, mechanically stronger materials, nanobiological devices, molecular electronics, quantum devices, carbon nanotubes, etc. have been made using nanotechnology. Even social scientists are debating ethical use of nanotechnology.

The two main approaches in order to explain nanotechnology to the general public have been oversimplified and have become known as the "top-down" approach. This involves fabrication of device structures via monolithic processing

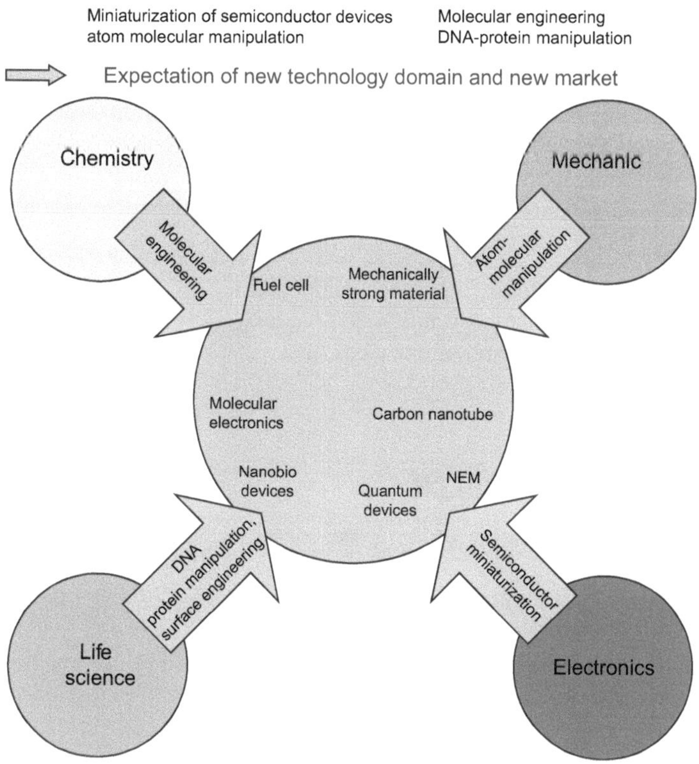

FIGURE 1.1

Multidisciplinary nature of nanotechnology.

on the nanoscale. This approach has been used with spectacular success in the semiconductor devices used in consumer electronics. The "bottom-up" approach involves the fabrication of device structures via systematic assembly of atoms, molecules, or other basic units of matter. This is the approach nature uses to repair cells, tissues, and organ systems in living things and indeed for life processes such as protein synthesis. Tools are evolving which will give scientists more control over the synthesis and characterization of novel nanostructures yielding a range of new products in the near future.

1.3 NANOTECHNOLOGY TO NANOMANUFACTURING

A huge amount of research is being carried out internationally, and governments and research organizations are spending large amounts of money and human resources into nanotechnology. This has generated interesting scientific output and potential commercial applications, some of which have been translated into products produced on a large scale. However, in order to realize commercial benefits far more from lab scale applications need to be commercialized and for that to happen nanotechnology needs to enter the realm of nanomanufacturing. This involves using the technologies available to produce products on a large scale which is economically viable. Regardless of whether a top-down or bottom-up approach is used, a nanomanufacturing/nanofabrication technology should

- be capable of producing components with nanometer precision,
- be able to create systems from these components,
- be able to produce many systems simultaneously,
- be able to structure in three dimensions, and
- be cost-effective.

1.3.1 TOP-DOWN APPROACH

The most successful industry utilizing the top-down approach is the electronics industry (Fig. 1.2).

This industry is utilizing techniques involving a range of technologies such as chemical vapor deposition (CVD), physical vapor deposition, lithography (photolithography, electron beam, and X-ray lithography), wet and plasma etching, and so on to generate functional structures at the micro- and nanoscale (Fig. 1.3).

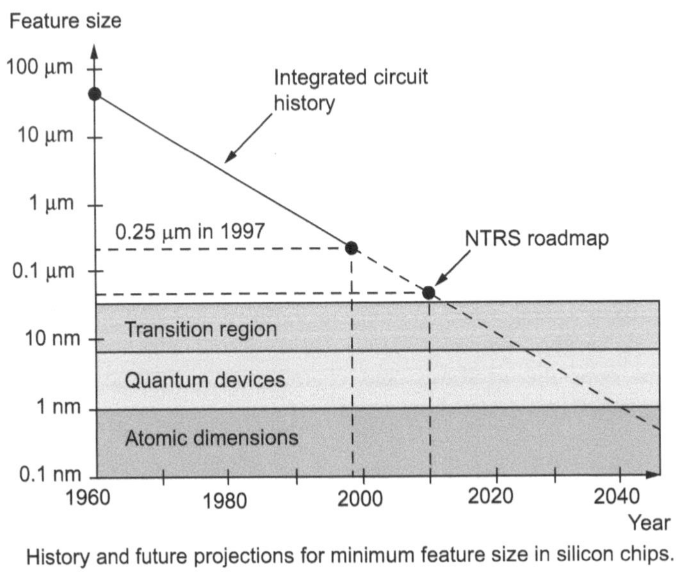

FIGURE 1.2

Features size evolution in silicon chips.

Evolution and development of these technologies have allowed emergence of numerous electronic products and devices that have enhanced the quality of life throughout the world. The feature sizes have shrunk continuously from about 75 μm to below 100 nm. This has been achieved by improvements in deposition technology and more importantly due to the development of lithographic techniques and equipment such as X-ray lithography and electron beam lithography.

Techniques such as electron beam lithography, X-ray lithography, and ion beam lithography have advantages in terms of resolution achieved; however, there are disadvantages associated with cost, "optics," and detrimental effects on the substrate. These methods are currently under investigation to improve upon current lithographic process used in the integrated circuits industry. With continuous developments in these technologies, it is highly likely that the transition from microtechnology to nanotechnology will generate a whole new generation of exciting products and features.

A demonstration of how several techniques can be combined together to form a "nano" wine glass (Fig. 1.4). In this example, a focused ion beam and CVD have been employed to produce this striking nanostructure.

The top-down approach is being used to coat various coatings to give improved functionality. For example, vascular stents are being coated using CVD technology with ultrathin diamond-like carbon coatings in order to improve biocompatibility and blood flow (Fig. 1.5). Graded a-Si × Cy:H interfacial layers result in greatly reduced cracking, enhanced adhesion.

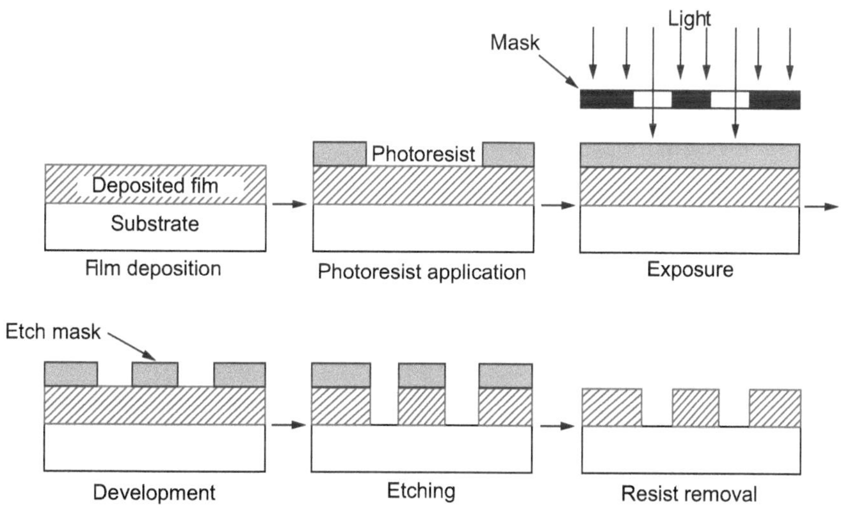

FIGURE 1.3

A typical process sequence employed in the electronics industry to generate functional devices at the micro- and nanoscale [3].

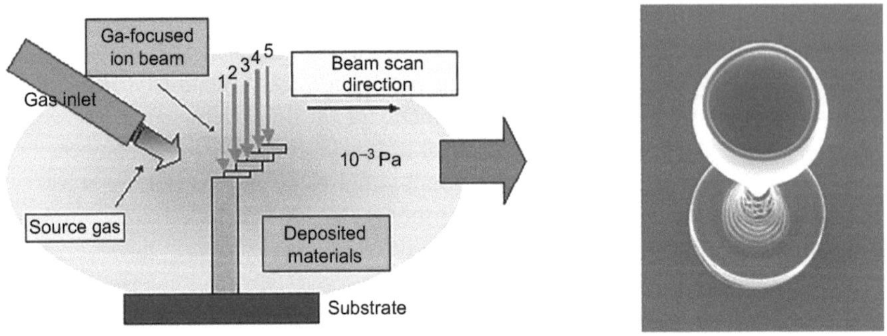

FIGURE 1.4

Demonstration of three-dimensional nanostructure fabrication [4].

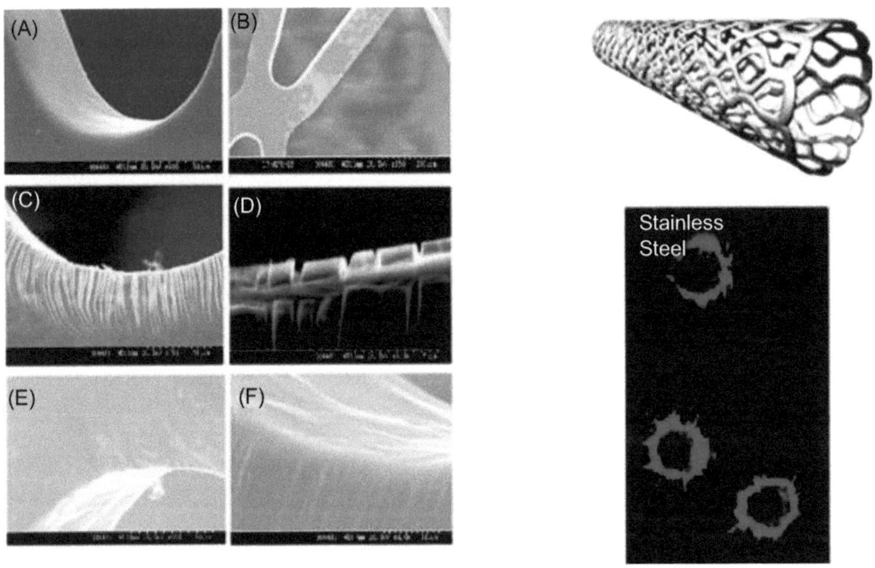

FIGURE 1.5

Examples of stents coated with diamond-like carbon using plasma-enhanced CVD.

Dr. T. Okpalugo, University of Paisley (2007) Private communication.

1.3.2 BOTTOM-UP APPROACH

The bottom-up approach involves making nanostructures and devices by arranging atom by atom. The scanning tunneling microscope (STM) has been used to build nanosized atomic features such as the letters IBM written using xenon atoms on nickel [5] (Fig. 1.6). While this is beautiful and exciting, it remains that the experiment was carried out under carefully controlled conditions, i.e., liquid

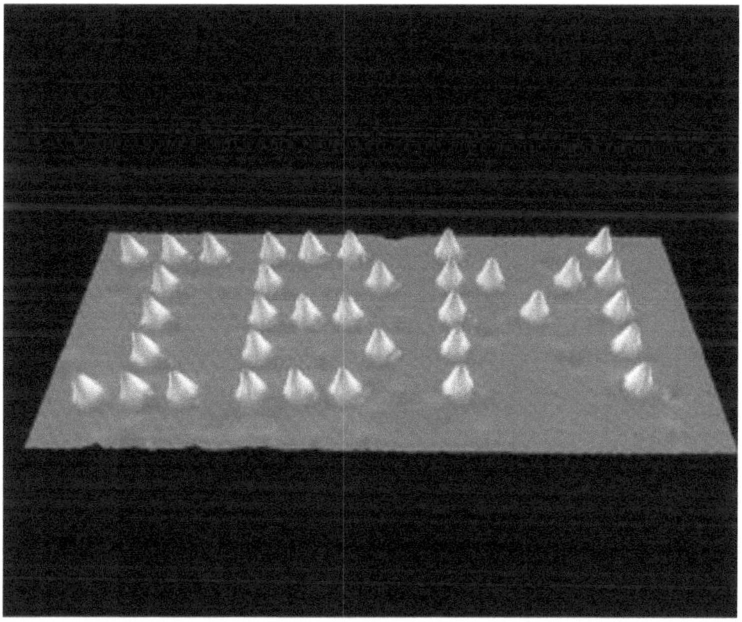

FIGURE 1.6

Positioning single atoms with a scanning tunneling microscope [5].

helium cooling, high vacuum, and it took something like 24 h to get the letters right. Also, the atoms are not bonded to the surface just adsorbed and a small change in temperature or pressure will dislodge them. Since this demonstration, significant advances have been made in nanomanufacturing.

The discovery of the STM's ability to image variations in the density distribution of surface-state electrons created a compulsion to have complete control of not only the atomic landscape but also the electronic landscape [6]. Here they have positioned 48 iron atoms into a circular ring in order to "corral" some surface-state electrons and force them into "quantum" states of the circular structure (Fig. 1.7). The ripples in the ring of atoms are the density distribution of a particular set of quantum states of the corral. The artists were delighted to discover that they could predict what goes on in the corral by solving the classic eigenvalue problem in quantum mechanics—a particle in a hard-wall box.

Probably the most publicized material in recent years has been carbon nanotubes. Carbon nanotubes, long, thin cylinders of carbon, were discovered in 1991 by Iijima. These are large macromolecules that are unique for their size, shape, and remarkable physical properties. They can be thought of as a sheet of graphite (a hexagonal lattice of carbon) rolled into a cylinder. These intriguing structures have sparked much excitement in the recent years and a large amount of research has been dedicated to their understanding. Currently, the physical properties are

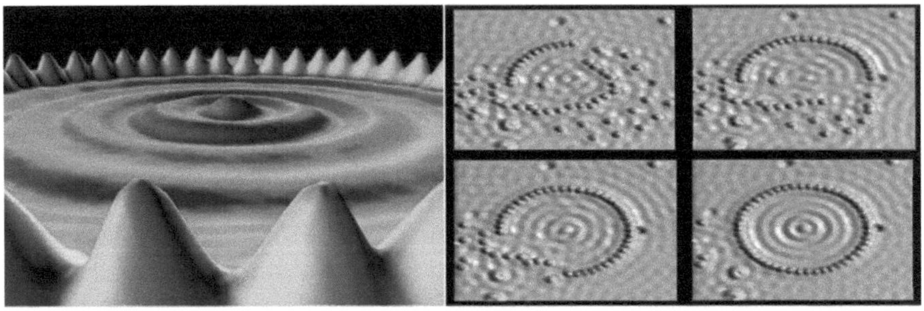

FIGURE 1.7

Confinement of electrons to quantum corrals on a metal surface [5].

still being discovered and disputed. What makes it so difficult is that nanotubes have a very broad range of electronic, thermal, and structural properties that change depending on the different kinds of nanotube (defined by its diameter, length, and chirality, or twist). To make things more interesting, besides having a single cylindrical wall, nanotubes can have multiple-walled cylinders inside the outer cylinder.

Bower et al. [7] have grown vertically aligned carbon nanotubes using microwave plasma-enhanced CVD system using a thin film cobalt catalyst at 825°C (Fig. 1.8). The chamber pressure used was 20 Torr. The plasma was generated using hydrogen which was replaced completely with ammonia and acetylene at a total flow rate of 200 sccm.

Lithographic methods are important for micro- and nanofabrication. Lithography (in Greek "lithos" means stone; "graphein" means to write) is a planographic printing technique using a plate or stone with a smooth surface. In micro- and nanofabrication, we mean pattern transfer. Due to limitations in current (and future) photolithographic processes there is a challenge to develop novel lithographic processes with better resolution for smaller features. One such development is that of *dip-pen nanolithography* (DPN). Dip-pen technology in which ink on a pointed object is transported to a surface via capillary forces is approximately 4000 years old. The difference with DPN is that the pointed object has a tip which has been sharpened to a few atoms across in some cases. DPN is a scanning probe nanopatterning technique in which an atomic force microscope (AFM) tip is used to deliver molecules to a surface via a solvent meniscus, which naturally forms in the ambient atmosphere. It is a direct-write technique and is reported to give high-resolution patterning capabilities for a number of molecular and biomolecular "inks" on a variety of substrates, such as metals, semiconductors, and monolayer functionalized surfaces.

DPN allows one to precisely pattern multiple patterns with good registration. It is both a fabrication and an imaging tool, as the patterned areas can be imaged with clean or ink-coated tips. The ability to achieve precise alignment of multiple

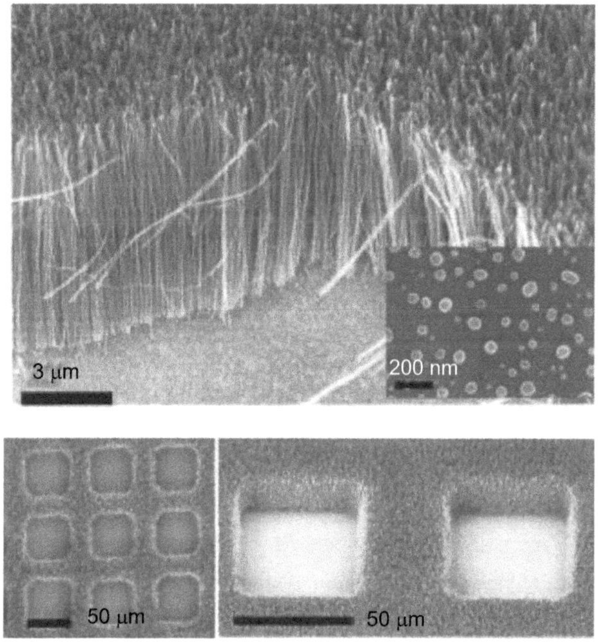

FIGURE 1.8

Multiwall carbon nanotubes with a diameter of 30 nm and length of 12 μm have been formed within 2 min [7].

patterns is an additional advantage earned by using an AFM tip to write, as well as read nanoscopic features on a surface. These attributes make DPN a valuable tool for studying fundamental issues in colloid chemistry, surface science, and nanotechnology. For instance, diffusion and capillarity on a surface at the nanometer level, organization and crystallization of particles onto chemical or biomolecular templates, monolayer etching resists for semiconductors, and nanometer-sized tethered polymer structures can be investigated using this technique. In order to create stable nanostructures, it is beneficial to use molecules that can anchor themselves to the substrate via chemisorption or electrostatic interactions. When alkanethiols are patterned on a gold substrate, a monolayer is formed in which the thiol head groups form relatively strong bonds to the gold and the alkane chains extend roughly perpendicular to surface. Creating nanostructures using DPN is a single step process which does not require the use of resists. Using a conventional AFM, DPN has been reported to achieve ultrahigh resolution features with line widths as small as 10–15 nm with ~5 nm spatial resolution. For nanotechnological applications, it is not only important to pattern molecules in high resolution, but also to functionalize surfaces with patterns of two or more components (Figs. 1.9–1.11).

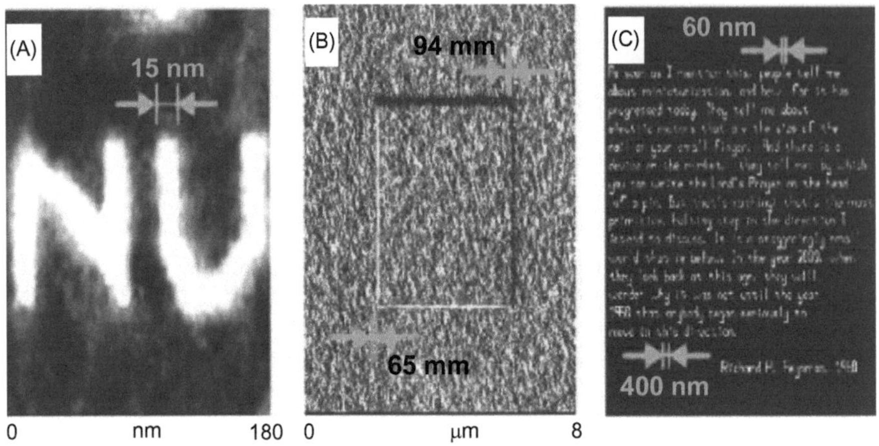

FIGURE 1.9

(A) Ultrahigh resolution pattern of mercaptohexadecanoic acid on atomically-flat gold surface. (B) DPN-generated multicomponent nanostructure with two aligned alkanethiol patterns. (C) Richard Feynmann's historic speech written using the DPN nanoplotter.

A. Byrne, Lecture at University of Ulster 2006 Private communications.

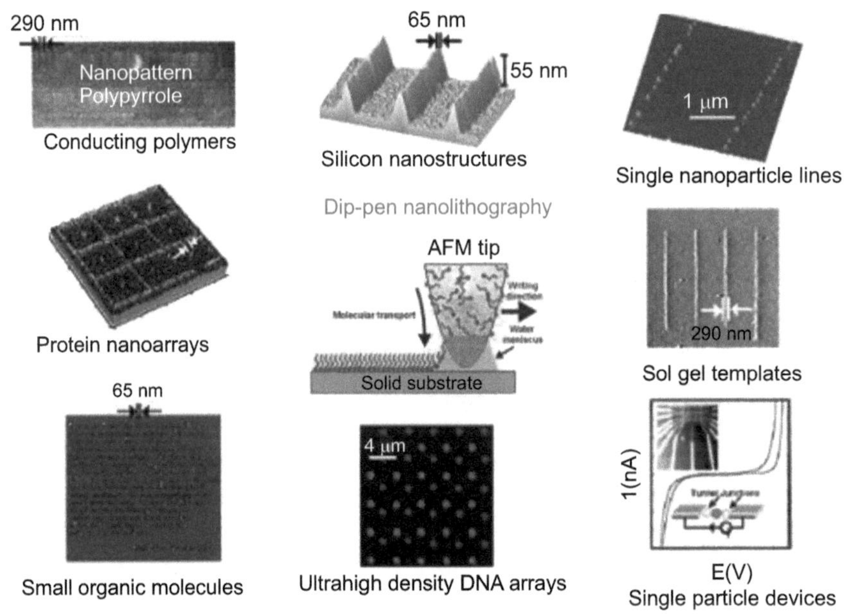

FIGURE 1.10

Some of the potential applications of DPN.

Dr. T. Okpalugo, University of Paisley (2007) Private communication.

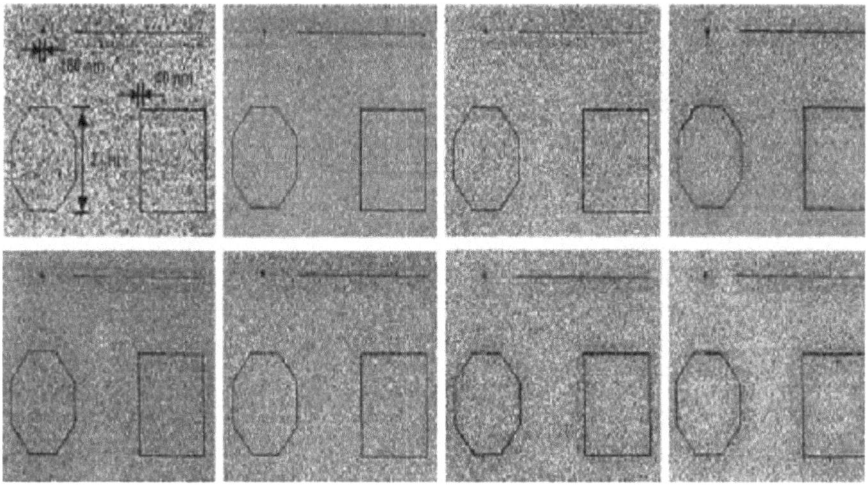

FIGURE 1.11

Lateral force microscopy (LFM) images of nanolithography patterns, which were formed using an eight-pen nanoplotter capable of doing parallel DPN.

Dr. T. Okpalugo, University of Paisley (2007) Private communication.

Fig. 1.12 shows the basic concept of nanomanufacturing. Individual atoms, which are given in the periodic table, form the basis from nanomanufacturing. These can be assembled into molecules and various structures using various methods including directed self-assembly, templating, etc. and may be positioned appropriately depending on the final requirements. Further along the devices architecture, integration, in situ processing may be employed culminating in nanosystems, molecular devices, etc.

1.4 NANODENTISTRY

Over the years, developments in dentistry have made many dental treatment procedures fast, reliable, safe, and much less painful. New technologies such as nanotechnology, dental implantology, cosmetic surgery, use of lasers, and digital dentistry have had great impact on dental treatment methodologies and recovery time. Although nanotechnology has always existed, its discovery is attributed to Feynman who won the Nobel Prize in 1959 about his theories regarding future nanosized devices. In the field of medicine, nanotechnology has been applied in diagnosis, prevention, and treatment of diseases. Nanotechnology offers considerable scope in dentistry to improve dental treatment, care, and prevention of oral diseases. The following chapters in this book discuss about the recent developments in this interdisciplinary field bridging nanotechnology and dentistry.

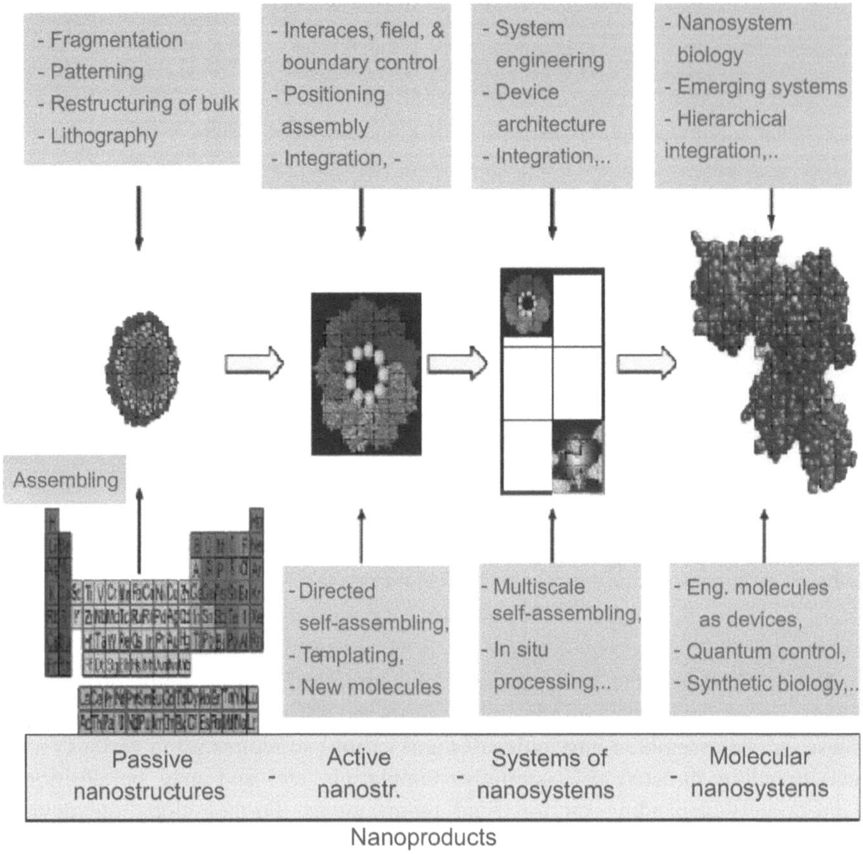

FIGURE 1.12

Summary of nanotechnology [8].

Nanotechnology has been used in dentistry for tooth sealants and fillers that use nanosized particles to improve their strength, luster, and resist wear. The application of nanoparticles in dental materials and their synthesis has been discussed in Chapter 2, Nanoparticles for dental materials: Synthesis, analysis, and applications. Antimicrobial nanoparticles in restorative composite materials are being used to prevent dental caries. For example, silver particles as antibacterial agents when used in fillers and toothpastes can retard bacterial growth and reduce tooth decay (Chapter 3: Antimicrobial nanoparticles in restorative composites). It is envisaged that in the long-term, biomimetic approaches and nanotechnology will be used to repair and rebuild damaged enamel. Composite materials are becoming popular due to their esthetic appearance and superior wear properties designed to replicate the properties of enamel (Chapter 4: Nanotechnology in operative dentistry: A perspective approach of history, mechanical behavior, and clinical application). The properties of these materials such as compressive strength, material flow, tensile strength, flexural strength have been improved using nanotechnology. Microfill composites are made using the top-down approach to nanotechnology where materials such as ceramics, quartz, and glasses start off as bulk materials and then they ground into particles sizes below 100 nm However, nanocomposites are made using a bottom-up approach where atoms and molecules combine to produce nanoparticles much smaller than those produced by the first approach.

The applications of nanoscale manufacturing have also been used in the field of dental implantology. Chapters 5–8 discuss briefly about nanotechnology applications in dental implantology, microscale to nanoscale surface modification techniques, titanium nanotubes as carriers of osteogenic growth factors and antibacterial drugs and cellular responses to surface modifications, and their applications in implantology and bone tissue engineering. Chapter 9, Corrosion resistance of Ti6Al4V with nanostructured TiO_2 coatings, discusses about improving biocompatibility and bioactivity of titanium alloy (Ti6Al4V) by electrodeposition of TiO_2 nanoparticles. Multiwalled carbon nanotubes/hydroxyapatite nanoparticles incorporated membranes have been explored for their applications in guided tissue regeneration of the periodontium (Chapter 10: Multiwalled carbon nanotubes/hydroxyapatite nanoparticles incorporated GTR membranes). Nanoapatitic composite scaffolds for stem cell delivery and bone tissue engineering have been discussed in Chapter 11, Nanoapatitic composite scaffolds for stem cell delivery and bone tissue engineering. The phenomenon of self-assembly of proteins and peptides and their applications in bionanotechnology and dentistry are discussed briefly in Chapter 12, Self-assembly of proteins and peptides and their applications in bionanotechnology and dentistry. In Chapter 13, Surface engineering of dental tools with diamond for enhanced life and performance, the application of CVD of Diamond Films onto Dental Burs and tools has been presented. Nanomechanical characterization of mineralized tissues in the oral cavity and nanoindentation techniques for the determination of mechanical properties of dental materials and dental implants have been discussed in the subsequent Chapters 14–16.

Nanoparticle-based drug delivery systems have been widely used in targeted treatment of various forms of cancer. In Chapter 17, Nanoparticulate drug delivery systems for oral cancer treatment, a general outline on cancer treatment techniques and the recent developments in nanoparticle-based drug delivery systems for oral cancer treatment has been discussed. The chapter also covers the limitations of such systems for cancer treatment. Carbon nanotubes are fast emerging as biomaterials. In Chapter 18, Carbon nanotubes: Applications in cancer therapy and drug delivery research, carbon nanotubes for drug delivery and cancer treatment are described. Chapter 19, Nanodiagnostics in microbiology and dentistry, discusses about nanodiagnostics in microbiology and dentistry. The chapter covers the recent developments and futuristic applications of nanotechnology in these fields. Chapter 20, Neurotoxicity of nanomaterials, addresses the toxicity of nanomaterials.

1.5 FUTURE DIRECTIONS AND CONCLUSIONS

Biomedical scientists and clinicians all over the world are working toward prevention and early delivery of care to maintain human health. It is envisaged that nanotechnology will have a great impact in dental research and improvement in current treatment methodologies leading to superior oral health care in the near future. Nanomaterials will be used far more widely and will yield superior properties and combined with biotechnology, laser, and digital guided surgery will thus provide excellent dental care. Smarter preventive measures and earlier interventions to avert craniofacial disorders using nanodiagnostics seems a reality. Nanotechnology research will definitely pave the way for development of tools which would allow clinicians to diagnose and treat oral malignancies at their earliest stage. Biomimetics and nanotechnology have given us the knowledge to bioengineer lost tooth and remineralization of carious lesions. This is one field which has stimulated immense interest among the dental and nanotechnology researchers. Salivary glands can be a gateway to the body for the delivery of precise molecular therapies using nanoparticle-based drug delivery systems with fewer side effects. Nanofillers have improved the esthetic, physical, and mechanical properties of dental composite materials.

Futuristic applications have been proposed on utilizing nanobots (nanoscale robots) to treat carious lesions, dentin hypersensitivity, induce dental anesthesia, teeth repositioning (using orthodontic nanobots that could directly manipulate periodontal tissues allowing rapid, painless movement). Dentifrobots (nanorobots in dentifrices) delivered through mouthwash or toothpaste could patrol supra- and subgingival surfaces of tooth performing continuous plaque/calculus removal and metabolize trapped organic matter into harmless and odorless vapor. These proposals may seemingly look outrageous, but inventions have always been the brainchildren of outrageous ideas of the scientific community. Predictive tools like "lab-on-a-chip" can utilize saliva as a media to diagnose dental and other physical anomalies of the human body. The transition from nanotechnology to nanomanufacturing is truly under way. Numerous products are now on the market and many new sophisticated and intelligent "nano" products are being developed and will become available in the near future.

REFERENCES

[1] R.P. Feynman, There is plenty of room at the bottom, Eng. Sci. 23 (1959) 22–36 and www.zyvex.com/nanotech/feynman.html (1960).

[2] N. Tanaguchi, On the basic concept of nanotechnology, Proc. ICPE, 1974.

[3] B. Bushan, Springer Handbook of Nanotechnology, Springer, Berlin, 2017.

[4] T. Fujii, K. Iwasaki, M. Munekane, T. Takeuchi, et al., J. Micromech. Microeng. 15 (2005) S286–S291.

[5] D.M. Eigler, E.K. Schweizer, Positioning single atoms with a scanning tunneling microscope, Nature 344 (1990) 524–526.
[6] M.F. Crommie, C.P. Lutz, D.M. Eigler, Confinement of electrons to quantum corrals on a metal surface, Science 262 (1993) 218–220.
[7] C. Bower, W. Zhu, S. Jin, O. Zhou, Appl. Phy. Lett. 77 (6) (2000) 830–832.
[8] M.C. Roco, NSF Nanoscale Science and Engineering Grantees Conference, Dec 12–15, 2005.

CHAPTER

Nanoparticles for dental materials: Synthesis, analysis, and applications

2

Sumita B. Mitra

Mitra Chemical Consulting LLC, St. Pete Beach, FL, United States

CHAPTER OUTLINE

2.1 Introduction: Why Use Nanoparticles? 18
2.2 Synthesis of Nanoparticles 19
2.2.1 Synthesis by Mechanical Attrition 19
2.2.2 Synthesis Through Sol–Gel Process 20
2.2.3 Synthesis of Silsesquioxane Nanoparticles 22
2.2.4 Synthesis of Polymer-Templated Nanoparticles 23
2.3 Examples of Dental Materials Using Nanoparticles 23
2.3.1 Nanocomposites Containing Oxide Nanoparticles 23
2.3.2 Silsequioxane-Based Composites 27
2.3.3 Calcium Phosphate and Calcium Fluoride Nanoparticles-Based Composites 28
2.3.4 Nanoparticles in Glass Ionomer Systems 28
2.3.5 Nanotechnology in Dental Adhesives 30
2.4 Selected Properties of Dental Materials Containing Nanoparticles 30
2.4.1 Optical Properties 30
2.4.2 Wear Properties 32
2.4.3 From B.D. Craig, S.B. Mitra, G.A. Kobussen, M.C. Doruff, H.L. Lechuga, M.R. Atkinson, Polish Retention Comparison of Experimental and Commercial Restorative Composite Materials, J. Dent. Res. 88, (2009) (Spec Issue A Abstract 1506). Mechanical Properties 33
2.5 Clinical Experience With Dental Materials Containing Nanoparticles 35
2.6 Conclusions 35
References 36

Emerging Nanotechnologies in Dentistry. DOI: http://dx.doi.org/10.1016/B978-0-12-812291-4.00002-9

2.1 INTRODUCTION: WHY USE NANOPARTICLES?

Nanotechnology is the production of functional materials and structures in the nanoscale using various physical and chemical methods [1]. Today, the revolutionary development of nanotechnology has become a highly energized discipline of science and technology. The US National Nanotechnology Initiative defines nanotechnology in terms of three requirements (http://www.nano.gov/nanotech-101/what/definition):

1. Technology development at the atomic, molecular or macromolecular levels, in the length scale of 1–100 nm range
2. Creating and using structures, devices and systems that have novel properties and functions because of their small and/or intermediate size
3. Ability to control or manipulate on the atomic/molecular scale

Thus, the term nanotechnology should imply a high degree of control and planning. The intense interest in using nanomaterials stems from the idea that they may be employed to manipulate the structure of materials to provide dramatic improvements in chemical, mechanical, and optical properties. During the last decade, the use of nanoparticles has become very popular in the design and development of many dental materials, since they can provide a unique combination of properties. By far the largest application has been in dental composites, although several unique adhesive systems containing nanoparticles have also been commercialized. Every property has a critical length scale, and by using building blocks smaller than the critical length scale—such as nanoparticles—one can capitalize on the manifestation of physics at small sizes. An example of this is in light scattering. Since nanoparticles have dimensions well below the wavelength of visible light (400–800 nm), they cannot scatter that particular light resulting in inability to detect the particles by naked eye. This has tremendous implications for controlling the optical properties of materials containing these particles.

Another important parameter to consider while using nanoparticles is that due to their extremely small size they have a high surface area to volume ratio. Thus, the precise control of the chemical composition of the surface of nanoparticles becomes a prerequisite to the reliability and reproducibility of the nanoparticles. This factor is of crucial importance since a key element in nanotechnology is the ability to deliberately control and manipulate the assembly of the nanoparticles to provide the desirable properties of the nanomaterial and the ultimate performance of the materials into which they are incorporated. The section in this chapter on the design of true nanofill composites will amplify this point. The other related factor is that the size of individual nanoparticles often approaches that of the host matrix materials and hence there can be a molecular-scale interaction between these nanoparticles and the materials comprising the matrix. This factor has been utilized in providing unique characteristics to dental adhesives, such as adhesion strength and radiopacity without adversely affecting other properties. Further details will be provided in the section on adhesives.

2.2 SYNTHESIS OF NANOPARTICLES

There are two approaches to nanotechnology, and these are termed bottom-up and top-down approaches. The bottom-up approach involves synthesis of nanostructures from atomic level or molecular level, whereas the top-down approach involves depositing, for example, thin films and then removing unwanted regions using lithography and etching leaving behind nanostructures. The latter has been used widely for making components for electronic devices at the nanoscale. Both approaches have been used widely and the success of the approaches is highly dependent on the particular requirements of the application. The essence of nanotechnology is the ability to control and manipulate the nanostructures to provide unique properties of materials.

2.2.1 SYNTHESIS BY MECHANICAL ATTRITION

The grinding or milling of large coarse-grained materials to produce smaller sized particles has long been a major component of ceramic processing and powder metallurgy. Accordingly the uses of similar attrition techniques in mechanical devices have also been tried in producing nanoparticles. This is a "top-down" manufacturing process and under certain conditions the resulting particulate powders can exhibit nanostructural characteristics. Although used extensively to prepare small particles this procedure is energy consuming and has limited success in providing true nanostructured components for dental materials.

The fundamental principle of size reduction in mechanical attrition devices such as ball mills lies in the energy imparted to the sample during impacts between the milling media. For brittle materials, particle fracture is described by Griffith theory [2]. The stress (σ_F) at which crack propagation occurs leading to catastrophic failure, and hence size reduction depends on the surface energy (γ), modulus of elasticity (E) and length of the initiated crack (c). K is an empirical constant that is usually determined experimentally for each type of material.

$$\sigma_F = K(\gamma E/c)^{1/2} \tag{2.1}$$

When stress at the crack tip equals the strength of cohesion between the atoms, the crack becomes unstable and propagates, leading to fracture. However, as fragments decrease in size, the tendency to aggregate increases and particle fineness approaches a limit as milling continues and maximum energy is expended [3]. Furthermore, the grinding media (i.e., the balls and inner walls of the ball mill) become coated with fine particles that cushion the microbed from impact. The foregoing factors limit the size reduction and hence provision of particles of nanometer scale becomes more difficult.

Another limitation of the milling process is that milling generally produces a normal size distribution. Hence, a substantial portion of the particles are higher than the required 100 nm size. The classification of particles to eliminate the larger sizes has been tried but is generally of limited success. Often, dental composites contain the unclassified milled particles. However, by virtue of the milling process although they contain a fraction of nanosized particles they cannot be classified as true nanofill composites.

2.2.2 SYNTHESIS THROUGH SOL–GEL PROCESS

Among various processes available for synthesis of nanomaterials, sol–gel synthesis is one of the most versatile and cost-effective method for production. A sol is defined as a stable colloidal suspension of solid particles within a liquid while a gel is defined as a porous three-dimensional semisolid network that expands in a stable fashion [4]. There are many definitions of sol–gel systems. For dental fillers it is useful to consider the definition by Segal [5] who defined the sol–gel process as the production of inorganic oxides, either from colloidal dispersions or from metal oxides. The starting point of a sol–gel preparative method generally involves the mixing of different precursor chemicals, usually in a liquid form. The precursors generate a solution containing a stable colloidal dispersion of the particles, known as the sol. The solubility of the precursors in the solutions depends on the Gibbs free energy of mixing and solubility limit of the reactants. Two main groups of precursors are most common—metal alkoxides $M(OR)_n$ and metallic salts M_mX_n where X is an anionic group. For mixed sols judicious choice of solvents is critical. The most common nanoparticles for dental materials are oxides; water is therefore always present as a major reactant in order to transform the precursor. Precursors undergo hydrolysis and condensation reactions to generate metal-oxy or metal-hydroxy polymers. Thus, the sol evolves toward the formation of a gel-like diphasic system containing both a liquid phase and a solid phase that comprise morphologies ranging from discrete particles to continuous polymeric networks. Since the sol–gel method generates materials from atomic levels, it is often referred to as a "bottom-up" process. By controlling the purity of the precursor solution(s), it is possible to control the chemistry of the end product.

The basic steps involved in sol–gel synthesis of oxide nanoparticles consists of the hydrolysis and condensation of metal alkoxide $M(OR)_n$, where M is a metal atom, and R is an organic group, according the following scheme:

Hydrolysis

$$M(OR)_4 + H_2O \rightarrow M(OH)_4 + 4ROH$$

Condensation

$$M(OH)_4 + M(OH)_4 \rightarrow (OH)_3M - O - M(OH)_3 + H_2O$$

The kinetics of hydrolysis and condensation reactions is governed mainly by the ratio of molar concentrations of water and alkoxide. In general, larger R values (groups containing >3 carbon atoms) generate powder particles, whereas lower R values are more suitable for depositing thin films or fibers [6].

There are three regimes in the generation of monodispersed particles in the nanometers range. The first is the induction period during which the reaction slowly generates chemical complexes which serve as the growth units. The second is the nucleation period during which the concentration of the solution builds up until a critical supersaturation is reached and nucleation occurs [7]. This is followed by the growth period. The shape of the oxide particles by the sol–gel process has been modeled by Pierre [8] and is dependent on the growth mechanisms of the particles. The oxide particles grow by mononuclear or polynuclear mechanisms. During the initial stages, the growth occurs by a mononuclear mechanism whereby successive layers are deposited on the nucleated particle surface. The resultant particles form in well-defined shapes governed by the particular crystal structure of the oxide. As the particles grow the area of the layers increases and growth mechanism progressively changes to a polynuclear growth mechanism. The nucleation of a new layer takes place before the completion of the deposition of the previous layer. The particle surface becomes less smooth at the molecular scale but less faceted at the macroscopic scale. The particular crystallographic shape is lost and the particle becomes spherical. As the particle size of the polynuclear species increases at a certain point the supply of chemical complexes becomes insufficient because they are consumed by the growing particles. At this point, the transport of chemical complexes by diffusion becomes rate controlling. Conditions such as sol concentration, rate of addition, mixing, pH, and nature of the counter ion of metallic salts are some of the factors that determine the size and shape of the final particles. An extensive treatment is provided in Ref. [8].

Control of the structure of the nanoparticles is a special challenge since as the size nanoparticles increases, diffusion becomes limited and agglomeration of the particles takes place. Most often this leads to precipitation of oxide particles. Although this is not a problem in classical ceramic processing, it becomes detrimental to creating stable nanoparticles needed to control the nanostructure and provide the desired properties. By controlling the sol–gel process small clusters of controlled sizes can be obtained. However, it is still important to prevent agglomeration of the nanoclusters beyond a critical length. The best way to prevent undue agglomeration and control the size of the nanoclusters is to chemically modify their surfaces. Since the nanosized building blocks have to be incorporated into a host matrix to provide a desired end product, it is convenient to functionalize the surface of the inorganic oxide nanoparticle or nanocluster with reactive organic group that can be later tethered to the organic host matrix by strong covalent or ionic bonding.

The sol–gel method has been used with much success in designing and manufacturing nanoparticles for dental composites. Nanoparticles of silica and zirconia derived by the sol–gel method are used in the commercial nanofill composites, glass ionomers and adhesives. The zirconia nanoparticles provide radiopacity as well as serve as reinforcement for the matrix without sacrificing the optical properties. A more detailed description is given in Sections 2.3 and 2.4 of this chapter. An aqueous sol–gel method was also used for the synthesis of nanostructured SiO_2–BaO powder for use in dental composite resins [9]. Tantalum oxide nanoparticles prepared by this technique have been used for imparting radiopacity to dental composites [10].

2.2.2.1 Functionalization of oxide nanoparticles

There are two strategies for the functionalization of the surface of nanoparticles. In one method, the organic group can be grafted after the isolation of the nanoparticle or nanocluster. In the second method the organic group can be introduced in situ during the synthesis of the nanoparticle. The most common type postsynthesis modification involves the use of silane coupling agents. The terminal O– or –OH groups at the surface of the oxide nanoparticles are reacted with silicon halides or alkoxides with organic functionality. Often the organic functional group contains a reactive group that can be used for anchoring the derivatized particle to the host matrix. Common reactive groups are vinyl, acryloyl, methacryloyl, epoxy, and amino groups.

In another method of functionalization and stabilization of the nanoparticles, the groups at the surface of preformed clusters are exchanged with treatment agents within solution. Surface oxide ions can be exchanged with multidentate ions to prevent agglomeration. Electrostatic interaction between a tin oxide nanocluster and methacrylic acid was used to attach a polymerizable group [11].

2.2.3 SYNTHESIS OF SILSESQUIOXANE NANOPARTICLES

The term silsesquioxane refers to all structures with the empirical formula $RSiO_{1.5}$, where R is hydrogen or any alkyl, alkylene, aryl, or arylene group or organofunctional derivatives of the aforementioned groups. The silsesquioxanes include random structures, ladder structures, cage structures, or partial cage structures. The oxygen to silicon ratio of 1.5 in these materials in intermediate that of silica ($RSiO_2$) and silicones (R_2SiO), thus making them interesting hybrid compositions with unique physical properties. They are known to exist in the form of polycyclic oligomers and polymers. The polyhedral oligomeric silsesquioxane (POSS) structures are often referred to as POSS materials. The majority of POSS compounds have highly symmetrical, fully condensed silicone–oxygen frameworks with as equivalent oxygen functionality on each silicone atom. POSS compounds are physically large with respect to polymer dimensions and fall in the nanoscale range. A POSS molecule contains covalently bonded reactive functionalities suitable for polymerization or grafting them to polymer chains of the host matrix. The nonreactive organic groups are chosen to modify the solubility and compatibility of the POSS segments with the various polymeric segments. Incorporation of the POSS into a polymer matrix can result in significant improvements in a variety of physical and mechanical properties due to the reinforcement at the molecular level and the inorganic framework's ceramic-like properties.

The most common method of synthesis of silsesquioxanes involves the condensation of reaction of alkyl silanes containing three hydrolysable functionalities. The reactions for the synthesis of POSS compounds may be divided into two major groups depending on the nature of the starting materials [12]. Monomers of the general formula $RSiX_3$ ($X =$ alkyl or alkoxide groups) react under appropriate conditions by controlled hydrolysis and condensation giving rise to Si–O–Si bonds with the subsequent formation of polyhedral cage framework.

$$n\,RSiX_3 + 1.5\,n\,H_2O \rightarrow (RSiO_{1.5})_n + 3n\,HX$$

The second major class of reactions involves the manipulation of the substituents of the silicon atom without affecting the silicon–oxygen skeleton of the molecules. A large number of substituents have been appended to the silicon–oxygen cages. Such substituents include alcohols, phenols, alkoxysilanes, chlorosilanes, epoxides, esters, fluoroalkyls, halides, isocyanates, acrylates, and methacrylates [13].

2.2.4 SYNTHESIS OF POLYMER-TEMPLATED NANOPARTICLES

Supramolecular assemblies of polymers that are ordered and structured in their alignment have been utilized to create an environment for the synthesis of ordered materials on a nanoscale regime. These environments can also be used to template the materials synthesized within them. Examples of suitable self-assembly molecules are natural and synthetic polypeptides, lipids, synthetic functional polymers, and surfactants. High molecular weight polyacrylic acid has been used to synthesize needle-like nanoapatite crystals with diameters smaller than 100 nm [14]. Naturally occurring polymers such as chitosan and collagen have also been used as porous scaffold to precipitate nanosized hydroxyapatite crystals in situ [15,16]. Moreover, surfactant molecules, because of their tendency to form micelles, can serve as templates for the synthesis of nanoparticles. Thus, poly(methyl methacrylate)-grafted nanoclays were prepared through the free radical polymerization of the monomer in the presence of surfactants. These nanoparticles can be used to reinforce dental adhesives [17]. An excellent review on the recent advances in synthetic strategies for preparing polymer/inorganic hybrid nanoparticles was published recently [18].

2.3 EXAMPLES OF DENTAL MATERIALS USING NANOPARTICLES

2.3.1 NANOCOMPOSITES CONTAINING OXIDE NANOPARTICLES

To date oxide nanoparticles have been the most prevalent types of nanomaterials used in dentistry. This section will focus on the use of nanoparticles in dental composite filling materials. At present there are two distinct types of dental nanocomposites available, which are nanofills and nanohybrids [19,20].

- Nanofills—these contain nanometer-sized particles (1–100 nm) throughout the resin matrix, with no other large primary particles being included.
- Nanohybrids—these consist of large particles (0.4 to 5 μm) with added nanometer-sized particles.

One of the most significant contributions in dentistry has been the development of resin-based composite technology containing inorganic oxides or glass fillers. Adhesively bonded composites have the advantage of conserving sound tooth structure with the potential of tooth reinforcement, while at the same time providing cosmetically acceptable restorations. However for many years no one composite material has been able to meet all the requirements of both posterior and anterior restorations [21]. Prior to the advent of nanotechnology in dental materials a class of composites called "microfills" were used for anterior applications because of their high initial gloss and superior polish retention. Unfortunately microfill composites are not suitable for high stress-bearing areas (e.g., Class I, II, and IV restorations) of the dentition. On the other hand, the "hybrid" composite materials containing blends of micron and submicron fillers were of sufficient strength and wear resistance to be used for posterior teeth but did not have the long-term polish retention desirable for anterior teeth [22]. The use of nanoparticles addresses the aforementioned difficulty by combining high mechanical strength with long-term polish retention in one material.

2.3.1.1 Nanofill composites

As mentioned earlier, the nanofills are dental composites in which all the fillers are in the 1–100-nm range. Two types of nanoparticles have been synthesized and utilized for preparing the nanofill dental composites [23]. The first of these is the most common and are nanomeric particles which are essentially monodispersed nonaggregated and nonagglomerated particles of silica. The surface of the nanoparticles, are treated with silane coupling agents (e.g., 3-methacryloxypropyl-trimethoxysilane, MPTS) using a proprietary method. MPTS is a bifunctional material, also known as a coupling agent. It contains a silica ester function on one end for bonding to the inorganic surface and a methacrylate group on the other end to make the filler compatible with the resin before curing to prevent any agglomeration or aggregation. MPTS also allows for chemical bonding of the nanomers to the resin matrix during curing. More recently stabilized nanomers of radiopaque materials have been synthesized and used in dental materials. Thus, zirconia nanoparticles of average size of less than 50 nm have been stabilized with carboxylate were prepared [24] and used for as fillers in dental composites.

Nanomers usually start from sols and hence have a very high monodispersity (i.e., narrow size distribution). Because of this, if nanomeric particles alone are used to make highly filled composites, the rheological properties are rather poor. Attempts have been made to mix nanomers of different sizes but with little improvement in handling without sacrificing the optical properties.

To overcome the disadvantage of using nanomeric particles only and to avoid the disadvantages of nanohybrid composites (see Section 2.3.1.2) researchers at 3M Company have designed a second type of nanofillers that are called nanoclusters. The nanoclusters are made by lightly sintering nanomeric oxides to form clusters of a controlled particle size distribution so as to provide composites with good rheological properties [23]. Nanoclusters from silica sols only [25] as well as from mixed oxides of silica and zirconia [26] have been synthesized. The primary particle size of the nanomers used to prepare the clusters range from 5 to 75 nm, while the clusters have a wider distribution ranging from 100 nm to submicron level and have an average size of 0.6 μs. Fig. 2.1 shows an SEM image of a nanocluster of silica in the composite 3M ESPE Filtek Supreme Plus (3M ESPE Filtek Supreme Plus Universal Restorative System, 3M ESPE Dental Products, 3M Center, St. Paul MN, 55144-1000) after the resin matrix was removed by washing with acetone. In this material the surface of the nanoclusters are treated with MPTS to provide compatibility and chemical bonding with the resin system. The differences in particle architecture of nanomers, nanoclusters and conventional microhybrid fillers are readily apparent from transmission electron micrographs (TEMs) of composites prepared from these fillers. Fig. 2.2A shows a nanocomposite filled with 75 nm diameter nanoparticles only, Fig. 2.2B shows a nanocomposite filled with a mixture of nanoclusters alone and Fig. 2.2C shows a conventional microhybrid composite (3M ESPE Filtek Z250).

In the commercial dental composite, Filtek Supreme Plus nanoclusters have been used with nanomers to provide nanocomposites called *nanofills* in which all the primary particles throughout the resin matrix are in the nanometer size. The Filtek Supreme Plus system from 3M ESPE was the only commercial nanofill

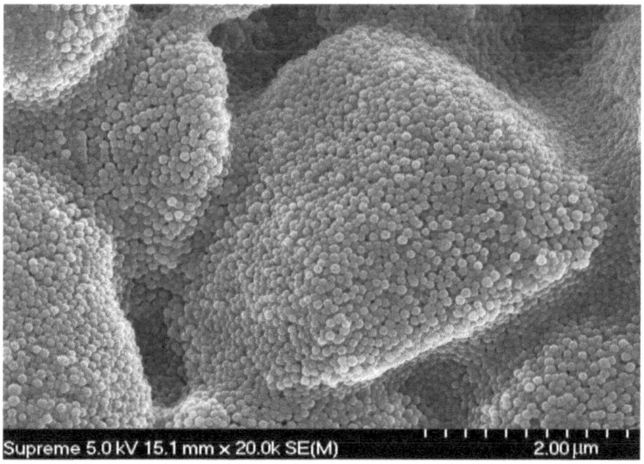

FIGURE 2.1

SEM image of silica nanoclusters in Filtek Supreme (3M ESPE Dental Products) after removal of resin by acetone extraction.

Image provided by J. Perdigao, University of Minnesota. Reproduced with permission.

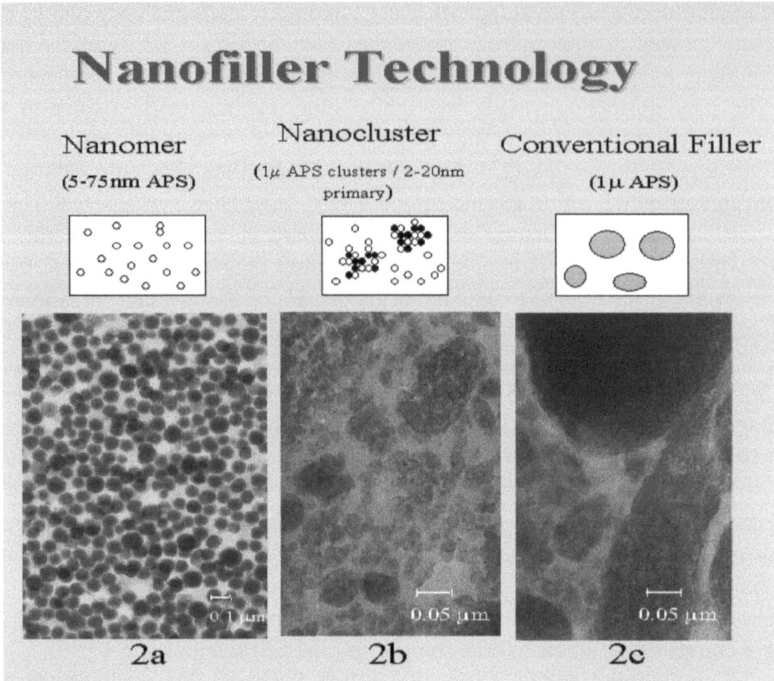

FIGURE 2.2

Schematics and transmission electron micrographic images of composites studied. (A) Composite with nanoparticles (×60,000 magnification). (B) Composite with nanocluster particles (×300,000 magnification). (C) Composite with hybrid filler (×300,000 magnification).

available [20] until its updated product Filtek Supreme Ultra system became available more recently (Filtek Supreme Ultra Universal Restorative, 3M Oral Care Division, St. Paul, MN 55144-1000, USA) [27]. A schematic diagram of the nanofill composite Filtek Supreme Plus is shown in Fig. 2.3. The use of spheroidal nanocluster fillers with a broad particle size distribution results in high filler loading, desirable handling characteristics and excellent mechanical properties. The properties of the nanofill composite will be compared with those of other dental composites as will be shown in later in this chapter.

2.3.1.2 Nanohybrid composites

As mentioned earlier, the use of unagglomerated nanomeric particles only often lead to undesirable rheological properties in a highly filled composite. Some manufacturers have added larger particles in the submicron and micron range or incorporated prepolymerized organic fillers to improve the handling properties of the composites where nanomers were included. The resulting composites are called

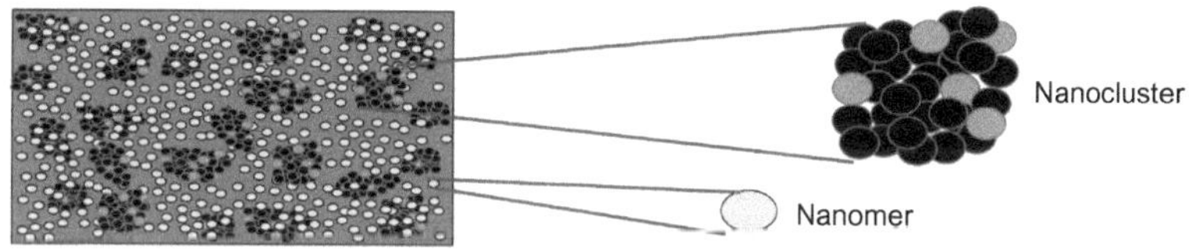

FIGURE 2.3

Schematic diagram of a nanofill composite (3M ESPE Filtek Supreme Plus).

nanohybrids [20]. These materials have improved esthetic properties compared to the traditional hybrids and microhybrids but since polish retention is governed by the size of the largest the nanohybrid approach still suffer from the loss of larger particles and the potential loss of initial gloss [28]. Almost all major dental manufacturers have now introduced commercial products in the nanohybrid category. The properties of the nanohybrid composite will be compared with those of the nanofills and other conventional dental composites in Section 2.4.

2.3.2 SILSEQUIOXANE-BASED COMPOSITES

There have been several research publications regarding the incorporation of silsesquioxanes in dental composites to improve a variety of properties including optical properties and shrinkage reduction of dental composites. Monomethacrylate functionalized POSSs was copolymerized with other methacrylate monomers and evaluated as dental restorative materials [29]. In this study, the incorporation of only 5.0% of the functionalized POSS monomer by weight was reported to significantly reduce the shrinkage of the dental composite while improving the mechanical properties. However, these results could not yet be reproduced in other laboratories [30]. In another study, monomethacrylated and dimethacrylated POSS were incorporated in dental resins and their properties studied. Monofunctional POSS showed high tendency toward aggregation and crystallization leading to weakening of the physical properties of the composites. Dimethacrylated POSS showed better miscibility at low concentrations ($<3\%$) but led to marginal reduction of shrinkage and increase in flexural strength [31]. Experimental methacrylate and epoxy functionalized nanocomposites based on silsesquioxane cores for dental applications have been reported by Soh et al. [32]. Silsesquioxane-based composites with varying amounts of methacrylate and/or epoxide groups have been prepared via Pt-catalyzed hydrosilylation of di (propylene glycol) allyl ether methacrylate, propargyl methacrylate and 4-vinyl-cyclohexene oxide combinations. The resultant composites were clear viscous liquids at room temperature and exhibited lower shrinkage upon curing than traditional monomers used for dental applications. However, the physiomechanical properties such as degree of conversion, hardness, modulus and shrinkage of a series of polyhedral silsesquioxane methacrylate monomers developed for dental applications have been found to be lower than those of conventional materials of proven clinical performance [33].

2.3.3 CALCIUM PHOSPHATE AND CALCIUM FLUORIDE NANOPARTICLES-BASED COMPOSITES

Materials that release calcium, fluoride, or phosphate ions have been shown to provide remineralization to the tooth structure [34]. Accordingly, the use of nanoparticles based on calcium phosphates and calcium fluoride in nanocomposites have been investigated. Xu et al. [35] have reported the synthesis of anhydrous dicalcium phosphate nanoparticles incorporated in composites. However, it was necessary to incorporate nanosilica-fused whiskers into the composites for the mechanical strength to be adequate. Similar results have been seen for composites into which nanocalcium fluoride particles were incorporated [36]. In an interesting approach, nanohydroxyapatites (HAP) having a particle size of 20 nm were synthesized to mimic the natural building blocks of human enamel and were found to provide anticaries repair effect [37]. It is hypothesized in the report that nano-HAP with a size of 20 nm shares the same characteristics of natural building block of enamel and can thus serve as repair material. Conventional HAP was not found to perform similarly in control experiments. In another study, an experimental nanocomposite containing amorphous calcium phosphate was shown to provide remineralization action on demineralized enamel although controlled study comparing with conventional material under similar conditions was not reported in this paper [38].

2.3.4 NANOPARTICLES IN GLASS IONOMER SYSTEMS

Conventional and resin-modified glass ionomers (RMGIs) are important classes of dental materials due to their ability to release fluoride in the oral environment and remineralize tooth structure. Additionally, RMGIs provide excellent moisture compatibility and ability to control postoperative sensitivity [39]. However, their optical properties are not ideal for situations where highly esthetic restorations are desirable. In order to preserve the beneficial properties of these materials and at the same time improve the esthetic properties, a novel RMGI material has been developed in which nanoclusters of silica and zirconia as well as unagglomerated zirconia nanomers of tailored refractive index and surface functionality were added along with a unique fluoroaluminosilcate glass filler to provide a two paste RMGI (3M ESPE Ketac Nano) (Ketac Nano Light Curing Glass Ionomer

Restorative, 3M ESPE Dental Products, 3M Co, St. Paul, MN 55144-100. http://solutions.3m.com/wps/portal/3M) [40,41]. The setting mechanism and interfacial behavior of this paste–paste nanoionomer as reported in a recent publication was found to be comparable to those of classical powder-liquid types of conventional and RMGIs with the added advantage of extremely low visual opacity [42].

An SEM of a section of the cured nanomer is displayed in Fig. 2.4 in which the nanoparticles are clearly visible. The polish and abrasion resistance of the nanoionomer as well as the fluoride release and recharge behavior compared to conventional systems will be described in the properties section. The visual opacity of the resultant nanoionomer is very low compared to conventional materials yet the mechanical properties, fluoride release, remineralization ability, adhesion strength, and mechanical properties have not been compromised [43–45]. Further detail is shown in Section 2.4.

Titania nanoparticles have also been dispersed into conventional glass ionomer systems to improve their optical translucency without compromising the mechanical properties [46]. In another approach, nanohydroxyapatite and nanofluorapatite were synthesized using an ethanol-based sol–gel method, mixed with commercial fluoroaluminosilicate powders and incorporated in conventional glass ionomer systems to improve their mechanical properties [47]. Nanoparticles of ytterbium fluoride and barium sulfate have been added to glass ionomers with an aim to improving their radiopacity [48]. While improvement in this property was observed other properties such as compressive strength and setting time suffered indicating that using nanoparticles is not a guarantee of a superior product and careful design and optimization of materials are needed.

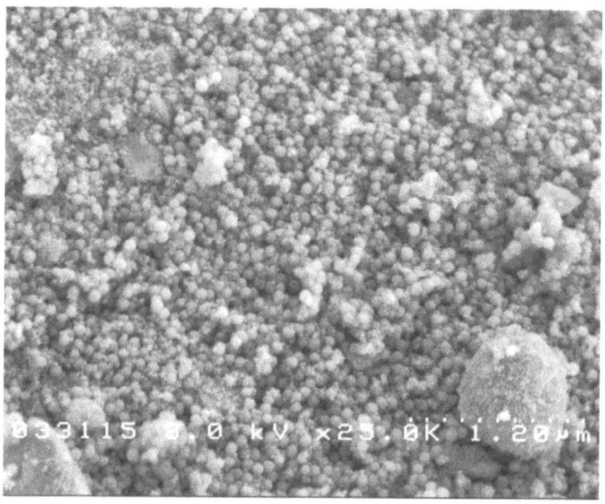

FIGURE 2.4

Scanning electron micrograph of a section of the cured nanoionomer Ketac Nano Glass Ionomer (3M ESPE Dental Products). The nanoparticles are clearly visible.

2.3.5 NANOTECHNOLOGY IN DENTAL ADHESIVES

In order to increase the cohesive strength of dental adhesives, it is common to use filler particles which have been treated with a polymerizable silane. However, since the adhesive liquids are not very viscous the filler particles tend to settle out during storage which may lead to inconsistency in their performance. To overcome this disadvantage discrete silane-treated nanoparticles of silica or zirconia in the size range of 5–7 nm have been added to dental adhesives. These particles are too small to be affected by gravity or to be visible by the bare eye and can provide transparent adhesive liquids. Examples of such adhesives are 3M ESPE Single Bond Plus and 3M ESPE Scotch Bond SE adhesive. The latter adhesive contains zirconia nanoparticles [49] to impart radiopacity to the adhesive. The morphology at the interface of a nanofilled adhesive with dentin along with bond strengths were studied and reported recently [50]. TEM analysis showed a homogeneous distribution of nanofillers within the interfibrillar spaces of the hybrid layer interface formed between this adhesive (Adper Single Bond 2) and dentin. Weakly agglomerated nanoparticles of tantalum oxide and silica with primary particle size of 10 nm have been prepared by flame pyrolysis and incorporated into experimental dental adhesives. Stable suspensions were obtained with no decrease in bond strength [51].

2.4 SELECTED PROPERTIES OF DENTAL MATERIALS CONTAINING NANOPARTICLES

2.4.1 OPTICAL PROPERTIES

A primary reason for using nanoparticles in dental materials is to improve their optical properties without sacrificing mechanical strength and wear resistance. Since the size of individual nanoparticles are smaller than the wavelength of visible light nanocomposites made from these particles exhibit extremely low visual opacity—a property highly desirable for making esthetic dental restorations. A wide variety of shades and opacities can be constructed increasing the clinician's latitude to accurately match the natural dentition. As shown in Fig. 2.5 disks of composites made from hybrid and microfill fillers are quite opaque whereas that made from nanofillers is almost glass-like in translucency [23]. In addition, when placed on a dark background, as would be the case in the oral cavity, the nanoparticles preferentially scatter blue light, giving the composite a desirable opalescent effect. The ability to create a nanocomposite with a very low opacity provides the ability to formulate a vast range of shades and opacity options from the very translucent shades needed for the incisal edge and for the final layer in multilayered restorations to the more opaque shades desired in the enamel, body, and dentin shades.

Although the average size of the clusters in nanofill composite is similar to that in conventional hybrid or microhybrid fillers, the nanoclusters are

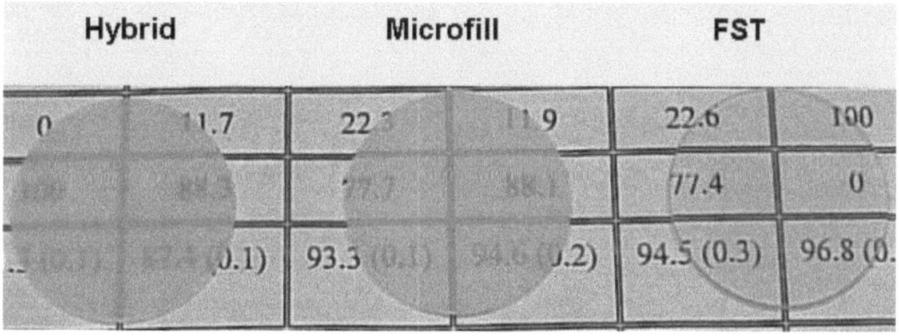

FIGURE 2.5

Optical effect of nanocomposite material FST (Filtek Supreme translucent shades) vs other types of composites: hybrid and microfill.

fundamentally different from hybrid filler particles. Typically, hybrid fillers are large dense particles of an average size of about 1 μm. These particles cannot be further subdivided under normal abrasive forces in the mouth. Similar remarks also apply to microhybrids, which are only slightly smaller than hybrids in average particle size (0.4–0.6 μm). By contrast the nanoclusters consist of loose agglomerates of primary particles clearly seen in the cluster domain in Figs. 2.1 and 2.2B. The addition of nanomers reduces the interstitial spacing between the nanoclusters thus providing a very smooth surface in a nanofilled composite. During normal abrasion the wear rate and pattern of the nanoclusters closely match those of the surrounding matrix, resulting in increased polish retention and surface gloss when compared to traditional hybrid types of composites. In contrast, the resin matrix of hybrids and microhybrids wear away by abrasive force overtime exposing the large filler particles. The average particle size of these fillers is typically between 0.4 and 0.6 μm. As these materials are abraded further, loss of the large filler particles can occur, resulting in the formation of pits or voids, leaving a rough, less-polished surface. Fig. 2.6 shows scanning electron micrographs of toothbrush-abraded surfaces of a nanofill composite (Fig. 2.6A) compared to that of a conventional microhybrid composite (Fig. 2.6B). The difference in surface topography is readily apparent. In the latter case the abraded surface is quite rough and shows large particles protruding from the surface due to the wearing away of the resin matrix and possibly formation of pits due to loss of filler. On a macro scale, the abrasion causes a loss of optical gloss of the composite, which appears dull to the unaided eye. Gloss retention of nanofill composites is much superior to that of hybrids and microhybrids. The nanohybrids have somewhat better gloss retention than hybrids but since the optical properties are dominated by the largest particles present in a system these are not much better than microhybrids and are inferior to true nanofill composites. In a study reported by Craig and co-workers [52] cured composite blocks of a commercial nanofill composite, two nanohybrid composites and a microhybrid composite were

FIGURE 2.6

Scanning electron micrographic images of toothbrush-abraded surfaces of (A) a nanofill composite Filtek Supreme and (B) a hybrid composite Filtek Z250. (3M ESPE Dental Products.)

subjected to 6000 strokes of toothbrush abrasion under controlled load. Gloss of the samples initially and after the abrasion was measured using a micro trigloss meter at 60° observation angle. Although after initial polish all composites exhibited high gloss the nanofill composites showed significantly higher polish retention than any of the microhybrid and nanohybrid composites studied. Fig. 2.7 shows the results of the gloss retention study initially and after 600 strokes. A study by Yap et al. [53] investigated the polish retention of aged composites including a nanofill composite, an oromocer composite and conventional composites over a 6-month aging period in distilled water at elevated temperatures. At all-time intervals, composite materials based on oromocer and nanomer technology were significantly smoother than all other materials investigated. Several studies on the polish retention of nanoionomers also show the superior gloss and polish retention of nanoionomers [44].

2.4.2 WEAR PROPERTIES

The wear resistance of dental composites is of paramount importance. Wear of a restoration may result from physical or pathological conditions. Mastication during chewing of food or bruxism can cause attrition of the restorative by the opposite dentition and results in the loss of material and ultimately the original anatomical shape. Several in vitro test protocols to accelerate and predict the wear of dental materials have been developed [54]. Typically hybrid composites have clinically shown excellent wear resistance. The in vitro wear resistance of the nanofill composite by a 3-body abrasion test using a modification of the method developed by the Academisch Centrum Tandheelkunde Amsterdam (ACTA method) [55], Filtek Supreme Plus demonstrated superior wear resistance compared with conventional hybrid and microhybrid composites [56]. The wear facets of the samples after 117,000 cycles of wear was imaged by atomic force

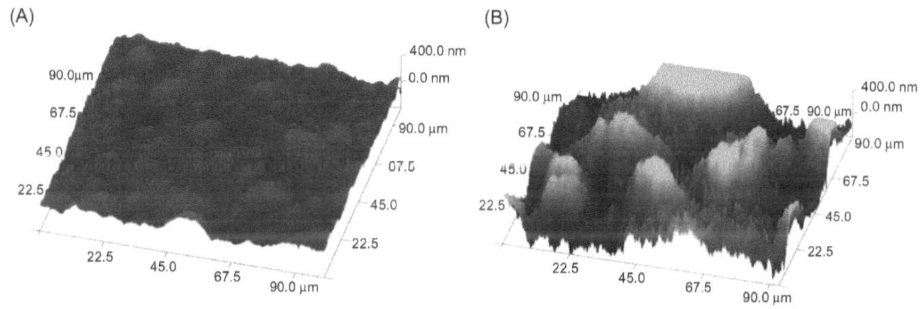

FIGURE 2.7

AFM images in tapping mode of toothbrush-abraded surfaces (A) of a nanofilled composite Filtek Supreme Ultra compared to (B) a conventional composite Premise (B).

microscopy (AFM) using a Digital Instruments Dimension 5000 SPM. The nanocomposite sample showed a much smoother wear facet and roughness compared to conventional composites [57]. Cha et al. have also reported excellent wear resistance of the same nanofill composite compared to other restoratives studied using an oral simulator and a pin-on-disc methodology [58]. Several clinical studies have verified the in vitro results and are discussed in Section 2.5. The newer generation of nanofilled composite similarly shows excellent wear resistance as shown by AFM, optical scanning Wyko analysis and gloss measurements [59]. Thus, the AFM image in tapping mode of brushed samples of the nanoionomer Filtek Supreme Ultra vs a conventional composite Premise as shown in Fig. 2.7 illustrates the superior wear resistance of nanofilled composites.

2.4.3 FROM B.D. CRAIG, S.B. MITRA, G.A. KOBUSSEN, M.C. DORUFF, H.L. LECHUGA, M.R. ATKINSON, POLISH RETENTION COMPARISON OF EXPERIMENTAL AND COMMERCIAL RESTORATIVE COMPOSITE MATERIALS, J. DENT. RES. 88, (2009) (SPEC ISSUE A ABSTRACT 1506). MECHANICAL PROPERTIES

In general, the commercial nanocomposites show mechanical properties very comparable to the clinically proven hybrid and microhybrid composites as reported in several studies. Fig. 2.8 shows a comparative summary of the results of diametral tensile strength, compressive strength, flexural strength and fracture resistance of the nanocomposite Filtek Supreme (FST = standard shades, FST = translucent shades with several control materials). The values of all the mechanical properties were equal to or better than the control composites and yet the nanocomposite materials had superior optical properties vs the comparative materials [23]. The spatial and cure-time distribution of dynamic-mechanical properties of this dimethacrylate nanocomposite was studied by Illie et al. [60].

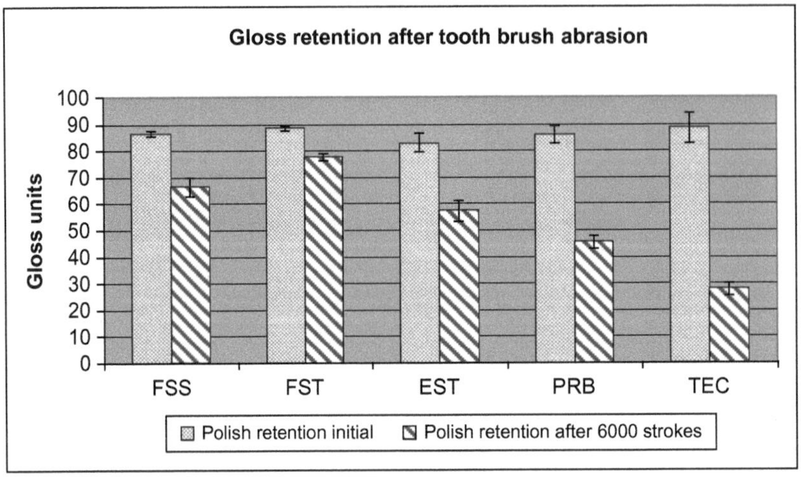

FIGURE 2.8

Gloss retention of composites. FSS, Filtek Supreme nanofill composite standard shade A2B; FST, Filtek Supreme nanofill composite translucent shade CT (3M ESPE Dental Products); EST, EsthetX microhybrid composite (Dentsply Caulk); PRB, Premise nanohybrid composite body shade A3 (Kerr Dental); TEC, Tetric Evoceram nanohybrid composite A3 (Ivoclar Vivadent).

From D. Wu, B.N. Holmes, S.B. Mitra, B.U. Kolb, W. Thompson, N.J. Johnson, Wear resistance and mechanical properties of novel dental nanocomposites, J. Dent. Res. 81 (2002) (Special Issue A-37:Abstract 64).

Nanodynamic mechanical parameters (complex and loss modulus and tan δ) as well as micromechanical properties (hardness, modulus of elasticity, creep, and elastic/plastic deformation) were investigated. The influence of short and medium-term immersion in water up to 1 year at 37°C was comparable for conventional microhybrid composite (Z250) and the nanofill composite Filtek Supreme. The influence of nanosized filler particles and nanoclusters materials on the biaxial flexural strength of resin-based nanofill and nanohybrid composites has been studied by Curtis et al. [61]. The composite containing nanoclusters (Filtek Supreme) demonstrated distinctive and unique patterns of response to preloading. Cyclic preloading at 20 and 50 N significantly increased the Weibull modulus of both the standard shades and translucent shades of this material. The biaxial flexural strength of this nanofill composite was maintained or significantly increased compared to other composites used in this study following 20 and 50 N cyclic preload. Thus, the nanoclusters were found to provide a distinct reinforcing mechanism compared with the microhybrids, microfills or nanohybrid composite systems, resulting in significant improvements to the strength and reliability, irrespective of the environmental and storage conditions. It was hypothesized that silane infiltration within the interstices of the nanoclusters may modify the response to preloading induced stresses, thereby enhancing damage tolerance and providing the potential for improved clinical performance.

2.5 CLINICAL EXPERIENCE WITH DENTAL MATERIALS CONTAINING NANOPARTICLES

Several clinical investigations on the performance of restorations using the nanofill composite Filtek Supreme have demonstrated that the material is suitable for long-term universal application to restore the dentition providing both excellent esthetics, mechanical function and wear resistance. In a clinical evaluation of this material in anterior teeth after 3 years the surface polish was reported to remain high throughout the study and the composite appeared to display a "self-polishing" effect [62]. The performance of the same nanofill resin composite was found to be efficacious for clinical use in stress-bearing posterior restorations as reported by Ernst et al. [63] in a 2-year in vivo study. A randomized clinical trial to evaluate the clinical performance and wear of the nanocomposite Filtek Supreme was performed vs the control hybrid composite [64]. The nanocomposite was found to be more polishable than the hybrid material. After 3 years the nanocomposite was found to meet the American Dental Association acceptance guidelines for tooth colored restorative materials for posterior teeth and its performance was comparable to that of human enamel. A follow-up report of the same clinical study after 5 years in vivo continued to show good clinical performance and enamel-like vertical wear resistance [65,66]. A 2-year clinical study comparing conventional, nanohybrid and nanofill composites showed the nanocomposites to have acceptable clinical performance similar to the control microhybrid material [67]. Similar results have been reported in another clinical study at 18 months [68]. Extended clinical study at 6 years and 10 years by other authors verify the long-term performance of nanocomposites [69,70]. A few studies have also reported on the clinical experience with the nanoionomer Ketac Nano. Burgess et al. reported this material to provide improved finish, resulting in a more esthetic restoration [71]. Good clinical results have also been reported at 2 years by Croll and Berg [72] in pediatric restorations, which were found to be superior esthetic properties and maintained their clinical function.

2.6 CONCLUSIONS

The use of nanoparticle technology allows the formulation of dental materials with high translucency, excellent initial polish as well as retention of gloss while maintaining mechanical properties and wear resistance equivalent to other clinically proven materials. Additional benefits in terms of imparting high radiopacity and strength in dental adhesives is described. Finally, several clinical investigations have documented the excellent performance of these materials in a variety of oral restorations.

REFERENCES

[1] R.E. Kirk, D.F. Othmer, J. Kroschwitz, M. Howe-Grant, Nanotechnology, Encyclopedia of Chemical Technology, fourth ed, Wiley, Supplement volume: New York, 1998, pp. 397–437.

[2] E. Kuhn, Powder Metallurgy, ASM Handbook, Vol. 7, ASM International, Materials Park, OH, 1984, pp. 56–70.

[3] Y.D. Kim, J.Y. Chung, J. Kim, H. Jeon, Formation of nanocrystalline Fe–Co powders produced by mechanical alloying, Mater. Sci. Eng. A 291 (2000) 17–21.

[4] P.C. Hiemenz, Principles of Colloid and Surface Chemistry, Marcel Dekker, New York, 1997.

[5] D.L. Segal, Sol–gel processing: routes to oxide ceramics using colloidal dispersions of hydrous oxides and alkoxide intermediate, J. Non-Cryst. Solids 63 (1984) 183–191.

[6] V. Nagpal, R. Davis, J. Riffle, Synthesis of ultra-fine sterically stabilized titania particles by sol–gel process, Proc. ACS Div. Polym. Mater. Eng. 67 (1992) 235–236.

[7] S. Shukla, S. Seal, Sol–gel-derived oxide and sulfide nanoparticles, in: M.-I. Baraton (Ed.), Synthesis, Functionalization and Surface Treatment of Nanoparticles, American Scientific Publishers, California, 2003, pp. 36–38.

[8] A.C. Pierre, Introduction to sol–gel processing, Boston, Kluwer Academic Publishers, MA, 1998.

[9] N.N. Ghosh, Development of aqueous sol–gel method for synthesis of nanostructured SiO_2–BaO powders for use in dental composites, Ceram. Trans. 172 (2006) 207–212.

[10] D.C.N. Chan, H.W. Titus, K. Chung, H. Dixon, S.T. Wellinghoff, H.R. Rawls, Radiopacity of tantalum oxide nanoparticles filled resins, Dent. Mater. 15 (3) (1999) 219–222.

[11] F. Ribot, F. Banse, F. Dieter, C. Sanchez, Stabilization of sol–gel derived tin oxide particles, New J. Chem. 19 (1995) 1145–1153.

[12] G. Li, L. Wang, H. Ni, C.U. Pittman, Polyhedral oligomeric silsesquioxane (POSS) polymers and copolymers: a review, J. Inorg. Organomet. Polym. 11 (2002) 123–149.

[13] J.D. Lichtenhan, Silsesquioxane-basedpolymers, in: J.C. Salamone (Ed.), Polymeric Materials Encyclopedia, Vol. 10, CRC Press Boca Raton, FL, USA, 1996, pp. 7768–7778.

[14] C. Moisescu, Z. Vroom, G. DeWith, Synthesis of nanoapatite crystals, Phosphorus Res. Bull. 13 (2002) 193–196.

[15] L. Kong, Y. Gao, W. Cao, Y. Gong, N. Zhao, X. Zhang, Preparation and characterization of nano-hydroxyapatite/chitosan composite scaffolds, J. Biomed. Mater. Res. A 75 (2) (2005) 275–282.

[16] C. Liu, Z. Han, J.T. Czernuska, Gradient collagen/nanohydroxyapatite composite scaffold: development and characterization, Acta Biomater. 5 (2) (2009) 661–669.

[17] M. Atai, L. Solhi, A. Nodehi, S.M. Mirabedini, S. Kasraei, K. Akbari, et al., PMMA-grafted nanoclay as novel filler for dental adhesives, Dent. Mater. 25 (3) (2009) 339–347.

[18] M.A. Hood, M. Mari, R. Munoz-Espi, Synthetic strategies in the preparation of polymer/inorganic hybrid nanoparticles, Materials 7 (2014) 4057–4087.

[19] The Dental Advisor, Composite Update 26 (08) (2009) 1–3.

[20] CRA Foundation Newsletter, Clinicians' Guide to dental products and techniques, Understanding current resin-based composites relative to filler *CR Newsletter*, 31(10) (2007), 1–2.

[21] D. Fortin, M.A. Vargas, The spectrum of composites: new techniques and materials, J. Am. Dent. Assoc. 131 (6) (2000) 26S–30S.

[22] G.E. Denehy, A direct approach to restore anterior teeth, Am. J. Dent. 13 (2000) 55–59 (special issue).

[23] S.B. Mitra, D. Wu, B.N. Holmes, An application of nanotechnology in advanced dental materials, J. Am. Dent. Assoc. 134 (10) (2003) 1382–1390.

[24] R.S. Davidson, B.U. Kolb, D.B. Anderson, J.A. Higgins, M.J. Hendrickson, J.T. Brady, Zirconia particles, US Patent 7,241,437, 2007.

[25] D. Wu, B.U. Kolb, S.B. Mitra, B.D. Craig, B.N. Holmes, R.P. Rusin, et al., Dental Compositions containing nanofillers and related methods, US Patent 7,393,882, 2008.

[26] M.S. Windisch, X. Zhang, R.P. Rusin, S.B. Mitra, Clustered particle dental fillers, US Patent 6,730,156, 2004.

[27] The Dental Advisor Award Issue, 2013, https://www.dentaladvisor.com/evaluations/3m-espe-filtek-supreme-ultra-2013-award-issue.

[28] E.J. Swift, Ask the experts—nanocomposites, J. Esthet. Restor. Dent. 17 (1) (2005) 3–4.

[29] F. Gao, B.M. Culbertson, Y. Tong, S.R. Schricker, Evaluation of multi-methacryl-POSS for potential organic-inorganic hybrid dental restorative materials, Polym. Prep. Div. Polym. Chem. Amer. Chem. Soc. 41 (2000) 580–581.

[30] N. Mozner, S. Klapdohr, Nanotechnology for dental composites, Int. J. Nanotechnol. 1 (1/2) (2004) 130–156.

[31] W.G. Wang, X. Sun, L. Huang, Y. Gao, J.H. Ban, L.J. Shen, et al., Structure–property relationships in hybrid dental nanocomposite resins containing monofunctional and multifunctional polyhedral oligomeric silsequioxanes, Int. J. Nanomed. 9 (1) (2013) 841–852.

[32] M.S. Soh, U.J. Yap, A. Sellinger, Methacrylate and epoxy functionalized nanocomposites based on silsequioxane cores for use in dental applications, Eur. Polym. J. 43 (2) (2007) 315–327.

[33] M.S. Soh, U.J. Yap, A. Sellinger, Effect of chain modifications on the physicomechanical properties of silsequioxane-based dental nanocomposites, J. Biomed. Mater. Res. B, Appl. Biomater. 85B (1) (2008) 75–86.

[34] M. Vollenweider, T.J. Brunner, S. Knecht, R.N. Grass, M. Zehnder, T. Imfeld, et al., Remineralization of human dentin using ultrafine bioactive glass particles, Acta Biomater. 3 (6) (2007) 936–943.

[35] H.H.K. Xu, M.D. Weir, L. Sun, Nanocomposites with Ca and PO_4 release: effects of reinforcement, dicalcium phosphate particle size and silanization, Dent. Mater. 23 (12) (2007) 1482–1491.

[36] H.H.K. Xu, J.L. Moreau, L. Sun, L.C. Chow, Strength and fluoride release characteristics of a calcium fluoride based dental nanocomposite, Biomaterials 29 (32) (2008) 4261–4267.

[37] L. Li, H. Pan, J. Tao, X. Xu, C. Mao, X. Gu, et al., Repair of enamel by using hydroxyapatite nanoparticles as the building blocks, J. Mater. Chem. 18 (34) (2008) 4079–4084.

[38] M.D. Weir, L.C. Chow, H.H.K. Xu, Remineralization of demineralized enamel via calcium phosphate nanocomposite, J. Dent. Res. 91 (10) (2012) 979–984.

[39] S.B. Mitra, Glass ionomer and related filling materials, in: V.S. Dhuru (Ed.), Contemporary Dental Materials, Oxford University Press, New Delhi, India, 2004, pp. 66–80.

[40] J. Blaes, Pearls for your practice, Dent. Econ. 97 (9) (2007) 154–156.

[41] A. Falsafi, V.J. Madsen, S.B. Mitra, J.D. Oxman, T.T. Ton, A novel nano-ionomeric restorative with improved polish and wear-resistance, J. Dent. Res. 87 (2008) (Special Issue B: Abstract no, 0244).

[42] A. Falsafi, S.B. Mitra, J.D. Oxman, H.T. Bui, T.T. Ton, Mechanisms of setting reactions and interfacial behavior of a nano-filled resin-modified glass ionomer, Dent. Mater. 30 (6) (2014) 632–642.

[43] H.T. Bui, A. Falsafi, S. Mitra, J. Oxman, T. Ton, Fluoride release of a new nano-ionomer restorative material, J. Dent. Res. 87 (2008) (Special Issue A: Abstract no. 0987).

[44] J.D. Oxman, A. Falsafi, S. Mitra, T.T. Ton, V.J. Madsen, H.T. Bui, Improved polish, wear-resistance and esthetics of a nanoionomer restorative materials, J. Dent. Res. 87 (2008) (Special Issue C PEF IADR Abstract 0039).

[45] S.B. Mitra, J.D. Oxman, A. Falsafi, T. Ton, Fluoride release and recharge behavior of a nano-filled resin-modified glass ionomer compared with that of other fluoride releasing materials, Amer. J. Dent. 24 (6) (2011) 372–378.

[46] B.D. Craig, B.U. Kolb, J.D. Oxman, R.F. Peez, S.A. Frank, Use of nanoparticles to adjust refractive index of dental composites, US Patent 7,090,721, 2006.

[47] A. Moshverinia, S. Ansari, Z. Movasaghi, R.W. Billington, J.A. Daar, I.U. Rehman, Modification of conventional glass-ionomer cements with *N*-vinylpyrrolidone containing polyacids, nano-hydroxy and fluorapatite to improve mechanical properties, Dent. Mater. 24 (2008) 1381–1390.

[48] L.H. Prentice, M.J. Tyas, M.F. Burrow, The effect of ytterbium fluoride and barium sulphate nanoparticles on the reactivity and strength of glass-ionomer cement, Dent. Mater. 22 (2006) 746–751.

[49] R.S. Davidson, B.U. Kolb, D.B. Anderson, J.A. Higgins, M.J. Hendrickson, J.T. Brady, Zirconia particles, US Patent 7,241,437, 2007.

[50] V. Di Hipolito, A.F. Reis, S.B. Mitra, M.F. de Goes, Interaction morpohology and bond strength of nanofilled simplified-step adhesives to acid etched dentin, Eur. J. Dent. 6 (4) (2012) 349–360.

[51] N. Silikas, K. Masouras, J. Satterthwaite, D.C. Watts, Effect of nanofillers in adhesives and aesthetic properties of dental resin composites, Int. J. Nano Biomater. 1 (2) (2007) 116–127.

[52] B.D. Craig, S.B. Mitra, G.A. Kobussen, M.C. Doruff, H.L. Lechuga, M.R. Atkinson, Polish retention comparison of experimental and commercial restorative composite materials, J. Dent. Res. 88 (2009) (Special issue A: Abstract no. 1506).

[53] S.-H. Yap, A.U.J. Yap, C.-K. Teo, J.-J. Ng, Polish retention of new aesthetic restorative materials over time, Singapore Dent. J. 26 (1) (2004) 39–43.

[54] S.D. Heintze, A. Cavalleri, M. Forjanic, G. Zellweger, V. Rousson, A comparison of three different methods for the quantification of the in vitro wear of dental materials, Dent. Mater. 22 (2006) 1051–1062.

[55] A.J. deGee, P. Pallav, C.L. Davidson, Effect of abrasion medium on wear of stress-bearing composites and amalgam *in-vitro*, J. Dent. Res. 65 (1986) 654–658.

[56] D. Wu, B.N. Holmes, S.B. Mitra, B.U. Kolb, W. Thompson, N.J. Johnson, Wear resistance and mechanical properties of novel dental nanocomposites, J. Dent. Res. 81 (2002) (Special Issue A-37:Abstract 64).

[57] D. Wu, V.W. Jones, S.B. Mitra, J.S. Fuller, Correlation of wear and surface roughness of dental materials, J. Dent. Res. 85 (2004) (Special Issue A:Abstract 3274).

[58] H.-S. Cha, Y.K. Lee, B.S. Lim, S.H. Rhee, H.C. Yang, Evaluation of wear resistance of dental resin composites with a 3D profilometer, J. Biomed. Mater. Res. B: Appl. Biomater. 71 (2004) 414–420.

[59] B.D. Craig, S.B. Mitra, G.A. Kobussen, M.C. Doruff, H.L. Lechuga, M.R. Atkinson, Polish retention comparison of experimental and commercial restorative composite materials, J. Dent. Res. 88 (2009) (Spec Issue A Abstract 1506).

[60] N. Illie, R. Hickel, D.C. Watts, Spatial and cure-time distribution of dynamic-mechanical properties of a dimethacrylate nano-composite, Dent. Mater. 25 (2009) 411–418.

[61] A.R. Curtis, W.M. Palin, G.J. Fleming, A.C. Shortall, P.M. Marquis, The mechanical properties of nanofilled resin-based composites: the impact of dry and wet cyclic pre-loading on bi-axial flexure strength, Dent. Mater. 25 (2009) 188–197.

[62] J. Dunn, C. Munoz, A. Wilson, M. Arambula, R. Randall, Three year clinical evaluation of Filtek Supreme in anterior teeth, J. Dent. Res. 83 (2004) (Special Issue A: Abstract No., 0359).

[63] C.P. Ernst, M. Brandenbusch, K. Canbek, G. Meyer, F. Gottschalk, B. Witllershausen, Clinical study of a nanofiller resin composite: 2 year results, J. Dent. Res. 84 (2005) (Special Issue A: Abstract No., 0578).

[64] S. Palaniappan, D. Bharadwaj, D.L. Mattar, M. Peumans, B. Van Meerbeek, P. Lambrechts, Three-year randomized clinical trial to evaluate the clinical performance and wear of a nanocomposite versus a hybrid composite, Dent. Mater. 25 (2009) 1302–1314.

[65] S. Palaniappan, M. Peumans, B. Van Meerbeek, P. Lambrechts, Clinical and in-vitro evaluation of posterior composite wear: five-year RCT, J. Dent. Res. 87 (2008) (Special issue B, Abstract No., 0240).

[66] S. Palaniappan, D. Bhardwaj, D.L. Mattar, M. Peumans, B. Van Meerbeek, P. Lambrechts, Nanofilled and microhybrid composite restorations: five-year clinical wear performances, Dent. Mater. 27 (7) (2011) 692–700.

[67] S.H. Mahmoud, A.E. El-Embaby, A.M. Abdallah, H.H. Hamama, Two-year clinical evaluation of ormocer, nanohybrid and nanofilled composite restorative systems in posterior teeth, J. Adhes. Dent. 19 (4) (2008) 315–322.

[68] Z. Ergucu, L.S. Turkun, Clinical performance of novel resin composites in posterior teeth: 18 month results, J. Adhes. Dent. 9 (2) (2009) 209–216.

[69] J.W.V. van Dijken, U. Pallesen, A six-year prospective randomized study of a nano-hybrid and a conventional hybrid resin composite in Class II restorations, Dent. Mater. 29 (2) (2013) 191–198.

[70] J.W.V. van Dijken, U. Pallesen, A randomized 10-year prospective follow-up of Class II nanohybrid and conventional hybrid resin composite restorations, J. Adhes. Dent. 16 (6) (2014) 585–592.

[71] J.O. Burgess, Fluoride-releasing materials and their adhesive characteristics, Compend. Contin. Educ. Dent. 29 (2008) 82–91.

[72] T.P. Croll, J.H. Berg, Nanoionomer restorative cement: observation after 2 years of use, Inside Dent. 5 (2009) 60–67.

CHAPTER

Antimicrobial nanoparticles in restorative composites

3

Nurit Beyth[1], Ira Yudovin-Farber[1], Arijit Basu[1,2,3], Ervin I. Weiss[1] and Abraham J. Domb[1]

[1]The Hebrew University of Jerusalem, Jerusalem, Israel [2]Koch Institute for Integrative Cancer Research, Cambridge, MA, United States [3]Massachusetts Institute of Technology, Cambridge, MA, United States

CHAPTER OUTLINE

3.1 Introduction 41
3.2 Antibacterial Restorative Composites 42
3.2.1 Filler Phase Modification 43
3.2.2 Matrix Phase Modification 44
3.3 Antimicrobial Macromolecules 45
3.3.1 Polycationic Disinfectants 46
3.3.2 Polyethyleneimine 46
3.4 Nanoparticles 47
3.4.1 Polyethylenimine Nanoparticles 48
3.5 Conclusions 53
References 54

3.1 INTRODUCTION

Restorative composite materials are commonly used in dentistry as hard tissue substitute, i.e., enamel and dentin. In the past 35 years, numerous studies investigated the antibacterial properties of various restorative composites as well as their constituents. Regrettably, restorative composite materials fail to display any inhibition after being cured [1–4]. Moreover, restorative composites were reported to accumulate more dental plaque in the long run compared to enamel and other restorations. The lack of antibacterial properties of cured composites means there is no inhibitory effect against plaque accumulation on the surface, and cariogenic bacteria such as *Streptococcus mutans* can easily grow on composites. The result is deterioration of the restoration material while infecting the neighboring soft and hard tissues, including the enamel, dentin, and gingiva. Consequently, recurrent caries evolve around these restorations, which as a matter of course is treated

Emerging Nanotechnologies in Dentistry. DOI: http://dx.doi.org/10.1016/B978-0-12-812291-4.00003-0

by restoration replacement, resulting in additional tissue loss. A diverse management strategy comprises the abolition of the cause for dental caries in terms of antimicrobial treatment [5–7].

Antimicrobial agents are chemical compounds capable of killing pathogenic microorganisms [8]. Several studies have provided restorative composites that possess antibacterial activity [9–12]; some of which were affixed with a soluble antimicrobial agent, and some carry a stationary antibacterial agent. The approach for preparing antibacterial restorative composites with a soluble agent is to impregnate them with an antibacterial agent, which is gradually released over time. However, discharge of the antibacterial agent from the materials' bulk has several disadvantages; adverse influence on mechanical properties of the base material; the release of the agent may generate a porous structure; the efficacy is time limited, and possible toxicity to the adjacent tissues may occur given that the rate of diffusion is difficult to monitor. Several of the soluble antibacterial agents that have been introduced are low molecular weight agents, such as antibiotics, fluoride, chlorhexidine, silver ions, iodine, and quaternary ammonium compounds.

Residual toxicity of low molecular antibacterial agents may be solved by generating antibacterial macromolecules. The later may be accomplished by polymerization of biologically active monomers [13,14] or by immobilization of disinfectants/antibiotics onto polymeric surfaces [15]. Current literature suggests that restorative material might enhance development of secondary caries [16].

Farrugia and Camilleri reviewed the need for developing dental composite materials containing antimicrobial agents. They highlighted that the reoccurrence of dental caries after restorative treatment is increased. Secondary caries formation is the main limiting factor for composites and glass ionomer cements. Composites are more prone toward recurrent decay and plaque accumulation compared to amalgam. Consequently, demineralization of the tooth occurs by the attached acid producing bacteria (*S. mutans*) in the presence of sugars. Moreover, these bacteria adhere to these restorative materials. Therefore, materials reinforced with bactericidal agents are now the material of choice for dental composites [17]. This would in theory prevent recurrent decay.

3.2 ANTIBACTERIAL RESTORATIVE COMPOSITES

Restorative composite materials typically involve a dispersed phase of filler particles that are distributed within a continuous phase (matrix phase). Usually the inorganic filler particles, which are zirconium/silica based, are dispersed in an organic matrix of resin components, such as bisphenol A glycidyl methacrylate, urethane dimethacrylate, and triethylene glycol dimethacrylate that are cured during application. Addition of an antibacterial component can be achieved through modifications made to the filler phase [10–12] or the matrix phase [7,18,19] of the restorative composite.

3.2.1 FILLER PHASE MODIFICATION

Restorative composites consist 70%–90% (w/w) filler; various modifications of the filler components to achieve antibacterial composites have been reported, for example, using silver.

3.2.1.1 Released antibacterial agents

Silver has been used in medicine for centuries because of its antimicrobial properties [20]. Silver has a long history in medicine and pharmacology as an antibacterial agent. Consequently, addition of a silver component into the filler or replacement of the filler with silver-supported materials has been evaluated. Silver-containing composites demonstrated strong antimicrobial effect in in vitro, in vivo, and in clinical practice. Such composites prevented and arrested recurrent caries. However, they caused tooth discoloration. Silver composites inhibit the demineralization process and show antibacterial effect by interfering with bacterial cell membranes, cytoplasmic enzymes, and DNA replication [21]. Pure silver ions implanted in SiO_2 filler particles exhibited an antibacterial effect on oral streptococci [11,22]. Moreover, leaching of silver ions from restorative composites loaded with high concentrations of silver-containing fillers resulted in antibacterial activity attributed to the antiadherence activity of the silver-supported substratum [12,23]. Furthermore, antibacterial activity was demonstrated in Ag–silica glass prepared by the sol–gel method [24].

Silver nanoparticles incorporated into biocompatible polymers prevent biofilm formation on the tooth surface. Corrêa et al. reviewed the use of silver nanoparticles used in dentistry, highlighting their antimicrobial effect, mechanical properties, safety, and effectiveness over long period of time [25].

Silver-supported material Novaron in acrylic resin poly-(methylmethacylic acid) (PMMA) composites with zirconium dioxide nanoparticles (nano-ZrO_2) and aluminum borate whiskers were reported. Results showed that the silanized nanocomposite improve the flexural strength. Novaron synergistically improved the flexural strength (increase of 44%). Composites show a significant enhancement of antimicrobial properties, without compromising the biocompatibility as shown in in fibroblast cells model [26].

An additional component used as an anticariogenic agent is fluoride. The anticariogenic effect of fluoride is attributed to various mechanisms, such as reduction of the demineralization; enhancement of the remineralization; interference with pellicle and plaque formation; and the inhibition of microbial growth and metabolism. For this purpose, fluoride-releasing filler systems are used, such as strontium fluoride (SrF_2) or ytterbium trifluoride (YbF_3) and leachable glass fillers [27–29]. The filler systems release fluoride by means of an exchange reaction of water diffusion into the composite and fluoride release from the particles. One of the major disadvantages of fluoride release is that it may cause voids in the matrix as the fluoride leaches out of the material. Furthermore, most of the fluoride is already released during the setting reaction, followed by a smaller amount of long-term fluoride release. Recently, Bakri et al. reported fluoride release studies from fluoridated glass fillers. The fluoridated fillers were loaded at various concentrations in PMMA. They release fluoride ion in a sustained manner, but the flexural strength decreased as loading increased [30]. Acrylate-based resin containing hollow silica microsphere loaded with NaF were reported. These NaF reservoirs showed higher cumulative fluoride release compared to traditional acrylate resins. Moreover, fluoride release could be tailored by changing the shell thickness [31].

A recent review compiled the dental use of metal oxide nanoparticles, emphasizing on zinc and titanium dioxide. ZnO and TiO_2 nanoparticles (NPs) have shown significant antimicrobial activity against oral bacteria; proved safe in in vivo toxicity assays. The nanoparticles have been produced at an industrial scale for food industry. Thus, ZnO and TiO_2 NPs are also commercially viable [32,33].

3.2.1.2 Nonreleased antibacterial agents

Filler modification may also produce a nonreleased antibacterial agent [34]. For example, Yoshida et al. [12] reported that an experimental restorative composite containing silver-supported fillers inhibits the growth of *S. mutans*, when in direct contact with the silver composites, with no release of silver ions. Contact inhibition of bacteria was also achieved by a restorative composite containing bactericide-immobilized filler [35]. The antibacterial component utilized was 12-methacryloyloxydodecylpyridinium bromide (MDPB). MDPB was found advantageous over agent-releasing materials in terms of maintaining favorable mechanical properties after aging; and over silver immobilized agents in terms of color stability (an imperative feature in restorative composites). The MDPB activity is assumed to be by contact inhibition. It is presumed that the agent cannot penetrate fully through the cell wall or membrane unlike free antimicrobials. Hence, the effect of the filler-modified MDPB-containing composites is not as intensive as the materials that release antibacterial agents.

3.2.2 MATRIX PHASE MODIFICATION

Matrix phase modification is basically easier to accomplish than filler modification, for the reason that it is less complicated to alter resins. Similarly to filler modification, matrix modification involves two approaches: (1) addition of a released antibacterial agent and (2) addition of nonreleased antibacterial agents.

3.2.2.1 Released antibacterial agents

Released agents are soluble components capable of diffusing in an aqueous environment. One example for soluble antibacterial agents is organic fluoride components added to the matrix. Matrix-bound fluorides include acrylic-amine-HF salts, methacryloyl acid-fluoride, and acrylic-amine-BF3 [36]. Regrettably, fluoride levels leached from composites are much lower and not as effective compared to levels released from conventional or resin-modified glass-ionomers. Furthermore, it is not proven by prospective clinical studies whether the incidence of secondary caries can be significantly reduced by the fluoride release of restorative materials [28]. An additional soluble antimicrobial used as an added agent in restorative composites is chlorhexidine [37]. It has been demonstrated that chlorhexidine inhibits bacterial growth around restorative composites. However, resins such as poly(ethyl methacrylate) and tetrahydrofurfuryl methacrylate, incorporating chlorhexidene or chlorhexidine-diacetate-based systems, release up to 50% of the agent within 14 days [38,39]. An additional agent added to restorative composites was benzalkonium chloride. Benzalkonium chloride, a quaternary ammonium compound, was added to an orthodontic composite based on the concept that minute amounts could provide antibacterial properties without significantly affecting the materials' physical properties. This addition resulted in enhanced antimicrobial properties [40].

3.2.2.2 Nonreleased antibacterial agents

A different approach relates to usage of insoluble disinfectants that can inactivate target microorganisms by contact without leaching from the carrier material. This mechanism is based on immobilized bactericides, which have the advantage of being nonvolatile and chemically stable [41]. One example was the usage of triclosan that induced bacterial death presumably by acting on the cell wall of the bacteria [34]. Consequently, triclosan incorporation in a restorative composite might be better when compared with chlorhexidine, fluoride, and benzalkonium chloride in the long run because of its nonreleasing nature.

To overcome the disadvantages of antibacterial materials based on releasing of antiseptic agents or small-molecule antimicrobial agents, a possible solution is the usage of polymeric macromolecules with antimicrobial groups.

A 5-hydroxymethylthiazole (nonquaternary ammonium) antibacterial methacrylate monomer was synthesized [42]. The 5-hydroxymethylthiazole was reacted with the active acrylate isocyanate. The resulting polymer showed enhanced antimicrobial activity without release of the active components.

3.3 ANTIMICROBIAL MACROMOLECULES

An alternative approach gaining interest in recent years is the development of macromolecular materials possessing antimicrobial properties without releasing an agent into the solution. As compared with small-molecule antibacterial agents, macromolecular materials have some advantages, such as being nonvolatile, chemically stable, and sustaining long-term antibacterial activity [8,34,43–45]. Furthermore, insoluble polymeric contact disinfectants may inactivate or remove target microorganisms by contact without releasing any biocide from the bulk phase. For example, Imazato et al. reported of an antibacterial monomer covalently bonded to the polymer network of a restorative composite. This MDPB was found to have a long-lasting antibacterial effect without altering the mechanical properties and the curing behavior [18,19].

Polymeric antimicrobials may also possess high antimicrobial ability due to high local density of active groups. Particularly, polycationic antimicrobials, which bear quaternary ammonium, have high charge density and excellent process ability, exhibiting high antimicrobial activity [46].

3.3.1 POLYCATIONIC DISINFECTANTS

Cationic polymer disinfectants include ion-exchange fibers [47], alkoxysilanes [48,49], insoluble pyridinium-type polymers [50–53], polyionenes [54], polymer surfaces derivatized with poly(vinyl-*N*-pyridinium) [55,56], and immobilized *N*-alkylated polyethyleneimine (PEI) [57–59]. Specifically, cationic polymers bearing quaternary ammonium groups, mainly crosslinked anion-exchange resins, including quaternary ammonium-type resins, exhibit high antibacterial potency.

In view of the fact that bacterial cell surface is regularly negatively charged, polymers comprising cations or retaining positive charges on their surfaces may have a contact disinfectant effect. A number of polymers that exhibit antibacterial properties were developed for this purpose, including soluble and insoluble pyridinium-type polymers [49,58]. Water-soluble pyridinium-type polycations had been previously described as having bactericidal properties. Furthermore, hexyl-poly(vinyl-*N*-pyridinium) surface coating has proven to kill 90%–99% of Gram-negative and Gram-positive bacteria, through either air or water [55]. Most of these compounds, bearing quaternary ammonium groups, appear to act primary by interacting with the cell wall and disrupting negatively charged bacterial cell membrane. A structure–activity relationship analysis has revealed that in order to create antibacterial surfaces, immobilized long polymeric chains have to be hydrophobic (but not excessively so) and positively charged [59].

3.3.2 POLYETHYLENEIMINE

Numerous publications have demonstrated the antimicrobial utility of cationic polymers with quaternary ammonium groups [60]. In particular, it was reported that quaternary ammonium PEI (QPEI) possesses excellent antibacterial activities [61–64]. Polyethyleneimine is a branched water-soluble polyamine in which the ratio of primary, secondary, and tertiary amine groups is equal to 1:2:1, approximately (Fig. 3.1).

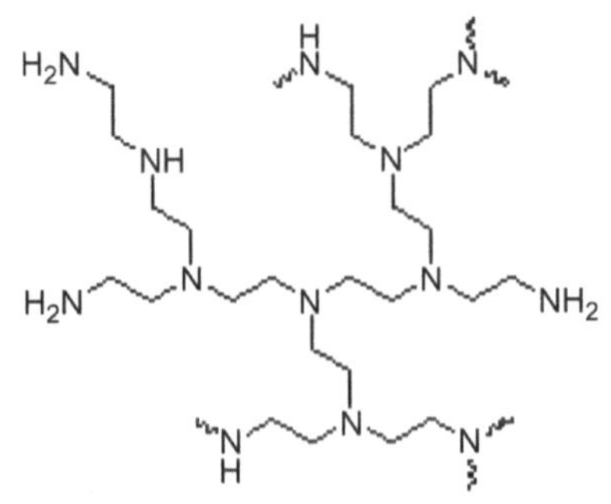

FIGURE 3.1

Chemical structure of polyethyleneimine.

The effectual antibacterial activity of QPEI is attributed to the greater density of quaternary ammonium groups along its backbone. Compared to small-molecular-weight quaternary ammonium salts, QPEI has a stronger antibacterial effectiveness because of the macromolecular effect of concentrating antibacterial groups.

The mechanism of action of antibacterial quaternary ammonium compounds is believed to be a sequence of events beginning with cationic binding and electrostatic interaction between the QPEI to the microorganism cell wall components, causing a strong adsorption, which later disrupts the membrane function, resulting in leakage of constituents, such as K^+ ions, DNA, and RNA, and subsequently causing cell death [65]. Accordingly, quaternary ammonium causes lysis of bacterial cells and shows strong bactericidal effect. Thus, the antibacterial mechanism of QPEI is a sterilizing process.

QPEI was reported to have a 100% antibacterial ratio against *Escherichia coli* cells following contact time of 4 min [57]. Lin et al. determined that *N*-alkylated poly(ethyleneimine) may be effective against a variety of Gram-positive and Gram-negative bacteria [53,59]. Its antibacterial activity was found to be dependent on molecular weight of the conjugate. The results showed, for example, that *N*-alkylated PEI of 2 and 0.8 kDa had a weak bactericidal activity. Furthermore, a structure–activity relationship analysis has revealed that in order to create antibacterial surfaces immobilized long polymeric chains have to be hydrophobic (but not excessively so) and positively charged. This design strategy has been validated not only with macroscopic surfaces but also with nanoparticles.

A detailed study against the oral pathogens was reported for QPEI-based nanoparticles. The oral pathogens include *Enterococcus faecalis*, *S. mutans*, *Actinomyces viscosus*, *Lactobacillus casei*, and whole saliva. The nanoparticles showed long-lasting microbial protection against all these pathogens [66].

Hoshika et al. demonstrated that 10% quaternary compounds increased the degree of conversion of methacrylate. It lowered wet toughness and 15% lowered the mechanical strength below acceptable levels [67]. We believe this lowering of mechanical strength is subjective and should vary from different quaternary compounds.

We reported QPEI nanoparticles by PEI crosslinking and alkylation with octyliodide followed by quaternization by methyl iodide. The particles formed were uniform (160–190 nm) and the method is reproducible. These nanoparticles incorporated in dental restorative resin (2% w/w) demonstrated strong antibacterial activity [61,68,69].

3.4 NANOPARTICLES

Nanoparticles carrying antimicrobial agents have been reported to control oral infections. Most of the time such particles were synthesized by traditional methods and can be classified as particles sized not greater than 100 nm [70]. The main advantage (as per the review of Allaker and Memarzadeh) of nanoparticles is their effectiveness against biofilms formed by bacteria within the oral cavity.

Øilo and Bakken provided an overview of the scientific literature regarding the dental biomaterials dealing with biofilms. They outlined the association among the chemical, mechanical, and physical properties of dental biomaterials and oral biofilm formation [71].

Usage of nanoparticles is advantageous as active antibacterial groups, since their surface area is exceedingly outsized relative to their size and scope. Thus, nanoparticles may provide high activity even in cases when only a small dosage of the particles is added. Particles should be homogeneously distributed in a restorative composite in order to generate an outer surface concentration of about 1 particle per sq. μm, thereby achieving effective antibacterial activity against bacteria averagely sized 1−3 μm.

Branched polymers are preferred for crosslinking as small amount of crosslinking is required to form insoluble nanoparticles, which are essential in materials anticipated to withstand exposure to an aqueous environment, such as restorative composites. Additionally, crosslinking prevents the unfolding of the polymer and separation of the various polymeric chains that form the particle. Furthermore, incorporation of nanoparticles is considered to be beneficial in durability over utilization of restorative composites containing leachable antibacterial agents. As opposed to restorative composites containing a water-soluble antimicrobial antibacterial nanoparticles, when incorporated are not released, consequently resulting in additional stability in a wet environment. He et al. recently demonstrated graphene oxide nanoparticles are effective against dental pathogens like *S. mutans*, *Porphyromonas gingivalis*, and *Fusobacterium nucleatum* [72].

3.4.1 POLYETHYLENIMINE NANOPARTICLES

According to previous reports, antibacterial activity of cationic biocides can be achieved due to hydrophobic nature and positive charge. *N*-alkylation of PEI crosslinked nanoparticles with alkyl halides of various lengths results in hydrophobic nanoparticles, and by subsequent *N*-methylation PEI's primary, secondary, and tertiary amino groups become quaternary ammonium groups.

3.4.1.1 Synthesis

QPEI particles can be prepared in several steps using 1,5-dihalidopentane as crosslinking agent, which causes intramolecular crosslinking [51]. Subsequently, the crosslinked PEI particles are *N*-alkylated with linear alkyl halides of different chain lengths varying from butyl to hexadecyl followed by neutralization with sodium bicarbonate and further *N*-methylation is carried out with methyl iodide to obtain QPEI nanoparticles. In order to increase the rate of quaternization of PEI,

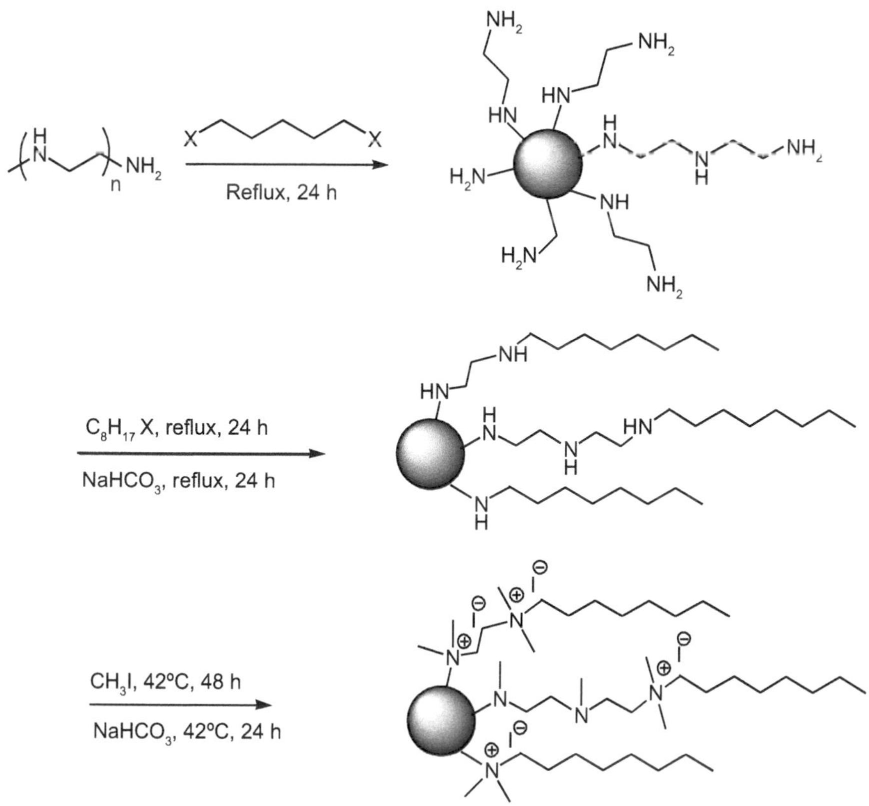

SCHEME 3.1

Synthesis of quaternary ammonium polyethyleneimine nanoparticles.

neutralization should be repeated with an additional portion of sodium bicarbonate and the quaternization step should be continued for additional 24 h (Scheme 3.1) [73–76].

Antibacterial activity of PEI is dependent on the quality and quantity of quaternary amines. Therefore, in order to increase the degree of amino group substitution, it is beneficial to increase the alkylation with alkyl halides of various lengths followed by methylation. Alkylation agents impart hydrophobic nature to the hydrophilic PEI. This *N*-alkylation makes the polymer more hydrophobic and methylation may also raise its positive charge by converting the PEI's primary, secondary, and tertiary amino groups into quaternary amino groups. Studies on PEI nanostructured samples, modified with various alkylation agents based on alkyl bromides, evaluated their antibacterial properties as a function of the alkyl group length [50]. The alkyl halides of eight methylene groups were found preferred.

The effect of quaternization degree on the antibacterial activity of QPEI was also studied. It was reported that the antibacterial effect was strengthened with the increase of the quaternization degree [77].

Theoretically, positively charged quaternary ammonium groups strongly attract the negatively charged microorganisms' bodies. Thus, higher electrostatic interaction will determine stronger adsorption toward microorganisms, resulting in enhanced antibacterial activity.

3.4.1.2 Characterization

The antibacterial activity of QPEI nanoparticle incorporated in restorative composites was studied with respect to molecular weight of polyethyleneimine, degree of crosslinking, *N*-alkyl chain length and *N*-methylation [78,79].

- *Crosslinking degree*: three QPEI derivatives crosslinked at various degrees were prepared and tested for their antibacterial activity being incorporated in dental composite resin at 1% w/w. The starting polyamine was polyethyleneimine (750 kDa) crosslinked at 1:0.01, 1:0.04, and 1:0.2 (monomer units of PEI/dihalidopentane) mole ratios. As expected, high degree of crosslinking resulted in a reduced yield of octyl substitution (6.04 carbon/nitrogen), while lower degree of crosslinking [1:0.01 and 1:0.04 (monomer units of PEI/dihalidopentane) mole ratios] resulted in increase of the carbon/nitrogen content (6.53 and 6.85, respectively). The antibacterial efficiencies of QPEI prepared from low degree of crosslinking resulted only in a slight inhibition of the bacterial growth, whereas QPEI nanoparticles prepared from high degrees of crosslinking inhibited more effectively bacterial growth, but less successfully than moderate crosslinked QPEI. The most effective compound embedded within the matrix of restorative composite resin was octyl-alkylated QPEI crosslinked at 1:0.04 (monomer units of PEI/dihalidopentane) mole ratio. The reason for the reduced activity of the low crosslinked compound can be attributed to the insufficient crosslinking degree of the nanoparticles, which might result in separation of the various polymeric chains that form the particle. On the contrary, crosslinking at 1:0.2 (monomer units of PEI/dihalidopentane) mole ratio resulted in more compact particles in comparison with low degree of crosslinking which might be responsible for the reduced access of the hydrophobic chains to the bacterial membrane that might be critical for the effectiveness of the compound.
- *N-methylation effect*: unlike QPEI-based nanoparticles, nonmethylated octyl-PEI-based nanoparticles showed reduced antibacterial activity with bacterial recovery reduced to 34% compared to the negative control, in which restorative composite resins were not treated with QPEI particles. Thus, *N*-methylation step is essential to the antibacterial activity of the particles. This behavior can be explained by the fact that quaternary methylation converts remained secondary and tertiary amines to quaternary amino groups. Therefore, use of foregoing alkylation and methylation methodology elevates

antibacterial efficiency of the octyl-alkylated QPEI being incorporated within the matrix of the clinically used dental composite materials.

- *Molecular weight of starting polyethyleneimine*: QPEI nanoparticles prepared from crosslinked polyethyleneimine of various molecular weights (25 and 750 kDa) *N*-alkylated with octyl halide followed by quaternization with methyl iodide, were embedded in dental composite resin at 1% w/w and tested for their antibacterial activity. High antibacterial effect was obtained with QPEIs having average molecular weights of 25 and 750 kDa. The results show that QPEIs prepared from high molecular weight polyethyleneimine are efficient in inhibition of bacterial growth probably due to better access of the hydrophobic polymeric flexible chains to the bacterial surface.
- *Effect of particle size*: Dental composite resin embedded with 1% w/w QPEI microparticles was tested for its antibacterial effect in comparison with resin containing QPEI nanoparticles. QPEI particles up to 3.4 μm were found to be highly effective in inhibition of *S. mutans* growth indicating in minor effect resulted from surface density differences between nano- and microtested particles. Experiments to prepare larger microparticles of QPEI were failed.
- *Effect of counter ion*: Nitrate, acetate, and iodide form QPEI nanoparticles demonstrated similar efficiency in bacterial growth inhibition. This behavior can be explained by the fact that antibacterial activity of the QPEI particles is depended on the hydrophobic chain and positive charge of the derivative and not on the counter ion. Thus, all tested materials similarly inhibited bacterial growth. Another explanation for this behavior is the fact that counter ion can affect antibacterial properties where it alters the solubility of the biocides, whereas QPEI nanoparticles are crosslinked. Thus, counter ions showed minor effect on the antibacterial activity of the QPEI nanoparticles. Based on foregoing data, it was decided to focus on the iodide form quaternary ammonium polyethylenimine (QA-PEI) due to simplicity of the synthesis and further study physical, chemical, and biological properties of the restorative composite resins incorporating QPEI particles.

Surface chemical analysis of the restorative composites containing QPEI depicted surface modification of higher hydrophobicity and presence of quaternary amino groups on the surface of the modified restorative composites compared to the corresponding commercial material although only 1% of the particles was added. In particular, the water contact angles were increased following the addition of the QPEI nanoparticles, raising the hydrophobicity of the material surface [77]. The presence of active antibacterial components on the surface of the restorative composite materials may also offer an additional explanation for the long-lasting antibacterial properties of the materials following incorporation of QPEI. These findings add another aspect to the belief that the effective antibacterial outcome of these components is through lethal direct contact with bacteria.

3.4.1.3 Incorporation of polyethyleneimine nanoparticles

The combination of high antimicrobial potency of QPEI polycations together with the advantages of nanoparticles may be used to produce improved restorative composites' properties.

The polymeric particles may be physically entrapped within the matrix, chemically bound thereto, or both. In case the particles are to be chemically bound to the polymeric host, the particles have functional groups that are capable of reacting with the host polymer, or with monomers thereof.

The antibacterial effect of incorporated QPEI nanoparticles in restorative composites was investigated in detail by many in vitro tests [77–80].

Various QPEI nanoparticles were tested for antibacterial effect in long-lasting studies. *N*-alkylation was performed with a long-alkyl chain followed by *N*-methylation. Although some of the tested nanoparticles were found to be effective in inhibition of bacterial growth, the octyl-alkylated QPEI nanoparticles showed the most potent antibacterial properties when incorporated in a restorative composite [78,79].

The antibacterial activity of octyl-alkylated QPEI nanoparticles incorporated at 1% w/w in various restorative composite systems was further investigated. Bonding, flowable, and hybrid restorative composites incorporating the nanoparticles were tested after being cured by light initiation. Interestingly, although the nanoparticles were incorporated at a low concentration, a strong antibacterial activity was evident upon contact without leach-out of the nanoparticles and without compromising the restorative composite's mechanical properties [80]. Furthermore, the scanning electron microscopy (SEM) of *S. mutans*, which was the test organism in this study, revealed bacterial debris and no streptococcal chains within 24 h of bacterial contact. Although the detailed mechanism of the antibacterial effect of these materials has not been fully determined, it was suggested that quaternary ammonium compounds cause lysis of the bacterial cells as was seen in this study in the SEM micrographs.

The antibacterial spectrum was further assayed on *Staphylococcus aureus*, *Staphylococcus epidermidis*, *E. faecalis*, *Pseudomonas aeruginosa*, and *E. coli* [80]. Interestingly, 2% w/w of incorporated nanoparticles was sufficient to inhibit the growth of all bacterial strains tested. Reducing the amount of the nanoparticles to 1% w/w resulted in complete inhibition of *S. aureus* and *E. faecalis*, and decreased growth of *S. epidermidis*, *P. aeruginosa*, and *E. coli*. The bacterial sensitivity of the tested strains in this study to the added QPEI nanoparticles was differential: in some bacteria, low concentrations resulted in complete inhibition, whereas in others complete inhibition was obtained only at higher concentrations. It appears that the Gram-positive bacterial growth was affected to a greater extent by the added QPEI, compared with that of the Gram-negative strains tested. This can be accredited to the substantial dissimilarity in the cell wall structure. In the Gram-negative bacteria such as *P. aeruginosa* and *E. coli* the presumed antimicrobial mechanism involves displacement of the divalent cations that hold together the negatively charged surface of the lipopolysaccharide network, resulting in outer membrane disruption

[81,82]. Furthermore, damage to the outer membrane permeability barrier, may cause further penetration of the cationic groups into the inner membrane, causing leakage. Alternatively, Gram-positive bacteria like *S. aureus*, *E. faecalis*, and *S. epidermidis* have a simpler cell wall structure consisting only of a rigid peptidoglycan layer. This layer, though thick, has numerous pores, which allow different molecules to readily penetrate the thick cell wall and reach the cytoplasmatic membrane. Although it is not known how the initial damage to the outer and/or cytoplasmatic membrane ultimately kills the bacteria, these nanoparticles presumably actively participate by inducing autolysis.

An additional virtue fundamental in bioactive materials is biocompatibility. Incorporation of the QPEI nanoparticles at various concentrations showed no change in cell viability or of specific alterations in cell activity when compared to the base restorative composite materials, thus indicating similar in vitro biocompatibility properties [80]. Moreover, in vivo toxicity tests, which were assessed on Wistar rats by the implantation of modified composite specimens, revealed no inflammation response 1 week after the implantation of restorative composite resin that was embedded with up to 2% w/w QPEI [54]. Furthermore, the release of QPEI from the restorative composites was also studied using various biological and chemical methods. In these tests, no residual particles or polymers were detected in the eluted supernatants; bacterial growth was similar to the growth that was obtained from the eluted supernatants of the base restorative composites. Consequently, it may be deduced that the bioactivity of restorative composites incorporating QPEI nanoparticles is not through the release of compounds to the medium and that the activity is associated with surface contact.

3.5 CONCLUSIONS

Incorporation of QPEI nanoparticles in a resin composite had a long-lasting antimicrobial effect against a wide range of bacteria with no measured effect on biocompatibility.

QPEI nanoparticles were found to have a strong bactericidal activity against *S. mutans* and a wide variety of microorganisms rapidly killing bacterial cells when incorporated at small concentrations into restorative composites. QPEI nanoparticles have a potential to be incorporated into dental resin-based restorative materials to provide bactericidal activity without causing adverse effect on physiologic properties or on biocompatibility. Further in vivo tests are needed to clarify the clinical significance of the addition of QPEI nanoparticles into restorative composites. QPEI nanoparticles are suitable candidates as additives for the restorative composite materials possessing antibacterial activity.

We observed that there is a growing interest in the use of nanomaterials (either nanomedicines or dental materials/devices) in the clinics. Nanoparticles are now investigated to strengthen dentine or regenerate pulp tissue. However, they are mostly explored as antibacterials and as nanofillers to improve the mechanical strength. Overall, the clinical benefits clearly outweigh the toxicity aspects of using nanoparticles.

According to the recent review by Besinis et al., there are several areas where knowledge can be improved [83]. Briefly, these include the following:

1. Formation of protein corona and alteration of salivary gland functioning by nanomaterials.
2. Targeted research on the oral cavity biofilm is needed in context to drug delivery, antimicrobial functions, and penetration to the dentine.
3. Mechanistic investigations about how nanomaterials strengthen the tooth structure (enamel, dentine, pulp, and cementum).
4. Possibility of stem-cell treatments for regenerating the pulp.

REFERENCES

[1] K.K. Skjöland, Plaque accumulation on different dental filling materials, Eur. J. Oral. Sci. 81 (1973) 538–542.

[2] K.K.R. Skjørland, T. Sønju, Effect of sucrose rinses on bacterial colonization on amalgam and composite, Acta Odontol. Scand. 40 (1982) 193–196.

[3] R.T. Weitman, W.B. Eames, Plaque accumulation on composite surfaces after various finishing procedures, J. Am. Dent. Assoc. 91 (1975) 101–106.

[4] M. Svanberg, I.A. Mjor, D. Orstavik, Mutans streptococci in plaque from margins of amalgam, composite, and glass-ionomer restorations, J. Dent. Res. 69 (1990) 861–864.

[5] L. Cheng, K. Zhang, M.D. Weir, M.A.S. Melo, X. Zhou, H.H. Xu, Nanotechnology strategies for antibacterial and remineralizing composites and adhesives to tackle dental caries, Nanomedicine 10 (2015) 627–641.

[6] I. Mehdawi, A. Young, Antibacterial composite restorative materials for dental applications, in: L. Barnes, I. Cooper (Eds.), Biomaterials and Medical Device-Associated Infections, Elsevier Ltd. 2014, p. 199.

[7] S. Imazato, S. Ma, J.-H. Chen, H.H. Xu, Therapeutic polymers for dental adhesives: loading resins with bio-active components, Dent. Mater. 30 (2014) 97–104.

[8] E.-R. Kenawy, Biologically active polymers. IV. Synthesis and antimicrobial activity of polymers containing 8-hydroxyquinoline moiety, J. Appl. Polym. Sci. 82 (2001) 1364–1374.

[9] J.R. Jedrychowski, A.A. Caputo, S. Kerper, Antibacterial and mechanical properties of restorative materials combined with chlorhexidines, J. Oral. Rehabil. 10 (1983) 373–381.

[10] T. Syafiuddin, In vitro inhibition of caries around a resin composite restoration containing antibacterial filler, Biomaterials 18 (1997) 1051–1057.

[11] K. Yamamoto, S. Ohashi, M. Aono, T. Kokubo, I. Yamada, J. Yamauchi, Antibacterial activity of silver ions implanted in SiO_2 filler on oral streptococci, Dent. Mater. 12 (1996) 227–229.

[12] K. Yoshida, M. Tanagawa, M. Atsuta, Characterization and inhibitory effect of antibacterial dental resin composites incorporating silver-supported materials, J. Biomed. Mater. Res. 47 (1999) 516–522.
[13] S. Thamizharasi, J. Vasantha, B.S.R. Reddy, Synthesis, characterization and pharmacologically active sulfamethoxazole polymers, Eur. Polym. J. 38 (2002) 551–559.
[14] W.-S. Moon, K.-H. Chung, D.J. Seol, E.-S. Park, J.-H. Shim, M.-N. Kim, et al., Antimicrobial effect of monomers and polymers with azole moieties, J. Appl. Polym. Sci. 90 (2003) 2933–2937.
[15] R.S. Nohr, J. Gavin Macdonald, New biomaterials through surface segregation phenomenon: new quaternary ammonium compounds as antibacterial agents, J. Biomater. Sci., Polym. Ed. 5 (1994) 607–619.
[16] I. Nedeljkovic, W. Teughels, J. De Munck, B. Van Meerbeek, K.L. Van Landuyt, Is secondary caries with composites a material-based problem?, Dent. Mater. 31 (2015) e247–e277.
[17] C. Farrugia, J. Camilleri, Antimicrobial properties of conventional restorative filling materials and advances in antimicrobial properties of composite resins and glass ionomer cements—a literature review, Dent. Mat. 31 (2015) e89–e99.
[18] S. Imazato, M. Torii, Y. Tsuchitani, J. McCabe, R. Russell, Incorporation of bacterial inhibitor into resin composite, J. Dent. Res. 73 (1994) 1437–1443.
[19] S. Imazato, J. McCabe, Influence of incorporation of antibacterial monomer on curing behavior of a dental composite, J. Dent. Res. 73 (1994) 1641–1645.
[20] J.-Y. Peng, M. Botelho, J. Matinlinna, Silver compounds used in dentistry for caries management: a review, J. Dent. 40 (2012) 531–541.
[21] J.J.Y. Peng, M.G. Botelho, J.P. Matinlinna, Silver compounds used in dentistry for caries management: a review, J. Dent. 40 (2012) 531–541.
[22] N. Talebian, E. Zare, Structure and antibacterial property of nano-SiO_2 supported oxide ceramic, Ceram. Int. 40 (2014) 281–287.
[23] M. Tanagawa, K. Yoshida, S. Matsumoto, T. Yamada, M. Atsuta, Inhibitory effect of antibacterial resin composite against *Streptococcus mutans*, Caries Res. 33 (1999) 366–371.
[24] M. Kawashita, S. Tsuneyama, F. Miyaji, T. Kokubo, H. Kozuka, K. Yamamoto, Antibacterial silver-containing silica glass prepared by sol–gel method, Biomaterials 21 (2000) 393–398.
[25] J.M. Corrêa, M. Mori, H.L. Sanches, A.Dd Cruz, E. Poiate, I.A.V.P. Poiate, Silver nanoparticles in dental biomaterials, Int. J. Biomater. 2015 (2015) 485275.
[26] Z. Han, B. Zhu, R. Chen, Z. Huang, C. Zhu, X. Zhang, Effect of silver-supported materials on the mechanical and antibacterial properties of reinforced acrylic resin composites, Mater. Des. 65 (2015) 1245–1252.
[27] J. Arends, G.E.H.M. Dijkman, A.G. Dijkman, Review of fluoride release and secondary caries reduction by fluoridating composites, Adv. Dent. Res. 9 (1995) 367–376.
[28] X. Xu, J.O. Burgess, Compressive strength, fluoride release and recharge of fluoride-releasing materials, Biomaterials 24 (2003) 2451–2461.
[29] A. Yap, E. Khor, S. Foo, Fluoride release and antibacterial properties of new-generation tooth-colored restoratives, Oper. Dent. 24 (1998) 297–305.
[30] I.A. Al-Bakri, M.V. Swain, S.J. Naoum, W.M. Al-Omari, E. Martin, A. Ellakwa, Fluoride release, recharge and flexural properties of polymethylmethacrylate containing fluoridated glass fillers, Aust. Dent. J. 59 (2014) 208–214.

[31] Y. Ji, J. Gao, H. Yin, A. Wang, T. Jiang, G. Wu, et al., Fluoride release from hollow silica microsphere-containing dental restorative acrylate resin, J. Nanosci. Nanotechnol. 15 (2015) 3967–3974.
[32] S.T. Khan, A.A. Al-Khedhairy, J. Musarrat, ZnO and TiO_2 nanoparticles as novel antimicrobial agents for oral hygiene: a review, J. Nanopart. Res. 17 (2015) 1–16.
[33] O. Bondarenko, K. Juganson, A. Ivask, K. Kasemets, M. Mortimer, A. Kahru, Toxicity of Ag, CuO and ZnO nanoparticles to selected environmentally relevant test organisms and mammalian cells in vitro: a critical review, Arch. Toxicol. 87 (2013) 1181–1200.
[34] E.-R. Kenawy, S. Worley, R. Broughton, The chemistry and applications of antimicrobial polymers: a state-of-the-art review, Biomacromolecules 8 (2007) 1359–1384.
[35] S. Imazato, Antibacterial activity of bactericide-immobilized filler for resin-based restoratives, Biomaterials 24 (2003) 3605–3609.
[36] J. Hicks, F. Garcia-Godoy, K. Donly, C. Flaitz, Fluoride-releasing restorative materials and secondary caries, Dent. Clin. N. Am. 46 (2002) 247–276.
[37] A. Wiegand, W. Buchalla, T. Attin, Review on fluoride-releasing restorative materials—fluoride release and uptake characteristics, antibacterial activity and influence on caries formation, Dent. Mater. 23 (2007) 343–362.
[38] M.P. Patel, A.T. Cruchley, D.C. Coleman, H. Swai, M. Braden, D.M. Williams, A polymeric system for the intra-oral delivery of an anti-fungal agent, Biomaterials 22 (2001) 2319–2324.
[39] D. Leung, D.A. Spratt, J. Pratten, K. Gulabivala, N.J. Mordan, A.M. Young, Chlorhexidine-releasing methacrylate dental composite materials, Biomaterials 26 (2005) 7145–7153.
[40] V. Sehgal, V.S. Shetty, S. Mogra, G. Bhat, M. Eipe, S. Jacob, et al., Evaluation of antimicrobial and physical properties of orthodontic composite resin modified by addition of antimicrobial agents—an in-vitro study, Am. J. Orthodon. Dentofacial Orthop. 131 (2007) 525–529.
[41] E.-R. Kenawy, S.D. Worley, R. Broughton, The chemistry and applications of antimicrobial polymers: a state-of-the-art review, Biomacromolecules 8 (2007) 1359–1384.
[42] W. Luo, Q. Huang, F. Liu, Z. Lin, J. He, Synthesis of antibacterial methacrylate monomer derived from thiazole and its application in dental resin, J. Mech. Behav. Biomed. Mater. 49 (2015) 61–68.
[43] A. Muñoz-Bonilla, M. Fernández-García, Polymeric materials with antimicrobial activity, Progr. Polym. Sci. 37 (2012) 281–339.
[44] I. Francolini, G. Donelli, F. Crisante, V. Taresco, A. Piozzi, Antimicrobial polymers for anti-biofilm medical devices: state-of-art and perspectives, Biofilm-based Healthcare-associated Infections, Springer, Cham, Switzerland, 2015, pp. 93–117.
[45] N. Beyth, S. Farah, A.J. Domb, E.I. Weiss, Antibacterial dental resin composites, Reactive Funct. Polym. 75 (2014) 81–88.
[46] I. Cakmak, Z. Ulukanli, M. Tuzcu, S. Karabuga, K. Genctav, Synthesis and characterization of novel antimicrobial cationic polyelectrolytes, Eur. Polym. J. 40 (2004) 2373–2379.
[47] T. Nonaka, E. Noda, S. Kurihara, Graft copolymerization of vinyl monomers bearing positive charges or episulfide groups onto loofah fibers and their antibacterial activity, J. Appl. Polym. Sci. 77 (2000) 1077–1086.

[48] A. Isquith, E. Abbott, P. Walters, Surface-bonded antimicrobial activity of an organo-silicon quaternary ammonium chloride, Appl. Microbiol. 24 (1972) 859–863.

[49] A.A. Scheie, Modes of action of currently known chemical anti-plaque agents other than chlorhexidine, J. Dent. Res. 68 (1989) 1609–1616.

[50] G. Li, J. Shen, Y. Zhu, A study of pyridinium-type functional polymers. III. Preparation and characterization of insoluble pyridinium-type polymers, J. Appl. Polym. Sci. 78 (2000) 668–675.

[51] G. Li, J. Shen, A study of pyridinium-type functional polymers. IV. Behavioral features of the antibacterial activity of insoluble pyridinium-type polymers, J. Appl. Polym. Sci. 78 (2000) 676–684.

[52] J. Nagaya, A. Tanioka, K. Miyasaka, Preparation and characterization of membranes with positively charged main chain from polyethylenimine, J. Appl. Polym. Sci. 48 (1993) 1441–1447.

[53] J. Lin, S. Qiu, K. Lewis, A.M. Klibanov, Mechanism of bactericidal and fungicidal activities of textiles covalently modified with alkylated polyethylenimine, Biotechnol. Bioeng. 83 (2003) 168–172.

[54] T. Ikeda, H. Yamaguchi, S. Tazuke, Phase separation in phospholipid bilayers induced by biologically active polycations, Biochim. Biophys. Acta (BBA)—Biomembr. 1026 (1990) 105–112.

[55] J.C. Tiller, S.B. Lee, K. Lewis, A.M. Klibanov, Polymer surfaces derivatized with poly(vinyl-*N*-hexylpyridinium) kill airborne and waterborne bacteria, Biotechnol. Bioeng. 79 (2002) 465–471.

[56] J.C. Tiller, C.J. Liao, K. Lewis, A.M. Klibanov, Designing surfaces that kill bacteria on contact, Proc. Natl. Acad. Sci. 98 (2001) 5981–5985.

[57] B. Gao, X. Zhang, Y. Zhu, Studies on the preparation and antibacterial properties of quaternized polyethyleneimine, J. Biomater. Sci., Polym. Ed. 18 (2007) 531–544.

[58] K. Lewis, A.M. Klibanov, Surpassing nature: rational design of sterile-surface materials, Trends Biotechnol. 23 (2005) 343–348.

[59] J. Lin, S. Qiu, K. Lewis, A.M. Klibanov, Bactericidal properties of flat surfaces and nanoparticles derivatized with alkylated polyethylenimines, Biotechnol. Prog. 18 (2002) 1082–1086.

[60] Y. Ge, S. Wang, X. Zhou, H. Wang, H.H. Xu, L. Cheng, The use of quaternary ammonium to combat dental caries, Materials 8 (2015) 3532–3549.

[61] S. Farah, O. Aviv, N. Laout, S. Ratner, N. Beyth, A.J. Domb, Antimicrobial silica particles loaded with quaternary ammonium polyethyleneimine network, Polym. Adv. Technol. 25 (2014) 689–692.

[62] J. Barros, M. Silva, M. Rodrigues, F. Alves, M. Lopes, I. Pina-Vaz, et al., Antibacterial, physicochemical and mechanical properties of endodontic sealers containing quaternary ammonium polyethylenimine nanoparticles, Int. Endod. J. 47 (2014) 725–734.

[63] J. Barros, M.G. Silva, I.N. Rôças, L.S. Gonçalves, F.F. Alves, M.A. Lopes, et al., Antibiofilm effects of endodontic sealers containing quaternary ammonium polyethylenimine nanoparticles, J. Endod. 40 (2014) 1167–1171.

[64] G.M. De Souza, Nanoparticles in restorative materials, Nanotechnology in Endodontics, Springer, Cham, Switzerland, 2015, pp. 139–171.

[65] N. Kawabata, M. Nishiguchi, Antibacterial activity of soluble pyridinium-type polymers, Appl. Environ. Microbiol. 54 (1988) 2532–2535.

[66] D.K. Shvero, N. Zatlsman, R. Hazan, E.I. Weiss, N. Beyth, Characterisation of the antibacterial effect of polyethyleneimine nanoparticles in relation to particle distribution in resin composite, J. Dent. 43 (2015) 287–294.

[67] T. Hoshika, Y. Nishitani, M. Yoshiyama, W.O. Key Iii, W. Brantley, K.A. Agee, et al., Effects of quaternary ammonium-methacrylates on the mechanical properties of unfilled resins, Dent. Mater. 30 (2014) 1213–1223.

[68] S. Farah, W. Khan, I. Farber, D. Kesler-Shvero, N. Beyth, E.I. Weiss, et al., Crosslinked QA-PEI nanoparticles: synthesis reproducibility, chemical modifications, and stability study, Polym. Adv. Technol. 24 (2013) 446–452.

[69] S. Farah, O. Aviv, N. Laout, S. Ratner, N. Beyth, A.J. Domb, Quaternary ammonium poly(diethylaminoethyl methacrylate) possessing antimicrobial activity, Colloids Surf. B: Biointerf. 128 (2015) 608–613.

[70] R.P. Allaker, K. Memarzadeh, Nanoparticles and the control of oral infections, Int. J. Antimicrob. Agents 43 (2014) 95–104.

[71] M. Øilo, V. Bakken, Biofilm and dental biomaterials, Materials 8 (2015) 2887–2900.

[72] J. He, X. Zhu, Z. Qi, C. Wang, X. Mao, C. Zhu, et al., Killing dental pathogens using antibacterial graphene oxide, ACS Appl. Mater. Interf. 7 (2015) 5605–5611.

[73] I. Yudovin-Farber, N. Beyth, E.I. Weiss, A.J. Domb, Antibacterial effect of composite resins containing quaternary ammonium polyethyleneimine nanoparticles, J. Nanopart. Res. 12 (2010) 591–603.

[74] S. Farah, O. Aviv, N. Laout, S. Ratner, N. Beyth, A.J. Domb, Quaternary ammonium polyethylenimine nanoparticles for treating bacterial contaminated water, Colloids Surf. B: Biointerfaces 128 (2015) 614–619.

[75] W. Zheng, J. Hu, Z. Han, Z. Wang, Z. Zheng, J. Langer, J. Economy, Synthesis of porous carbon fibers with strong anion exchange functional groups, Chem. Commun. 51 (2015) 9853–9856.

[76] S. Farah, O. Aviv, N. Laout, S. Ratner, N. Beyth, A.J. Domb, Quaternary ammonium poly(diethylaminoethyl methacrylate) possessing antimicrobial activity, Colloids Surf. B: Biointerfaces 128 (2015) 608–613.

[77] B. Gao, X. Zhang, J. Wang, Preparation and antibacterial characteristic of water-insoluble antibacterial material QPEI/SiO_2, J. Mater. Sci.: Mater. Med. 19 (2008) 3021–3028.

[78] I. Yudovin-Farber, N. Beyth, A. Nyska, E.I. Weiss, J. Golenser, A.J. Domb, Surface characterization and biocompatibility of restorative resin containing nanoparticles, Biomacromolecules 9 (2008) 3044–3050.

[79] I. Yudovin-Farber, N. Beyth, E.I. Weiss, A.J. Domb, Antibacterial effect of composite resins containing quaternary ammonium polyethyleneimine nanoparticles, J. Nanopart. Res. 12 (2009) 591–603.

[80] N. Beyth, Y. Houri-Haddad, L. Baraness-Hadar, I. Yudovin-Farber, A.J. Domb, E.I. Weiss, Surface antimicrobial activity and biocompatibility of incorporated polyethylenimine nanoparticles, Biomaterials 29 (2008) 4157–4163.

[81] M. Vaara, Agents that increase the permeability of the outer membrane, Microbiol. Rev. 56 (1992) 395–411.

[82] N. Beyth, I. Yudovin-Farber, R. Bahir, A.J. Domb, E.I. Weiss, Antibacterial activity of dental composites containing quaternary ammonium polyethylenimine nanoparticles against *Streptococcus mutans*, Biomaterials 27 (2006) 3995–4002.

[83] A. Besinis, T. De Peralta, C.J. Tredwin, R.D. Handy, Review of nanomaterials in dentistry: interactions with the oral microenvironment, clinical applications, hazards, and benefits, ACS Nano 9 (2015) 2255–2289.

CHAPTER

4 Nanotechnology in operative dentistry: A perspective approach of history, mechanical behavior, and clinical application

Eduardo G. Mota[1] and Karthikeyan Subramani[2]

[1]Pontifical Catholic University of Rio Grande do Sul (PUCRS), Porto Alegre, RS, Brazil

[2]Roseman University of Health Sciences, Henderson, NV, United States

CHAPTER OUTLINE

4.1 Introduction 60
4.2 Historical Review: Nanotechnology Applications in Operative Dentistry 60
4.3 Biomimetics 61
4.4 Nanotechnology in CAD/CAM 63
4.5 Fillers in Composite Resins 65
4.6 SEM and EDS Evaluations 66
4.7 Filler Weight Content (wt%) 67
4.8 Water Sorption 67
4.9 Mechanical Behavior 69
4.9.1 Compressive Strength 70
4.9.2 Diametral Tensile Strength 71
4.9.3 Flexural Strength and Flexural Modulus 72
4.9.4 Microhardness 73
4.9.5 Nanohardness 74
4.9.6 Wear Resistance 75
4.10 Clinical Application 79
4.11 Conclusions 81
Acknowledgments 81
References 81

Emerging Nanotechnologies in Dentistry. DOI: http://dx.doi.org/10.1016/B978-0-12-812291-4.00004-2

4.1 INTRODUCTION

Since the pioneering research in the 1950s, numerous innovations have been applied in operative dentistry. Composite resin technology has continuously advanced since its inception by Bowen as a reinforced Bis-GMA (bisphenol A-glycidyl methacrylate) system [1]. This color-based filling material is composed by a triad: organic phase (dimethacrylates), inorganic phase (silanated fillers), and initiator/activator system. One of the most important breakthroughs in composite resin technology was the development of photocurable resin material. This was followed by the development of reduced filler particle size and increased filler loading which significantly improved the universal applicability of light-cured composite resins.

Resin composites are widely used in dentistry and have become one of the most common esthetic restorative material due to their mechanical strength, excellent esthetics properties, moderate cost (when compared to ceramics), ability to bond with the tooth, improved formulation, simplified clinical procedures, and the decline in amalgam usage due to mercury toxicity. During the last few decades, the increasing demands in esthetic dentistry have led to the development of resin composites for direct restorations with improved physical and mechanical properties, esthetics, and clinical longevity. The latest innovation in the field has been the introduction of nanofilled materials, by combining nanometer scale particles and nanoclusters in a conventional resin matrix.

4.2 HISTORICAL REVIEW: NANOTECHNOLOGY APPLICATIONS IN OPERATIVE DENTISTRY

The term "nanotechnology" was coined by the researcher named Norio Taniguchi in 1974. Nanotechnology aims the creation and utilization of materials and devices at the level of atoms, molecules, and supramolecular structures, and in the exploitation of unique properties of particles with size ranging from 0.1 to 100 nm ($1\ nm = 10^{-9}\ m$). Nanofilled composite resin materials are believed to offer excellent wear resistance, strength, and ultimate esthetics due to their exceptional polishability and luster retention [2]. In operative dentistry, nanofillers constitute spherical silicon dioxide (SiO_2) particles with an average size of 5–40 nm (0.005 to 0.04 μm). However, this 0.04-μm scale filler is not new in dentistry. In the 1970s, the minifilled composites (0.04 μm, i.e., 40 nm) were launched in the marketplace. The nanometer scale filler particles have been already applied in composite resins. So, the following question can be asked: What is the difference between this resin and the actual nanofilled resins? The answer lies in the manufacturing procedure. While the colloidal silica from the 1970s obtained by pyrogenic method allows a maximum load of 55 wt%, the ordered growth of nanofillers allows up to 87 wt% of inorganic phase (for details of filler weight content, please refer Table 4.2).

It was only in 2003 that the primary results of a nanofilled composite were published in the Journal of American Dental Association by Mitra et al. [3]. The authors reported improvements in optical properties, similar mechanical behavior when compared to microhybrid material, and same polishability of minifilled resins.

4.3 BIOMIMETICS

The early stages of dental materials as a science were characterized by answering how the natural and artificial material behaves. The development and discovery of new materials were based in fortuitous observations, i.e., the acid etch surface observed by Buonocore and then applied to dental enamel, or mixing materials in order to improve properties which were shown to be effective as seen in published manuscripts of Wilson and Kent since 1968 up to 1972 that culminate the development of glass ionomer cement.

The following decades of 1980s and 1990s were comprised by testing and discussing which dental material behaves similarly to dental hard tissues. While exhaustion of this subject was reached, the new nanotechnology field has gained attention. Therefore, a new question was asked: How nanotechnology can be used to create materials capable of simulating the physical and mechanical behavior of enamel/dentin?

Biomimetics has been the answer to this question. Biomimetics has been described as the science that applies nanotechnology to create materials and devices that imitate nature. According to Saunders [4], the biomimetics is supporting numerous improvements on nanofillers used in restorative materials in dentistry. The following clinical case (Figs. 4.1–4.4) presents the association of minimally invasive dentistry with nanohybrid composite veneers achieving high anterior esthetic.

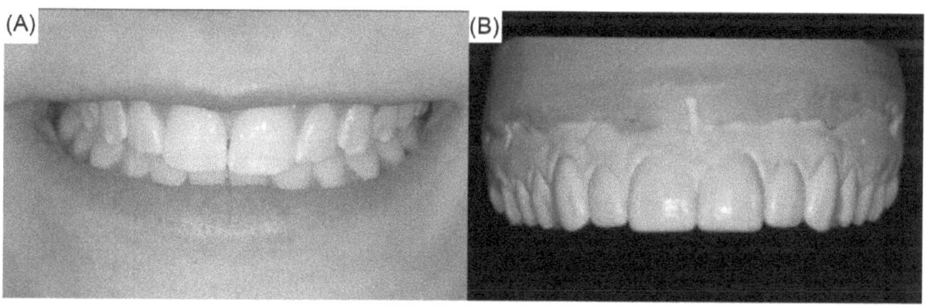

FIGURE 4.1

The preoperatory photo shows the challenge condition of color and shape (A) and the wax-up proposal (B) for the case.

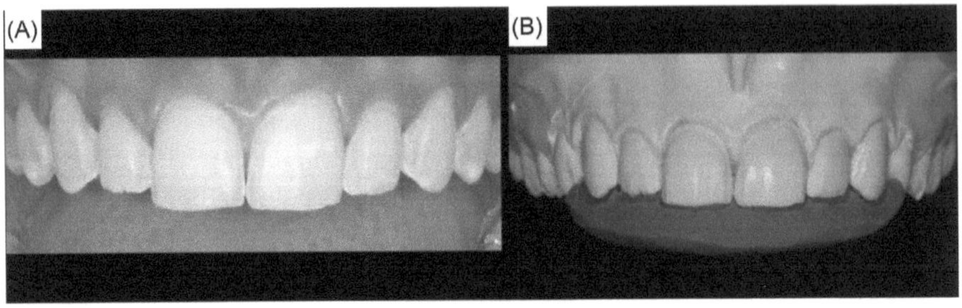

FIGURE 4.2

Minimally invasive preparation (A) and working model with prep guide (B).

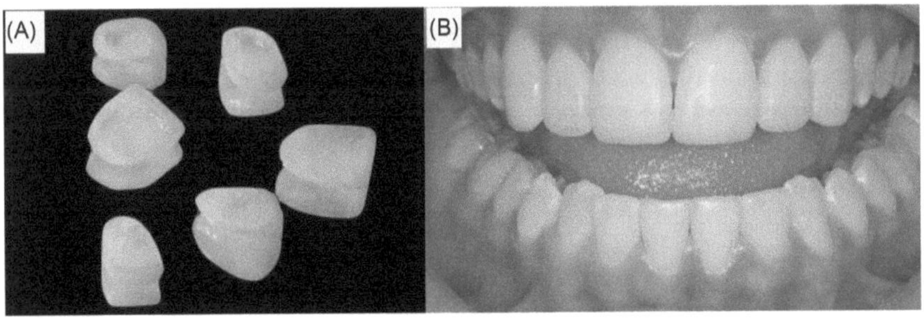

FIGURE 4.3

Nanohybrid composite veneers (A) and postoperatory condition (B) after luting.

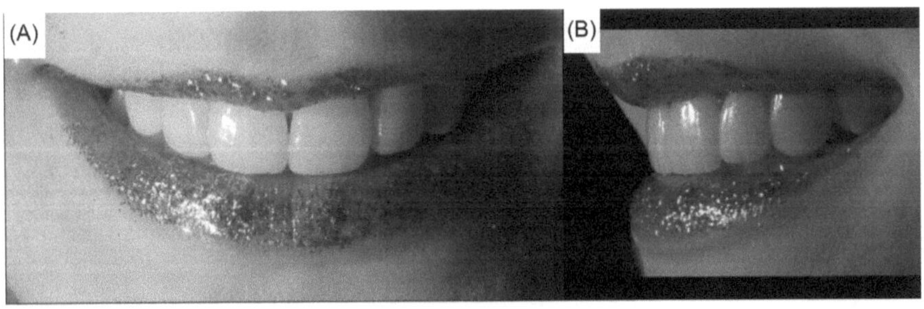

FIGURE 4.4

High esthetic biomimetic in front (A) and lateral (B) views of nanohybrid composite veneers.

Courtesy of H. Costa.

FIGURE 4.5

Scanning electron microscopy (SEM) of nanoclusters.

Besides the size that allows a higher amount of inorganic phase into the composite, the shape, i.e., nanoclusters (Fig. 4.5), nanorods, nanotubes, and nanofibers, has a significant and positive influence on resin bonded composites mechanical property.

The nanosized fillers inserted into resin composites lead to better physical and mechanical characteristics. An optical advantage of high esthetic potential, due to nanofiller size of 20 nm smaller than visible light wavelength that range from 400 to 800 nm, enhances the translucency and color mimetic. Significant augmentation of microhardness, diametral tensile, compressive, and flexural strengths is recorded when nanofillers are incorporated. Further composition changes as $CaPO_4$ nanostructures [5], caries preventive CaF_2 nanoparticles, and recombinant amelogenins are being added in order to synthesize biomimetic composites. Therefore, in the near future, nanofilled composites shall not be used as simple bioinert way of enamel/dentin replacement, but as a complex, biomimetic, and biointeractive material.

4.4 NANOTECHNOLOGY IN CAD/CAM

In 2012, a new nanoceramic Lava Ultimate (3MEspe, St. Paul, MN, EUA) was presented for CAD/CAM purposes (Fig. 4.6). The polymerized blocks contain 80 wt% of silica (20 nm diameter) and zirconia (4 to 11 nm diameter) treated with a silane coupling agent in a 95.05% polymerized organic phase. The nanofiller is produced through a bottom-up process although the restoration is milled top down. It results in fine marginal adaptation, improved mechanical behavior with high luster retention (Figs. 4.7 and 4.8).

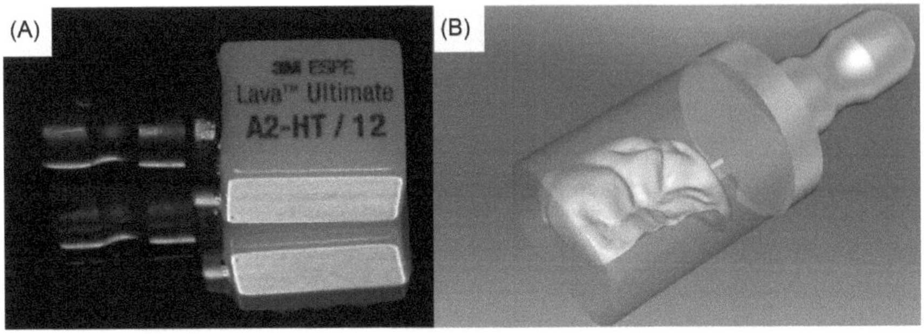

FIGURE 4.6

Block of nanofilled composite Lava Ultimate (3MEspe, St. Paul, MN, EUA) for CEREC (Sirona, Bensheim, Germany) (A) and computer-assisted designed (CAD) onlay proposal (B).

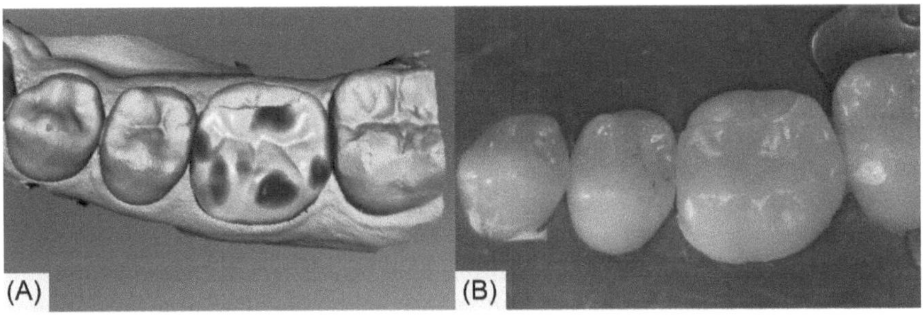

FIGURE 4.7

Intraoral scanning and CAD design (A) and computer-assisted manufactured cemented onlay (B).

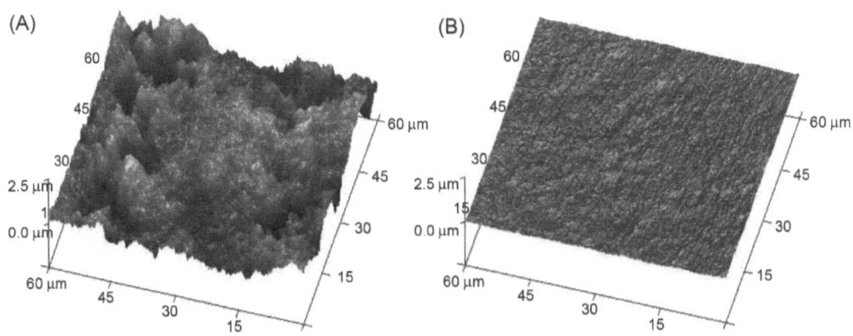

FIGURE 4.8

Atomic force microscopy of milled (A) and polished (B) surfaces of CAD/CAM nanoceramic.

Courtesy of Nunes LS and Fracasso L.

4.5 FILLERS IN COMPOSITE RESINS

Since the first formulation of composite resins in the 1960s, a basic triad is used: monomer, silane treated filler, and initiators. The filler used by Bowen [1], in 1963, consisted of milled quartz particles with average size ranging from 8 to 12 μm (8000–12,000 nm) as shown in Fig. 4.9.

Due to the esthetic limitations of macrofilled composites (lack of surface gloss), the minifilled composites were introduced in the 1970s. The filler material was produced by a pyrogenic method allowing a maximum load of 55 wt%, with better polishability, however with a significantly lower mechanical strength.

It was only in the 1980s and 1990s when mixtures of previous filler materials were tested. These hybrid fillers (600–2000 nm) were commercialized as hybrid, microhybrid, and condensable (whisker shaped) composites. Improvement in the mechanical strength was achieved; however, the polishability was still a limitation. A maximum load of filler from 70 to 77 wt% was recorded then. Still, the particle size of conventional composites was not similar to the size of the hydroxyapatite crystal, dentinal tubule, and enamel rod and that there was a potential for compromise in adhesion between the macroscopic restorative material and the nanoscopic (1–10 nm in size) tooth structure [6].

However, the milling procedure cannot normally reduce the filler particle size below 100 nm. Thus, nanotechnology has been used as an innovative way to manufacture fillers from the controlled growth of an initial molecule (bottom-up approach), resulting in homogeneity in shape and size (in the range between 5 and 20 nm).

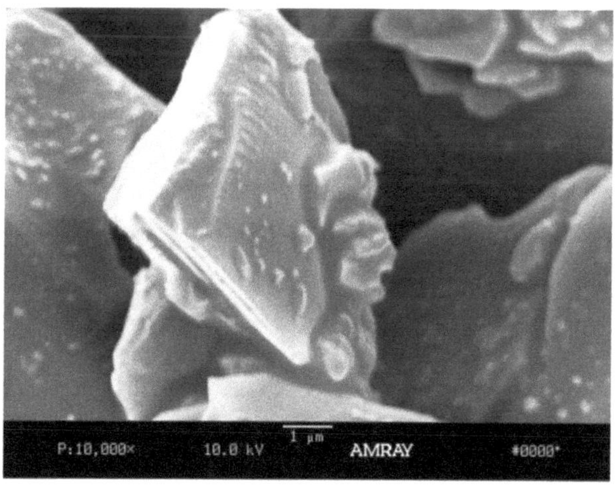

FIGURE 4.9

SEM of irregular macrofiller.

The Filtek Supreme (3MEspe, St. Paul, MN, USA) was the commercial milestone of nanotechnology application in operative dentistry in the beginning of this century. This composite resin was a combination of aggregated zirconia/silica cluster filler with a primary particle size of 5–20 nm and nanoagglomerated 20 nm silica filler in 78.5 wt%.

Over the last few years, a combination of microhybrid and nanofilled composites has been commercialized. This new combination has increased the filler weight content up to 87% by filling spaces between larger particles with smaller ones and has retained optical and mechanical characteristics which are known to be exclusive to nanofilled composites.

4.6 SEM AND EDS EVALUATIONS

SEM has been used to analyze the shape and size of the filler particles and as a qualitative method to classify the composite resins [7]. Comparative SEM images of nanofilled composites are presented in Fig. 4.10.

Morphological comparisons show that there are similar patterns between Esthet-X (Dentsply, Caulk, Milford, USA) and Grandio (Voco, GMBH Cuxhaven, Germany) [8]. High amount of irregular fillers is seen with additional load of small nanofillers around and/or over them occupying the empty spaces. The same pattern was registered to both enamel and dentin shades. A completely different pattern can be seen with Filtek Supreme XT (3MEspe, St. Paul, MN, USA) with high amount of rounded, uniform size fillers in both enamel and dentin shades. Following the classification proposed by Lutz and Phillips in 1983 [8], all evaluated composites could be classified as microhybrid, except the Filtek Supreme XT which is an exclusive nanofilled composite in accordance to Mitra et al. [3] and Beun et al. [9].

When fillers were analyzed with energy dispersive spectroscopy (EDS) [10], the following components were observed: aluminum (Al), barium (Ba), silicates (Si), and other heavy metals used to increase the radio opacity (Table 4.1). For

FIGURE 4.10

SEM: (A) Nanoclusters of Filtek Supreme XT, the exclusive nanofilled composite. (B) Esthet-X filler, a microhybrid composite with additional load of nanofillers. (C) Grandio, a high filled microhybrid composite with additional load of nanofillers.

Table 4.1 Descriptive Table of Components (wt%) Recorded from Nanofilled Composites Submitted to EDS

Composite Resins	Ba	Al	O	Si
Esthet-X enamel	37.56	6.15	23.37	32.92
Esthet-X dentin	37.00	6.05	17.42	39.53
Filtek supreme XT enamel	–	–	40.79	59.26
Filtek supreme XT dentin	–	–	37.38	62.62
Grandio enamel	–	10.38	36.93	52.70
Grandio dentin	5.18	10.12	33.27	51.42

–, *Not recorded.*

clinical inference, the composites with Ba and Al (Esthet-X) shall be more evident in X-ray and for clinical longevity of the filling. For Grandio, the dentin shade must be inserted in deeper increments which is more radiopaque when compared to the enamel shade (as seen in the X-ray in Fig. 4.33B).

4.7 FILLER WEIGHT CONTENT (WT%)

Thermogravimetric analysis is used to evaluate filler concentration in percentage by weight. The specimens of composite were inserted in a platinum container and subjected to 20°C/min temperature increase until 700°C (TGA 2050, TA Instruments, New Castle, DW, USA) in a nitrogen saturated atmosphere. The temperature of organic matrix degradation and filler weight percent (wt%) were recorded. The amount of inorganic residues was established by stabilization of the weight of the sample. The determination of the inorganic content was performed by weighing the mass of the composite specimen before and after the entire elimination of the organic phase [11]. The filler weight content of common nanofilled composites ranged from 75.75 to 87 wt% (see Fig. 4.11 and Table 4.2) [10].

In the relevant literature, it can be seen that there is a strong positive correlation between filler content and mechanical properties. Xu [12] concluded that there was a uniform improvement of elastic modulus and hardness when filler load was enhanced. Similarly, Sabbagh et al. [13] recorded a strong correlation ($r = 0.82$) between the filler weight content and elastic modulus of composite resins. Thus, the following section will describe the mechanical properties of nanofilled composites.

4.8 WATER SORPTION

Water absorption by composite materials is a diffusion-controlled process, and the uptake of water occurs largely in the resin matrix [14]. The water sorbed by

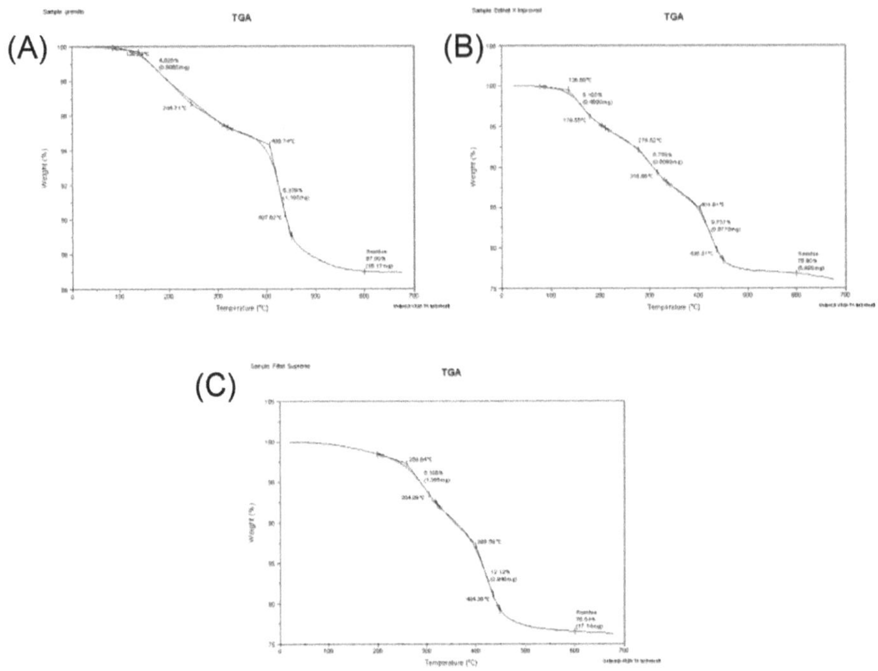

FIGURE 4.11

Comparative charts of thermogravimetry analysis of nanofilled enamel shade composites. (A) Grandio, (B) Esthet-X, and (C) Filtek Supreme XT.

Table 4.2 Descriptive Table of Filler Weight Content (wt%) of Nanofilled Composites Evaluated by Thermogravimetric Analyses

Composite Resin	wt%
Esthet-X enamel	76.8
Esthet-X dentin	75.75
Filtek supreme XT enamel	76.22
Filtek supreme XT dentin	76.54
Grandio enamel	87
Grandio dentin	86.89

the polymer matrix can cause filler/matrix debonding (Fig. 4.26) or even hydrolytic degradation of the fillers [15] and may reduce the mechanical properties of the composite materials [16]. Water sorption (WS) also results in increased overall weight.

Solubility, leaching, and hydrolytic degradation are results of either the breaking of chemical bonds in the resin or softening through the plasticizing action of water [17]. When resin samples are immersed in water, some of the components, such as unreacted monomers or fillers, dissolve and leach out of the sample. The release of these components may influence the initial dimensional changes of the composite, the clinical performance, the esthetic aspect of the restoration, or the biocompatibility of the material [17]. Solubility therefore results in weight reduction.

WS and water solubility (WSL) measurements were performed as described by Oysaed and Ruyter [18]. Ten disk specimens were used for each material. The diameter and the thickness of the specimens were measured and the volume (V) was calculated. The discs were conditioned in a desiccator for 3 days, containing calcium sulfate at 37°C until a constant weight had been achieved (w0). The disks were placed in a glass vial containing 100 mL of distilled water. The vials were wrapped by aluminum foil to protect from light and placed in an incubator at 37°C and at intervals removed, blot dried, and weighed. This was continued until the weight change/week was less than 0.32 μg (constant weight—w1). The discs were removed from the water and replaced in a desiccator for 24 h and then reweighed for the last time (w2). These steps were carried out to evaluate WS and WSL, in μg/cm^3.

$$\mathrm{WS} = \mathrm{w1} - \mathrm{w2/V}$$
$$\mathrm{WSL} = \mathrm{w0} - \mathrm{w2} = \mathrm{V}$$

where w0 is the sample weight before immersion, w1 is the sample weight after immersion, and w2 is the sample weight after immersion and desiccation.

For WS and WSL, statistical differences were recorded when nanofilled composites were compared. The exclusive nanofilled composite, Supreme XT, was more susceptible to water uptake. This can be explained by the higher contact surface of nanosized fillers. Samples cured with halogen device were less sensitive in water (Figs. 4.12 and 4.13).

4.9 MECHANICAL BEHAVIOR

In 1983, the composite resin materials were classified according to the average size of inorganic filler that became a standard. Based on the size, composite resin is classified as macroparticulated or traditional (up to 50 μm), hybrid (8–30 μm), microhybrid (0.7–3.6 μm), and microparticulated or minifilled (0.04–0.2 μm) [8]. Due to many differences between materials classified in the same group, other methods for classification were suggested based on mechanical properties [7,19].

Compressive strength, diametral tensile strength, flexural strength, and hardness have also been evaluated [20–22], due to a direct influence of composition in the mechanical behavior [23]. However, different composites are available in the market and are classified just by inorganic filler size, suggesting that resins of the same group would have similar mechanical behavior. Therefore, the following described mechanical properties and results will show a dissimilar behavior.

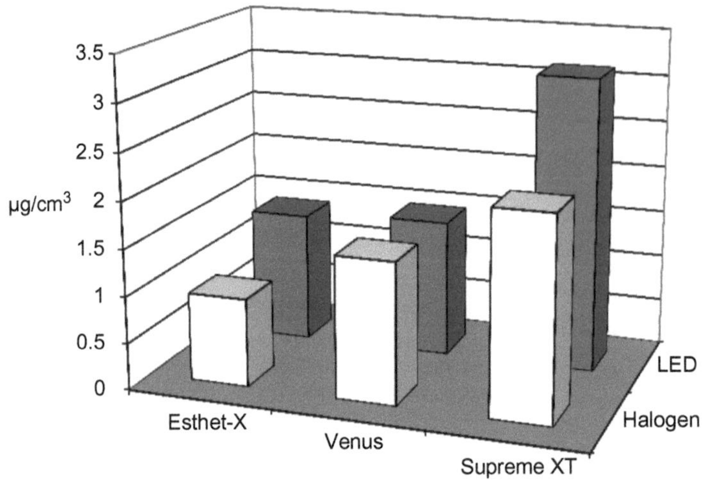

FIGURE 4.12

Comparison of water sorption (µg/cm^3) between nanofilled composites when photocured by two different sources.

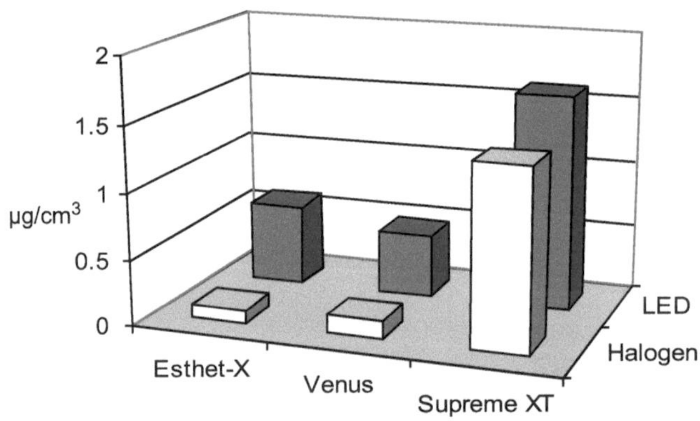

FIGURE 4.13

Comparison of solubility (µg/cm^3) between nanofilled composites when photocured by two different sources.

4.9.1 COMPRESSIVE STRENGTH

The compressive measurement test induces stress into the sample tested (in order to quantify the maximum strength, the filler and polymeric resin can withstand). This mechanical property results in higher cohesive strength of the material. Resin-based restorations in functional areas are submitted to up to 800 N. The relation between the maximum strength (N) applied on a sample divided by the sectional area (mm^2) results in the maximum compressive strength in megapascal

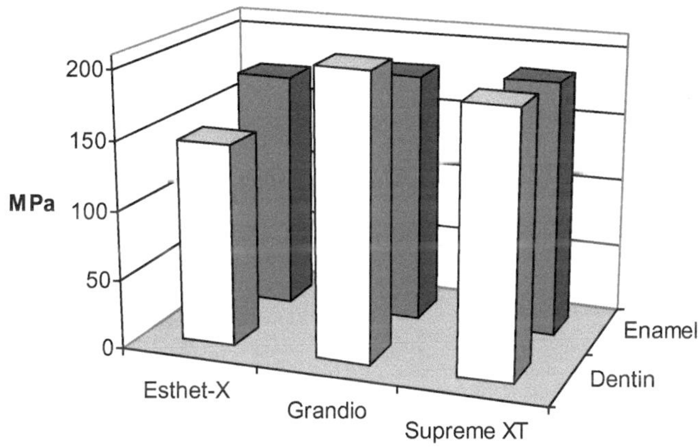

FIGURE 4.14

Graphic representation of compressive strength (MPa) of nanofilled composites in dentin and enamel shades.

(MPa). A comparative view in Fig. 4.14 describes the compressive strength [22] of nanofilled composites. The compressive strength ranges from 146.63 to 206.08 MPa. There was no statistical difference between them and in both shades (Anova and Tukey, $P = 0.18$).

4.9.2 DIAMETRAL TENSILE STRENGTH

This in vitro methodology was developed to test the cohesion of materials when subjected to tensile stress. However, when brittle materials are exposed to tensile stress, they are most likely to break easily. The diametral tensile test applies a compressive load into a cylindrical-shaped sample, resulting in elongation as an indirect of tensile strength. This mechanical property is calculated (MPa) as a relation of two times the maximum strength recorded under the relation of π, diameter, and height of the sample. The diametral tensile strength [22] ranged from 34.87 to 50.26 (Fig. 4.15). There was a significant difference between the composites ($P = 0.02$). Supreme XT enamel shade was statistically higher then Esthet-X Dentin. The other composites and shades did not differ from both resins studied [24].

Besides the results after 24 h of storage, the effect of artificial accelerated aging on mechanical properties has also been recorded [25]. Samples of nanofilled composites were submitted to 300,000 mechanical cycles of 80 N at 37°C. Fig. 4.16 presents the effect of this load on nanofilled composites. The results (MPa) ranged from 41.6 (Esthet-X) to 53.39 (Supreme XT). No differences were recorded between composites submitted up to 300,000 mechanical cycles and the samples which were not submitted to the mechanical loading cycle. This means that nanofilled composites might support the masticatory load better than other microhybrid composites.

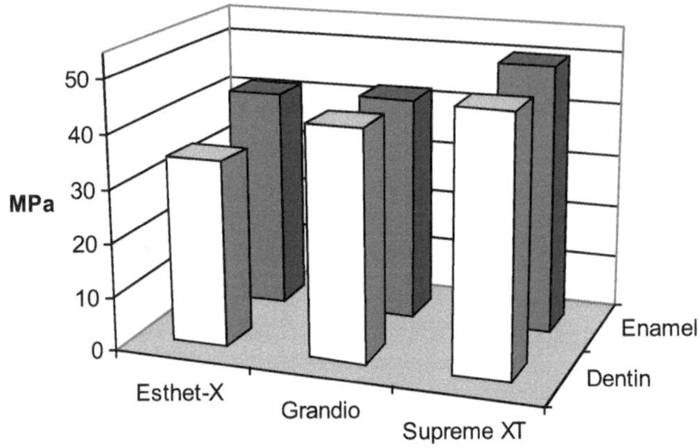

FIGURE 4.15

Graphic representation of diametral tensile strength (MPa) of nanofilled composites in dentin and enamel shades.

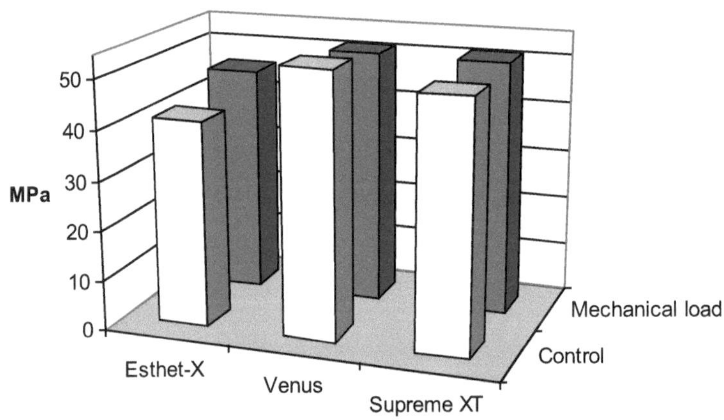

FIGURE 4.16

Comparative graph of diametral tensile strength of nanofilled composites submitted up to 300.000 cycles of 80 N.

4.9.3 FLEXURAL STRENGTH AND FLEXURAL MODULUS

The flexural strength [22] (Fig. 4.17) was carried out according to the International Standardization Organization specification 4049 due to the capability to test complex stress (compressive, tensile, and shear) simultaneously using one sample.

The flexural modulus (GPa) is the relation of flexural strength to the deformation of materials. The higher the flexural or elastic modulus of a material, the higher the energy required to deform it. In literature, 12–14 GPa has been reported as elastic modulus of human dentin. Restorative materials are expected to perform like natural dentin; however, different behavior [22] is presented in Fig. 4.18.

4.9.4 MICROHARDNESS

The hardness test measures the ability of a material to be indented or abraded by another material. In dentistry, microhardess is commonly used due to smaller sample dimensions and multiple compositions of the materials. Two main methods are described in the literature: Vickers and Knoop. Vickers indenter is associated to brittle materials and use a square-shaped indenter tip (Fig. 4.19A). However, some researchers classify composite resins as elastic materials due to the organic phase and use Knoop point with "V" shape (Fig. 4.19B).

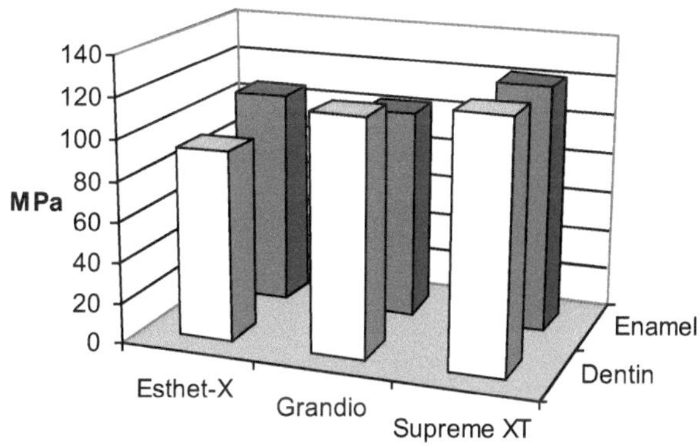

FIGURE 4.17

Comparative graph of flexural strength (MPa) of nanofilled composites in enamel and dentin shades.

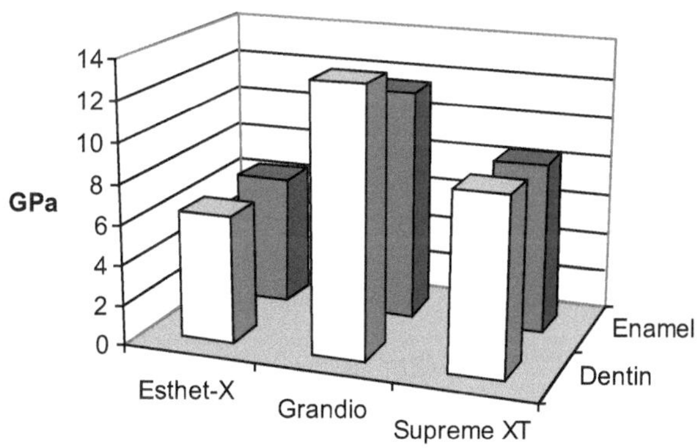

FIGURE 4.18

Comparative graph of flexural modulus (GPa) of nanofilled composites in enamel and dentin shades.

The potential resistance to wear is mainly explained by hardness. Resin composites fillings in functional areas are more susceptible to abrasion. In Fig. 4.20, the exclusively nanofilled composite Supreme XT showed the highest average microhardness when compared to Esthet-X and Venus. However, there was no significant effect of 300,000 load cycles on the hardness. This is a completely different behavior when compared to minifilled and microhybrid composites [25]. At the same time, this gives an indication of higher clinical longevity which must be considered.

4.9.5 NANOHARDNESS

Since the use of nanofillers started, the term "nanohardness" has been introduced. The use of a nanoindenter with a Berkovich diamond tip and with a nominal radius of 5000 nm has been widely used. A nanoindenter measures the hardness

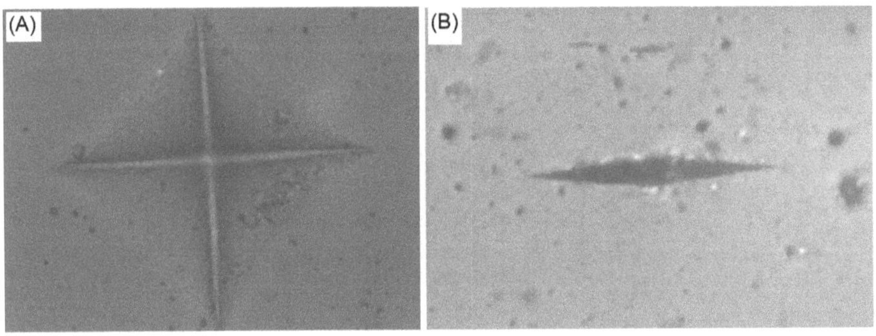

FIGURE 4.19

Microhardness impressions using Vickers (A) and Knoop (B) indenters of Supreme XT and Grandio, respectively.

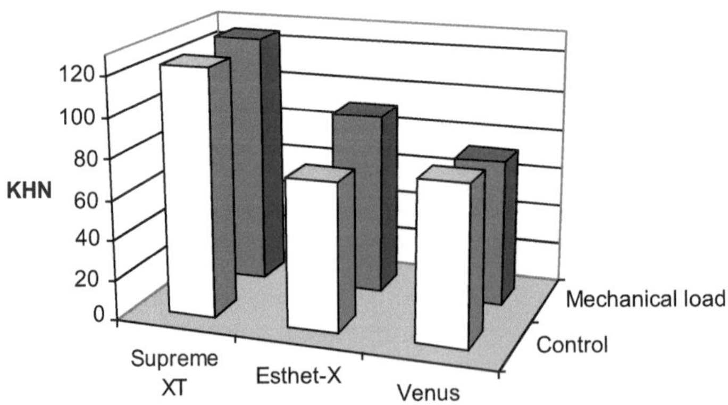

FIGURE 4.20

Comparative graph of Knoop microhardness of nanofilled composites when submitted to 300,000 cyclic loads.

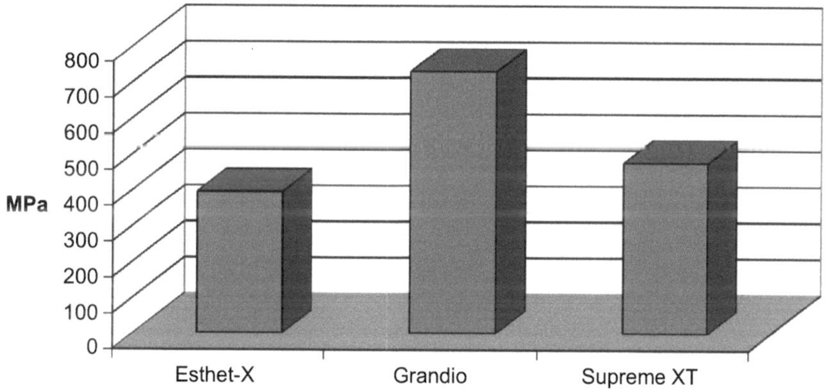

FIGURE 4.21

Comparison of nanohardness.

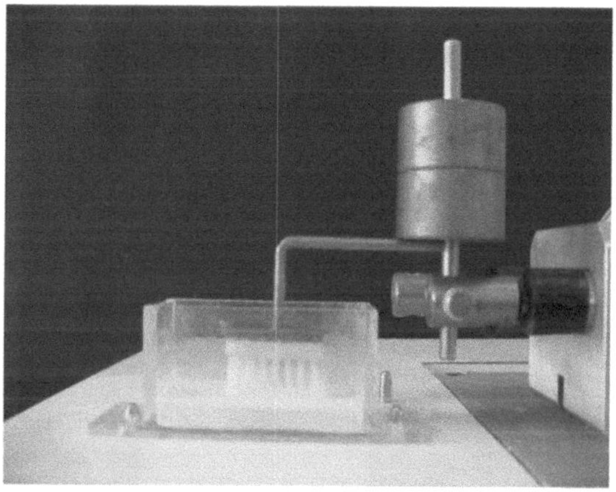

FIGURE 4.22

Sample exposed to three-body wear test.

and elastic modulus of a material on an extremely smaller surface scale of 50 nm [26]. Nanohardness results [22] ranged from 392.94 to 474.79 MPa (Fig. 4.21), far less than 4350 from enamel recorded by Machado [26].

4.9.6 WEAR RESISTANCE

According to O'Brien and Yee [27], clinical wear mechanisms of composite resin are described as follows: (1) organic matrix wear, (2) loss of filler bond to the matrix, (3) shear of filler, (4) matrix cohesive failure, and (5) exposure of air blisters. Some of these phenomena still occur in nanofilled composites and can be seen in Figs. 4.22–4.28. Morphological aspects observed in SEM in different simulated periods of three-body wear show rounded fillers exposed by the organic phase removal, small cracks, and loss of fillers due to debonding.

The wear resistance depends upon the filler weight content (as described in Section 4.7), organic phase, and silanization. In 2002, Kim et al. [28] suggested

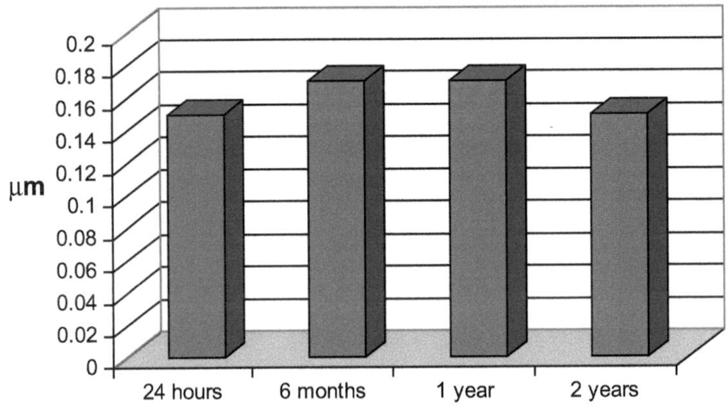

FIGURE 4.23

Comparative wear Ra (μm) of Supreme XT after 24 h, 6 month, 1 year, and 2 years of three-body simulated test.

Courtesy of Braun AP and Oshima HMS.

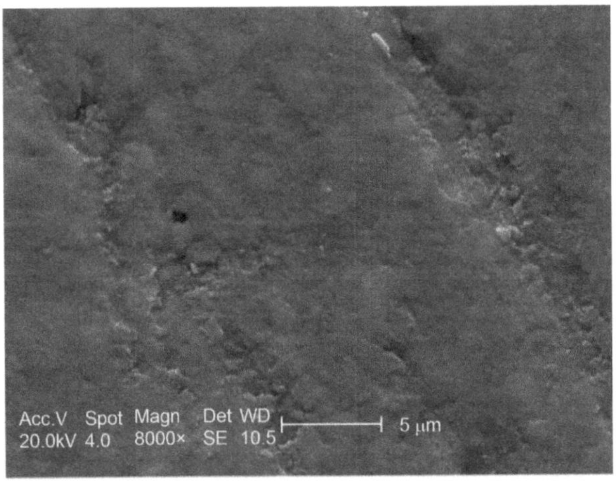

FIGURE 4.24

Morphological aspect of Filtek Supreme XT after finishing and polishing (8000 ×). The organic phase covers entirely the fillers.

Courtesy of Braun AP and Oshima HMS.

FIGURE 4.25

Morphological aspect of Filtek Supreme XT after 1 year of simulated wear (8000×). It is possible to observe the removal of superficial organic phase exposing rounded particles and nanoclusters.

Courtesy of Braun AP and Oshima HMS.

FIGURE 4.26

Morphological aspect of Filtek Supreme XT after 2 years of simulated wear (8000×). Small cracks and filler debonding can be observed.

Courtesy of Braun AP and Oshima HMS.

FIGURE 4.27

Morphological aspect of Filtek Supreme XT after 4 years of simulated wear (8000 ×).

Courtesy of Braun AP and Oshima HMS.

FIGURE 4.28

Morphological aspect of Filtek Supreme XT after 6 years of simulated wear (8000 ×). A similar pattern can be observed compared to 1, 2, and 4 years suggesting maintenance of roughness.

Courtesy of Braun AP and Oshima HMS.

the use of smaller particles in order to improve the filler load, reducing the exposure of organic resin to wear that was confirmed later by Yesil et al. [29]. The effect of this wear (Fig. 4.22) in a nanofilled composite (Supreme XT) is measured by surface roughness (μm) and is presented in Fig. 4.28.

4.10 CLINICAL APPLICATION

The following clinical report (Figs. 4.29–4.33) illustrates an interproximal restorative procedure in a 20-year-old male patient with nanofilled composite resin.

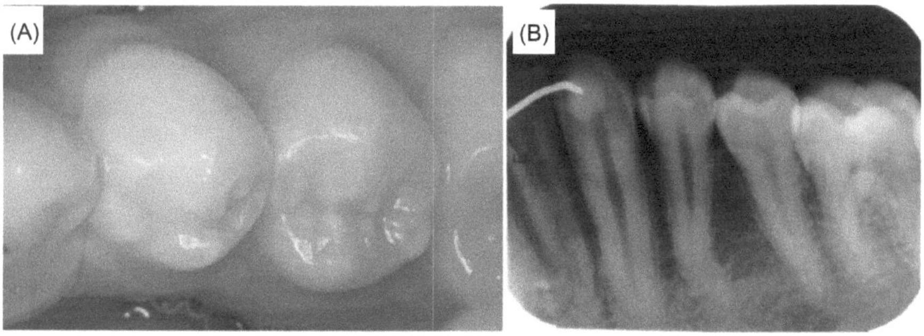

FIGURE 4.29

The clinical protocol followed all conservative procedures, removing only the affected hard tissues. The beginning aspect of the interproximal cavity (A) and the X-ray confirmation (B).

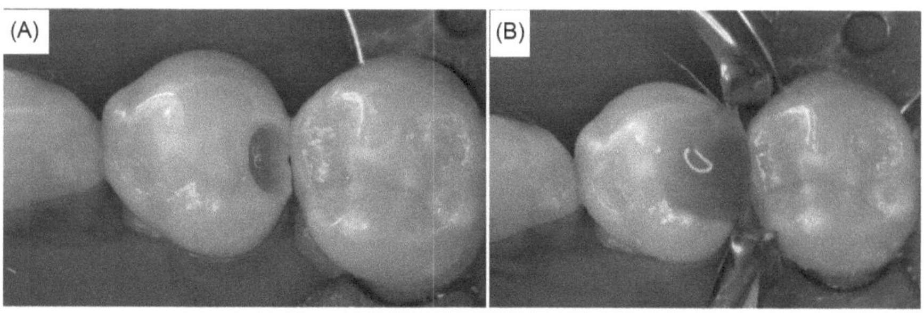

FIGURE 4.30

After complete removal of decayed tissue (A), the adhesive protocol was total acid etch with 37% phosphoric acid (B) for 15 s in dentin and 20 s in enamel followed by water rinse and gently air blow; two consecutive layers of one-bottle adhesive and light curing for 20 s with 1000 mW/cm^2 LED light source.

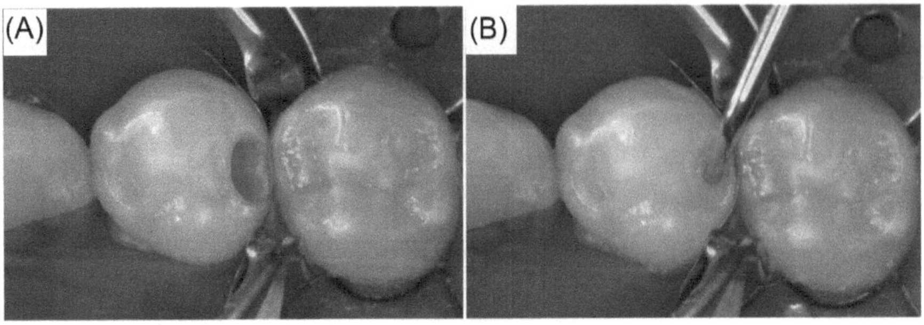

FIGURE 4.31

Hybridization aspect of preparation (A). The nanofilled composite resin Grandio A2 dentin and A1 enamel shades were used, incrementally inserted (B), and light cured for 20 s each increment with 1000 mW/cm^2 LED light source.

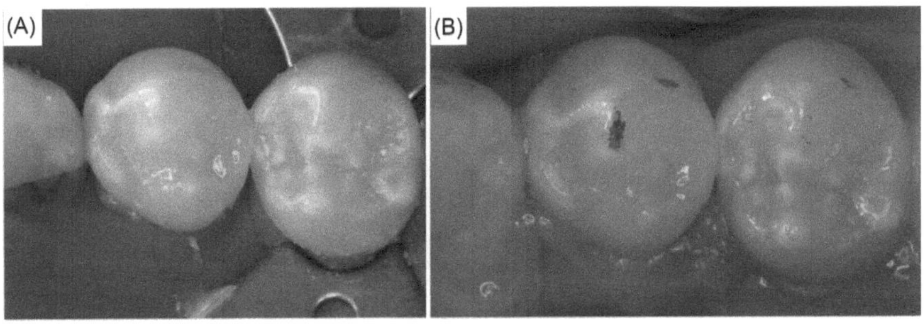

FIGURE 4.32

The filling aspect immediately after matrix removal (A) and the immediate aspect after finishing and polishing (B).

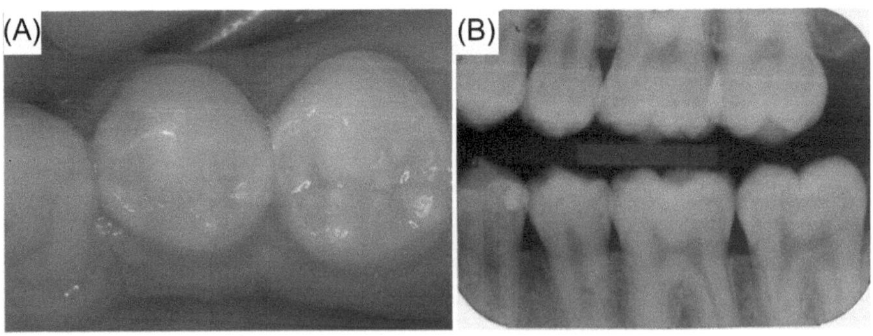

FIGURE 4.33

Part (A) shows the filling after 1 week and the final X-ray (B) with radiopacity for further radiographic control.

4.11 CONCLUSIONS

In dentistry, nanofiller could be considered an inorganic particle with average size of 40 nm or 0.04 μm. This size, however, is not an innovation in dental composites, because the minifilled composites had the same 0.04 μm (40 nm) filler particles since the 1970s. The real innovation is the nanofiller manufacture and the possibility of improving the load of inorganic phase [22]. The effect of this high filler load is widely recorded in terms of mechanical properties. Microhybrid composites with additional load of nanofillers are the best clinical choice. The next research frontier in operative dentistry is the manufacture of a filler material with shape and composition that can closely mimic the optical and mechanical characteristics of the natural hard tissues (enamel and dentin).

ACKNOWLEDGMENTS

The authors would like to thank the associated professors and students of Dentistry Graduate Program of Pontifical Catholic University of Rio Grande do Sul, Brazil.

REFERENCES

[1] R.L. Bowen, Properties of a silica-reinforced polymer for dental restorations, J. Am. Dent. Assoc. 66 (1963) 57–64.

[2] Y. Xia, F. Zhang, H. Xie, N. Gu, Nanoparticle-reinforced resin-based dental composites, J. Dent. 36 (6) (2008) 450–455.

[3] S.B. Mitra, D. Wu, B.N. Holmes, An application of nanotechnology in advanced dental materials, J. Am. Dent. Assoc. 134 (2003) 1382–1390.

[4] S.A. Saunders, Current practicality of nanotechnology in dentistry. Part 1: Focus on nanocomposite restoratives and biomimetics, Clin. Cosm. Inv. Dent. 1 (2009) 47–61.

[5] H.H. Xu, L. Sun, M.D. Weir, S. Takagi, L.C. Chow, B. Hockey, Effects of incorporating nanosized calcium phosphate particles on properties of whisker-reinforced dental composites, J. Biomed. Mater. Res. B Appl. Biomater. 81 (1) (2007) 116–125.

[6] D.A. Terry, Direct applications of a nanocomposite resin system: part 1—The evolution of contemporary composite materials, Pract. Proced. Aesthet. Dent. 16 (2004) 417–422.

[7] H. Hosoda, T. Yamada, S. Inokoshi, SEM and elemental analysis of composite resins, J. Prosthet. Dent. 64 (1990) 669–676.

[8] F. Lutz, R.W. Phillips, A classification and evaluation of composite resin systems, J. Prosthet. Dent. 50 (1985) 480–488.

[9] S. Beun, T. Glorieux, J. Devaux, J. Vreven, G. Leloup, Characterization of nanofilled compared to universal and microfilled composites, Dent. Mater. 23 (1) (2007) 51–59.

[10] E.G. Mota, L. Hörlle, H.M. Oshima, L.M. Hirakata, Evaluation of inorganic particles of composite resins with nanofiller content, Stomatologija 14 (4) (2012) 103–107.

[11] J. Sabbagh, L. Ryelandt, L. Bachérius, J.J. Biebuyck, J. Vreven, P. Lambrechts, et al., Characterization of the inorganic fraction of resin composites, J. Oral. Rehab. 31 (11) (2004) 1090–1101.
[12] H.H.K. Xu, Dental composite resins containing silica-fused ceramic single-crystalline whiskers with various filler levels, J. Dent. Res. 78 (1999) 1304–1311.
[13] J. Sabbagh, J. Vreven, G. Leloup, Dynamic and static moduli of elasticity of resin-based materials, Dent. Mater. 18 (2002) 64–71.
[14] M. Braden, R.L. Clarke, Water absorption characteristics of dental microfine composite filling materials, Biomaterials 5 (1984) 369–372.
[15] K.J. Söderholm, M. Zigan, M. Ragan, W. Fischlschweiger, M. Bergman, Hydrolytic degradation of dental composites, J. Dent. Res. 63 (1984) 1248–1254.
[16] A. El-Hadary, J.L. Drummond, Comparative study of water sorption, solubility, and tensile bond strength of two soft lining materials, J. Prosthet. Dent. 83 (2000) 356–361.
[17] J.L. Ferracane, Elution of leachable components from composites, J. Oral. Rehab. 21 (1994) 441–452.
[18] H. Oysaed, I.E. Ruyter, Composites for use in posterior teeth: mechanical properties tested under dry and wet conditions, J. Biomed. Mater. Res. 20 (1986) 261–271.
[19] B.R. Lang, M. Jaarda, R.F. Wang, Filler particle size and composite resin classification systems, J. Oral. Rehab. 19 (1998) 569–584.
[20] T. Brosh, Y. Ganor, I. Belov, R. Pilo, Analysis of strength properties of light-cured resin composites, Dent. Mater. 15 (1999) 174–179.
[21] D.S. Cobb, K.M. MacGregor, M.A. Vargas, G.E. Denehey, The physical properties of packable and conventional posterior resin-based composites: a comparison, J. Am. Dent. Assoc. 131 (2000) 1610–1615.
[22] R.S. Rosa, C.E. Balbinot, E. Blando, E.G. Mota, H.M. Oshima, L. Hirakata, et al., Evaluation of mechanical properties on three nanofilled composites, Stomatologija 14 (4) (2012) 126–130.
[23] E.C. Say, A. Civelek, A. Nobecourt, M. Ersoy, C. Guleryuz, Wear and microhardness of different resin composite materials, Oper. Dent. 28 (2003) 628–634.
[24] E.G. Mota, H.M. Oshima, L.H. Burnett Jr, L.A. Pires, R.S. Rosa, Evaluation of diametral tensile strength and Knoop microhardness of five nanofilled composites in dentin and enamel shades, Stomatologija 8 (3) (2006) 67–69.
[25] P.N. Gomes, S.C. Dias, M.R. Moyses, L.J. Pereira, B.G. Negrillo, J.C. Ribeiro, Effect of artificial accelerated aging on Vickers microhardness of composite resins, Gen. Dent 56 (7) (Nov–Dec 2008) 695–699.
[26] C. Machado, W. Lacefield, A. Catledge, Human enamel nanohardness, elastic modulus and surface integrity after beverage contact, Braz. Dent. J 19 (1) (2008) 68–72.
[27] W.J. O'Brien, J. Yee Jr, Microstructure of posterior restorations of composite resin after clinical wear, Oper. Dent. 5 (3) (1980) 90–94 Summer.
[28] K. Kim, J.L. Ong, O. Okuno, The effect of filler loading and morphology on the mechanical properties of contemporary composites, J. Prosthet. Dent. 87 (2002) 642–649.
[29] Z.D. Yesil, S. Alapati, W. Johnston, R.R. Seghi, Evaluation of the wear resistance of new nanocomposite resin restorative materials, J. Prosthet. Dent 99 (6) (2008) 435–443.

CHAPTER

Impact of nanotechnology on dental implants

5

Karthikeyan Subramani[1], Sandrine Lavenus[2], Julie Rozé[2], Guy Louarn[2] and Pierre Layrolle[2]

[1]*Roseman University of Health Sciences, Henderson, NV, United States*

[2]*University of Nantes, Nantes, France*

CHAPTER OUTLINE

5.1 Introduction 83

5.2 Nanoscale Surface Modifications 86

5.3 Interactions of Surface Dental Implants With Blood 87

5.4 Interactions Between Surfaces and MSCs 89

5.4.1 Origin of MSCs 89

5.4.2 Migration, Adhesion, and Proliferation 89

5.4.3 Differentiation 90

5.5 Tissue Integration 91

5.6 Conclusion 93

Acknowledgments 93

References 93

5.1 INTRODUCTION

Implants are commonly used in dental surgery for restoring teeth. One of the challenges in implantology is to achieve and maintain the osseointegration as well as the epithelial junction of the gingival with implants. An intimate junction of the gingival tissue with the neck of dental implants may prevent bacteria colonizations leading to peri-implantitis, while direct bone bonding may ensure a biomechanical anchoring of the artificial dental root (Fig. 5.1).

The first step of the osseointegration of implants is called primary stability and is related to the mechanical anchorage, design of implants and bone structure [1]. This primary interlock decreases with time at the benefit of the secondary anchorage, which is characterized by a biological bonding at the interface between bone tissues and implant surface. Between the primary mechanical and secondary biological anchorage, a decreased of implant stability could be observed. Many studies have attempted to enhance the osseointegration of

Emerging Nanotechnologies in Dentistry. DOI: http://dx.doi.org/10.1016/B978-0-12-812291-4.00005-4

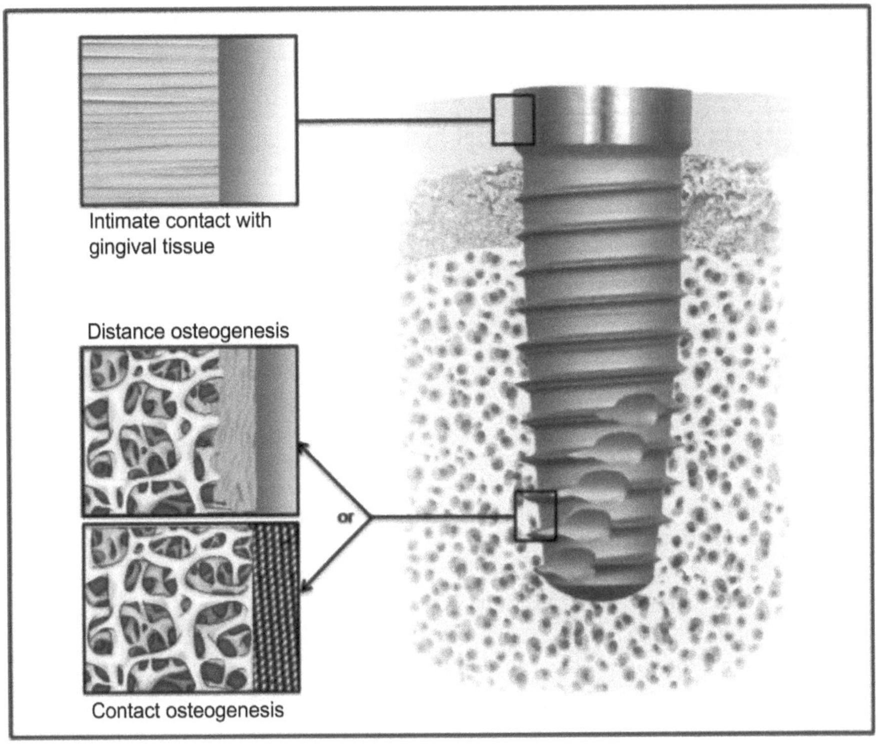

FIGURE 5.1

Tissue integration of dental implant. Note the intimate contact with gingival tissue in the upper part and the desired contact osteogenesis in the tapered lower part rather than distance osteogenesis.

implants by various surface modifications. The aim is to provide metal implants with surface biological properties for the adsorption of proteins, the adhesion and differentiation of cells and tissue integration. These biological properties are related to chemical composition, wettability and roughness of metal implants surfaces. However, the control of these surface properties at the protein and cell levels, thus in the nanometer range, remains a challenge for researchers and dental implants manufacturers.

Nanotechnologies may produce surfaces with controlled topography and chemistry that would help understanding biological interactions and developing novel implant surfaces with predictable tissue-integrative properties [2–4]. Various processing methods derived from the electronic industry such as lithography, ionic implantation, anodization, radio frequency plasma treatments may be applied to the surfaces of dental implants to produce controlled features at the nanometer scale. These surfaces may then be screened by using high throughput biological assays in vitro. For instance, specific protein adsorption, cell adhesion and differentiation of stem cells should be studied in relation to the surface

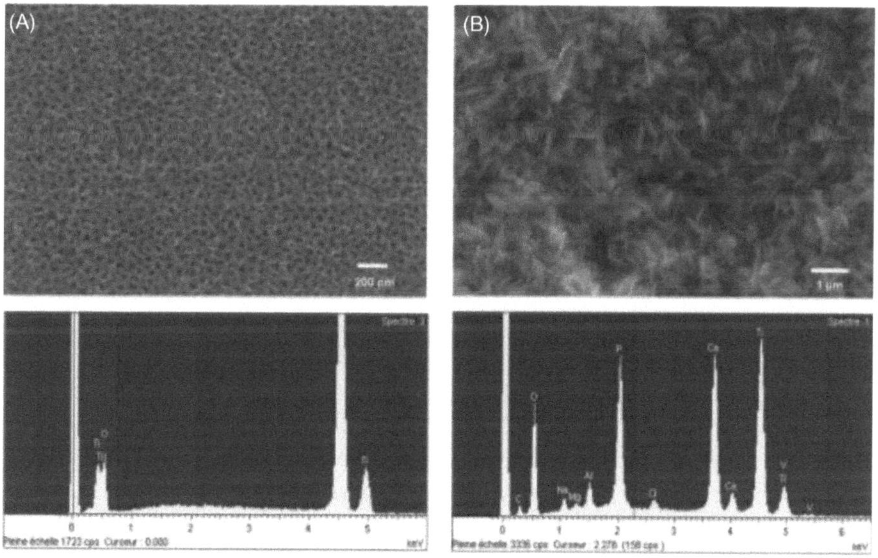

FIGURE 5.2

Scanning electron micrographs and energy dispersive analysis for X-ray of (A) nanostructured titanium surface obtained by anodization and (B) nanosized thin calcium phosphate (CaP) coating on titanium produced by electrochemical deposition. Note the regular array of TiO_2 nanopores of approximately 100 nm in diameter and the nanosized CaP crystals on titanium surfaces.

properties. This approach may define the ideal surface for a specific biological response. Following in vitro screening, nanostructured surfaces may then be tested in animal models to validate hypothesis in a complex in vitro environment.

New coating technologies have also been developed for applying hydroxyapatite (HA) and related calcium phosphates (CaP), the mineral of bone, onto the surface of implants (Fig. 5.2). Many studies have demonstrated that these CaP coatings provided titanium implants with an osteoconductive surface [5,6]. Following implantation, the dissolution of CaP coatings in the peri-implant region increased ionic strength and saturation of blood leading to the precipitation of biological apatite nanocrystals onto the surface of implants. This biological apatite layer incorporates proteins and promotes the adhesion of osteoprogenitor cells that would produce the extracellular matrix of bone tissue. Furthermore, it has been also shown that osteoclasts, the bone resorbing cells, are able to degrade the CaP coatings through enzymatic ways and created resorption pits on the coated surface [6]. Finally, the presence of CaP coatings on metals promotes an early osseointegration of implants with a direct bone bonding as compared to noncoated surfaces. The challenge is to produce CaP coatings that would dissolve at a similar rate than bone apposition in order to get a direct bone contact on implant surfaces.

This chapter reviews the different steps of the interactions between biological fluids, cells, tissues and surfaces of implants. Recent nanoscale surface modifications and CaP coating technologies of dental implants are discussed. The sequence of biological events in relation to surface properties is related. Mechanisms of interaction with blood, platelets, hematopoietic and mesenchymal stem cells (MSCs) on the surface of implants are described. These early events have shown to condition the adhesion, proliferation, and differentiation of cells as well as the osseointegration of implants. Future implant surfaces may improve the tissue-integrative properties and long-term clinical success for the benefits of patients.

5.2 NANOSCALE SURFACE MODIFICATIONS

Surfaces properties play a determinant role in biological interactions. In particular, the nanometer sized roughness and the chemistry have a key role in the interactions of surfaces with proteins and cells. These early interactions will in turn condition the late tissue integration. In this prospect, different methods have been reported for enhancing bone healing around metal implant [2,7].

Modifying surface roughness has been shown to enhance the bone to implant contact and improve their clinical performance [2,8]. Grid-blasting, anodization, acid etching, chemical grafting, and ionic implantation were the most commonly used methods for modifying surface roughness of metal implants. Combinations of these techniques could be used such as acid-etching after grit-blasting in order to eliminate the contamination by blasting residues on implant surfaces. This grit-blasting residue may interfere with the osseointegration of the titanium dental implants [9–11]. It has been shown that grit-blasting with biphasic calcium phosphate (BCP) ceramic particles gave a high average surface roughness and particle-free surfaces after acid etching of titanium implants. Studies conducted both in vitro and in vivo have shown that BCP grit-blasted surfaces promoted an early osteoblast differentiation and bone apposition as compared to mirror-polished or alumina grit-blasted titanium [12,13]. Anodization is a method commonly used to obtain nanoscale oxides on metals including titanium [14,15]. By adjusting the anodization condition such as voltage, time and shaking, nanoscale properties could be controlled. Shankar et al. [16] have reported that the diameters of the nanotubes could be modified to a range from 20 to 150 nm in modifying voltage conditions. On the other hand, Kang et al. [17] found that TiO_2 nanotube arrays were more uniform on electropolished than machined titanium. Moreover, TiO_2 nanotubes on Ti improved the production of alkaline phosphatase (ALP) activity by osteoblastic cells. In particular, nanotubes with a diameter of 100 nm up-regulated level of ALP activity as compared to 30–70-nm-diameter nanotube surfaces [18]. Since ALP is a marker of osteogenic differentiation, these surfaces may demonstrate enhanced bone tissue integrative properties.

Another approach for improving osseointegration of dental implants is to apply a CaP coating having osteoconductive properties [19–21]. Different methods have been developed to coat metal implants with CaP layers such as plasma spraying, biomimetic and electrophoretic deposition. Nevertheless, plasma-sprayed HA coated dental implants have been related to clinical failures due to coating delimitation and heterogeneous dissolution rate of deposited phases. An electrochemical process which consists of depositing CaP crystals from supersaturated solutions has been proposed for coating titanium implants with CaP layers [22,23]. Upon implantation, these CaP coatings dissolve and release Ca^{2+} and HPO_4^{2-} increasing saturation of blood in the peri-implant region. This dissolution led to the precipitation of biological apatite nanocrystals with the incorporation of various proteins. This biological apatite layer will promote cell adhesion, differentiation into osteoblast and the synthesis of mineralized collagen, the extracellular matrix of bone tissue. In addition to dissolution, osteoclast cells are also able to resorb the CaP coatings and activate osteoblast cells to produce bone tissue. As the result, these CaP coatings promote a direct bone-implant contact without an intervening connective tissue layer leading to a proper biomechanical fixation of dental implants.

5.3 INTERACTIONS OF SURFACE DENTAL IMPLANTS WITH BLOOD

During surgery, blood vessels are injured and, thus, dental implant surfaces interact with blood components (Fig. 5.3). Various plasma proteins get adsorbed on

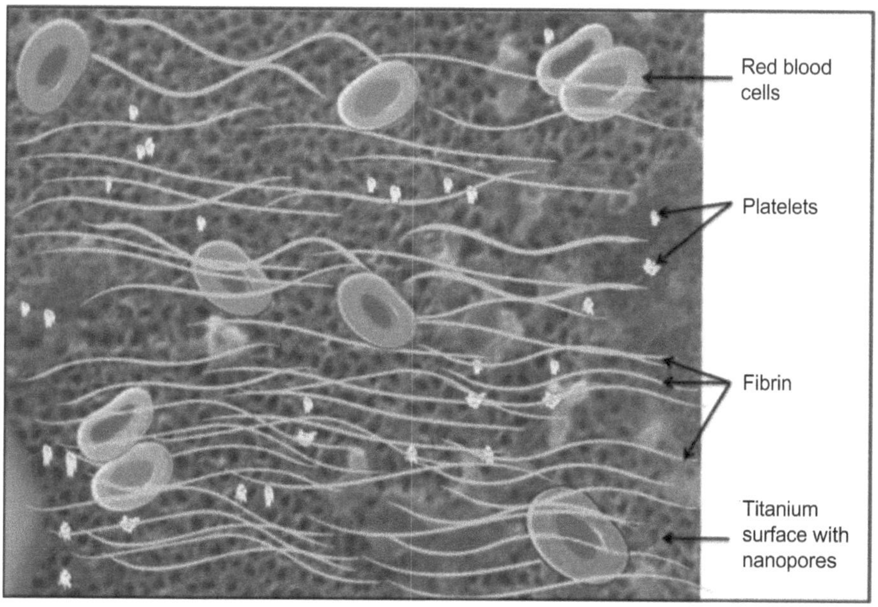

FIGURE 5.3

Interactions of surface of dental implants with blood. Note the numerous proteins, red blood cells and activated platelets that lead to blood clotting on implants.

the material surface within a minute. Platelets from blood interact also with the implant surface. Plasma proteins modified the surface while activated platelets are responsible of thrombus formation and blood clotting. Subsequently, the migrations of various cell types interact with the surface through membrane integrin receptors. These early events occur prior to peri-implant tissue healing.

Plasma contains dissolved substances such as glucose, amino acids, cholesterols, hormones, urea, and various ions (Fig. 5.4). Most of these components are needed for the viability of cells and tissues. All of these blood substances could interact with implant surface thus modifying their chemical properties like charge or hydrophobicity.

Blood interactions with implants lead to protein adsorption, which is dependent on the surface properties of the material and occurs through a complex series of adsorption and displacement steps known as the Vroman effect [24]. A hydrophilic surface is better for blood coagulation than a hydrophobic surface. Consequently, dental implants manufacturers have developed high hydrophilic and rough implant surfaces which in turn exhibited better osseointegration than conventional ones [25]. Adsorption of proteins such as fibronectin, vitronectin on surface of dental implants could promote cell adhesion by cell-binding RGD domain (arg-gly-asp). This RGD sequence interacts with integrin present on the cell membrane [26]. Interactions between cell membrane integrins and proteins coated onto implant surface play a key role in adhesion of many cells types. After proteins absorption, the osseointegration is characterized by platelets adhesion

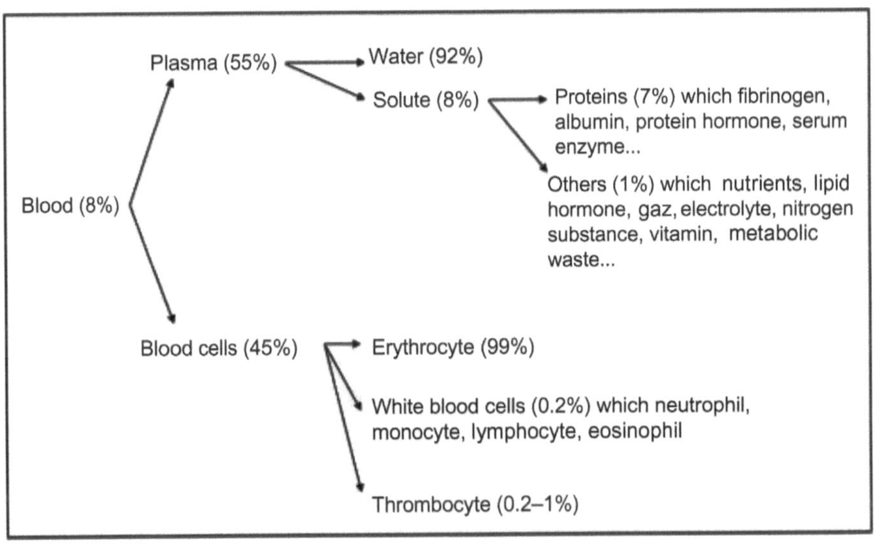

FIGURE 5.4

Scheme showing blood composition and components that primarily interact with surface of dental implants.

and fibrin clots formation at the injured blood vessels site. It has been shown that implants in contact with platelet-rich plasma (PRP) with a platelet concentration of approximately 1000,000 protein/μL have a positive effect on osseointegration. At lower concentrations of PRP, the effect was not optimal, while higher concentrations resulted in a paradoxically inhibitory effect of bone regeneration. Other studies were not in agreement with this PRP beneficial effect on the osseointegration of dental implants [27]. The assessment of bio activity of surface treated dental implants should be tested in vitro using biological fluids containing blood components [2].

5.4 INTERACTIONS BETWEEN SURFACES AND MSCs

Following blood clotting around dental implants, several cells interact with surfaces for tissue healing. MSCs attracted to the injured site by chemotactic factors have a determinant role in peri-implant tissue healing.

5.4.1 ORIGIN OF MSCs

MSCs are stem cells derived from somatic tissue which can be differentiated into mesenchymal lineages such bone, cartilage, fat, and skin. In addition, MSCs are present in many connective tissues and blood at low concentrations serving as a sort of internal repair system. MSCs are distinguished from other cell types by two important characteristics. First, they are unspecialized cells able to renew themselves through cell division, sometimes after long periods of inactivity. Second, under certain physiologic or experimental conditions, they can be induced to become tissue- or organ-specific cells with special functions. MSCs have high proliferative and multi potent capacity leading to differentiated cells under the guidance of various cues or niches. MSCs are conventionally defined as adherent, nonhematopoietic cells expressing markers such as CD13, CD29, CD44, CD54, CD73, CD90, CD105, and CD166, and being negative for CD14, CD34, and CD45 [28,29]. While originally identified in the bone marrow [30]. MSC have been extracted from numerous tissues including adipose [31,32], heart [33], dental pulp [34], peripheral blood [35], cord blood [36]. One of the major properties of MSCs is their ability to differentiate into various cells like adipocytes [37], chondrocytes [31], osteoblasts [38], neurons [39,40], muscles [40,41], hepatocytes [42] in vitro after treatment with induction agents.

5.4.2 MIGRATION, ADHESION, AND PROLIFERATION

The integration of implant with neighboring bone and gingival tissue depends on successful crosstalk between old tissue and implant surface. The challenge in dental implant research is the capability of the surface to guide cells colonization and

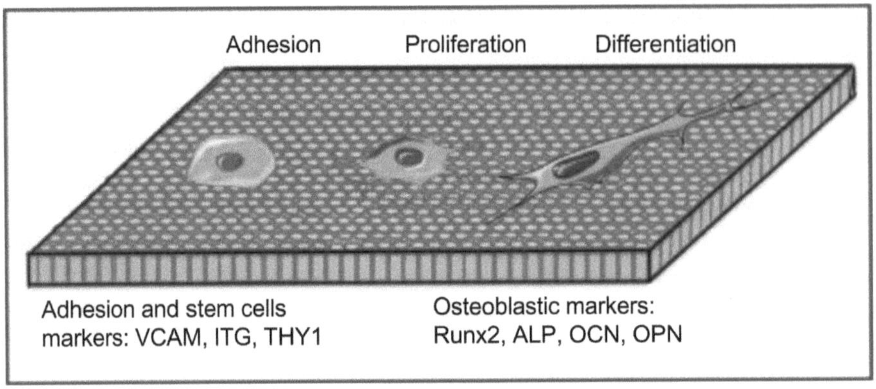

FIGURE 5.5

Scheme the adhesion, proliferation and differentiation of mesenchymal stem cells on nanostructured surfaces. The adhesion of stem cells is characterized by the expression of cell surface markers (VCAM, ITG, THY1), while phenotypic markers (Runx2, ALP, OCN, OPN) are specific to their osteoblastic differentiation. (OCN: osteocalcin; OPN: osteopontin).

differentiation. Cell migration, adhesion and proliferation on implant surfaces are a prerequisite to initiate the tissue regeneration (Fig. 5.5). Authors have shown that some factors present in tissues and secreted during the inflammatory phase are able to attract MSCs to the injured site [43,44]. MSCs migration and proliferation were stimulated in vitro by many growth factors including PDGF [45,46], EGF [46,47], VEGF [48], TGF-β [45,49], BMP-2 and BMP-4 [45,48], etc. These factors are certainly released in the injured sites by cells involved in tissue healing. Furthermore, plasma clot serve as storage to fibrin molecules and release system for a variety of bioactive factors including growth factors that attract and differentiate MSCs into specific lineages [50–52]. The platelet factors are well-known to stimulate the proliferation of MSCs [53]. The formation of a clot matrix with a potent chemo-attractive factor like PDGF, EGF, or fibrin may further enhance MSCs numbers and peri-implant tissue healing surface. Moreover, the plasma clot in contact with implant surface represents a three-dimensional microporous structure that allows diffusion of regulatory factors [54,55] and is involved in the migration, proliferation and differentiation of MSCs. After MSC recruitment in the injured site, cells adhere on the local extracellular matrix as well as on the implant surface beginning an extensive proliferation in order to build up new tissue. Again, surface modifications of implants in the nanometer range condition the biological responses.

5.4.3 DIFFERENTIATION

In the microenvironment, MSCs are stimulated by some specific factors to differentiate into the adequate cell line. Under the influence of these factors, MSCs switch to osteoblastic cells in contact to bone tissue while they differentiate into fibroblastic lineage in the gingival tissue region. These two differentiation pathways are in concurrence around dental implants. In some cases, implants are encapsulated by fibrous tissue due to the proliferation and differentiation of MSCs into fibroblastic cells. In response to cytokine, fibroblasts migrate and generate a capsule of collagen, the first step in generation of gingival tissue or rejection on contact to bone. This fibrous capsule prevents bonding between implant surface and juxtaposed bone and caused a failure of the implant [56]. On the other hand, both the differentiation of MSCs into fibroblastic lineage and the fibroblastic adhesion are desired in the gingival upper part of dental implants. Fibroblasts adhesion has been shown to be lower on nanoscale surface compared to conventional surfaces [57]. Moreover, nanometer-sized features have been shown to decrease fibroblast adhesion and proliferation [58,59]. The micro and nanoscale surface properties of metal implant including chemistry, roughness, wettability, could affect bone formation [60]. Numerous treatment such as machining, grit-blasting, Ti/HA plasma spray, chemical etching, anodization are available to modify the implant surface. Research has specifically demonstrated that nanorough Ti [61] and nanostructured Ti can enhance osteoblast adhesion and differentiation compared to their nanosmooth control [62]. Furthermore, surface with micro- and nanopores have shown to enhance greatly osseointegration [63,64]. Surface properties may control the steps of adhesion, proliferation, and differentiation of MSCs and, thus, condition tissue integration.

5.5 TISSUE INTEGRATION

Brånemark et al. [65] described the osseointegration as a direct structural and functional bone to implant contact under load. As previously discussed, the biological events occurring at the tissue-implant interface are influenced by the chemistry, topography, and wettability of dental implant surfaces. The challenge in developing new implant surface consist in increasing the clinical success rate as well as decreasing the tissue healing time for immediate loading of implants, particularly in aesthetic situations [66–68]. One of the objectives is to develop implant surface having predictable, controlled, and guided tissue healing. For instance, surfaces that promote contact osteogenesis rather than distance osteogenesis would be desired in bony site while intimate fibrous tissue healing in gingival tissue (Fig. 5.1). In order to enhance this intimate contact between tissues and implant, surface treatments at the nanometer scale have been performed on metal implants and tested in various animal models. Implant surface with various roughness have been used to increase the total area available for osteoapposition. Kubo et al. [66] observed a substantial increase by 3.1 times in bone–titanium interfacial strength by Ti nanotube (300 nm) at 2 weeks of implantation in femur rats. These results suggest the establishment of nanostructured surfaces for improved osteoconductivity. Moreover, Ogawa et al. [69] have prepared Ti nanostructure by physical vapor deposition and tested their osseointegration in femur of rats. They found an increased surface area by up to 40% and a greater strength of osseointegration for the nanostructured compared to an acid-etched surface. Some authors have correlated the initial events in bone formation adjacent to surface with the long-term tissue response to these materials in humans [70,71].

By mimicking the chemical composition of natural bone, HA and CaP coatings on Ti greatly enhance osseointegration. As shown in Fig. 5.6, a greater direct bone apposition was observed on CaP coated than on bare Ti implants. During the bone healing process, calcium and phosphate ions are released from the CaP coating in the peri-implant region and saturate body fluids to precipitate a biological apatite, which serves as a substrate for osteoblastic cells producing bone tissue. Several authors have shown the benefit of using CaP-coated titanium

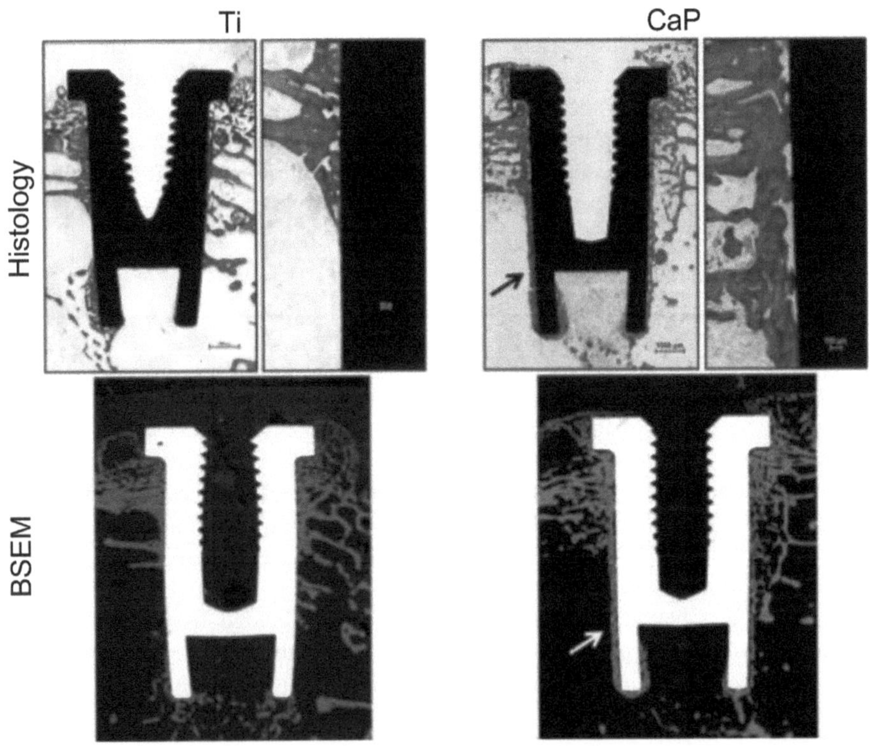

FIGURE 5.6

Micrographs showing the osseointegration of bare titanium (Ti) and calcium phosphate (CaP) coated implants after implantation in femoral condyles of rabbits for 4 weeks. Note the direct bone apposition on CaP coated implants (arrows) on both histology (basic fuchsine, toluidine blue staining) and back-scattered electron microscopy (BSEM) images.

implants for improving the osseointegration [72,73]. In particular, Le Guehennec et al. [21] have studied the osseointegration of four implant surfaces in the femoral epiphyses of rabbits after 2 and 8 weeks of healing. In this study, the bone-implant contact and bone growth inside the chambers were compared for four different implant surfaces and shown that biomimetic coating method may enhance the bone apposition onto titanium. In order to prevent coating delamination and implant loosening, the CaP coating should dissolve or degrade under osteoclastic activity at a similar rate than bone apposition. The final result should be a direct bone-implant coating without the presence of fibrous tissue. Another advantage of these CaP coatings is related to their preparation by biomimetic methods at physiological temperature and pH from simulated body fluids. CaP crystals have characteristics that resemble bone mineral in term of size and composition. Furthermore, it is possible to incorporate biologically active drugs such as antibiotics or growth factors during the precipitation of CaP coatings on Ti implants [74]. These molecules could be locally and gradually released in the peri-implant region for either preventing bacterial infections or stimulating bone growth.

5.6 CONCLUSION

Many reports have shown that nanometer-controlled surfaces have a great effect on early events such as the adsorption of proteins, blood clot formation and cell behaviors occurring upon implantation of dental implants. These early events have an effective impact on the migration, adhesion and differentiation of MSCs. Nanostructured surfaces may control the differentiation pathways into specific lineages and ultimately direct the nature of peri-implant tissues. Despite an active research in dental implants, the ideal surface for predictive tissue integration remains a challenge.

ACKNOWLEDGMENTS

The authors are grateful to Jean-Charles Ricquier for contribution in figures preparation. We acknowledge the pharmaceutical company SERVIER for using some drawings taken from their website.

REFERENCES

[1] J. Rozé, et al., Correlating implant stability to bone structure, Clin. Oral Implants Res. 20 (2009) 1140–1145.

[2] L. Le Guéhennec, A. Soueidan, P. Layrolle, Y. Amouriq, Surface treatments of titanium dental implants for rapid osseointegration, Dent. Mater. 23 (2007) 844–854.

[3] S. Lavenus, J.-C. Ricquier, G. Louarn, P. Layrolle, Cell interaction with nanopatterned surface of implants, Nanomedicine (Lond) 5 (2010) 937–947.

[4] S. Lavenus, G. Louarn, P. Layrolle, Nanotechnology and dental implants, Int. J. Biomater. 2010 (2010) 915327.

[5] R.G.T. Geesink, Osteoconductive coatings for total joint arthroplasty, Clin. Orthop. Relat. Res. (2002) 53–65.

[6] S. Leeuwenburgh, et al., Osteoclastic resorption of biomimetic calcium phosphate coatings in vitro, J. Biomed. Mater. Res. 56 (2001) 208–215.

[7] R.G. Geesink, K. de Groot, C.P. Klein, Chemical implant fixation using hydroxyl-apatite coatings. The development of a human total hip prosthesis for chemical fixation to bone using hydroxyl-apatite coatings on titanium substrates, Clin. Orthop. Relat. Res. (1987) 147–170.

[8] M.M. Shalabi, J.G.C. Wolke, J.A. Jansen, The effects of implant surface roughness and surgical technique on implant fixation in an in vitro model, Clin. Oral Implants Res. 17 (2006) 172–178.

[9] M. Esposito, J.M. Hirsch, U. Lekholm, P. Thomsen, Biological factors contributing to failures of osseointegrated oral implants. (I). Success criteria and epidemiology, Eur. J. Oral Sci. 106 (1998) 527–551.

[10] M. Esposito, J.M. Hirsch, U. Lekholm, P. Thomsen, Biological factors contributing to failures of osseointegrated oral implants. (II). Etiopathogenesis, Eur. J. Oral Sci. 106 (1998) 721–764.

[11] W.-D. Müeller, et al., Evaluation of the interface between bone and titanium surfaces being blasted by aluminium oxide or bioceramic particles, Clin. Oral Implants Res. 14 (2003) 349–356.

[12] L. Le Guehennec, et al., Osteoblastic cell behaviour on different titanium implant surfaces, Acta Biomater. 4 (2008) 535–543.

[13] A. Citeau, et al., In vitro biological effects of titanium rough surface obtained by calcium phosphate grid blasting, Biomaterials 26 (2005) 157–165.

[14] S. Oh, et al., Stem cell fate dictated solely by altered nanotube dimension, Proc. Natl. Acad. Sci. U.S.A. 106 (2009) 2130–2135.

[15] L. Zhang, Y. Han, Effect of nanostructured titanium on anodization growth of self-organized TiO_2 nanotubes, Nanotechnology 21 (2010) 55602.

[16] K. Shankar, et al., Highly-ordered TiO_2 nanotube arrays up to 220 μm in length: use in water photoelectrolysis and dye-sensitized solar cells, Nanotechnology 18 (2007) 065707.

[17] S.H. Kang, H.S. Kim, J.-Y. Kim, Y.-E. Sung, An investigation on electron behavior employing vertically-aligned TiO_2 nanotube electrodes for dye-sensitized solar cells, Nanotechnology 20 (2009) 355307.

[18] K.S. Brammer, et al., Improved bone-forming functionality on diameter-controlled TiO(2) nanotube surface, Acta Biomater. 5 (2009) 3215–3223.

[19] N.C. Geurs, R.L. Jeffcoat, E.A. McGlumphy, M.S. Reddy, M.K. Jeffcoat, Influence of implant geometry and surface characteristics on progressive osseointegration, Int. J. Oral Maxillofac. Implants 17 (2002) 811–815.

[20] J.E. Davies, Understanding peri-implant endosseous healing, J. Dent. Educ. 67 (2003) 932–949.

[21] L. Le Guehennec, et al., Histomorphometric analysis of the osseointegration of four different implant surfaces in the femoral epiphyses of rabbits, Clin. Oral Implants Res. 19 (2008) 1103–1110.

[22] M.A. Lopez-Heredia, P. Weiss, P. Layrolle, An electrodeposition method of calcium phosphate coatings on titanium alloy, J. Mater. Sci. Mater. Med. 18 (2007) 381–390.
[23] R.Z. LeGeros, Properties of osteoconductive biomaterials: calcium phosphates, Clin. Orthop. Relat. Res. (2002) 81–98.
[24] R. Miller, Z. Guo, E.A. Vogler, C.A. Siedlecki, Plasma coagulation response to surfaces with nanoscale chemical heterogeneity, Biomaterials 27 (2006) 208–215.
[25] T. Sawase, et al., Photo-induced hydrophilicity enhances initial cell behavior and early bone apposition, Clin. Oral Implants Res. 19 (2008) 491–496.
[26] G. Balasundaram, M. Sato, T.J. Webster, Using hydroxyapatite nanoparticles and decreased crystallinity to promote osteoblast adhesion similar to functionalizing with RGD, Biomaterials 27 (2006) 2798–2805.
[27] G. Weibrich, T. Hansen, W. Kleis, R. Buch, W.E. Hitzler, Effect of platelet concentration in platelet-rich plasma on peri-implant bone regeneration, Bone 34 (2004) 665–671.
[28] W. Richter, Mesenchymal stem cells and cartilage in situ regeneration, J. Intern. Med. 266 (2009) 390–405.
[29] T.E. Ichim, et al., Mesenchymal stem cells as anti-inflammatories: implications for treatment of Duchenne muscular dystrophy, Cell. Immunol. 260 (2010) 75–82.
[30] A.J. Friedenstein, K.V. Petrakova, A.I. Kurolesova, G.P. Frolova, Heterotopic of bone marrow. Analysis of precursor cells for osteogenic and hematopoietic tissues, Transplantation 6 (1968) 230–247.
[31] A.C.W. Zannettino, et al., Multipotential human adipose-derived stromal stem cells exhibit a perivascular phenotype in vitro and in vivo, J. Cell. Physiol. 214 (2008) 413–421.
[32] M.Q. Wickham, G.R. Erickson, J.M. Gimble, T.P. Vail, F. Guilak, Multipotent stromal cells derived from the infrapatellar fat pad of the knee, Clin. Orthop. Relat. Res. (2003) 196–212. Available from: http://dx.doi.org/10.1097/01.blo.0000072467.53786.ca.
[33] M.J. Hoogduijn, et al., Human heart, spleen, and perirenal fat-derived mesenchymal stem cells have immunomodulatory capacities, Stem Cells Dev. 16 (2007) 597–604.
[34] Y.-Y. Jo, et al., Isolation and characterization of postnatal stem cells from human dental tissues, Tissue Eng. 13 (2007) 767–773.
[35] Q. He, C. Wan, G. Li, Concise review: multipotent mesenchymal stromal cells in blood, Stem Cells 25 (2007) 69–77.
[36] W. Oh, D.-S. Kim, Y.S. Yang, J.K. Lee, Immunological properties of umbilical cord blood-derived mesenchymal stromal cells, Cell. Immunol. 251 (2008) 116–123.
[37] D.L. Morganstein, et al., Human fetal mesenchymal stem cells differentiate into brown and white adipocytes: a role for ERRalpha in human UCP1 expression, Cell Res. 20 (2010) 434–444.
[38] L. Marinucci, et al., Effects of Hydroxyapatite and Biostite((R)) on Osteogenic Induction of hMSC, Ann. Biomed. Eng. 38 (2010) 640–648.
[39] G. Lepski, et al., Limited Ca2+ and PKA-pathway dependent neurogenic differentiation of human adult mesenchymal stem cells as compared to fetal neuronal stem cells, Exp. Cell Res. 316 (2010) 216–231.
[40] A.J. Engler, S. Sen, H.L. Sweeney, D.E. Discher, Matrix elasticity directs stem cell lineage specification, Cell 126 (2006) 677–689.
[41] Y. Liu, et al., Flk-1 + adipose-derived mesenchymal stem cells differentiate into skeletal muscle satellite cells and ameliorate muscular dystrophy in mdx mice, Stem Cells Dev. 16 (2007) 695–706.

[42] M. Chivu, et al., In vitro hepatic differentiation of human bone marrow mesenchymal stem cells under differential exposure to liver-specific factors, Transl. Res. 154 (2009) 122–132.
[43] H. Agis, B. Kandler, M.B. Fischer, G. Watzek, R. Gruber, Activated platelets increase fibrinolysis of mesenchymal progenitor cells, J. Orthop. Res. 27 (2009) 972–980.
[44] J.P. Vogel, et al., Platelet-rich plasma improves expansion of human mesenchymal stem cells and retains differentiation capacity and in vivo bone formation in calcium phosphate ceramics, Platelets 17 (2006) 462–469.
[45] Y. Mishima, M. Lotz, Chemotaxis of human articular chondrocytes and mesenchymal stem cells, J. Orthop. Res. 26 (2008) 1407–1412.
[46] Y. Ozaki, et al., Comprehensive analysis of chemotactic factors for bone marrow mesenchymal stem cells, Stem Cells Dev. 16 (2007) 119–129.
[47] S.A. Kuznetsov, A.J. Friedenstein, P.G. Robey, Factors required for bone marrow stromal fibroblast colony formation in vitro, Br. J. Haematol. 97 (1997) 561–570.
[48] J. Fiedler, F. Leucht, J. Waltenberger, C. Dehio, R.E. Brenner, VEGF-A and PlGF-1 stimulate chemotactic migration of human mesenchymal progenitor cells, Biochem. Biophys. Res. Commun. 334 (2005) 561–568.
[49] H. Jian, et al., Smad3-dependent nuclear translocation of beta-catenin is required for TGF-beta1-induced proliferation of bone marrow-derived adult human mesenchymal stem cells, Genes Dev. 20 (2006) 666–674.
[50] I. Catelas, J.F. Dwyer, S. Helgerson, Controlled release of bioactive transforming growth factor beta-1 from fibrin gels in vitro, Tissue Eng. Part C Methods 14 (2008) 119–128.
[51] C. Wong, E. Inman, R. Spaethe, S. Helgerson, Fibrin-based biomaterials to deliver human growth factors, Thromb. Haemost. 89 (2003) 573–582.
[52] M.W. Mosesson, Fibrinogen and fibrin structure and functions, J. Thromb. Haemost. 3 (2005) 1894–1904.
[53] G. Rock, D. Neurath, M. Lu, A. Alharbi, M. Freedman, The contribution of platelets in the production of cryoprecipitates for use in a fibrin glue, Vox Sang. 91 (2006) 252–255.
[54] I. Catelas, et al., Human mesenchymal stem cell proliferation and osteogenic differentiation in fibrin gels in vitro, Tissue Eng. 12 (2006) 2385–2396.
[55] T. Schildhauer, D. Seybold, J. Gebmann, G. Muhr, M. Koller, Fixation of porous calcium phosphate with expanded bone marrow cells using an autologous plasma clot, Mater. Sci. Eng. Technol. 38 (2007) 1012–1014.
[56] J.A. Hobkirk, Progress in implant research, Int. Dent. J. 33 (1983) 341–349.
[57] E. Eisenbarth, J. Meyle, W. Nachtigall, J. Breme, Influence of the surface structure of titanium materials on the adhesion of fibroblasts, Biomaterials 17 (1996) 1399–1403.
[58] A. Cohen, P. Liu-Synder, D. Storey, T. Webster, Decreased fibroblast and increased osteoblast functions on ionic plasma deposited nanostructured Ti coatings, Nanoscale Res. Lett. 2 (2007) 385–390.
[59] D. Miller, R. Vance, A. Thapa, T. Webster, K. Haberstroch, Comparison of fibroblast and vascular cell adhesion to nano structured poly(lactic co glycolic acid) films, Appl. Bionics Biochem. 2 (2005) 1–7.
[60] R.M. Streicher, M. Schmidt, S. Fiorito, Nanosurfaces and nanostructures for artificial orthopedic implants, Nanomedicine (Lond) 2 (2007) 861–874.

[61] S. Puckett, R. Pareta, T.J. Webster, Nano rough micron patterned titanium for directing osteoblast morphology and adhesion, Int. J. Nanomed. 3 (2008) 229–241.

[62] C. Yao, E.B. Slamovich, T.J. Webster, Enhanced osteoblast functions on anodized titanium with nanotube like structures, J. Biomed. Mater. Res. A 85 (2008) 157–166.

[63] K.-H. Frosch, et al., Growth behavior, matrix production, and gene expression of human osteoblasts in defined cylindrical titanium channels, J. Biomed. Mater. Res. A 68 (2004) 325–334.

[64] Oh Sh, R.R. Finõnes, C. Daraio, L.-H. Chen, S. Jin, Growth of nano-scale hydroxyapatite using chemically treated titanium oxide nanotubes, Biomaterials 26 (2005) 4938–4943.

[65] P.I. Brånemark, et al., Osseointegrated titanium fixtures in the treatment of edentulousness, Biomaterials 4 (1983) 25–28.

[66] K. Kubo, et al., Cellular behavior on TiO_2 nanonodular structures in a micro-to-nanoscale hierarchy model, Biomaterials 30 (2009) 5319–5329.

[67] R.Z. LeGeros, R.G. Craig, Strategies to affect bone remodeling: osteointegration, J. Bone Miner. Res. 8 (Suppl 2) (1993) S583–596.

[68] R.M. Pilliar, Cementless implant fixation—toward improved reliability, Orthop. Clin. N. Am. 36 (2005) 113–119.

[69] T. Ogawa, L. Saruwatari, K. Takeuchi, H. Aita, N. Ohno, Ti nano-nodular structuring for bone integration and regeneration, J. Dent. Res. 87 (2008) 751–756.

[70] B.D. Boyan, Z. Schwartz, J.C. Hambleton, Response of bone and cartilage cells to biomaterials in vivo and in vitro, J. Oral Implantol. 19 (1993) 116–122; discussion 136–137

[71] D. Kohavi, et al., Effect of titanium implants on primary mineralization following 6 and 14 days of rat tibial healing, Biomaterials 13 (1992) 255–260.

[72] M.K. Jeffcoat, E.A. McGlumphy, M.S. Reddy, N.C. Geurs, H.M. Proskin, A comparison of hydroxyapatite (HA)-coated threaded, HA-coated cylindric, and titanium threaded endosseous dental implants, Int. J. Oral Maxillofac. Implants 18 (2003) 406–410.

[73] E.A. McGlumphy, L.J. Peterson, P.E. Larsen, M.K. Jeffcoat, Prospective study of 429 hydroxyapatite-coated cylindric omniloc implants placed in 121 patients, Int. J. Oral Maxillofac. Implants 18 (2003) 82–92.

[74] Y. Liu, K. de Groot, E.B. Hunziker, BMP-2 liberated from biomimetic implant coatings induces and sustains direct ossification in an ectopic rat model, Bone 36 (2005) 745–757.

CHAPTER

6 Titanium surface modification techniques for dental implants—From microscale to nanoscale

Karthikeyan Subramani[1], Reji T. Mathew[2] and Preeti Pachauri[3]

[1]Roseman University of Health Sciences, Henderson, NV, United States [2]Midwestern University, Downers Grove, IL, United States [3]Rama Dental College, Hospital & Research Center, Kanpur, India

CHAPTER OUTLINE

6.1 Introduction 99
6.2 Titanium Surface Modification Methods 100
6.2.1 Mechanical Modification of Titanium Surface 101
6.2.2 Physicochemical Modification of Titanium Surface 101
6.2.3 Biochemical Modification of Titanium Surface 104
6.2.4 Physical Modification of Titanium Surface 114
6.3 Recent Techniques 115
6.3.1 Discrete Crystalline Deposition (DCD) 115
6.3.2 Laser Ablation 115
6.3.3 Titanium Oxide Blasted and Acid-Etched Implants 115
6.3.4 Photofunctionalization 115
6.4 Limitations & Conclusion 116
Acknowledgments 116
References 117

6.1 INTRODUCTION

Titanium was once considered a rare metal, but nowadays it is one of the most important metals in the industry. The element was first discovered in England by Gregor in 1790, although it did not receive its name until Klaproth named it after the mythological first sons of the earth, the Titans, in 1795. Chemically, titanium is one of the transition elements in group IV and period 4 of Mendeleef's periodic table. It has an atomic number of 22 and an atomic weight of 47.9. Being a transition element, titanium has an incompletely filled shell in its electronic structure [1]. The

Emerging Nanotechnologies in Dentistry. DOI: http://dx.doi.org/10.1016/B978-0-12-812291-4.00006-6

incomplete shell enables titanium to form solid solutions with most substitutional elements. In the elemental form, titanium has a high melting point (1668 8C) and possesses a hexagonal closely packed crystal structure (hcp) α up to a temperature of 882.5 8C. Titanium transforms into a body centered cubic structure (bcc) β above this temperature [2]. Titanium (Ti) and its alloys are widely used to manufacture dental implants, maxillofacial, and orthopedic prostheses. In the field of orthopedics, more than 500,000 total hip and knee replacement surgeries with titanium implants are performed in the United States every year and the number is increasing [3]. Replacement of missing tooth structure and restoration of oral function with surface modified titanium implants is the ultimate goal in implant dentistry. Titanium and its alloys have been used extensively due to their excellent mechanical properties, low specific weight, high resistance to corrosion, and high biocompatibility [4,5]. Commercially, pure titanium is widely used in dental implants and titanium alloys with iron, aluminum, vanadium, molybdenum, and other elements (e.g., Ti−6Al−4V and Ti−6Al−7Nb) are used to manufacture orthopedic reconstructive prostheses like hip implants, knee implants, bone screws, and plates [6,7]. Even though titanium has excellent biomechanical properties, the long-term failure of titanium-based hip implants after 10−15 years remains unresolved. This has been attributed to insufficient early osseointegration and corrosion wear particles at a later stage [8−10]. Titanium dental implants fail due to various factors like insufficient early osseointegration, bacterial infection, surgical trauma, premature overloading, micro movements caused by inadequate prosthesis design, improper surgical placement, metal fatigue, and inadequate quality and quantity of bone surrounding the implant [11,12]. Over the past decade, various techniques of titanium surface modification have been employed to fabricate implant surfaces in order to promote osseointegration, faster healing time, higher bone-to-implant-contact (BIC) ratio, and longevity of titanium implants. These approaches are reviewed in the following sections of this chapter.

6.2 TITANIUM SURFACE MODIFICATION METHODS

To formulate novel titanium implant surface that could enhance early osseointegration, an overall knowledge of the research work done so far is essential. Titanium known as a "valve metal," when exposed to air, water, or other oxygen containing atmosphere it forms an oxide layer having 2−5 nm thickness [13]. The spontaneously formed oxide layer (TiO_2) provides resistance to corrosion and protection of the underlying metal [14]. Titanium being relatively inert, cannot directly bind to the bone and therefore osseointegration via the oxide layer (TiO_2) becomes relatively a long process [15,16]. In an effort to enhance cell-implant surface interaction, many surface modification methods have been investigated in the past to optimize surface topography and accelerate osseointegration with the host tissue [17−22]. The process of new bone formation at the implant-bone interface has been summarized in Fig. 6.1 [18]. It is well established that physical

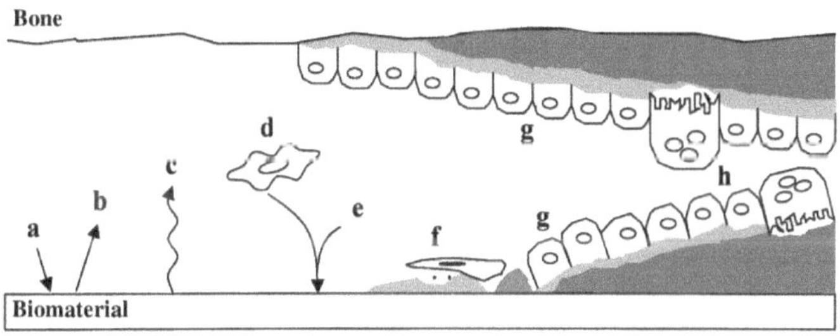

FIGURE 6.1

Representation of events at the bone-implant interface. (a) Protein adsorption from blood and tissue fluids, (b) protein desorption, (c) surface changes and material release, (d) inflammatory and connective tissue cells approach the implant, (e) possible targeted release of matrix proteins and selected adsorption of proteins such as Bone Sialoprotein (BSP) and Osteopontin (OPN), (f) formation of lamina limitans and adhesion of osteogenic cells, (g) bone deposition on both the exposed bone and implant surfaces, (h) remodeling of newly formed bone [18].

or chemical changes of titanium surface can influence bone cell (osteoblast) attachment [23–26]. There are three major ways [27] to modify titanium implant surfaces, which are

- Mechanical
- Chemical
- Physical

6.2.1 MECHANICAL MODIFICATION OF TITANIUM SURFACE

Alteration in titanium surface morphology has been used to influence osteoblast attachment, differentiation, proliferation, and migration. Common mechanical surface modification methods, such as micromachining [28–31], grinding, polishing [32], and blasting [33–38], involve physical treatment, shaping, or removal of the materials surface. The typical objective of mechanical modification is to obtain specific surface topographies and roughness, remove surface contamination, and/or improve adhesion in subsequent bonding steps. For a more detailed discussion on mechanical methods used in the surface modification of titanium and its alloys, the readers are referred to Ref. [39] (Table 6.1).

6.2.2 PHYSICOCHEMICAL MODIFICATION OF TITANIUM SURFACE

Physicochemical characteristics like surface free energy, surface charge, chemical composition, and surface wettability are fundamental parameters that influence osteoblast attachment. Physicochemical properties of titanium substrates can be altered primarily through manufacturing processes [40,41]. Chemical methods

Table 6.1 Overview of Surface Modification Methods for Titanium and its Alloys Implants

Surface Modification Methods	Modified Layer	Objective
1. Mechanical methods	Rough or smooth surface formed by subtraction process	Produce specific surface topographies, clean, and roughen surface, improve adhesion in bonding
a. Micromachining	<10 nm of surface oxide layer	Remove oxide scales and contamination
b. Grinding	~1 μm of sodium titanate gel	Improve biocompatibility, bioactivity, or bone conductivity
c. Polishing	~5 nm of dense inner oxide and porous outer layer	Improving biocompatibility, bioactivity, or bone conductivity
d. Blasting	~10 mm of thin film, such as calcium phosphate, TiO_2, and silica	Improve biocompatibility, bioactivity, or bone conductivity
2. Chemical methods	~10 to 40 mm of TiO_2 layer, adsorption, and incorporation of electrolyte anions	Produce specific surface topographies; improved corrosion resistance; improve biocompatibility, bioactivity, or bone conductivity
a. Chemical treatment	~1 μm of TiN, TiC, TiCN, diamond, and diamond-like carbon thin film	Improve wear resistance, corrosion resistance, and blood compatibility
i. Acidic Treatment	Modification through silanized titania, photochemistry, self-assembled monolayers, protein-resistance, etc.	Induce specific cell and tissue response by means of surface-immobilized peptides, proteins, or growth factors
ii. Alkaline treatment	~30 to ~200 mm of coatings, such as titanium, HA, calcium silicate, Al_2O_3, ZrO_2, TiO_2	Improve wear resistance, corrosion resistance, and biological properties
b. Hydrogen peroxide treatment	~1 μm of TiN, TiC, TiCN, diamond, and diamond-like carbon thin film	Improve wear resistance, corrosion resistance, and blood compatibility
c. Sol–gel	~10 nm of surface modified layer and/or ~μm of thin film	Modify surface composition; improve wear, corrosion resistance, and biocompatibility
d. Anodic oxidation	~1 to ~100 nm of surface modified layer	Clean, sterilize, oxide, nitride surface; remove native oxide layer
e. Chemical vapor deposition (CVD)		
f. Biochemical methods		
i. Osteoinductive biomolecular cues [cell adhesive proteins and growth factors like bone morphogenetic protein-2 (BMP-2)]		
ii. Microscale and nanoscale coating of hydroxyapatite/calcium phosphate/alumina coatings with bioactive molecular cues, osteoinductive growth factors, and antibacterial drugs		
iii. Organic nanoscale self-assembled monolayers (SAMs)		
iv. Bioactive, biodegradable hydrogels		
v. Antibacterial agents or antibacterial drug delivery directly from titanium surface		

3. Physical methods a. Thermal Spray i. Flame spray ii. Plasma spray iii. High velocity oxy-fuel (HVOF) iv. Detonation-gun spray (DGUN) b. Physical vapor deposition (PVD) i. Evaporation ii. Ion Plating iii. Sputtering c. Ion implantation and deposition i. Bean-line ion implantation ii. Plasma Immersion Ion Implantation (PIII) d. Glow discharge Plasma Treatment		

described here include chemical treatment, electrochemical treatment (anodic oxidation), sol–gel, chemical vapor deposition (CVD), and biochemical modification. During the chemical treatment, electrochemical treatment, and biochemical modification, chemical, electrochemical, or biochemical reactions occur, respectively, at the interface between titanium and a solution. CVD is a process involving chemical reactions between chemicals in the gas phase and the sample surface resulting in the deposition of a nonvolatile compound on the substrate. On the other hand, in the sol–gel process, chemical reactions do not occur at the interface between the sample surface and solution or gel, but rather in the solution. This is often combined with biochemical modification of titanium surface. This recent trend in titanium surface engineering is aimed to design biologically inspired surfaces that have the potential to mimic natural bone architecture and stimulate osteoblast adhesion, differentiation, proliferation, migration, and thereby enhancing bone formation and osseointegration. A biologic organic matrix or biomolecular layer covalently bound to titanium surface can effectively modulate osteoblast response. The potential advantage of delivering organic molecules directly to the bone-implant interface is that it could be used to control orientation of biomolecules to cells, therefore inducing favorable intracellular and extracellular responses to accelerate bone formation.

6.2.3 BIOCHEMICAL MODIFICATION OF TITANIUM SURFACE

Biochemical modification of titanium surface has been achieved through the following ways:

- Osteoinductive biomolecular cues (cell adhesive proteins and peptides) passively adsorbed or covalently bound to titanium surface.
- Microscale and nanoscale coating of hydroxyapatite/calcium phosphate/alumina coatings with bioactive molecular cues, osteoinductive growth factors, and antibacterial drugs.
- Organic nanoscale self-assembled monolayers (SAMs) with chemically linked biomolecular cues, osteoinductive growth factors, and drugs.
- Bioactive, biodegradable hydrogels with incorporated biomolecules and osteoinductive growth factors.
- Antibacterial agents or antibacterial drug delivery directly from titanium surface to decrease primary bacterial adhesion and prevention of biofilm formation in order to reduce the risk of peri-implantitis-induced implant failure.

6.2.3.1 Osteoinductive biomolecular cues

Osteoinductive biological macromolecules have been delivered from titanium surface to study their role in modulating osteoblast response and enhancing osseointegration. These biomolecules have been passively adsorbed or covalently bound to the titanium surface. In vitro studies with passively adsorbed extracellular matrix proteins like Fibronectin and Vitronectin on titanium surface showed increased osteoblast adhesion, proliferation, and differentiation when compared with bare titanium surfaces and those coated with Vitronectin [42]. Fibronectin and collagen type I and III adsorbed onto Ti–6Al–4V surface have significantly influenced the behavior of rat osteoblasts in terms of alkaline phosphatase activity and collagen synthesis [43,44]. Since these passively bound biomolecules can be washed off or removed easily, covalent binding on to the titanium surface was advocated. Collagen has been covalently bound through acrylic acid grafting onto a hydrocarbon plasma film deposited on titanium surface. An in vivo study on rabbit femur model showed enhanced bone healing with the collagen modified titanium surface [45].

Another approach utilized anodic oxidation (Anodization) to fabricate oxide layer (TiO_2) on titanium surface. Collagen type was immobilized by partial incorporation onto this oxide layer [46]. In vitro studies with stimulated body fluid (SBF) showed spherical calcium phosphate formed on immobilized fibrillar collagen (FC) type 1 on titanium specimen with air-formed passive layer (AFP) (Fig. 6.2) [46].

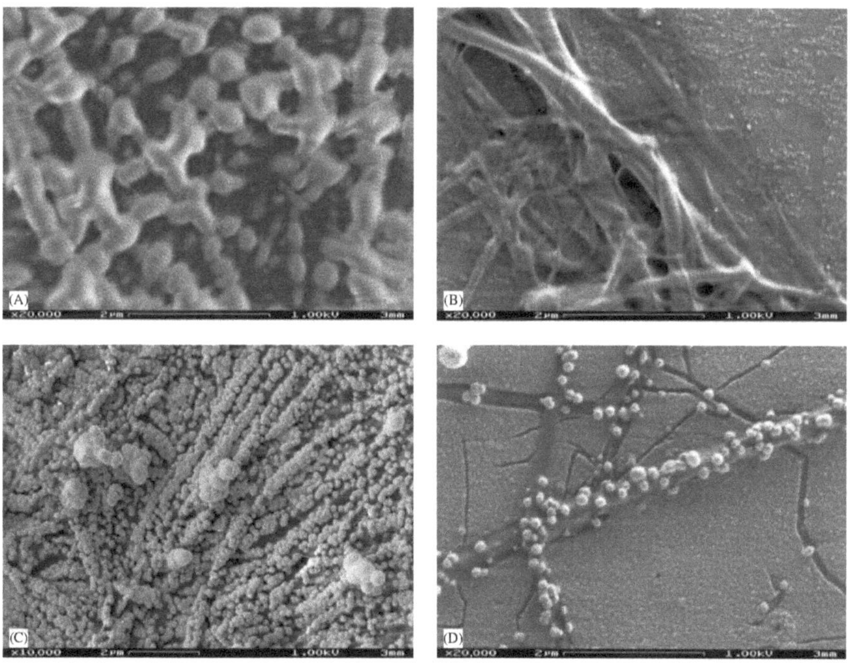

FIGURE 6.2

SEM images of spherical calcium phosphate formed on immobilized fibrillar collagen (FC) type 1 on titanium specimen with air-formed passive layer (AFP). The surface states FC/AFP (left) and FC/Anodic oxidation (AO) (right) after exposure to SBF for 1 h (top) and 24 h (bottom) [46].

The limitations of delivering whole proteins or protein fragments on titanium surface include the risk of 3D structure deformation leading to protein denaturation, protein conjugation, or protein-induced immunological reactions [47]. In order to address these limitations, recent approaches have focused on using short synthetic or natural peptides that present the bioadhesive motif. The most frequently used bioadhesive peptide motif is the arginine–glycine–aspartic acid (RGD) sequence which mediates bone cell attachment to several extracellular matrix proteins like fibronectin, vitronectin, bone sialoprotein, and osteopontin [48].

"Silanization" is a simple technique employed to immobilize RGD peptide sequence covalently onto titanium surface. The technique involves coating the surface with silane molecules through chemical treatment and self-assembly. This silane molecule acts as a "bridging molecule" between the substrate and the biomolecules. In a study, using 3-aminopropyltriethoxysilane (APTES) chemistry, RGD sequence was bound to Ti–6Al–4V surface [49]. In vivo study in canines showed improved mechanical fixation with RGD-coated implants [50]. In a similar study, RGD–cysteine (RGDC) peptide sequence was immobilized on titanium surface through covalent addition by silanization technique using APTES chemistry [51]. Primary calvarial osteoblasts cultured on RGDC-coated titanium surface showed increased cell attachment, morphology, proliferation, and osteocalcin messenger RNA expression than the bare titanium surface [51]. These studies demonstrate that silanization technique is not only simple and cost-effective but also can be successfully used for covalent binding of cell adhesive peptide motifs or other osteoinductive growth factors on titanium surface for implant applications.

Bone morphogenetic protein-2 (BMP-2) is a potent osteogenic growth factor which has been studied extensively for bone regeneration. BMP-2 has been delivered directly from the titanium implant surface or through a carrier scaffold/membrane for such purposes. It was shown that titanium implant can be made "osteoconductive" by coating the surface with BMP-2 [52]. Preclinical results revealed higher bone regeneration around implant surfaces coated with BMP-2 than the uncoated surfaces. Ectopic bone formation was reported in another preclinical study using BMP-2 delivered from titanium implant surface [53].

6.2.3.2 Microscale and nanoscale coating of hydroxyapatite/calcium phosphate/alumina

A percentage of 70 human bone is made up of inorganic mineral—hydroxyapatite [$Ca_{10}(PO4)_6(OH)_2$]. Modifying titanium surface biochemistry at the micro and nanoscale level includes fabrication of highly adherent hydroxyapatite and calcium phosphate. Aluminum oxide (alumina) deposited on titanium surface can serve as a template on which hydroxyapatite or calcium phosphate can be coated. For such purpose, plasma spray technique has been utilized [54,55]. Alumina deposited on titanium substrate by electron beam evaporation and subsequent anodization produced highly adherent nanoporous coatings. In vitro studies with primary human osteoblast cells showed that the nanopores significantly enhanced the adhesion of cells (Fig. 6.3) [56]. These coatings can be further functionalized

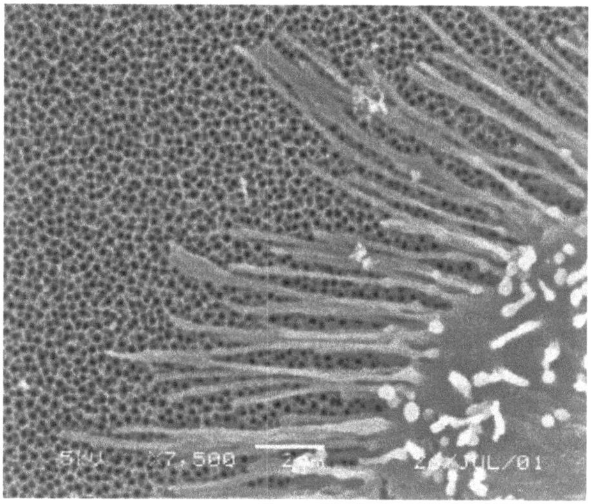

FIGURE 6.3

SEM micrograph of human osteoblast cells on nanoporous alumina showing cell filopodia attaching to the pores of the material [56].

with osteoinductive biomolecules. Nanoscale hydroxyapatite coating up to 25–40 nm were fabricated on titanium surface using sol–gel dip-coating [57,58], electrochemical deposition, [59] or electrostatic atomization spray techniques [60]. In vitro study with rabbit mesenchymal stem cells on nanostructured hydroxyapatite coatings indicated enhanced osteoblast response and biocompatibility of such surfaces [61]. Recent in vivo study in rabbits showed increased bone formation on nanohydroxyapatite coated titanium implants and significant influence on early bone formation compared to uncoated implants after 4 weeks of healing [62]. These coatings can be further functionalized with osteoinductive biomolecules. In vitro study with SaOS-2 human osteoblast-like cells on nanohydroxyapatite coatings functionalized with collagen by sol–gel technique showed increased osteoblast activity [63]. Silanization using APTES chemistry has also been utilized in covalent attachment of RGD peptide on hydroxyapatite coated titanium substrates. In vitro study with osteoprogenitor cells isolated from human bone marrow stroma cells showed enhanced osteoblast adhesion and function on such surfaces [64].

Different forms of calcium phosphate (octa, di, tri, and tetra calcium phosphate) have been coated on titanium surface [65,66]. Plasma spraying technique has been utilized to coat calcium phosphate on titanium cylinders. In vivo implantation in dog femur and histological evaluation revealed that these coatings enhanced bone formation with excellent bone to implant surface contact [67]. Pulsed laser deposition (PLD) technique has also been employed to fabricate calcium phosphate coatings on titanium substrates (Fig. 6.4). These types of coatings

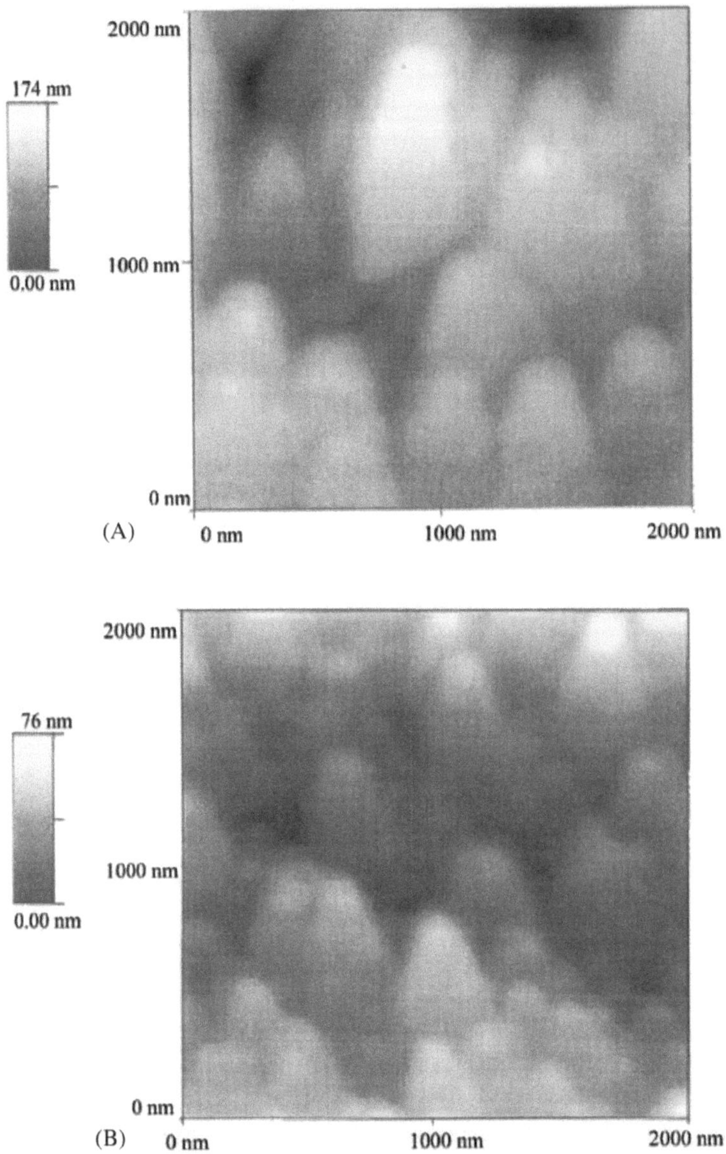

FIGURE 6.4

AFM images of the coating surface showing tips of B100 nm high (A) crystalline and (B) amorphous calcium phosphate coatings produced by PLD [68].

exhibited much higher adhesive strength with substrates than conventional plasma-sprayed coatings [68]. In vitro study with human osteoblasts showed that the coatings enhanced osteoblast adhesion, proliferation, and differentiation.

Titanium has also been coated with biphasic calcium phosphate particles using grid-blasting technique to increase the roughness of titanium implants and enhance the attachment of mouse osteoblasts and increase the alkaline phosphatase activity [69]. A recent approach involved placing of Ti−6Al−4V discs in concentrated SBF forming a dense layer of amorphous calcium phosphate [70]. This layer served as seeding surface and the discs were placed in supersaturated calcium phosphate solution that produced a coating up to 100 μm thick. BMP-2, a potent osteoinductive agent was passively absorbed on to the coating. The titanium discs with calcium phosphate-BMP-2 coating were implanted subcutaneously in rats. The results of this study demonstrated ectopic bone formation and sustained osteogenic activity. Recent in vivo study with a similar setup of calcium phosphate-BMP-2 coating and titanium implants with BMP-2 passively deposited showed increased volume of bone deposition [52]. Bone interface coverage was highest for coated implants bearing no BMP-2, followed by coated implants bearing either incorporated only, or incorporated and adsorbed BMP-2; it was however lowest for uncoated implants bearing adsorbed BMP-2. This study showed the importance of using calcium phosphate coating to deliver BMP-2 at the bone-implant interface. A recent study in vivo has shown that BMP-2 adsorbed biomimetic calcium phosphate coating was osteoinductive, but not highly efficacious [71]. The use of such biomimetic coating has been reviewed in the literature for their applications in clinical orthopedics and dentistry [72]. Such attempts have shown that titanium surface can be modified at the micro and nanometer scales with hydroxyapatite/calcium phosphate coating which can deliver osteoinductive factors and effectively induce osseointegration at the bone-implant interface.

6.2.3.3 Organic nanoscale self-assembled monolayers (SAMs)

Another way of biochemical modification of titanium surface is through SAMs. This technique involves adsorption and self-assembly of single layers of molecules on a substrate. Molecular self-assembly of alkane phosphate SAMs on metal oxides like TiO_2 and Al_2O_3 have been reported [73,74]. The hydrophilicity of these alkane phosphate SAMs can be modified with a hydroxy-terminated end group. When smooth and rough titanium surfaces coated with hydroxy-terminated (hydrophilic) and methyl-terminated (hydrophobic) alkane phosphate SAMs were exposed to human fibroblasts, more fibroblasts were found on smooth surface. The researchers concluded that surface wettability was much less important than surface roughness [75]. But recent studies have concentrated on the role of SAMs in the formation of hydroxyapatite on titanium substrate. Titanium wafers with carboxyl (−COOH), hydroxyl (−OH), phosphate ($-PO_4H_2$), and sulfonic acid ($-SO_3H$) end group SAMs were placed in SBF. The sulfonic acid ($-SO_3H$) and carboxyl (−COOH) end groups produced a predominantly crystalline HA. Furthermore, the carboxyl (−COOH) end group yielded the thickest layer which possessed crystalline characteristics very similar to that of the human bone [76]. A similar in vitro study on titanium coated with amine ($-NH_2$), thiol ($-SH$), and sulfonic acid ($-SO_3H$) end group SAMs showed that thiol ($-SH$) end group favored the formation of HA (Fig. 6.5), resulting in a layer with a thickness of

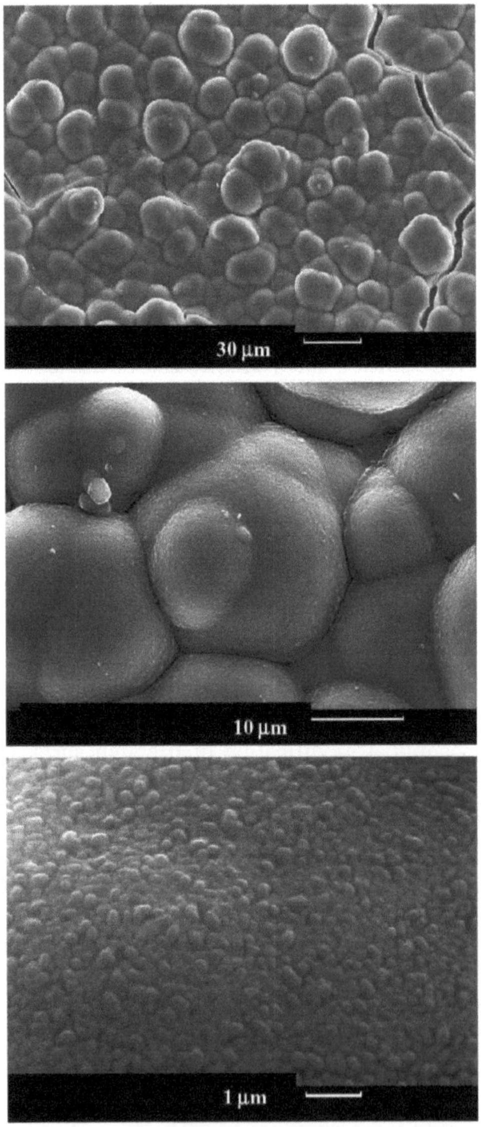

FIGURE 6.5

SEM images of hydroxyapatite (HA) on titanium coated with thiol ($-SH$) end group [77].

about 10 μm [77]. In vivo study was performed with titanium surface with three coatings: self-assembled monolayer of phosphonate molecules (SAMP), SAMP + RGD peptide, or hydroxyapatite. Histological analysis showed abundant new bone formation around all of the three groups, though higher enhanced formation was apparent in the two SAMP groups [78]. Self-assembled RGD peptide-modified surfaces have also been highlighted as one of the most promising and novel methods to enhance osteoblast function. In vitro study using rat calvarial osteoblasts on gold-coated titanium surfaces with self-assembled RGDC monolayer showed enhanced cell attachment, increased cell spreading, and significantly greater cell proliferation on the RGDC-coated surfaces [79]. Such studies showed that SAMs could be used to control the physicochemical properties (e.g., wettability) and biochemical composition of titanium surface to influence the response of host tissue.

6.2.3.4 Hydrogels on titanium surface

Hydrogel is a network of polymer chains that swell in aqueous solution. Hydrogel is composed of long polymer chains connected by crosslinks. The crosslinks may be biodegradable or nonbiodegradable and are ionic interactions between polyelectrolyte chains. Crosslinking of polymer molecules or polymerization can be achieved by photopolymerization, changes in temperature, radiation, self-assembly, or by crosslinking enzymes. Hydrogels undergo responsive swelling by absorbing solvent when placed in an aqueous solution (solvation). Swollen hydrogels can absorb many times their own weight in water and can switch between swollen and collapsed forms. The key properties of hydrogels for biomedical applications are high biocompatibility, biodegradability, and ability to incorporate biomolecular cues (due to high permeability for oxygen, nutrients, and water-soluble metabolites). These key characteristics along with the ease of in situ fabrication have made hydrogels a biomaterial of choice in in vitro studies for analyzing cell-biomaterial interactions and in biomedical applications [80]. Hydrogels are extensively used in cosmetic and reconstructive surgery [81], as a matrix for the fabrication of artificial organs in tissue engineering [82], and as "intelligent" stimuli sensitive drug delivery systems [83]. Hydrogels have also been extensively used to fabricate contact lenses [84], breast implants [85], tissue engineering scaffolds [86], delivery vehicles for bioactive drug molecules [87], coatings for biosensors [88], dressings for wound healing and burn injuries [89], and as carrier scaffolds for Guided bone regeneration using osteogenic growth factors [90]. Although many polymer hydrogels have been studied, Poly(ethylene glycol) hydrogel (PEG) is one of the most widely investigated systems. PEG hydrogel can be fabricated into three-dimensional microstructures to study the response of cells for their applications in tissue engineering. PEG hydrogel offers the simplicity and advantage of incorporating bioactive molecules into the hydrogel matrix passively or by covalent linking with the PEG monomers. Time-dependent release of these biomolecular cues from the PEG hydrogel micropatterns serves as an excellent platform for studying cell response in tissue culture.

PEG hydrogel micropatterns have been fabricated by photopolymerization through a micropatterned photomask (Photolithography) [91].

Poly(ethylene glycol) PEG hydrogel has also been used as a carrier matrix for the delivery of cell adhesive peptides [92,93] and growth factors due to its biocompatibility, biodegradability, and tissue-like conformational properties [94,95]. PEG hydrogel has also been used to fabricate "nonfouling" or "antifouling" surfaces (surfaces resistant to protein adsorption) on titanium. Many techniques like molecular assembly [96] and formation of interpenetrating networks [97] have been used to immobilize PEG onto titanium surface as they are compatible with titanium surface chemistry. Photolithography combined with silanization technique has been utilized to produce cell adhesive and antiadhesive PEG hydrogel micropatterns on a silicon substrate [98]. In this study, vascular endothelial growth factor was tested for its osteoinductive property. Silanization of titanium and its alloys has already been reported for KRG peptide grafting onto Ti–6Al–4V surface [99] and RGD-containing peptides delivered from PEG hydrogel coated titanium surface [100]. Such biomimetic surface modifications to gain control over nonspecific protein adsorption and introducing specific ligands like cell adhesive peptides (RGD) linked covalently to the PEG layer [101] would give control over surface chemistry of titanium implants made for commercial applications.

6.2.3.5 Antibacterial titanium surfaces

As an alternative to the previous biochemical modification methods of titanium and its alloys, recent approaches have focused on delivering antibacterial drugs to prevent biofilm formation. Biofilm (plaque) formation on titanium dental implants hinders early bone formation and subsequently leads to peri-implantitis and implant failure. Peri-implantitis is the inflammatory process affecting the tissues around an osseointegrated implant in function, thereby resulting in loss of supporting bone. The signs and symptoms of peri-implantitis include bleeding on probing, increased probing pocket depth ($>$3 mm; generally noninfected implants allow the probe to penetrate approximately 3mm), mobility of the implant, suppuration, pain along with radiological evidence of vertical destruction of crestal bone (which is often saucer-shaped) (Figs. 6.6 and 6.7).

In a comparative study with different metals used in the fabrication of implants, titanium showed considerable antibacterial activity when compared with other metals like gold, cobalt, vanadium, aluminum, chromium, and iron [102]. This study concluded that the rank order of antibacterial activity expressed by these metals was gold $>$ titanium $>$ cobalt $>$ vanadium $>$ aluminum $>$ chromium $>$ iron. After gold, titanium exhibited a high degree of antibacterial properties. By contrast, a few other in vitro and in vivo studies have reported that titanium does not have a significant bacteriostatic effect on strains of oral bacteria [103,104]. Numerous approaches have focused on modifying the titanium surface to render it resistant to bacterial attachment, thereby reducing infection rate and the risk of inflammation with antibacterial surfaces [105]. A recent study

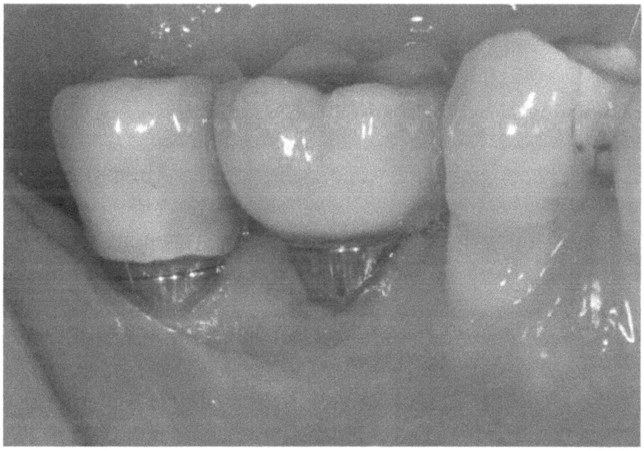

FIGURE 6.6

Clinical presentation of peri-implantitis.

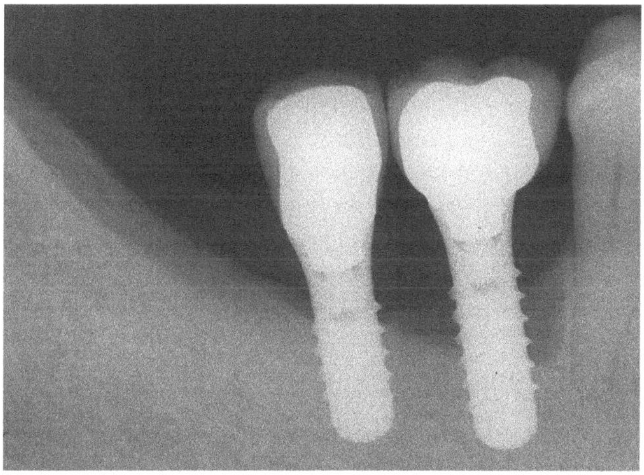

FIGURE 6.7

Radiographic image of peri-implantitis. Note bone loss around the mandibular first molar implant.

compared the adhesive and antimicrobial properties of nanoscale and microscale samples of zinc oxide (ZnO) and titanium oxide (TiO_2) to staphylococcal cells and osteoblasts [106]. ZnO was selected in this study because of its antimicrobial properties and TiO_2 was selected because it typically forms on titanium implants in the body. The authors concluded that, when compared to their microscale counterparts, the nanoscale ZnO and TiO_2 led to reduced staphylococcal cell adhesion and increased osteoblast adhesion. Such nanoscale coatings of ZnO on titanium implants may play a vital role in decreasing the formation of biofilm and thereby the subsequent peri-implant pathology. Photocatalytic bactericidal nanoscale TiO_2 when illuminated under ultraviolet radiation releases antibacterial active-oxygen species like OH, HO_2, HO_2^-, and H_2O_2 [107,108]. Such TiO_2 nanocoatings on titanium dental implants can decrease early bacterial colonization.

Silver-based antimicrobials are of interest due to their specific antibacterial activity and the nontoxicity of the active Ag^+ to human cells. Antimicrobial silver-coated titanium surface (TiAg) when evaluated for its specific antimicrobial activity showed significant antimicrobial potency against *Staphylococcus epidermis* and *Klebsiella pneumoniae* strains and at the same time, no cytotoxic effects on osteoblast and epithelial cells [109]. Due to similar mechanical performance when compared to pure titanium, TiAg coatings could be suitable to provide antimicrobial activity on load-bearing implant surfaces. In addition to silver-based antimicrobial coatings, antibiotic drug molecules delivered from titanium surface can inhibit bacterial colonization, prevent biofilm formation, and avert late-stage infection. Vancomycin-modified titanium surface have been shown to be effective against in vitro bacterial colonization on implant surfaces [110]. Covalently bound vancomycin-titanium surfaces proved to be bactericidal on *S. aureus* [111] and *S. epidermidis* [112]. Recent studies have also shown that the vancomycin-modified bactericidal titanium surface is stable and demonstrated superior inhibition of bacterial attachment, colonization, and proliferation [113–115]. In vitro studies on bacterial cells and fibroblasts on titanium surface modified Ca^+, N^+, F^+ ions by ion implantation method showed marked antibacterial effect with no deleterious effects on fibroblast proliferation [116]. These approaches should be carefully considered toward modifying and reactivating the titanium surface in vitro and in vivo to produce an osteoinductive and antibacterial surface for enhanced osseointegration and prolonged success rate of the implant.

6.2.4 PHYSICAL MODIFICATION OF TITANIUM SURFACE

During some surface modification processes, such as thermal spraying and physical vapor deposition, chemical reactions do not occur. In this case, the formation of surface modified layer, films, or coatings on titanium and its alloys are mainly attributed to the thermal, kinetic, and electrical energy. In the thermal spraying process, the coating materials are thermally melted into liquid droplets and coated to the substrate at a high speed (kinetic energy). Physical vapor deposition leads to film growth by reaction between a substrate surface and an adjacent vapor which supplies the coating material in the form of atoms, molecules, or ions, generated from a target and transported to the substrate surface on which condensation and reaction with atoms of the surface lattice take place. The generation of atoms, molecules, or ions from targets can be accomplished by resistance heating, electron beam, laser, or electrical discharge in vacuum. Glow discharge plasma treatment and ion implantation are also categorized as physical methods.

6.3 RECENT TECHNIQUES

6.3.1 DISCRETE CRYSTALLINE DEPOSITION (DCD)

The Osseotite surface of the dual acid-etched titanium alloy implant has been altered with a nanometer-scale manufacturing technique. Calcium phosphate (CaP) particles of 20–100 nm are deposited on a double acid-etched surface by a sol–gel process named discrete crystalline deposition (DCD). These CaP particles make up roughly 50% of the surface area [117] and exert a higher adhesive force to the implant surface than former techniques of CaP deposition [118,119]. Bacterial adhesion to the NanoTite surface has been shown to be lower compared to the predecessor Osseotite surface [120].

6.3.2 LASER ABLATION

This technique focuses on improving the integration of dental implants in the surrounding soft tissue. Therefore, nanoscale surface manufacturing techniques have been transferred to the implant collar. The neck of the dental implant has been processed in a laser micromachining step to generate a pattern of micro- and nanoscale microchannels. These microchannels act as a biologic seal by eliciting the attachment of connective tissue and bone and inhibiting epithelial down growth [121].

6.3.3 TITANIUM OXIDE BLASTED AND ACID-ETCHED IMPLANTS

Titanium oxide blasting produces the microscale surface roughness. The subsequent etching with hydrofluoric acid shapes the nanostructure of the implant [122]. A pleiotropic manufacturing effect is the accumulation of fluoride on the surface [123]. Fluoride containing surfaces have been hypothesized to propagate the host-to-implant reaction in early osseointegration [123]. Cell studies have demonstrated that the OsseoSpeed implant surface promotes a branched cell morphology of osteoblasts and an osteogenic gene expression profile as well as osteoinduction and osteogenesis in mesenchymal stem cells compared to TiOblast implants (DENTSPLY Implants, Mannheim, Germany), the titanium oxide blasted precursor [122].

6.3.4 PHOTOFUNCTIONALIZATION

UV treatment of dental implant surfaces enhances bioactivity and osseointegration by altering the titanium dioxide on the surface. By promoting interactions of cells and proteins to the implant on a molecular level, UV light is believed to enhance the osteoconductivity [124]. UV treatment reduces the degree of surface hydrocarbon and increases surface energy and wettability [125–128]. UV light has been suggested to raise the level of protein absorption and cellular attachment to titanium surfaces and has been shown to restore bioactivity caused by age-related degradations [129,130].

6.4 LIMITATIONS & CONCLUSION

Even though the titanium surface can be modified using the above mentioned techniques, these techniques do not produce a highly controllable, morphologically uniform nanoscale bone-like surface topography and chemistry. It is well known that HA coated titanium hip prostheses fail due to poor bond strength at the HA/Ti interface, nonuniformity in coating density, microcracks, and improper microstructural surface architecture [116,131–133]. Passively bound biomolecules can be washed off/removed from the implant surface during the process of surgical implantation. The release rate of covalently bound biomolecules from titanium surface cannot be controlled. There may be a rapid release or a very slow delayed release [134]. Hydrogel coatings can be too thin and fragile to withstand the surgical implantation procedure. Organic SAMs do share the same limitations. Moreover, the use of noble metal like gold coatings to enhance molecular self-assembly will not be of interest to the implant manufacturers due to higher cost. Owing to these limitations, and taking into consideration the need for revision surgery of orthopedic implants after 10–15 years, there is a real need for fabricating titanium surface with controlled surface features that mimic the organized nanoscale surface architecture of human bone. Novel research efforts should be directed toward designing a simple, cost-effective, and commercializable technique with which a "bone-mimicking" titanium surface could be fabricated for implant applications.

Micro- and nanoscale modification of titanium surface using physicochemical, morphological, and biochemical approaches have resulted in higher BIC ratio and improved osseointegration. Anodization of titanium has been effectively applied to fabricate nanotubular surfaces which closely mimic the nanoscale architecture of human bone. Such surfaces can be fabricated for implant applications and thus serve as an effective carrier for osteoinductive growth factors to promote bone formation and osseointegration. Antibacterial and antiinflammatory drugs can also be physically adsorbed on such surfaces to produce antibacterial implant surfaces in order to prevent infection and inflammation. These recent advances in utilizing TiO_2 nanotubes have to be evaluated in future clinical trials to understand the potential benefits and disadvantages of nanotubular implant surfaces.

ACKNOWLEDGMENTS

The authors are grateful for the kind contribution of Figs. 6.6 and 6.7, which were kindly provided by Dr. Christian Ramel, Prof. Christoph Hammerle, and Dr. Ronald Jung from the Clinic for Fixed and Removable Prosthodontics and Dental Material Science, University of Zurich, Plattenstrasse 11, Zurich, Switzerland.

REFERENCES

[1] H.R. Ogden, in: A.H. Clifford (Ed.), Rare Metals Handbook, Rinhdd Publishing Corporation, Chapman & Hall Ltd, London, 1961, pp. 559–579.
[2] E.W. Collings, The physical metallurgy of titanium alloys, in: H.L. Gegel (Ed.), ASM Series in Metal Processing, Edward Arnold Publications, Cleveland, Metals Park, OH, 1984.
[3] Improving medical implant performance through retrieval information: challenges and opportunities, in: NIH Technology Assessment Conference, 2000.
[4] T. Albrektsson, P.I. Branemark, H.A. Hansson, J. Lindstrom, Osseointegrated titanium implants. Requirements for ensuring a long-lasting, direct bone-to-implant anchorage in man, Acta Orthop. Scand. 52 (4) (1981) 155.
[5] J. Black, in: G. Hasting (Ed.), Handbook of Biomaterial Properties, Chapman and Hall, London, 1998, p. 179.
[6] M.A. Imam, A.C. Fraker, Titanium alloys as implant materials, ASTM Spec. Tech. Publ. 1272 (1996) 3–16.
[7] J. Katz, Applications of materials in medicine and dentistry: orthopedic applications, in: B.D. Ratner, F.J. Schoen, J.E. Lemons (Eds.), Biomaterials Science: An Introduction to Materials in Medicine, Academic Press, San Diego, 1996, pp. 335–346.
[8] R. Barbucci, Integrated Biomaterials Science, Kluwer, New York, 2002, pp. 189–689.
[9] M.A. Khan, R.L. Williams, D.F. Williams, Conjoint corrosion and wear in titanium alloys, Biomaterials 20 (8) (1999) 765.
[10] L.M. Rabbe, J. Rieu, A. Lopez, P. Combrade, Fretting deterioration of orthopaedic implant materials: search for solutions, Clin. Mater. 15 (4) (1994) 221.
[11] E.S. Rosenberg, J.P. Torosian, J. Slots, Microbial differences in 2 clinically distinct types of failures of osseointegrated implants, Clin. Oral Implants Res. 2 (3) (1991) 135.
[12] S. Schou, P. Holmstrup, E. Hjorting-Hansen, N.P. Lang, Plaque-induced marginal tissue reactions of osseointegrated oral implants: a review of the literature, Clin. Oral Implants Res. 3 (4) (1992) 149.
[13] M. Textor, V. Frauchiger, S. Tosatti, D.M. Brunnete, Properties and biological significance of natural oxide films on titanium and its alloys, in: D.M. Brunnete, M. Textor, P. Thomsen (Eds.), Titanium in Medicine, Springer, New York, 2001, pp. 171–230.
[14] J. Choi, R.B. Wehrspohn, J. Lee, U. Gösele, Anodization of nanoimprinted titanium: a comparison with formation of porous alumina, Electrochim. Acta 49 (16) (2004) 2645.
[15] P. Brånemark, Osseointegration and its experimental background, J. Prosthet. Dent. 50 (3) (1983) 399.
[16] L. Linder, T. Albrektsson, P.I. Branemark, Electron microscopic analysis of the bone-titanium interface, Acta Orthop. Scand. 54 (1) (1983) 45.
[17] M. Morra, Biomolecular modification of implant surfaces, Expert Rev. Med. Dev. 4 (3) (2007) 361.
[18] D.A. Puleo, A. Nanci, Understanding and controlling the bone-implant interface, Biomaterials 20 (23–24) (1999) 2311.
[19] M. Morra, Biochemical modification of titanium surfaces: peptides and ECM proteins, Eur. Cell Mater 12 (2006) 1.

[20] C. Yao, T.J. Webster, Anodization: a promising nano-modification technique of titanium implants for orthopedic applications, J. Nanosci. Nanotechnol. 6 (9–10) (2006) 2682.

[21] G. Mendonca, D.B. Mendonca, F.J. Aragao, L.F. Cooper, Advancing dental implant surface technology-from micron- to nanotopography, Biomaterials 29 (28) (2008) 3822.

[22] T.I. Kim, J.H. Jang, H.W. Kim, J.C. Knowles, Y. Ku, Biomimetic approach to dental implants, Curr. Pharm. Des. 14 (22) (2008) 2201.

[23] R.K. Sinha, F. Morris, S.A. Shah, R.S. Tuan, Surface composition of orthopaedic implant metals regulates cell attachment, spreading, and cytoskeletal organization of primary human osteoblasts *in-vitro*, Clin. Orthop. Relat. Res. 305 (1994) 258.

[24] K.T. Bowers, J.C. Keller, B.A. Randolph, D.G. Wick, C.M. Michaels, Optimization of surface micromorphology for enhanced osteoblast responses *in-vitro*, Int. J. Oral Maxillofac. Implants 7 (3) (1992) 302.

[25] G. Gronowicz, M.B. McCarthy, Response of human osteoblasts to implant materials: integrin-mediated adhesion, J. Orthop. Res. 14 (6) (1996) 878.

[26] N. Lumbikanonda, R. Sammons, Bone cell attachment to dental implants of different surface characteristics, Int. J. Oral Maxillofac. Implants 16 (5) (2001) 627.

[27] Y. Ito, M. Kajihara, Y. Imanishi, Material for enhancing cell adhesion by immobilization of cell-adhesive peptide, J. Biomed. Mater. Res. 25 (1991) 1325.

[28] P.J. Henry, Comparative surface analysis of two osseointegrated implant systems, Int. J. Oral. Maxillofac. Implants 2 (1987) 23.

[29] J. Lausmaa, B. Kasemo, H. Mattsson, Surface spectroscopic characterization of titanium implant materials, Appl. Surf. Sci. 44 (1990) 133–146.

[30] D.S. Sutherland, P.D. Forshaw, G.C. Allen, I.T. Brown, K.R. Williams, Biomaterials 14 (12) (1993) 893; X. Liu et al., Mater. Sci. Eng. R 47 (2004) 49–121.

[31] J.P. Lucchini, J.L. Aurelle, M. Therin, K. Donath, W. Becker, Clin. Oral. Implants Res. 7 (4) (1996) 397.

[32] B. Hignett, T.C. Andrew, W. Downing, E.J. Duwell, J. Belanger, E.H. Tulinski, Surface cleaning, finishing and coating, in: W.G. Wood (Ed.), Metals Handbook, vol. 5, American Society for Metals, Metals Park, OH, 1987, pp. 107–127.

[33] D. Buser, T. Nydegger, T. Oxland, D.L. Cochran, R.K. Schenk, H.P. Hirt, et al., J. Biomed. Mater. Res. 45 (2) (1999) 75.

[34] M. Baleani, M. Viceconti, A. Toni, Artif. Organs 24 (4) (2000) 296.

[35] I. Degasne, M.F. Basle, V. Demais, G. Hure, M. Lesourd, B. Grolleau, et al., Calcif. Tissue Int. 64 (6) (1999) 499.

[36] M. Kern, V.P. Thompson, J. Dent. 22 (5) (1994) 300.

[37] A. Wennerberg, T. Albrektsson, C. Johansson, B. Andersson, Biomaterials 17 (1) (1996) 15.

[38] A. Wennerberg, T. Albrektsson, J. Lausmaa, J. Biomed. Mater. Res. 30 (2) (1996) 251.

[39] J. Lausmaa, in: D.M. Brunette, P. Tengvall, M. Textor, P. Thomsen (Eds.), Titanium in Medicine, Springer, Berlin, 2001, pp. 231–266.

[40] R.E. Baier, Surface properties influencing biological adhesion, in: R.S. Manly (Ed.), Adhesion in Biological Systems, Academic Press, New York, 1970, pp. 15–48.

[41] J.M. Williams, R.A. Buchanan, Ion implantation of surgical Ti–6Al–4V alloy, Materials Sci. Eng. A 69 (1) (1985) 237.

[42] Y. Ku, C.P. Chung, J.H. Jang, The effect of the surface modification of titanium using a recombinant fragment of fibronectin and vitronectin on cell behaviour, Biomaterials 26 (25) (2005) 5153.
[43] S. Bierbaum, R. Beutner, T. Hanke, D. Scharnweber, U. Hempel, H. Worch, Modification of Ti6Al4V surfaces using collagen I, III, and fibronectin. I. Biochemical and morphological characteristics of the adsorbed matrix, J. Biomed. Mater. Res. A 67 (2) (2003) 421.
[44] S. Bierbaum, U. Hempel, U. Geissler, T. Hanke, D. Scharnweber, K.W. Wenzel, et al., Modification of Ti6Al4V surfaces using collagen I, III, and fibronectin. II. Influence on osteoblast responses, J. Biomed. Mater. Res. A 67 (2) (2003) 431.
[45] M. Morra, C. Cassinelli, G. Cascardo, P. Cahalan, L. Cahalan, M. Fini, R. Giardino, Surface engineering of titanium by collagen immobilization. Surface characterization and *in-vitro* and *in-vivo* studies, Biomaterials 24 (25) (2003) 4639.
[46] D. Scharnweber, R. Born, K. Flade, S. Roessler, M. Stoelzel, H. Worch, Mineralization behaviour of collagen type I immobilized on different substrates, Biomaterials 25 (12) (2004) 2371.
[47] M. Morra, Biochemical modification of titanium surfaces: peptides and ECM proteins, Eur. Cell. Mater. 12 (2006) 1–15.
[48] M. Dettin, M.T. Conconi, R. Gambaretto, A. Bagno, C. Di Bello, A.M. Menti, et al., Effect of synthetic peptides on osteoblast adhesion, Biomaterials 26 (22) (2005) 4507.
[49] H. Zreiqat, F.A. Akin, C.R. Howlett, B. Markovic, D. Haynes, S. Lateef, et al., Differentiation of human bone-derived cells grown on GRGDSP-peptide bound titanium surfaces, J. Biomed. Mater. Res. A 64 (1) (2003) 105.
[50] B. Elmengaard, J.E. Bechtold, K. Soballe, *In-vivo* effects of RGD-coated titanium implants inserted in two bone-gap models, J. Biomed. Mater. Res. A 75 (2) (2005) 249.
[51] S.J. Xiao, M. Textor, N.D. Spencer, M. Wieland, B. Keller, H. Sigrist, Immobilization of the cell-adhesive peptide Arg-Gly-Asp-Cys (RGDC) on titanium surfaces by covalent chemical attachment, J. Mater. Sci. Mater. Med. 8 (12) (1997) 867.
[52] Y. Liu, L. Enggist, A.F. Kuffer, D. Buser, E.B. Hunziker, The influence of BMP-2 and its mode of delivery on the osteoconductivity of implant surfaces during the early phase of osseointegration, Biomaterials 28 (16) (2007) 2677.
[53] J. Hall, R.G. Sorensen, J.M. Wozney, U.M. Wikesjo, Bone formation at rhBMP-2-coated titanium implants in the rat ectopic model, J. Clin. Periodontol. 34 (5) (2007) 444.
[54] E.P. Briggs, A.R. Walpole, P.R. Wilshaw, M. Karlsson, E. Pålsgård, Formation of highly adherent nano-porous alumina on Ti-based substrates: a novel bone implant coating, J. Mater. Sci.: Mater. Med. 15 (9) (2004) 1021.
[55] A.R. Walpole, E.P. Briggs, M. Karlsson, E. Pålsgård, P.R. Wilshaw, Nano-porous alumina coatings for improved bone implant interfaces, Materialwissenschaft und Werkstofftechnik 34 (12) (2003) 1064.
[56] M. Karlsson, E. Pålsgård, P.R. Wilshaw, L. Di Silvio, Initial *in-vitro* interaction of osteoblasts with nano-porous alumina, Biomaterials 24 (18) (2003) 3039.
[57] L. Guo, H. Li, Fabrication and characterization of thin nano-hydroxyapatite coatings on titanium, Surf. Coatings Technol. 185 (2–3) (2004) 268.
[58] J. Yang, L. Guo, H. Li, Sol–gel preparation of ultrathin nano-hydroxyapatite coating and its characterization, Sheng Wu Yi Xue Gong Cheng Xue Za Zhi 23 (5) (2006) 1075.
[59] R. Hu, C.J. Lin, H.Y. Shi, A novel ordered nano hydroxyapatite coating electrochemically deposited on titanium substrate, J. Biomed. Mater. Res. A 80 (3) (2007) 687.

[60] J. Huang, S.N. Jayasinghe, X. Su, Z. Ahmad, S.M. Best, M.J. Edirisinghe, et al., Electrostatic atomisation spraying: a novel deposition method for nano-sized hydroxyapatite, Key Eng. Mater. 309–311 (I) (2006) 635.
[61] F. Chen, W.M. Lam, C.J. Lin, G.X. Qiu, Z.H. Wu, K.D. Luk, et al., Biocompatibility of electrophoretical deposition of nanostructured hydroxyapatite coating on roughen titanium surface: *in-vitro* evaluation using mesenchymal stem cells, J. Biomed. Mater. Res. B. Appl. Biomater. 82 (1) (2007) 183.
[62] L. Meirelles, A. Arvidsson, M. Andersson, P. Kjellin, T. Albrektsson, A. Wennerberg, Nano hydroxyapatite structures influence early bone formation, J. Biomed. Mater. Res. A 87A (2) (2008) 299.
[63] X. Zhu, O. Eibl, L. Scheideler, J. Geis-Gerstorfer, Characterization of nano hydroxyapatite/collagen surfaces and cellular behaviors, J. Biomed. Mater. Res. A 79 (1) (2006) 114.
[64] M.C. Durrieu, S. Pallu, F. Guillemot, R. Bareille, J. Amédée, Ch Baquey, et al., RGD containing peptides onto hydroxyapatite to promote osteoblastic cells adhesion, J. Mater. Sci. Mater. Med. 15 (7) (2004) 779.
[65] J.F. Kay, Calcium phosphate coatings: understanding the chemistry and biology and their effective use, Compend. Suppl. 15 (1993) S520.
[66] C.P. Klein, P. Patka, J.G. Wolke, J.M. de Blieck-Hogervorst, K. de Groot, Long-term *in-vivo* study of plasma-sprayed coatings on titanium alloys of tetracalcium phosphate, hydroxyapatite and alpha-tricalcium phosphate, Biomaterials 15 (2) (1994) 146.
[67] C.P. Klein, P. Patka, H.B. van der Lubbe, J.G. Wolke, K. de Groot, Plasma-sprayed coatings of tetracalciumphosphate, hydroxyl-apatite, and alpha-TCP on titanium alloy: an interface study, J. Biomed. Mater. Res. 25 (1) (1991) 53.
[68] J.L. Arias, M.B. Mayor, J. Pou, Y. Leng, B. Leon, M. Perez-Amor, Micro- and nano-testing of calcium phosphate coatings produced by pulsed laser deposition, Biomaterials 24 (20) (2003) 3403.
[69] A. Citeau, J. Guicheux, C. Vinatier, P. Layrolle, T.P. Nguyen, P. Pilet, et al., *In-vitro* biological effects of titanium rough surface obtained by calcium phosphate grid blasting, Biomaterials 26 (2) (2005) 157.
[70] Y. Liu, K. de Groot, E.B. Hunziker, BMP-2 liberated from biomimetic implant coatings induces and sustains direct ossification in an ectopic rat model, Bone 36 (5) (2005) 745.
[71] Y. Liu, R.O. Huse, K. de Groot, D. Buser, E.B. Hunziker, Delivery mode and efficacy of BMP-2 in association with implants, J. Dent. Res. 86 (1) (2007) 84.
[72] Y. Liu, J.P. Li, E.B. Hunziker, K. de Groot, Incorporation of growth factors into medical devices via biomimetic coatings, Philos. Transact. A. Math. Phys. Eng. Sci. 364 (1838) (2006) 233.
[73] E. Gawalt, M.J. Avaltroni, N. Koch, J. Schwartz, Self-assembly and bonding of alkanephosphonic acids on the native oxide surface of titanium, Langmuir 17 (19) (2001) 5736.
[74] R. Hofer, M. Textor, N.D. Spencer, Alkyl phosphate monolayers, self-assembled from aqueous solution onto metal oxide surfaces, Langmuir 17 (13) (2001) 4014.
[75] M. Textor, S. Tosatti, M. Wieland, D.M. Brunnete, Use of Molecular Assembly Techniques for Tailoring The Chemical Properties on Smooth and Rough Titanium Surfaces, CRC Press, Boca Raton, 2003.
[76] D.P. Liu, P. Majewski, B.K. O'Neill, Y. Ngothai, C.B. Colby, The optimal SAM surface functional group for producing a biomimetic HA coating on Ti, J. Biomed. Mater. Res.—A 77 (4) (2006) 763.

[77] P.J. Majewski, G. Allidi, Synthesis of hydroxyapatite on titanium coated with organic self-assembled monolayers, Mater. Sci. Eng. A 420 (1−2) (2006) 13.
[78] F.J. Shannon, J.M. Cottrell, X.H. Deng, K.N. Crowder, S.B. Doty, M.J. Avaltroni, et al., A novel surface treatment for porous metallic implants that improves the rate of bony ongrowth, J. Biomed. Mater. Res. A 86 (1) (2008) 857.
[79] H. Huang, Y. Zhao, Z. Liu, Y. Zhang, H. Zhang, T. Fu, et al., Enhanced osteoblast functions on RGD immobilized surface, J. Oral. Implantol. 29 (2) (2003) 73.
[80] A.S. Hoffman, Hydrogels for biomedical applications, Ann. N.Y. Acad. Sci. 944 (2001) 62.
[81] S. Srouji, A. Rachmiel, I. Blumenfeld, E. Livne, A simple soft lithographic route to fabrication of poly(ethylene glycol) microstructures for protein and cell patterning, J. Cranio-Maxillofacial. Surg. 33 (2) (2005) 79.
[82] R. Landers, U. Hübner, R. Schmelzeisen, R. Mülhaupt, Rapid prototyping of scaffolds derived from thermoreversible hydrogels and tailored for applications in tissue engineering, Biomaterials 23 (23) (2002) 4437.
[83] K.S. Anseth, A.T. Metters, S.J. Bryant, P.J. Martens, J.H. Elisseeff, C.N. Bowman, In situ forming degradable networks and their application in tissue engineering and drug delivery, J. Contr. Rel. 78 (1−3) (2002) 199.
[84] M.G. Harris, M.D. Sarver, K.A. Polse, Patient response to thin hydrogel contact lenses: a comparative study, J. Am. Optom. Assoc. 48 (3) (1977) 295.
[85] G.F. Maillard, Carboxy-methyl cellulose hydrogels used to fill breast implants: a 15-year experience, Eur. J. Plast. Surg. 24 (4) (2001) 177.
[86] J.L. Drury, D.J. Mooney, Hydrogels for tissue engineering: scaffold design variables and applications, Biomaterials 24 (24) (2003) 4337.
[87] Y. An, J.A. Hubbell, Intraarterial protein delivery via intimally-adherent bilayer hydrogels, J. Control. Rel. 64 (1−3) (2000) 205.
[88] C.P. Quinn, C.P. Pathak, A. Heller, J.A. Hubbell, Photo-crosslinked copolymers of 2-hydroxyethyl methacrylate, poly(ethylene glycol) tetra-acrylate and ethylene dimethacrylate for improving biocompatibility of biosensors, Biomaterials 16 (5) (1995) 389.
[89] D. Eisenbud, H. Hunter, L. Kessler, K. Zulkowski, Hydrogel wound dressings: where do we stand in 2003? Ost. Wound. Manage. 49 (10) (2003) 52.
[90] L. Hong, Y. Tabata, M. Yamamoto, S. Miyamoto, K. Yamada, N. Hashimoto, et al., Comparison of bone regeneration in a rabbit skull defect by recombinant human BMP-2 incorporated in biodegradable hydrogel and in solution, J. Biomater. Sci. Polym. Ed. 9 (9) (1998) 1001.
[91] V.A. Liu, S.N. Bhatia, Three-dimensional photopatterning of hydrogels containing living cells, Biomed. Microdev. 4 (2002) 257.
[92] K.H. Park, M.H. Kim, S.H. Park, H.J. Lee, I.K. Kim, H.M. Chung, Synthesis of Arg-Gly-Asp (RGD) sequence conjugated thermo-reversible gel via the PEG spacer arm as an extracellular matrix for a pheochromocytoma cell (PC12) culture, Biosci. Biotechnol. Biochem. 68 (11) (2004) 2224.
[93] G.B. Schneider, A. English, M. Abraham, R. Zaharias, C. Stanford, J. Keller, The effect of hydrogel charge density on cell attachment, Biomaterials 25 (15) (2004) 3023.
[94] S.A. DeLong, J.J. Moon, J.L. West, Covalently immobilized gradients of bFGF on hydrogel scaffolds for directed cell migration, Biomaterials 26 (16) (2005) 3227.
[95] D. Seliktar, A.H. Zisch, M.P. Lutolf, J.L. Wrana, J.A. Hubbell, MMP-2 sensitive, VEGF-bearing bioactive hydrogels for promotion of vascular healing, J. Biomed. Mater. Res. A 68 (4) (2004) 704.

[96] J.L. Dalsin, L. Lin, S. Tosatti, J. Voros, M. Textor, P.B. Messersmith, Protein resistance of titanium oxide surfaces modified by biologically inspired mPEG-DOPA, Langmuir 21 (2) (2005) 640.
[97] T.A. Barber, L.J. Gamble, D.G. Castner, K.E. Healy, *In-vitro* characterization of peptide-modified p(AAm-*co*-EG/AAc) IPN-coated titanium implants, J. Orthop. Res. 24 (7) (2006) 1366.
[98] K. Subramani, M.A. Birch, Fabrication of poly(ethylene glycol) hydrogel micropatterns with osteoinductive growth factors and evaluation of the effects on osteoblast activity and function, Biomed. Mater. 1 (3) (2006) 144.
[99] M.C. Porte-Durrieu, F. Guillemot, S. Pallu, C. Labrugere, B. Brouillaud, R. Bareille, et al., Baquey Ch, Cyclo-(DfKRG) peptide grafting onto Ti−6Al−4V: physical characterization and interest towards human osteoprogenitor cells adhesion, Biomaterials 25 (2004) 4837.
[100] S. Tosatti, Z. Schwartz, C. Campbell, D.L. Cochran, S. VandeVondele, J.A. Hubbell, et al., RGD-containing peptide GCRGYGRGDSPG reduces enhancement of osteoblast differentiation by poly(L-lysine)-graft-poly(ethylene glycol)-coated titanium surfaces, J. Biomed. Mater. Res. 68A (2004) 458.
[101] T.A. Barber, S.L. Golledge, D.G. Castner, K.E. Healy, Peptide-modified p(AAm-*co*-EG/AAc) IPNs grafted to bulk titanium modulate osteoblast behavior *in-vitro*, J. Biomed. Mater. Res. A 64 (1) (2003) 38.
[102] C.W. Berry, T.J. Moore, J.A. Safar, C.A. Henry, M.J. Wagner, Antibacterial activity of dental implant metals, Implant. Dent. 1 (1) (1992) 59.
[103] A. Leonhardt, J. Olsson, G. Dahlén, Bacterial colonization on titanium, hydroxyapatite, and amalgam surfaces *in-vivo*, J. Dent. Res. 74 (9) (1995) 1607.
[104] A. Leonhardt, G. Dahlén, Effect of titanium on selected oral bacterial species *in-vitro*, Eur. J. Oral. Sci. 103 (6) (1995) 382.
[105] K. Subramani, R.E. Jung, A. Molenberg, C.H.F. Hammerle, Biofilm on dental implants: a review of the literature, Int, J. Oral Maxillofac. Impl. 24 (4) (2009) 616.
[106] G. Colon, B.C. Ward, T.J. Webster, Increased osteoblast and decreased Staphylococcus epidermidis functions on nanophase ZnO and TiO_2, J. Biomed. Mater. Res. A 78 (3) (2006) 595.
[107] M.F. Xu, S. Lin, X.M. Chen, Y.Z. Peng, Studies on characteristics of nanostructure of N-TiO_2 thin films and photo-bactericidal action, J. Zhejiang. Univ. Sci. B 7 (7) (2006) 586.
[108] N. Suketa, T. Sawase, H. Kitaura, M. Naito, K. Baba, K. Nakayama, et al., An antibacterial surface on dental implants, based on the photocatalytic bactericidal effect, Clin. Implant Dent. Relat. Res. 7 (2) (2005) 105.
[109] A. Ewald, S.K. Glückermann, R. Thull, U. Gbureck, Antimicrobial titanium/silver PVD coatings on titanium, Biomed. Eng. Online 5 (2006) 22.
[110] J. Parvizi, E. Wickstrom, A.R. Zeiger, C.S. Adams, I.M. Shapiro, J.J. Purtill, et al., Frank Stinchfield Award. Titanium surface with biologic activity against infection, Clin. Orthop. Relat. Res. 429 (2004) 33.
[111] B. Jose, V. Antoci Jr, A.R. Zeiger, E. Wickstrom, N.J. Hickok, Vancomycin covalently bonded to titanium beads kills *Staphylococcus aureus*, Chem. Biol. 12 (9) (2005) 1041.
[112] M.C. Lawson, C.N. Bowman, K.S. Anseth, Vancomycin derivative photopolymerized to titanium kills *S. epidermidis*, Clin. Orthop. Relat. Res. 461 (2007) 96.

[113] V. Antoci Jr, C.S. Adams, J. Parvizi, P. Ducheyne, I.M. Shapiro, N.J. Hickok, Covalently attached vancomycin provides a nanoscale antibacterial surface, Clin. Orthop. Relat. Res. 461 (2007) 81.

[114] V. Antoci Jr, S.B. King, B. Jose, J. Parvizi, A.R. Zeiger, E. Wickstrom, et al., Vancomycin covalently bonded to titanium alloy prevents bacterial colonization, J. Orthop. Res. 25 (7) (2007) 858.

[115] V. Antoci Jr, C.S. Adams, J. Parvizi, P. Ducheyne, I.M. Shapiro, N.J. Hickok, Covalently attached vancomycin provides a nanoscale antibacterial surface, Clin. Orthop. Relat. Res. 461 (2007) 81.

[116] M. Yoshinari, Y. Oda, T. Kato, K. Okuda, Influence of surface modifications to titanium on antibacterial activity *in-vitro*, Biomaterials 22 (14) (2001) 2043.

[117] E.A. Bonfante, R. Granato, C. Marin, et al., Biomechanical testing of microblasted, acid-etched/microblasted, anodized, and discrete crystalline deposition surfaces: an experimental study in beagle dogs, Int. J. Oral Maxillofac. Implants 28 (1) (2013) 136–142.

[118] M. Franchi, B. Bacchelli, D. Martini, et al., Early detachment of titanium particles from various different surfaces of endosseous dental implants, Biomaterials 25 (12) (2004) 2239–2246. View at Publisher.

[119] T. Kitsugi, T. Nakamura, M. Oka, Y. Senaha, T. Goto, T. Shibuya, Bone-bonding behavior of plasma-sprayed coatings of BioglassR, AW-glass ceramic, and tricalcium phosphate on titanium alloy, J. Biomed. Mater. Res. 30 (2) (1996) 261–269.

[120] R. Rodriguez, Y. Baena, C.R. Arciola, L. Selan, et al., Evaluation of bacterial adhesion on machined titanium, osseotite® and nanotite® discs, Int. J. Artif. Organs 35 (10) (2012) 754–761.

[121] M. Nevins, D.M. Kim, S.-H. Jun, K. Guze, P. Schupbach, M.L. Nevins, Histologic evidence of a connective tissue attachment to laser microgrooved abutments: a canine study, Int. J. Periodontics Restor. Dent. 30 (3) (2010) 245–255.

[122] P.G. Coelho, R. Jimbo, N. Tovar, E.A. Bonfante, Osseointegration: hierarchical designing encompassing the macrometer, micrometer, and nanometer length scales, Dent. Mater. 31 (1) (2015) 37–52.

[123] P.G. Coelho, J.M. Granjeiro, G.E. Romanos, et al., Basic research methods and current trends of dental implant surfaces, J. Biomed. Mater. Res. B: Appl. Biomater. 88 (2) (2009) 579–596.

[124] Y. Gao, Y. Liu, L. Zhou, et al., The effects of different wavelength UV photofunctionalization on micro-arc oxidized titanium, PLoS ONE 8 (7) (2013). Article ID e68086.

[125] B. Altmann, R.-J. Kohal, T. Steinberg, et al., Distinct cell functions of osteoblasts on uv-functionalized titanium- and zirconia-based implant materials are modulated by surface topography, Tissue Eng.—C: Methods 19 (11) (2013) 850–863.

[126] K.-H. Park, J.-Y. Koak, S.-K. Kim, C.-H. Han, S.-J. Heo, The effect of ultraviolet-C irradiation via a bactericidal ultraviolet sterilizer on an anodized titanium implant: a study in rabbits, Int. J. Oral Maxillof. Implants 28 (1) (2013) 57–66.

[127] H. Minamikawa, T. Ikeda, W. Att, et al., Photofunctionalization increases the bioactivity and osteoconductivity of the titanium alloy Ti6Al4V, J. Biomed. Mater. Res. A 102 (10) (2014) 3618–3630.

[128] A. Funato, M. Yamada, T. Ogawa, Success rate, healing time, and implant stability of photofunctionalized dental implants, Int. J. Oral Maxillof. Implants 28 (5) (2013) 1261–1271.

[129] N. Hori, T. Ueno, T. Suzuki, et al., Ultraviolet light treatment for the restoration of age-related degradation of titanium bioactivity,", Int. J. Oral Maxillof. Implants 25 (1) (2010) 49–62.
[130] H. Aita, N. Hori, M. Takeuchi, et al., The effect of ultraviolet functionalization of titanium on integration with bone, Biomaterials 30 (6) (2009) 1015–1025.
[131] J.P. Collier, V.A. Surprenant, M.B. Mayor, M. Wrona, R.E. Jensen, H.P. Surprenant, Loss of hydroxyapatite coating on retrieved, total hip components, J. Arthroplasty 8 (4) (1993) 389.
[132] S.D. Cook, K.A. Thomas, J.F. Kay, Experimental coating defects in hydroxylapatite-coated implants, Clin. Orthop. Relat. Res. 265 (1991) 280.
[133] A. Kar, K.S. Raja, M. Misra, Electrodeposition of hydroxyapatite onto nanotubular TiO_2 for implant applications, Surf. Coatings Technol. 201 (6) (2006) 3723.
[134] M. Schuler, D. Trentin, M. Textor, S.G. Tosatti, Biomedical interfaces: titanium surface technology for implants and cell carriers, Nanomedicine 1 (4) (2006) 449.

CHAPTER

7

Titanium nanotubes as carriers of osteogenic growth factors and antibacterial drugs for applications in dental implantology

Karthikeyan Subramani[1], Waqar Ahmed[2] and Preeti Pachauri[3]

[1]Roseman University of Health Sciences, Henderson, NV, United States [2]University of Lincoln, Lincoln, United Kingdom [3]Rama Dental College, Hospital & Research Center, Kanpur, India

CHAPTER OUTLINE

7.1 Introduction 125
7.2 Titanium Nanotubes 126
7.3 TiO_2 Nanotubes for Implant Fabrication 127
7.4 Functionalization of TiO_2 Nanotubes with Growth Factors and Antibacterial/ Antiinflammatory Drugs 129
7.5 Recent Advancements 132
7.6 Conclusions 132
References 133

7.1 INTRODUCTION

A nanotube (NT) is a tube-like structure at the nanometer-scale (10^{-9} m). Iijima [1] first synthesized carbon NTs in 1991. These contained at least two layers, or often many more, and measured about 3–30 nm in outer diameter. These bilayered NTs were invariably closed at both ends. In 1993, a new class of carbon NT was discovered with just a single layer [2]. These are known as single-walled NTs and are generally narrower than the multiwalled tubes with diameters typically in the range of 1–2 nm exhibiting extraordinary strength, unique electrical properties, and efficient conduction of heat. The advent of carbon NTs stimulated intense research activities worldwide due to its wide range of potential applications. These include potential applications in nanoelectromechanical systems,

Emerging Nanotechnologies in Dentistry. DOI: http://dx.doi.org/10.1016/B978-0-12-812291-4.00007-8

high-strength composite material synthesis, chemical sensors, nanotweezers, nanoprobes, hydrogen storage and energy conversion devices, field emission devices and their specific applications in the field of nanobiotechnology and biomaterials [3–5].

7.2 TITANIUM NANOTUBES

In the following years, successful chemical synthesis of titanium dioxide (TiO_2) NTs by anodization technique was reported [6–8]. Anodization is a widely used technique to produce protective oxide layer on valve metals like titanium. This technique has attracted great attention in recent years due to its simplicity as well as reproducibility of the results obtained [9,10]. Anodization is an electrolytic passivation technique used to increase the thickness of the natural oxide layer on metal surfaces. The schematic representation of TiO_2 NT fabrication by anodization is shown in Fig. 7.1 [11].

The thickness and structure of the oxide layers formed (amorphous or crystalline) depends on the applied potential between the electrodes, duration of anodization process, and the chemical composition of electrolyte used [12,13]. The structure of the oxide film formed on titanium is amorphous at low voltages

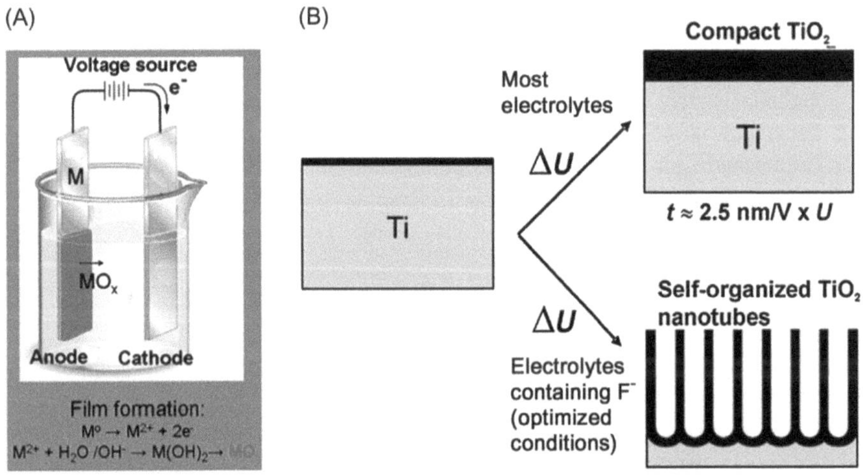

FIGURE 7.1

Schematic setup for anodization experiments. Anodization leads to an oxidation of metal species that form a solid oxide on the metal surface (A). (B) Depending on the anodization conditions (mainly potential, electrolyte, temperature), the solid oxide layer can be either compact, or nanotubular (nanoporous) [11].

(below 20 V) and crystalline at higher voltages [14]. Depending on the anodizing conditions, the crystal structure has been reported to be anatase, a mixture of anatase and rutile, or rutile (mineral forms of TiO_2) [15,16].

A completely different growth morphology leading to self-organized and ordered nanotubular, nanoporous structures of TiO_2 has been obtained when electrolytes containing fluoride ions and suitable anodization conditions were used [11]. Recent in vitro study of osteoblasts on anodized nanotubular titanium substrates has been reported to enhance cell adhesion and proliferation [17]. These NT-like pores possess higher surface energy and wettability compared to unanodized titanium [18]. Furthermore, it has been suggested that TiO_2 with a 3-D micro/nanoporous structure may enhance hydroxyapatite formability when compared to dense TiO_2 [19]. When the anodized titanium substrates were placed in simulated body fluid (SBF), bone-like apatite layer was formed on TiO_2 [20]. These studies indicate that nanotubular nanoporous TiO_2 structures can simulate the environment where bone formation/remodeling occur.

7.3 TIO_2 NANOTUBES FOR IMPLANT FABRICATION

Osteoblast adhesion and activity is enhanced on rougher titanium surfaces than on the smoother surfaces. Rougher surfaces enhance osteoblast activity as there is an increased surface area available for cell (osteoblast) interaction. The studies in the past have focused on modifying titanium surface at the microscale (10^{-6} m) and nanoscale (10^{-9} m) level. Therefore, fabrication of nanoscale structures (NTs) on the titanium substrates will offer more surface area than the microscale rough surface. Moreover, nanoscale features mimic the natural environment that the bone cells are accustomed to, specifically, osteoblasts consistently interact with hydroxyapatite crystals (20–40 nm long) uniquely patterned within a collagen matrix (type I collagen is a triple helix 300 nm in length, 0.5 nm in width, and periodicity of 67 nm) [21]. Numerous studies are being conducted to refine the nanoporous titanium substrate to ordered and controllable NTs with a surface that would closely mimic the nanoarchitectural environment of human bone.

The first generation of the TiO_2 NT arrays was grown in hydrofluoric acid (HF) electrolytes or acidic HF mixtures (Fig. 7.2) [11]. The thickness of the oxide layer was between 500 and 600 nm [22,23]. By substituting HF electrolytes with buffered neutral electrolytes containing sodium fluoride (NaF) or ammonium fluoride (NH_4F), self-organized TiO_2 NT layers with thicknesses higher than 2 μm were grown [24,25]. The third-generation NTs were grown in (almost) water-free electrolytes. Earlier studies in glycerol electrolytes showed tubes with extremely smooth walls and tube length exceeding 7 μm [26], while

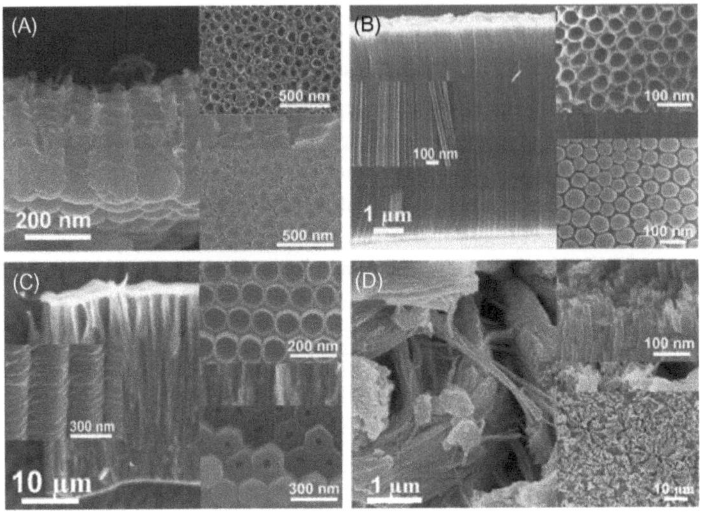

FIGURE 7.2

Scanning electron microscopy (SEM) images showing TiO_2 nanotube layers grown by different anodization processes of Ti. (A) Typical morphology obtained in acidic fluoride or HF electrolytes, (B) glycerol/fluoride electrolytes, (C) ethylene glycol/fluoride electrolytes. The insets show top views (open tubes), bottom views (closed ends), and side walls in detail. (D) Tubes grown by a different approach: rapid breakdown anodization (RBA); these tubes grow in disordered bundles within seconds at comparably high anodic potentials. The upper inset in (D) shows a side view of the layer, the lower inset shows a low magnification of the surface [11].

using acetic acid (CH_3COOH) electrolytes [27] remarkably smaller tube diameters were obtained. In aged ethylene glycol electrolytes and by further optimization of anodization parameters, the NT length reached 260 μm and the tubes with almost ideal hexagonal arrangement were grown [28]. In general, the morphology and the structure of NTs were strongly influenced by the electrochemical conditions, particularly the anodization voltage and the solution parameters (such as the HF concentration, the pH, and the water content in the electrolyte).

Using the same approach of controlled anodization in dilute fluoride electrolytes, NTs of intermetallic titanium compounds such as Ti–Al [29], binary alloys such as TiNb [30], TiZr [31], or on complex biomedical alloys such as Ti–6Al–7Nb [32] and Ti–29Nb–13Ta4.6Zr [33] were successfully fabricated (Fig. 7.3). Attempts have also been made to grow self-organized anodic NT layers on technologically relevant substrates such as Ti–6Al–4V and Ti–6Al–7Nb [32,34] for dental and orthopedic implant applications.

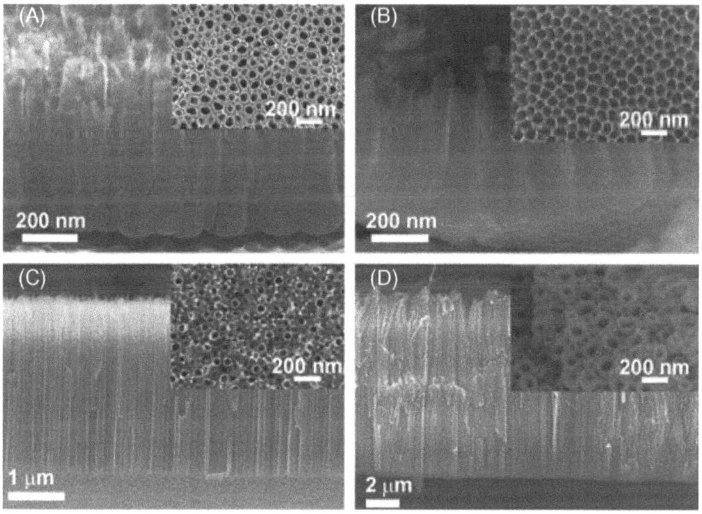

FIGURE 7.3

SEM images showing self-organized oxide nanotube layers grown on Ti–6Al–7Nb alloy (A), Ti–Al intermetallic compound (B), Ti–Nb alloy (C), and Ti–Zr alloy (D) by anodization in fluoride containing $(NH_4)_2SO_4$ electrolytes. The insets show top views of the nanotube layers [11].

7.4 FUNCTIONALIZATION OF TIO_2 NANOTUBES WITH GROWTH FACTORS AND ANTIBACTERIAL/ ANTIINFLAMMATORY DRUGS

TiO_2 nanotubular structures can be functionalized with osteogenic (bone-producing) growth factors which can play a vital role in implant osseointegration. In vivo study in rats reported that the adhesion of osteoblasts and bone regenerating ability were significantly accelerated by TiO_2 NTs when compared with clinically used hydroxy apatite-β tricalcium phosphate (HA-βTCP) inserts [35]. It was shown that the presence of chemically modified TiO_2 NTs have significantly accelerated the kinetics of hydroxyapatite growth by 7 times. Chemical modification of TiO_2 NTs with sodium hydroxide (NaOH) has resulted in surface with bioactive nanoscale sodium titanate [36], which can enhance osteogenesis (Fig. 7.4). A recent study on osteoblast adhesion on TiO_2 NTs (30–100 nm diameter) concluded that large diameter NTs, in the approximately 100 nm regime induced extremely elongated cellular shapes, which resulted in substantially enhanced upregulation of alkaline phosphatase activity, suggesting greater bone-forming ability than NTs with smaller diameters (30 nm) [37]. These NTs stimulate calcium phosphate crystal formation and deposition when exposed to 20 cycles of alternating immersion in saturated $Ca(OH)_2$ and 0.02 M $(NH_4)_2HPO_4$ (Fig. 7.5).

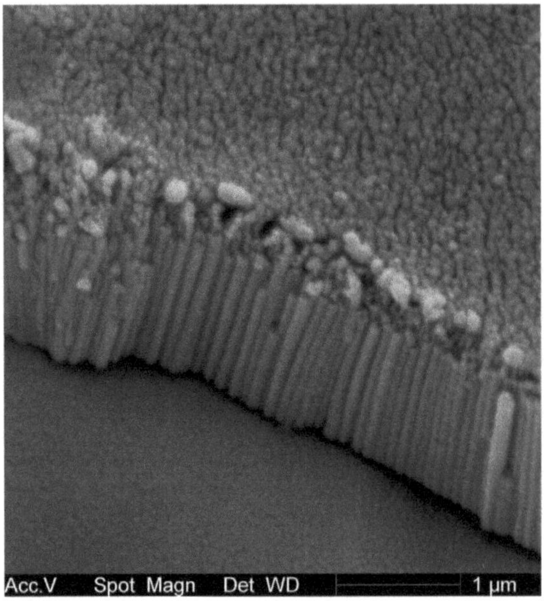

FIGURE 7.4

SEM image showing self-organized oxide nanotube layers grown cp-Ti surface. Anodization for 30 min using ethylene glycol electrolyte at 30 V and treated with 0.5 M NaOH for 15 min resulted in nanotubes covered with a layer of sodium titanate. Nanotubes—100 nm in diameter.

Such approaches prove that calcium phosphate coatings can be fabricated on these NTs, which may serve as carriers of osteogenic growth factors like bone morphogenetic protein-2 (BMP-2).

BMP-2 plays an important role in the formation of bone and recombinant human protein (rhBMP-2) is currently available for orthopedic use in the United States [38]. A recent study has shown that TiO_2 NTs functionalized with BMP-2 increases osteoblast adhesion [39]. Although BMP-2 is a potent osteoinductive growth factor, it is expensive. Such an approach has shown that TiO_2 NTs can be potentially used as a carrier for a wider range of inexpensive osteoinductive growth factors for implant applications in future. The most frequently used bioadhesive peptide motif is the arginine–glycine–aspartic acid (RGD) sequence which mediates bone cell attachment to several extracellular matrix proteins like fibronectin, vitronectin, bone sialoprotein, and osteopontin. RGD plays an important role in regulating cell migration, adhesion, spreading, and differentiation. TiO_2 NTs can be potentially used as carriers for such bioadhesive motifs. The delivery of such bioactive molecules from these nanoscale structures will closely mimic the natural bone architecture and would therefore stimulate osteoblast response and enhance osseointegration.

FIGURE 7.5

SEM image showing self-organized oxide nanotube layers grown cp-Ti titanium surface. Anodization for 30 min using ethylene glycol electrolyte at 30 V and treated with 20 cycles of alternating immersion in saturated $Ca(OH)_2$ and 0.02 M $(NH_4)_2HPO_4$ resulted in nanotubes covered with calcium phosphate crystals. Nanotubes—100 nm in diameter.

Peri-implantitis has become the common cause for dental implant failure. It is an inflammatory process affecting the tissues around an osseointegrated implant in function, thereby resulting in loss of supporting bone. There have been numerous mechanical and chemical agents used to clean the infected implant surface. But achieving complete reosseointegration with the implant surface is still a challenge. Recent studies have explored the use of TiO_2 NTs as antibacterial drug carriers to prevent bacterial adhesion, growth, and biofilm formation. TiO_2 NTs with 80 nm diameter and 400 nm lengths were loaded with gentamicin [40]. The results indicated that the NTs can be effectively filled with the drug and the drug-eluting NTs significantly reduced bacterial adhesion on the surface. This, further enhanced osteoblast differentiation on NTs filled with gentamicin. Another recent study was done on TiO_2 NTs coated with antibacterial drugs like penicillin/streptomycin and antiinflammatory drug (dexamethasone) using simple physical adsorption [41]. The results showed that these functionalized nanotubular surfaces increased osteoblast activity and proliferation. In an in vivo study in pigs, titanium implants covered with an ordered TiO_2 NT layer with an individual tube diameter of 30 nm and 25 commercially pure titanium (cp-Ti) implants were placed in the frontal skull [42]. Peri-implant bone formation and bone-implant contact (BIC) evaluation revealed that the nanotubular surface influenced bone formation and development by enhancing osteoblast function and the NT coatings resisted shearing forces evoked during implant insertion. Recently, an in vivo study in rabbit tibia on bone bonding of two titanium implant surfaces with TiO_2

NTs and TiO_2 grit-blasted surfaces has been conducted. After 4 weeks of implantation, pull-out testing indicated that TiO_2 NTs significantly improved bone-bonding strength by as much as ninefold when compared with TiO_2 grit-blasted surfaces. Histological analysis revealed greater BIC area, new bone formation, and higher calcium and phosphorus levels on the NT surfaces [43].

7.5 RECENT ADVANCEMENTS

Recently one study [44] determined antibacterial effects and biocompatibility of titania NTs (TNTs) with octenidine dihydrochloride (OCT)/poly(lactic-*co*-glycolic acid) (PLGA). In this study, OCT/PLGA was loaded into TNTs without completely filling the NTs using a simple solvent-casting technique to obtain sustained OCT release. OCT/PLGA-TNTs showed good antibacterial effects which could prevent postsurgical complications and even late cases of infection while osteogenic activity was enhanced. Moreover, the fabrication process of OCT/PLGA-TNTs is simple, economical, and versatile. Thus, OCT/PLGA-TNTs are highly attractive for biomedical implants because of their controlled drug release, long-term antibacterial efficacy, and promotion of osseointegration. Another study [45] demonstrated gentamicin-loaded NTs (NT-G) and NTs, especially those with 80 nm diameters, significantly promoted cell attachment, proliferation, spreading, and osteogenic differentiation. Therefore, NT modification and gentamicin loading can significantly improve the antibacterial ability and osteogenic activity of orthopedic implants. Several smart strategies including pH and temperature (thermal) [46,47], light [48–53], radiofrequency [54–63], magnet [64,65], and ultrasound [66] were used as triggers for drug release from TNTs, thus offering numerous opportunities for TNT applications

7.6 CONCLUSIONS

Anodization of titanium has been effectively applied to fabricate nanotubular surfaces which closely mimic the nanoscale architecture of human bone. Such surfaces can be fabricated for implant applications and thus serve as an effective carrier for osteoinductive growth factors to promote bone formation and integration. These recent advances in utilizing TiO_2 NTs have been evaluated in preclinical animal studies [42,43] with favorable results. Antibacterial and antiinflammatory drugs can also be physically adsorbed on such surfaces to produce antibacterial implant surfaces to prevent infection and inflammation following peri-implantitis.

REFERENCES

[1] S. Iijima, Helical microtubules of graphitic carbon, Nature 354 (6348) (1991) 56.
[2] S. Iijima, T. Ichihashi, Single-shell carbon nanotubes of 1-nm diameter, Nature 363 (1993) 603.
[3] P. Kim, C.M. Lieber, Nanotube nanotweezers, Science. 286 (1999) 2148.
[4] J.M. Planeix, N. Coustel, B. Coq, V. Brotons, P.S. Kumbhar, R. Dutartre, Application of carbon nanotubes as supports in heterogeneous catalysis, J. Am. Chem. Soc. 116 (17) (1994) 7935.
[5] M. Paradise, T. Goswami, Carbon nanotubes—production and industrial applications, Mater. Des 28 (5) (2007) 1477.
[6] P. Hoyer, Semiconductor nanotube formation by a two-step template process, Adv. Mater 8 (10) (1996) 857.
[7] T. Kasuga, M. Hiramatsu, A. Hoson, T. Sekino, K. Niihara, Formation of titanium oxide nanotube, Langmuir 14 (12) (1998) 3160.
[8] T. Kasuga, Formation of titanium oxide nanotubes using chemical treatments and their characteristic properties, Thin Solid Films 496 (1) (2006) 141.
[9] Y.T. Sul, The significance of the surface properties of oxidized titanium to the bone response: special emphasis on potential biochemical bonding of oxidized titanium implant, Biomaterials 24 (22) (2003) 3893.
[10] A.S. Karakoti, R. Filmalter, D. Bera, S.V. Kuchibhatla, A. Vincent, S. Seal, Spiral growth of one dimensional titania nanostructures using anodic oxidation, J. Nanosci. Nanotechnol 6 (7) (2006) 2084.
[11] J.M. Macak, H. Tsuchiya, A. Ghicov, K. Yasuda, R. Hahn, S. Bauer, TiO_2 nanotubes: self-organized electrochemical formation, properties and applications, Curr. Opin. Solid State Mater. Sci 11 (2007) 3.
[12] L. Young, Anodic Oxide Films, Plenum, New York, 1961.
[13] D. Vermilyea, Anodic Films Advances in Electrochemistry and Electrochemical Engineering, Wiley, London, 1963.
[14] J.W. Schultze, M.M. Lohrengel, D. Ross, Nucleation and growth of anodic oxide films, Electrochim. Acta. 28 (7) (1983) 973.
[15] J.L. Delplancke, R. Winand, Galvanostatic anodization of titanium—II. Reactions efficiencies and electrochemical behaviour model, Electrochim. Acta 33 (11) (1988) 1551.
[16] J.L. Delplancke, R. Winand, Galvanostatic anodization of titanium—I. Structures and compositions of the anodic films, Electrochim. Acta 33 (11) (1988) 1539.
[17] C. Yao, V. Perla, J.L. McKenzie, E.B. Slamovich, T.J. Webster, Anodized Ti and Ti6Al4V possessing nanometer surface features enhances osteoblast adhesion, J. Biomed. Nanotechnol. 1 (2005) 68.
[18] X. Zhu, J. Chen, L. Scheideler, R. Reichl, J. Geis-Gerstorfer, Effects of topography and composition of titanium surface oxides on osteoblast responses, Biomaterials 25 (18) (2004) 4087.
[19] B. Yang, M. Uchida, H.M. Kim, X. Zhang, T. Kokubo, Preparation of bioactive titanium metal via anodic oxidation treatment, Biomaterials 25 (6) (2004) 1003.
[20] H.M. Kim, H. Kaneko, M. Kawashita, T. Kokubo, T. Nakamura, Mechanism of apatite formation on anodically oxidized titanium metal in simulated body fluid, Key Eng. Mater 254 (2004) 741.

[21] J.Y. Rho, L. Kuhn-Spearing, P. Zioupos, Mechanical properties and the hierarchical structure of bone, Med. Eng. Phys 20 (2) (1998) 92.

[22] V. Zwilling, M. Aucouturier, E. Darque-Ceretti, Anodic oxidation of titanium and TA6V alloy in chromic media. An electrochemical approach, Electrochim. Acta 45 (6) (1999) 921.

[23] R. Beranek, H. Hildebrand, P. Schmuki, Self-organized porous titanium oxide prepared in H_2SO_4/HF electrolytes, Electrochem. Solid-State Lett 6 (3) (2003) B12.

[24] J.M. Macak, K. Sirotna, P. Schmuki, Self-organized porous titanium oxide prepared in Na_2SO_4/NaF electrolytes, Electrochim. Acta 50 (18) (2005) 3679.

[25] L.V. Taveira, J.M. Macak, H. Tsuchiya, L.F.P. Dick, P. Schmuki, Initiation and growth of self-organized TiO_2 nanotubes anodically formed in $NH_4F/(NH_4)_2SO_4$ electrolytes, J. Electrochem. Soc 152 (10) (2005) B405.

[26] J.M. Macak, H. Tsuchiya, L.V. Taveira, S. Aldabergerova, P. Schmuki, Smooth anodic TiO_2 nanotubes, Angew. Chem. Int. Ed 44 (2005) 7463.

[27] H. Tsuchiya, J.M. Macak, L. Taveira, E. Balaur, A. Ghicov, K. Sirotna, Self-organized TiO_2 nanotubes prepared in ammonium fluoride containing acetic acid electrolytes, Electrochem. Commun 7 (6) (2005) 576.

[28] S.P. Albu, A. Ghicov, J.M. Macak, P. Schmuki, 250 μm long anodic TiO_2 nanotubes with hexagonal self-ordering, Phys. Status Solidi (RRL) 1 (2) (2007) R65.

[29] H. Tsuchiya, S. Berger, J.M. Macak, A. Ghicov, P. Schmuki, Self-organized porous and tubular oxide layers on TiAl alloys, Electrochem. Commun 9 (9) (2007) 2397.

[30] A. Ghicov, S. Aldabergerova, H. Tsuchiya, P. Schmuki, $TiO_2-Nb_2O_5$ nanotubes with electrochemically tunable morphologies, Angew. Chem. Int. Ed 45 (42) (2006) 6993.

[31] K. Yasuda, P. Schmuki, Formation of self-organized zirconium titanate nanotube layers by alloy anodization, Adv. Mater 19 (13) (2007) 1757.

[32] J.M. Macak, H. Tsuchiya, L. Taveira, A. Ghicov, P. Schmuki, Self-organized nanotubular oxide layers on Ti−6Al−7Nb and Ti−6Al−4V formed by anodization in NH_4F solutions, J. Biomed. Mater. Res. A 75 (4) (2005) 928.

[33] H. Tsuchiya, J.M. Macak, A. Ghicov, Y.C. Tang, S. Fujimoto, M. Niinomi, Nanotube oxide coating on Ti−29Nb−13Ta−4.6Zr alloy prepared by self-organizing anodization, Electrochim. Acta 52 (1) (2006) 94.

[34] V. Zwilling, E. Darque-Ceretti, A. Boutry-Forveille, D. David, M.Y. Perrin, M. Aucouturier, Structure and physicochemistry of anodic oxide films on titanium and TA6V alloy, Surf. Interface Anal. 27 (7) (1999) 629.

[35] S. Kubota, K. Johkura, K. Asanuma, Y. Okouchi, N. Ogiwara, K. Sasaki, Titanium oxide nanotubes for bone regeneration, J. Mater. Sci. Mater. Med 15 (9) (2004) 1031.

[36] S.J. Seunghan Oh, Titanium oxide nanotubes with controlled morphology for enhanced bone growth, Mater. Sci. Engg. C 26 (2006) 1301.

[37] K.S. Brammer, S. Oh, C.J. Cobb, L.M. Bjursten, H. van der Heyde, S. Jin, Improved bone-forming functionality on diameter-controlled TiO_2 nanotube surface, Acta Biomater 5 (8) (2009) 3215.

[38] S.N. Khan, J.M. Lane, The use of recombinant human bone morphogenetic protein-2 (rhBMP-2) in orthopaedic applications, Expert. Opin. Biol. Ther 4 (5) (2004) 741.

[39] G. Balasundaram, C. Yao, T.J. Webster, TiO_2 nanotubes functionalized with regions of bone morphogenetic protein-2 increases osteoblast adhesion, J. Biomed. Mater. Res. A 84 (2) (2008) 447.

[40] K.C. Popat, M. Eltgroth, T.J. Latempa, C.A. Grimes, T.A. Desai, Decreased *Staphylococcus epidermis* adhesion and increased osteoblast functionality on antibiotic-loaded titania nanotubes, Biomaterials. 28 (32) (2007) 4880.

[41] G. Eaninwene, C. Yao, T.J. Webster, Enhanced osteoblast adhesion to drug-coated anodized nanotubular titanium surfaces, Int. J. Nanomed 3 (2) (2008) 257.

[42] C. von Wilmowsky, S. Bauer, R. Lutz, M. Meisel, F.W. Neukam, T. Toyoshima, In vivo evaluation of anodic TiO_2 nanotubes: an experimental study in the pig, J. Biomed. Mater. Res. B: Appl. Biomater 89 (1) (2009) 165.

[43] L.M. Bjursten, L. Rasmusson, S. Oh, G.C. Smith, K.S. Brammer, S. Jin, Titanium dioxide nanotubes enhance bone bonding in-vivo, J. Biomed. Mater. Res. A (2009).

[44] Z. Xu, Y. Lai, Wu Dong, W. Huang, S. Huang, L. Zhou, et al., Antibacterial effects and biocompatibility of titania nanotubes with octenidine dihydrochloride/poly(lactic-co-glycolic acid), BioMed Res. Int. (2015) 1–11.

[45] W.T. Lin, H.L. Tan, Z.L. Duan, B. Yue, R. Ma, G. He, et al., Inhibited bacterial biofilm formation and improved osteogenic activity on gentamicin-loaded titania nanotubes with various diameters, Int. J. Nanomed 9 (2014) 1215–1230.

[46] H. Jia, L.L. Kerr, Kinetics of drug release from drug carrier of polymer/TiO_2 nanotubes composite-pH dependent study, J. Appl. Polym. Sci 132 (2015) 41750.

[47] K. Cai, F. Jiang, Z. Luo, X. Chen, Temperature-responsive controlled drug delivery system based on titanium nanotubes, Adv. Eng. Mater 12 (2010) B565–B570.

[48] R. Guo, L. Zhang, H. Qian, R. Li, X. Jiang, B. Liu, Multifunctional nanocarriers for cell imaging, drug delivery, and near-IR photothermal therapy, Langmuir 26 (2010) 5428–5434.

[49] T. Sada, T. Fujigaya, Y. Niidome, K. Nakazawa, N. Nakashima, Near-IR laser-triggered target cell collection using a carbon nanotube-based cell-cultured substrate, ACS Nano 5 (2011) 4414–4421.

[50] J. Choi, J. Yang, D. Bang, et al., Targetable gold nanorods for epithelial cancer therapy guided by near-IR absorption imaging, Small 8 (2012) 746–753.

[51] R. Chen, X. Wang, X. Yao, X. Zheng, J. Wang, X. Jiang, Near-IR-triggered photothermal/photodynamic dual modality therapy system via chitosan hybrid nanospheres, Biomaterials 34 (2013) 8314–8322.

[52] P.P. Joshi, S.J. Yoon, Y.S. Chen, S. Emelianov, K.V. Sokolov, Development and optimization of near-IR contrast agents for immune cell tracking, Biomed. Opt. Express 4 (2013) 2609–2618.

[53] J.T. Robinson, K. Welsher, S.M. Tabakman, et al., High performance in vivo near-IR (>1 μm) imaging and photothermal cancer therapy with carbon nanotubes, Nano Res 3 (2010) 779–793.

[54] M.S. Aw, M. Kurian, D. Losic, Non-eroding drug-releasing implants with ordered nanoporous and nanotubular structures: concepts for controlling drug release, Biomater. Sci 2 (2014) 10–34.

[55] E.S. Glazer, S.A. Curley, Radiofrequency field-induced thermal cytotoxicity in cancer cells treated with fluorescent nanoparticles, Cancer 116 (2010) 3285–3293.

[56] S.A. Curley, P. Cherukuri, K. Briggsetal, Noninvasive radiofrequency field-induced hyperthermic cytotoxicity in human cancer cells using cetuximab-targeted gold nanoparticles, J. Pharmacol. Exp. Ther 7 (2008) 313–326.

[57] E.S. Glazer, K.L. Massey, C. Zhu, S.A. Curley, Pancreatic carcinoma cells are susceptible to noninvasive radio frequency fields after treatment with targeted gold nanoparticles, Surgery 148 (2010) 319–324.

[58] P. Cherukuri, S.A. Curley, Use of nanoparticles for targeted, noninvasive thermal destruction of malignant cells, Methods Mol. Biol 624 (2010) 359–373.

[59] P. Cherukuri, E.S. Glazer, S.A. Curley, Targeted hyperthermia using metal nanoparticles, Adv. Drug Deliv. Rev 62 (2010) 339–345.

[60] C.J. Gannon, P. Cherukuri, B.I. Yakobson, et al., Carbon nanotube-enhanced thermal destruction of cancer cells in a noninvasive radiofrequency field, Cancer 110 (2007) 2654–2665.

[61] C.J. Gannon, C.R. Patra, R. Bhattacharya, P. Mukherjee, S.A. Curley, Intracellular gold nanoparticles enhance non-invasive radiofrequency thermal destruction of human gastrointestinal cancer cells, J. Nanobiotechnol 6 (2008) 1–9.

[62] M. Raoof, S.A. Curley, Non-invasive radiofrequency-induced targeted hyperthermia for the treatment of hepatocellular carcinoma, Int. J. Hepatol 2011 (2011) 676957.

[63] J. Cardinal, J.R. Klune, E. Chory, et al., Noninvasive radiofrequency ablation of cancer targeted by gold nanoparticles, Surgery 144 (2008) 125–132.

[64] M.S. Aw, J. Addai-Mensah, D. Losic, Magnetic-responsive delivery of drug carriers using titania nanotube arrays, J. Mater. Chem 22 (2012) 6561–6563.

[65] N.K. Shrestha, J.M. Macak, F. Schmidt-Stein, et al., Magnetically guided titania nanotubes for site-selective photocatalysis and drug release, Angew. Chem. Int. Ed 48 (2009) 969–972.

[66] M.S. Aw, D. Losic, Ultrasound enhanced release of therapeutics from drug-releasing implants based on titania nanotube arrays, Int. J. Pharm 443 (2013) 154–162.

CHAPTER

8

Cellular responses to nanoscale surface modifications of titanium implants for dentistry and bone tissue engineering applications

Kaushik Subramanian[1,2], Daniel Tran[1,2] and Kytai T. Nguyen[1,2]

[1]*University of Texas at Arlington, Arlington, TX, United States*

[2]*University of Texas, Dallas, TX, United States*

CHAPTER OUTLINE

8.1 Introduction 138
8.2 Nanotopography Generated from Surface Modification of Ti Implants 138
8.2.1 Surface Modification of Ti Implants With Inorganic Materials/Nanoparticles 139
8.2.2 Surface Modifications of Ti Implants With Polymers 141
8.3 Nanotopography and Protein Absorption 142
8.4 Nanotopography Alters Osteoblast Responses 143
8.4.1 Cell Morphology 149
8.4.2 Cell Adhesion 149
8.4.3 Cell Proliferation 149
8.4.4 Bioactive Molecules 150
8.4.5 Osseointegration 150
8.5 Nanotopography and Stem Cell Responses 151
8.5.1 Effects of Nanotopography on Endothelial Progenitor Cells 151
8.5.2 Effects of Nanotopography on Bone Marrow Stem Cells 152
8.6 Conclusions 153
References 153

Emerging Nanotechnologies in Dentistry. DOI: http://dx.doi.org/10.1016/B978-0-12-812291-4.00008-X

8.1 INTRODUCTION

Biomaterials such as titanium (Ti) have been used for various purposes from treating injuries to replacing defects in the human body. Since their first use in the 1960s, biomaterials have been applied extensively in various medical applications [1], especially in the fields of orthopedics, dental, cardiovascular, wound healing, and drug delivery systems [2]. For bone implants, osseointegration (integration of bone cells with implant materials) has been a major goal while designing and fabricating implants; and toward this end, implant surfaces such as these of Ti have been greatly modified to improve bone cell adhesion and growth, leading to better osseointegration [3–9]. It has been shown previously that cell behaviors such as adhesion and osteoblast differentiation are greatly influenced by the nanoscale topography of the implant materials [10–15]. In the next sections, this chapter will review methods to create nanotopography on the Ti implants and responses of bone cells toward these modified surfaces.

8.2 NANOTOPOGRAPHY GENERATED FROM SURFACE MODIFICATION OF Ti IMPLANTS

Ti in its pure form has been tested and compared with other materials such as stainless steel and cobalt (Co)–chromium (Cr) alloys for their biocompatibility in vivo. Ti has commonly been used for dental implants over several decades. Commercially used Ti is pure Ti (ASTM F67) and Ti–6Al–4V (ASTM F136) [14]. Both pure Ti and its alloys exhibit superior mechanical properties, chemical stability, and in vivo biocompatibility [16–25]. Pure Ti is generally used when corrosion resistance is of greater importance than mechanical strength, whereas Ti–6Al–4V, an alpha–beta alloy, is used when the mechanical strength is required (Table 8.1). However, a major concern when Ti–6Al–4V is used is the leeching of aluminum (Al) and vanadium (V) ions from the alloy. Thus, other Ti alloys including Ti–6Al–7Nb are under investigation to search for materials with better corrosion properties [21,26,27]. Table 8.1 shows the mechanical properties and applications of different Ti alloys that are currently used.

Several modification methods have been developed to create distinct nanotopographic features because recent studies have shown that nanotopography may affect cell adhesion, growth, and differentiation. In addition, different surface structures and chemical functional groups can be added to the Ti surface so that they can be used to covalently bind to cell growth-promoting factors and bone-related factors such as fibroblast growth factors, bone morphogenic protein-2, and Isterix [5,34]. A few examples of surface modification methods include modification of Ti implants with inorganic materials such as hydroxyapatite (HA), chromium, cobalt, and V, and use of nanoparticles or incorporation of Ti powders with polymers such as poly(L-lactic acid) or poly(L-lactic-*co*-glycolic) acid

Table 8.1 Properties and Bone Tissue Engineering Applications of Different Ti Alloys [28–33]

Alloys	Elastic Modulus (GPa)	Ultimate Tensile Strength (MPa)	Elongation (%)	Applications
Pure Ti	100–110	240–550	15–24	Dental implants, maxillofacial and craniofacial implants, screws for spinal surgery
Ti–6Al–4V	110	930	10–15	Total joint replacement, primarily for hips, and knees
Ti–6Al–7Nb	105	860	10	Femoral hip stems, fracture fixation plates, spinal components, fasteners, nails, rods, screw, and wire
Ti–13Nb–13Zr	79–84	973–1037	10–16	Orthopedic implants

Ti, *titanium;* Al, *aluminum;* V, *vanadium;* Nb, *niobium;* Zr, *Zircon.*

(PLGA) to produce distinct indentations and nanoscale features. The modification techniques commonly used to create nanotopography on Ti implants will be briefly discussed in the next section.

8.2.1 SURFACE MODIFICATION OF Ti IMPLANTS WITH INORGANIC MATERIALS/NANOPARTICLES

Modification of Ti implants in the nanoscale dimension affects the topography and chemistry of the surface. It is important to note that the surface of an implant possesses three main properties, which are mechanical, topographic, and physicochemical properties, and that changing any of these properties will have an effect on the other two properties [5]. Although various methods have been developed to impart nanoscale modification to implant materials as shown in Table 8.2, there are only a few methods often used for surface modification due to their ease of use and reproducibility. Physical approaches for surface modification include compaction of nanoparticles and ion beam deposition. Of these methods, the compaction of nanoparticles and microparticles of titanium dioxide (TiO_2) is often used, and this modification yields implant surfaces with nanoscale and microscale features [24]. Since this method physically deposits the micro- and/or nanofeatures onto the implant surface, it has little impact on the bulk chemical properties of the material [41].

In addition to physical approaches, chemical methods including treatment of implants with active chemicals such as acids have also been developed to impart nanotopography onto the implant surfaces. Common chemicals used to produce nanostructures on Ti implant surfaces are sodium hydroxide (NaOH) and hydrogen

Table 8.2 Methods for Creating Nanofeatures on Ti Implants Using Inorganic Materials [24,35–40]

Methods	Characteristics
Physical approaches	*Surface chemistry of the implants will not be altered*
Compaction of micro-/nanoparticles	Localized only to certain parts of the implant surface
Possible weak bonds between particles and the implant surfaces	
Ion beam deposition	Create nanofeatures on the surface depending on materials used
Chemical approaches	*Surface chemistry of the implants might also be altered*
Acid etching	Often used with other methods
Imparts nanofeatures randomly	
Peroxidation	Produces a TiO_2 gel layer
Imparts both chemical and topographical properties	
NaOH treatment	Produces a sodium–titanate gel layer
Varies both chemical and topographical properties	
Can be used to deposit HA	
Others	*Surface chemistry of the implants might or might not be altered*
Sol–gel (colloidal particle adsorption)	Deposits nanoparticles
Creates a nanoscale thick film consisting of chemical properties	
Crystalline deposition	Deposits crystals to obtain a unique complex topography
Self-assembly of monolayers	Exposes functional end groups (e.g., RGD) that have specific functions
Lithography	Impart several microtopographic features on the surface
Expensive, time consuming	

peroxide (H_2O_2) [42]. NaOH treatment of Ti implants produces a sodium–titanate gel layer on the surface, whereas H_2O_2 treatment creates a titania gel layer. These gel layers can then be used to deposit osteoblast-promoting materials such as HA. Ti implants chemically treated with acids and NaOH have been shown to accelerate HA crystal growth in simulated body fluid [27]. Other chemical treatments such as peroxidation or acid oxidation (hydrofluoric acid) have also been used to create nanotopography [43]. It has been shown that treatment of Ti implant surfaces with H_2O_2/HCl increased adsorption of Arg-Gly-Asp (RGD) peptides onto the surface due to creation of an amorphous nanoscale feature on the implant surface [35]. In general, chemical approaches to generate nanotopography are popularly used in dental implants which might be due to their ease of use [41].

One novel approach to create nanofeatures on Ti implants is the deposition of nanoparticles onto the surface [36]. Nanoscale deposition of calcium phosphate has been achieved by the sol–gel coating method [37]. Other nanostructures made of metals such as alumina, titania, and zirconia have also been deposited onto implant surfaces to provide nanoscale ridges that aid cellular adhesion and differentiation [44,45]. Since the contacts are on the quantum scale, physical interaction between the nanoparticles and the implant surface has been shown to be very strong [46–48]. In addition, nanodeposition of calcium phosphate on acid-treated Ti has been shown to significantly increase mechanical interlocking with bone and accelerate healing of bone tissue [49]. Major concerns with this method include the agglomeration of nanoparticles during the coating process and the uneven distribution of nanoparticles onto the Ti implant surfaces, especially when nanoparticles are used at high concentrations.

A modern approach to create topography on Ti implants is the use of photolithography. However, this method is expensive, time consuming, and requires considerable development prior to actual use. Additionally, photolithography is rarely used to create nanoscale structures and mostly used to create microscale features [50]. The most widely used approach to create nanotopography on the implant surfaces is the sandblasting and acid-etching (SLA) method [51,52]. The SLA method has the advantages of both SLA techniques: creating both macro- to microscale structures (sandblasting) and micro- to nanoscale structures (acid etching). The major limitation of the SLA method is that it is a random process, thus it is hard to control the uniformity and distribution of nanostructures on the implant surfaces. To overcome the limitation of SLA methods, both plasma-spray coating of inorganic materials including HA and electron beam evaporation of calcium phosphate have also been commonly employed to generate nanostructures on the Ti implant surfaces with a better uniform distribution of nanofeatures on the implant surfaces. However, these methods are difficult to form the nanostructures on complex-shaped implants and inside the implant cavities.

Another method is the deposition of a preselected molecule onto the implant substrate by chemisorption. This method leads to the expression of selected functional groups at the cell implant interface [38] that may aid in cellular attachment onto the substrate or even lead to directed cell differentiation. One such popular molecule is the RGD peptide incorporated into a polyethylene glycol end group, which is adsorbed onto implant surfaces and aids in osseointegration [53]. Since this method involves chemisorption of functional groups, it leads to a localized modification of the chemical properties of the bulk substrate.

8.2.2 SURFACE MODIFICATIONS OF Ti IMPLANTS WITH POLYMERS

Coating the surface of metals such as Ti with functionalized polymer films has been shown as a useful strategy to improve osseointegration while retarding metal corrosion [54–60]. Polymers possess several advantages including ease of manipulation to vary their physical and chemical properties of the implants. Another major advantage of polymers is that they can be used for loading and controlled release of therapeutic reagents such as bone-promoting factors. However, they do possess some major disadvantages like the potential of releasing harmful compounds such as degraded products into the surrounding tissue and consisting of mechanical properties that are prone to wear and tear. Polymers also easily adsorb proteins, which causes an alteration in the surface chemistry of such polymers. Although these limiting factors prevent polymers from being more widely used in areas such as dental and bone replacement implants, several techniques have been developed to modify the Ti implant surfaces with various polymers to improve implant properties and enhance bone growth and integration.

Electrochemical polymerization is a technique that has been often employed to deposit polymer films on different metal surfaces. For example, homogenous passive polyacrylic acid films have been synthesized onto the surfaces of pure Ti and Ti–6Al–4V substrates to provide improved anticorrosion and new bioactive properties [55,61]. Various methods have also been devised to deliver biochemical factors at the interface between the implant surface and bone tissue. These methods include adsorption or covalent binding of bioactive molecules onto the polymer films [62,63]. Besides providing corrosion resistance and a bioactive surface, polymer coating of metal bone implants have also been shown to improve vascularization. Since metal implants show little or no support for revascularization, polymer coatings have been performed to aid in vascularization of the new bone tissue by entrapping and releasing biologically active molecules like vascular endothelial growth factor (VEGF), a key protein involved in vasculogenesis and angiogenesis [64]. This was supported by several studies which showed that VEGF aids in healing bone defects [65–68]. Coating Ti–6Al–4V implants with collagen and poly(2-hydroxyethyl methacrylate-2-methacryloyloxyethyl phosphate) loaded with VEGF has also been tested and shown to improve vascularization within the implant [69]. Moreover, immobilized VEGF on Ti alloy substrates coated with thin adherent polydopamine films induces the differentiation of human mesenchymal stem cells (hMSCs) into endothelial cells [70].

8.3 NANOTOPOGRAPHY AND PROTEIN ABSORPTION

Surface modification of Ti implants to create nanotopography on the implant surfaces has been shown to increase the adsorption of proteins, leading to an increase in cell adhesion, proliferation, and subsequently osseointegration. Adsorption of proteins produced by blood cells on the Ti implant is of critical importance, especially immediately after the implantation procedure. The proteins presented in blood plasma include albumins, fibrinogen, and immunoglobulins. Once the proteins have been adsorbed onto the implant surface, they act as the interface between the surface and the cells. As shown by a previous study, adsorption is not only affected by the texture of the Ti implant but also by the surface chemistry and hydrophobicity [71]. An implant made of Ti slows the process of blood clotting while reducing the adhesion of platelets when compared to other materials such as steel [72]. To increase the surface roughness of the Ti implant surfaces for protein adsorption, coating of the implant surface with carbon nanotubes has been used [73]. For example, F-actin more readily adsorbs onto a surface with a groove height of 1–2 nm than it does on a surface with a height of 4 nm or greater. The nanostructure can also affect the orientation of the adsorbed proteins such as fibronectin [74] and alter the conformation of the RGD containing proteins like fibronectin and vitronectin [41]. In general, nanotopography can alter protein interactions with an implant surface, leading to an increase in cell activities including osteoblast adhesion.

8.4 NANOTOPOGRAPHY ALTERS OSTEOBLAST RESPONSES

The cellular response of osteoblast cells to Ti surfaces depends on a variety of factors including structures and dimensions of the nanotopography as well as types of cells and materials. The potential impacts include changes in cell morphology, adhesion, proliferation, and production of other bioactive molecules (Fig. 8.1). Osteoblasts are cells that are primarily responsible for calcium deposition and formation of bone including teeth. They work in tandem with osteoclasts, which resorb bone, to constantly regulate bone formation. Cellular responses including cell morphology, adhesion, and proliferation toward the Ti-modified surfaces in vitro and in vivo are summarized in Tables 8.3 and 8.4, respectively.

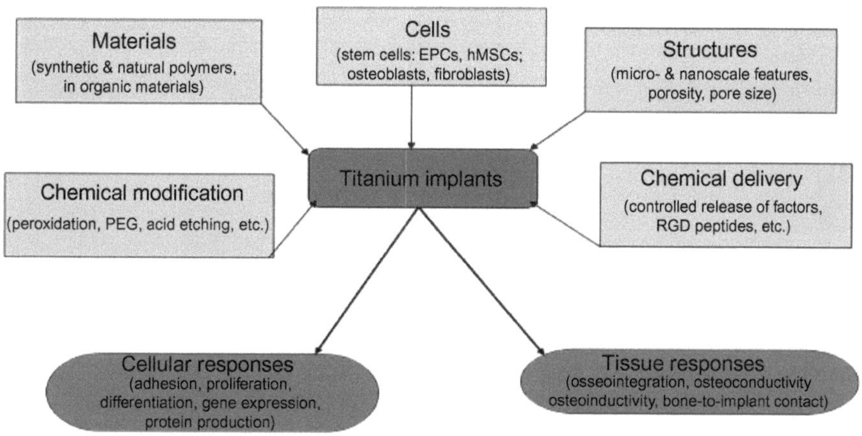

FIGURE 8.1

Methods of modification of Ti implants and their outcomes.

Table 8.3 Cellular Responses Towards Surface-Modified Titanium and its Alloys

Modified Structures/Chemicals	Modification Methods	Materials Used	Cell Type	Cell Responses	References
30–100 nm nanotubes	Anodization	Titanium dioxide (TiO_2)	hMSC, hSaOS-2	Enhanced adhesion for 30 nm induced ALP production, elongated morphology, differentiation for 70–100 nm	[75–79]
30–100 nm nanotubes	Anodization	TiO_2	BAEC	Enhanced focal adhesions upregulated antithrombotic cellular state (e.g., induced nitric oxide, endothelin-1)	[76]
Nanotubular	Anodization	Titania	RMSC	Enhanced cell adhesion, proliferation, ALP activity, bone matrix deposition	[80]
15, 55, and 100 nm nanopillars, 15–140 nm dot-, pillar-like structures	Anodization through a porous alumina mask	Titania (or TiO_2)	hMSC	Enhanced bone matrix nodule forming (15 nm)	[81,82]
Nanofibers	Surface coating	Silicon (Si), silicon oxide (SiO_2), and titanium oxide (TiO_2)	MC3T3-E1	Better cell proliferation for Si, SiO_2 fibers higher cellular differentiation for TiO_2 fibers	[83]
8–10 nm roughness	DC reactive magnetron sputtering	Titania	Primary rat osteoblasts	Enhanced cell adhesion, spreading, proliferation, and differentiation	[84]
30–50 nm roughness	Electrolytic deposition	TiO_2	Primary rat osteoblasts	Enhanced cell adhesion, spreading, proliferation	[85]
110 nm hemispherical	Electrostatic interactions of adsorbed colloidal particles	TiO_2	Primary human osteoblasts	No difference in cell adhesion and cytoskeletal formation accelerated PGE2 release, increased fiber length of F-actin and β-tubulin	[86]
Nanotopography	Acid treatment	H_2SO_4/H_2O_2	Rat calvarial osteoblasts	Induced cell proliferation, ALP activity favored BSP, osteopontin, fibronectin accumulation	[87]

100 nm, microtopography	Hydrofluoric acid (HF) treatment, TiO_2 grit blasting	TiO_2 grit blasted/HF	hMSC MC3T3-E1	Stimulated osteoblast differentiation more in acid-treated surface (nanotopography)	[88]
Enhanced osteoinductive transcription factors (RUNX-2, SMADs), growth factors (IGF-2, BMPs), and bone matrix proteins in acid treatment, increased levels of RUNX-2, osterix, BSP					[89]
Nano + micro, 10, 30, 100 μm hemispherical nano + micro	Mechanical polishing + acid etching, electrochemically microstructured (EM), EM + acid etching	N/A	MG63 cells	Cells more spread, longer and more filopods observed, vinculin positive focal contacts, little effect on cell morphology, a marked synergistic effect on cell proliferation	[90]
0.5–106 μm in diameter	Compaction	Metal powders: Ti, Ti–6Al–4V, CoCrMo	Human osteoblasts	Nanophase of Ti–6Al–4V and CoCrMo: more calcium and phosphorus deposition compared to microphase, no change in Ti microphase and nanophase	[91]
Micro-/nanotopography	Sand blasted and acid etched (SLA)	N/A	hMSCMG63	Induced levels of osteogenic markers SPP1, RUNX-2, BSP, WNT5A, enhanced cluster formation of osteoblasts, increased expression of bone-associated genes, ALP, osteocalcin, osteoprotegerin, collagen I	[92–94]
Nanotopography	HF treatment	N/A	MC3T3-E1	Improved biocompatibility	[95]
Micro-/nanotopography	Hydroxyapatite (HA)/ SLA by the ion beam assisted deposition (IBAD)	HA particles	hMSC	Induced adhesion	[96]

(Continued)

Table 8.3 Cellular Responses Towards Surface-Modified Titanium and its Alloys *Continued*

Modified Structures/Chemicals	Modification Methods	Materials Used	Cell Type	Cell Responses	References
Increased levels of ALP, OCN					
Improved cell proliferation and differentiation					
7, 15, 40 nm	Sol–gel coating	Nobium oxidation	MC3T3-E1	Optimal cell adhesion and spreading at 15 nm	[97]
Surface chemistry with Mg, Zn, CHAP	Ion beam implantation	Zinc, magnesium, alkoxide-derived hydroxy carbonate apatite (CHAP)	Human osteoblasts	Upregulated c-fos, MAP kinase pathways	[98]
60–70 nm thickness					
Surface chemistry with Ti alloys	Pulsed laser deposition	Ti–6Al–4V, TiNb30, and TiNb13Zr13	Osteoblast-like cells	Had positive effect on cell alignment	[99]
Influenced focal contact formation					
Surface chemistry with metals	Compaction	Ti, Ti–6Al–4V, CoCrMo	Human osteoblasts	Increased cell adhesion	[24]
Surface chemistry with HA, tricalcium phosphate (TCP), calcium titanate ($CaTiO_3$)	Ti coating/annealing	HA, TCP, $CaTiO_3$	Human osteoblasts	Enhanced cell adhesion for $CaTiO_3$, the most enhanced adhesion for TCP/$CaTiO_3$	[100,101]
Surface chemistry with RGD	H_2O_2 treatment/ adsorption	RGD	RMSC	High efficiency of RGD attachment	[35]
Rat calvarial osteoblasts					[57]
Calcified matrix formation					
Enhanced cell adhesion					
Surface chemistry with alumina, titania, HA	Compaction	Alumina, titania, HA powders	Rat osteoblasts	Increased vitronectin adsorption	[102]

Induced cell adhesion					
Surface chemistry with functional (e.g., carboxylic acid) groups	Grafting Ti with polymers	Pyrrole-3-acetic acid, chitosan	Mouse bone marrow cells	Enhanced protein binding	[54,63]
Induced cell attachment, proliferation, and ALP activity					
Surface chemistry with HA and titania	Sol–gel process	HA, TiO_2	Human osteoblast-like cells	Improved corrosion resistance	[39,103]
Increased ALP activity					
Enhanced cell adhesion					[103]
MC3T3-E1	Increased number of mineralized nodules				
Surface chemistry with titania composites	Nanophase titania, sonication disperse	PLGA, titania	Human osteoblasts	Increased cell adhesion	[104]
Enhanced bone functions (e.g., collagen, ALP activity, calcium, and mineral deposition)					
Surface chemistry with aluminum oxide (Al_2O_3)	Dipping	Al_2O_3	RMSC	Promoted stem cells to the osteoblast phenotype	[105]
Surface chemistry with nanoscale materials	Sol–gel-derived coating	Anatase (An), rutile (Ru), alumina (Al), and zirconia (Zr)	hMSC	Improved cell adhesion	[106]
Upregulated osteoblast differentiation genes, BSP, and OSX					
Surface chemistry with bioactive glass	Blasted Ti with granules	Bioactive glass particles	MC3T3-E1	Induced ALP activity	[107]
Enhanced osteogenic potential					

hMSC, *human mesenchymal stem cells;* hSaSO-2, *human osteoblast cell line;* BAEC, *bovine aortic endothelial cells;* hFOB, *human fetal osteoblast progenitor cell line;* MC3T3-E1, *mouse osteoblast cell line;* MG63, *human osteosarcoma immortalized cell line;* RMSC, *rat bone marrow stromal cells.*

Table 8.4 Tissue Responses Toward Surface-Modified Ti and Its Alloys in vivo [41]

Implant Materials	Modification	Animal Models	Tissue Responses	References
Ti, screw-shaped implants	Blasting with TiO_2, HF treatment, nano-HA	Rabbit	Increased removal torque values, demonstrated bone formation in contact with the implant surface	[108]
Ti implants	HA deposition by IBAD, SLA Ti implants, machined assisted implants	Rabbit, dog	Highest mean bone-to-metal contact ratio in HA-modified Ti implants, increased osseoactivity	[109–111]
Very smooth Ti implants	Nano-HA modification	Rabbit	Similar bone formation for untreated Ti and nano-HA Ti implants	[112]
Electropolished Ti implants	Nano-HA (10 nm in diameter) modification	Rabbit	Increased bone formation in nano-HA Ti implants, higher bone-to-implant contact for E-HA Ti implants compared to E Ti implants	[113]
Ti implants	Laser ablation followed by nano-HA deposition	Rabbit	Shorten implant healing, increased bone implant interactions	[114]
Ti cylinder implants	HA or titania nanostructures	Rabbit	Increased coverage area and feature density in nanotitania-coated implants, higher bone contact values (but not significant) for nanotitania implants	[115]
Ti implants	Calcium metaphosphate coated, anodic oxidized, HA deposition	Rabbit	Superior bone responses in all surface-modified implants, higher bone contact values	[116,117]
SLA titanium implants	Ti coated with calcium phosphate, Ti modified via anodic plasma-chemical treatment (APC), bisphosphonate-coated Ti (Ti + bisphos), and Ti coated with collagen containing chondroitin sulfate	Sheep	Surface Ti implants with bisphosphonate and collagen enhanced peri-implant bone formation	[118]
SLA zirconia implants				
Commercially pure titanium (cpTi) or Ti–6Al–4V	Discrete crystalline deposition of calcium phosphate nanoparticles	Rat	Increased in osteoconduction as a function of nanotopography generated by calcium phosphate nanoparticles, induced bone-to-implant contact (bone ingrowth), increased tensile test resistance	[119,120]
Ti implants	H_2SO_4/H_2O_2 mixture treatment (nanopits 5–50 nm)	Dog	Enhanced contact osteogenesis, more bone-to-implant contact, percentage of mineralized bone area	[121]
Ti implants	TiO_2 blasted with and without fluoride treatment	Rabbit, rat, dog	Fluoride-modified Ti implants demonstrated a firmer bone anchorage and greater bone integration, accelerated differentiation supported osteoinduction	[89,122–124]
Ti implants	Acid etching, dipping in an aluminum oxide solution	Rat	Increased bone-to-implant contact, increased bone-specific gene products (e.g., sialoprotein, osteocalcin, osteopontin, osterix, RUNX-2)	[105,106]

8.4.1 CELL MORPHOLOGY

Like most cells, osteoblast cells react by changing their morphology to suit the environment that they grow in. Osteoblasts maintain a rounded shape without extensions when they are on a flat Ti surface, whereas the Ti nanotube array causes the cells to become elongated, showing an increased number of extensions and filopodia [75]. This is correlated with an increase in the nanotube diameter from the 30 to 100 nm range. Changes in the morphology of osteoblasts growing on Ti nanotube arrays are also dependent on both the diameter of the nanotube and the spacing between the nanotubes [75]. Hemispherical cavities or indentations also affect the osteoblasts in a similar way. For instance, cavities in the 10 nm range cause the cells to become polygonal in shape with a high number of filopodia [125].

8.4.2 CELL ADHESION

The nanotopography surface can also affect how well the cells adhere; however, the level of cellular adhesion can be difficult to determine as the cells reach confluence. Ti nanotubes increase the number of cells that adhere to the surface [75,126] when compared to a flat Ti surface. This may be due to the increased hydrophilicity of the surface caused by the nanotube array. Nanophase surfaces of Ti have also been shown to increase cellular adhesion [24]. Nanoscale hemisphere structures, on the other hand, did not increase the number of adhered cells when compared to the flat Ti surface [86]. In addition, nanophase materials made up of grains that are compacted together and are smaller than 100 nm in diameter cause an increase in surface energy, leading to the enhanced cellular adhesion [127].

8.4.3 CELL PROLIFERATION

In addition to cell morphology and adhesion, the proliferation of bone cells on implant surfaces can also be altered by modifying the surface properties to create nanotopography on Ti implants as shown in Fig. 8.2. In the case of nanotubes

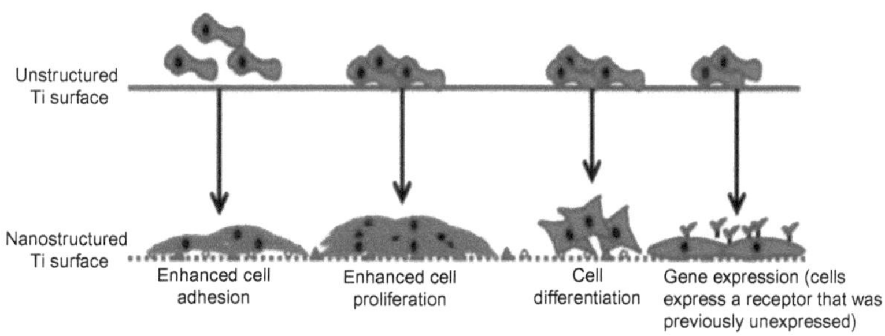

FIGURE 8.2

Varying effects of nanostructuring of Ti implants on cells.

that have self-assembled into an array, cellular response is determined by the nanotube dimension and the spacing between the nanotubes in the array. These factors can have a major effect on the osteoblast behavior. The spaces between nanotubes provide a physical pathway for the nutrients to reach the cells after the nanotubes have adhered to the surface, whereas nanostructures on the implant surfaces would improve the cellular adhesion [75]. Surface roughness of the nanotopography is also a major factor in the determination of the proliferative activity. Root-mean-square roughness values of 0.5–13 nm significantly decrease the proliferation of osteoblasts. Rougher surfaces lower proliferation while smoother surfaces promote proliferation [128]. A nanometric net of Ti filaments, on the other hand, can increase osteoblast proliferation [129].

8.4.4 BIOACTIVE MOLECULES

Besides cell responses such as cell morphology, adhesion, and proliferation, surface modification of bone implants with nanotopography also affects the cellular production of various bioactive and signaling molecules. For instance, hemispherical protrusions increased the levels of tumor necrosis factor-alpha (TNF-α) and prostaglandin E2 (PGE2) release from osteoblasts [86]. TNF-α has a role as an inflammatory mediator and assesses the ability to attract many cell types including osteoblasts, whereas PGE2 is the factor that stimulates osteoclasts to begin bone resorption. Prostaglandins are also important inflammatory mediators. The increased levels of TNF-α and PGE2 indicate an increase in osteoblast activity. On rougher nanotopography surfaces, cells release a higher level of PGE2 [130]. Osteoblasts also showed an increase in expression of osteopontin and bone sialoprotein (BSP) due to hemispherical indentations [125]. The presence of the proteins, osteopontin, and BSP indicates the early stages of bone formation and increased osteoblast differentiation generated from the nanostructure of the implant surface.

8.4.5 OSSEOINTEGRATION

Osseointegration, defined as the integration of bone with the implant, is a major factor that determines whether the transplant implantation is successful, and nanotopography can be used to enhance this integration process by selecting a specific nanoscale feature that is similar to the natural bone environment. The natural bone environment at the nanoscale level is made up of grains that are 10–50 nm in size and consist of mineralized osseous tissue. Ti surfaces that are smooth will cause the formation of a fibrous callus as part of the normal healing process after implantation. The fibrous callus will become calcified and remodeled into a bony callus. A fibrous callus is unwanted since it provides a boundary between the implant and bone and causes the implant to be much more loosely attached instead of being more directly incorporated into the bone. By mimicking the nanostructures of a healthy bone, this nanotopographical modification would lead to osteoblasts depositing calcium as if the implants were bony tissues meant to integrate the implant directly into the healthy bone tissues. Several nanostructures have been reported to increase osteoblast activities (Table 8.2). The Ti nanotube array previously mentioned also stimulates higher levels of alkaline phosphatase (ALP) activity [75]. Alkaline phosphate is a by-product of bone formation, and the level of alkaline phosphate can therefore be indicative of the activeness of the osseointegration. The addition of nanotubes has also been shown to increase the amount of calcium deposition on Ti surfaces [131].

8.5 NANOTOPOGRAPHY AND STEM CELL RESPONSES

Several studies have been shown that nanotopography is one of the major factors that determine the fate of stem cells, especially stem cell differentiation [132], and that stem cell biomaterial interactions, stem cell differentiation, and migration are in response to different nanotopographic features [133–137]. Stem cells normally remain in their mitotic cycle unless they are guided by an external cue which would then cause them to differentiate into more specialized cells [132]. One main factor that has been observed to guide stem cell differentiation is the presence and composition of extracellular matrix (ECM) components. Presence of collagen fibrils of diameter ranging from 15 to 300 nm provides cellular anchorage and presents cues for stem cell differentiation [135]. Cells attach to the ECM through attachment points such as integrins (transmembrane proteins) which form bonds with specific amino acid sequences found within the ECM network. These linkages cause a cascade of signaling pathways that affect different cellular processes such as migration, proliferation, and differentiation [137]. Several substrate characteristics such as size [138], lateral spacing [138], surface chemistry [139], and geometry [140] of the nanofeatures play an important role in integrin clustering and activation. For instance, it was shown that rat mesenchymal stem cells had maximum adhesion, spreading, and differentiation on TiO_2 nanotubes vertically aligned with a diameter of 15 nm (which is similar to the theoretical spacing of the integrin receptors on the cell surface). A change in the nanotube diameter showed a significant alteration in the cell behavior [138], e.g., focal adhesion kinase (FAK) activity was maximum on 15-nm diameter TiO_2 tubes in comparison to 100-nm diameter tubes.

8.5.1 EFFECTS OF NANOTOPOGRAPHY ON ENDOTHELIAL PROGENITOR CELLS

Endothelial progenitor cells (EPCs) have been observed to differentiate into endothelial cells [141], a major type of cells responsible for vasculature within the bone tissue, and to behave according to nanoscale features of a substrate [142–147]. For example, growth and differentiation of EPCs can be controlled by a ridge-groove type nanotopography generated within 1200 nm intervals and 600 nm in depth [141]. Although endothelial cellular biochemistry was not altered, EPCs cultured on substrate nanotopography formed aligned band structures after 6 days [141]. The study also showed that formation of long, thin-walled capillary tubes was much more prominent on nanostructured surfaces than unstructured surfaces. While reduced proliferation has been observed in cell types such as smooth muscle cells and human embryonic stem cells [144,148], EPCs have shown to express increased migration velocity and adhesion on linear substrate nanotopography [149]. The presence of nanotopographical features such as linear and long grooves in the substrate has also been shown to aid in maintaining morphology of elongated EPCs over a long period without causing any decrease in cell growth and to enhance the rate of capillary formation as compared to EPCs grown on flat surfaces. While nanotopography has been shown to affect cells at the protein level [150], which suggests a genetic level change in response to the nanoscale substrate features, it has been observed that no change in the characteristic endothelial-specific markers occurs in the EPCs grown on nanostructured surfaces [141].

8.5.2 EFFECTS OF NANOTOPOGRAPHY ON BONE MARROW STEM CELLS

Several studies have observed an obvious response from bone-marrow-derived stem cells to the topography of materials, and these cells elicit a large range of responses such as changes in adhesion, spreading, proliferation, and genomic responses [86,151–157]. For instance, nanotopography has a noticeable effect on the interactions between the hMSCs (or bone-marrow-derived stem cells) and the substrate by influencing the focal adhesion formation, organization of the cytoskeleton and cellular mechanical properties [158]. In addition, Berry et al. [159] have shown that the presence of nanometer-scale islands formed by polymer demixing elicited a response in cell morphology, resulting in a smaller cell body. Recent work employing bone-marrow-derived stem cells showed that these cells were responsive to 2D island nanotopography, showing a cellular adhesion and osteogenic differentiation. Bone marrow cells exposed to nanoislands have also been observed to express long filamentous processes which are believed to be fillipodia. These extensions are long and thin in proportion to the cell body and are believed to serve as points of interaction between the cells with the nanofeatures. Fillipodia are hypothesized to be one of the cells' main method of gathering information about their surrounding [160]. Fillipodia extensions have been observed in several other types of cells such as fibroblasts, endothelial cells, and macrophages when interacting with fabricated micro- and nanofeatures [154,161–163]. Once the fillipodia have found an appropriate feature, lamellipodia are set forth which causes cell movement toward the site [164]. The characterization of focal adhesion, namely FAK and vinculin, and actin rearrangement would provide evidence on how topography affects the cellular responses. For instance, Kim et al. [158] have observed that FAK and vinculin expressions in the elongated hMSCs differed from those grown on unstructured surfaces. Focal adhesion-recruited vinculin and phosphorylated Focal Adhesion Kinase (pFAK) are distributed evenly throughout hMSCs grown on unpatterned surfaces, but they are localized at the poles of the hMSCs grown on the nanostructures. In addition, Berry et al. [159] have reported that instead of adhering, differentiating, and proliferating, the cells exhibited clumped morphology and still emanated fillipodia. Transmembrane integrins bind to the substrate on one side and bind to different proteins associated with focal adhesions such as FAK, focal-adhesion-associated kinase, and actin stress fibers. These integrins play an important role in regulating cellular motility and behavior when bone marrow stem cells are exposed to nanostructures [159].

8.6 CONCLUSIONS

Different methods have been described to modify or to embellish Ti substrates with nanoscale features. These modifications can alter the chemistry and topography of the implant surface. These changes can be used to produce favorable changes in the cellular processes of the surrounding tissue for better osseointegration. Nanotopography has been used as cues for triggering specific biochemical and molecular cascades of cellular responses such as cell morphology, adhesion, and proliferation of bone cells as well as differentiation of stem cells on the Ti implants. The fields of tissue engineering and biomaterials have long been investigating the use of three-dimensional scaffolds; however, the use of intelligent scaffolds that present cues that closely resemble bone tissues in nature are yet to be fully explored. Thus, by conducting extensive research and clinical trials of nanostructured and nanomodified implants, it is expected that engineers and scientists will come up with implants that are more efficient in the future.

REFERENCES

[1] P.I. Branemark, R. Adell, U. Breine, B.O. Hansson, J. Lindstrom, A. Ohlsson, Intra-osseous anchorage of dental prostheses. I. Experimental studies, Scand. J. Plast. Reconstr. Surg. 3 (2) (1969) 81–100.

[2] L. Linder, T. Albrektsson, P.I. Branemark, H.A. Hansson, B. Ivarsson, U. Jonsson, Electron microscopic analysis of the bone–titanium interface, Acta Orthop. Scand. 54 (1) (1983) 45–52.

[3] R. Adell, B. Eriksson, U. Lekholm, P.I. Branemark, T. Jemt, Long-term follow-up study of osseointegrated implants in the treatment of totally edentulous jaws, Int. J. Oral Maxillofac. Implants 5 (4) (1990) 347–359.

[4] R. Adell, U. Lekholm, B. Rockler, P.I. Branemark, A 15-year study of osseointegrated implants in the treatment of the edentulous jaw, Int. J. Oral Surg. 10 (6) (1981) 387–416.
[5] T. Albrektsson, E. Dahl, L. Enbom, S. Engevall, B. Engquist, A.R. Eriksson, Osseointegrated oral implants. A Swedish multicenter study of 8139 consecutively inserted Nobelpharma implants, J. Periodontol. 59 (5) (1988) 287–296.
[6] T. Albrektsson, L. Sennerby, Direct bone anchorage of oral implants: clinical and experimental considerations of the concept of osseointegration, Int. J. Prosthodont. 3 (1) (1990) 30–41.
[7] C.J. Goodacre, J.Y. Kan, K. Rungcharassaeng, Clinical complications of osseointegrated implants, J. Prosthet. Dent. 81 (5) (1999) 537–552.
[8] D. Morton, R. Jaffin, H.P. Weber, Immediate restoration and loading of dental implants: clinical considerations and protocols, Int. J. Oral Maxillofac. Implants. 19 (Suppl) (2004) 103–108.
[9] G.A. Zarb, A. Schmitt, The longitudinal clinical effectiveness of osseointegrated dental implants: the Toronto study. Part III: problems and complications encountered, J. Prosthet. Dent. 64 (2) (1990) 185–194.
[10] C.A. Bain, Smoking and implant failure—benefits of a smoking cessation protocol, Int. J. Oral Maxillofac. Implants 11 (6) (1996) 756–759.
[11] L.F. Cooper, Biologic determinants of bone formation for osseointegration: clues for future clinical improvements, J. Prosthet. Dent. 80 (4) (1998) 439–449.
[12] J.P. Fiorellini, P.K. Chen, M. Nevins, M.L. Nevins, A retrospective study of dental implants in diabetic patients, Int. J. Periodontol. Restor. Dent. 20 (4) (2000) 366–373.
[13] R.A. Jaffin, C.L. Berman, The excessive loss of Branemark fixtures in type IV bone: a 5-year analysis, J. Periodontol. 62 (1) (1991) 2–4.
[14] A. Nanci, J.D. Wuest, L. Peru, P. Brunet, V. Sharma, S. Zalzal, Chemical modification of titanium surfaces for covalent attachment of biological molecules, J. Biomed. Mater. Res. 40 (2) (1998) 324–335.
[15] L. Tolstunov, Dental implant success–failure analysis: a concept of implant vulnerability, Implant Dent. 15 (4) (2006) 341–346.
[16] C. Aparicio, F.J. Gil, C. Fonseca, M. Barbosa, J.A. Planell, Corrosion behaviour of commercially pure titanium shot blasted with different materials and sizes of shot particles for dental implant applications, Biomaterials 24 (2) (2003) 263–273.
[17] S. Bruni, M. Martinesi, M. Stio, C. Treves, T. Bacci, F. Borgioli, Effects of surface treatment of Ti–6Al–4V titanium alloy on biocompatibility in cultured human umbilical vein endothelial cells, Acta Biomater. 1 (2) (2005) 223–234.
[18] V.M. Frauchiger, F. Schlottig, B. Gasser, M. Textor, Anodic plasma-chemical treatment of CP titanium surfaces for biomedical applications, Biomaterials 25 (4) (2004) 593–606.
[19] P. Linez-Bataillon, F. Monchau, M. Bigerelle, H.F. Hildebrand, *In-vitro* MC3T3 osteoblast adhesion with respect to surface roughness of Ti6Al4V substrates, Biomol. Eng. 19 (2–6) (2002) 133–141.
[20] T. Akahori, M. Niinomi, Fracture characteristics of fatigued Ti–6Al–4V ELI as an implant material, Mater. Sci. Eng. A 243 (1–2) (1998) 237–243.
[21] J.W. Park, K.B. Park, J.Y. Suh, Effects of calcium ion incorporation on bone healing of Ti6Al4V alloy implants in rabbit tibiae, Biomaterials. 28 (22) (2007) 3306–3313.

[22] D. Scharnweber, R. Beutner, S. Rossler, H. Worch, Electrochemical behavior of titanium-based materials—are there relations to biocompatibility?, J. Mater. Sci. 13 (12) (2002) 1215–1220.

[23] R. van Noort, D. Brown, B.E. Causton, E.C. Combe, A.M. Fletcher, C.H. Lloyd, Dental materials: 1984–85 literature review, Part 2. J. Dent. 15 (3) (1987) 93–115.

[24] T.J. Webster, J.U. Ejiofor, Increased osteoblast adhesion on nanophase metals: Ti, Ti6Al4V, and CoCrMo, Biomaterials 25 (19) (2004) 4731–4739.

[25] Y.L. Zhou, M. Niinomi, T. Akahori, H. Fukui, H. Toda, Corrosion resistance and biocompatibility of Ti–Ta alloys for biomedical applications, Mater. Sci. Eng. A 398 (2005) 28–36.

[26] M.A. Khan, R.L. Williams, D.F. Williams, The corrosion behaviour of Ti–6Al–4V, Ti–6Al–7Nb and Ti–13Nb–13Zr in protein solutions, Biomaterials 20 (7) (1999) 631–637.

[27] F.A. Muller, M.C. Bottino, L. Muller, V.A. Henriques, U. Lohbauer, A.H. Bressiani, *In-vitro* apatite formation on chemically treated (P/M) Ti–13Nb–13Zr, Dent. Mater. 24 (1) (2008) 50–56.

[28] S.R. Paital, N.B. Dahotre, Calcium phosphate coatings for bio-implant applications: materials, performance factors, and methodologies, Mater. Sci. Eng. R 66 (1–3) (2009) 1–70.

[29] M.A. Khan, R.L. Williams, D.F. Williams, The corrosion behaviour of Ti–6Al–4V, Ti–6Al–7Nb and Ti–13Nb–13Zr in protein solutions, Biomaterials 20 (7) (1999) 631–637.

[30] M.C. Bottino, P.G. Coelho, M. Yoshimoto, B. König, V.A.R. Henriques, A.H.A. Bressiani, et al., Histomorphologic evaluation of Ti–13Nb–13Zr alloys processed via powder metallurgy. A study in rabbits, Mater. Sci. Eng. C 28 (2) (2006) 223–227.

[31] S.L. Semiatin, V. Seetharaman, I. Weiss, Flow behavior and globularization kinetics during hot working of Ti–6Al–4V with a colony alpha microstructure, Mater. Sci. Eng. A 263 (2) (1999) 257–271.

[32] T. Seshacharyulu, S.C. Medeiros, W.G. Frazier, Y.V.R.K. Prasad, Hot working of commercial Ti–6Al–4V with an equiaxed α–β microstructure: materials modeling considerations, Mater. Sci. Eng. A 284 (1–2) (2000) 184–194.

[33] Y.G. Ko, W.S. Jung, D.H. Shin, C.S. Lee, Effects of temperature and initial microstructure on the equal channel angular pressing of Ti–6Al–4V alloy, Scr. Mater. 48 (2) (2002) 197–202.

[34] B.D. Boyan, Z. Schwartz, J.C. Hambleton, Response of bone and cartilage cells to biomaterials *in-vivo* and *in-vitro*, J. Oral Implantol. 19 (2) (1993) 116–122.

[35] F.K. Mante, K. Little, M.O. Mante, C. Rawle, G.R. Baran, Oxidation of titanium, RGD peptide attachment, and matrix mineralization rat bone marrow stromal cells, J. Oral Implantol. 30 (6) (2004) 343–349.

[36] B. Ben-Nissan, A.H. Choi, Sol–gel production of bioactive nanocoatings for medical applications. Part 1: an introduction, Nanomedicin 1 (3) (2006) 311–319.

[37] D.M. Liu, T. Troczynski, W.J. Tseng, Water-based sol–gel synthesis of hydroxyapatite: process development, Biomaterials 22 (13) (2001) 1721–1730.

[38] C.A. Scotchford, C.P. Gilmore, E. Cooper, G.J. Leggett, S. Downes, Protein adsorption and human osteoblast-like cell attachment and growth on alkylthiol on gold self-assembled monolayers, J. Biomed. Mater. Res. 59 (1) (2002) 84–99.

[39] H.W. Kim, Y.H. Koh, L.H. Li, S. Lee, H.E. Kim, Hydroxyapatite coating on titanium substrate with titania buffer layer processed by sol–gel method, Biomaterials 25 (13) (2004) 2533–2538.
[40] X.X. Wang, S. Hayakawa, K. Tsuru, A. Osaka, Bioactive titania-gel layers formed by chemical treatment of Ti substrate with a H_2O_2/HCl solution, Biomaterials 23 (5) (2002) 1353–1357.
[41] G. Mendonca, D.B. Mendonca, F.J. Aragao, L.F. Cooper, Advancing dental implant surface technology—from micron- to nanotopography, Biomaterials 29 (28) (2008) 3822–3835.
[42] J. Zhou, C. Chang, R. Zhang, L. Zhang, Hydrogels prepared from unsubstituted cellulose in NaOH/urea aqueous solution, Macromol. Biosci. 7 (6) (2007) 804–809.
[43] S.H. Oh, R.R. Finones, C. Daraio, L.H. Chen, S. Jin, Growth of nano-scale hydroxyapatite using chemically treated titanium oxide nanotubes, Biomaterials 26 (24) (2005) 4938–4943.
[44] S.H. Lee, H.W. Kim, E.J. Lee, L.H. Li, H.E. Kim, Hydroxyapatite–TiO_2 hybrid coating on Ti implants, J. Biomater. Appl. 20 (3) (2006) 195–208.
[45] K.T. Nguyen, K.P. Shukla, M. Moctezuma, A.R. Braden, J. Zhou, Z. Hu, Studies of the cellular uptake of hydrogel nanospheres and microspheres by phagocytes, vascular endothelial cells, and smooth muscle cells, J. Biomed. Mater. Res. 88 (4) (2009) 1022–1030.
[46] J.L. Arias, M.B. Mayor, J. Pou, Y. Leng, B. Leon, M. Perez-Amor, Micro- and nano-testing of calcium phosphate coatings produced by pulsed laser deposition, Biomaterials 24 (20) (2003) 3403–3408.
[47] A.H. Choi, B. Ben-Nissan, Sol–gel production of bioactive nanocoatings for medical applications. Part II: current research and development, Nanomedicin 2 (1) (2007) 51–61.
[48] L.D. Piveteau, B. Gasser, L. Schlapbach, Evaluating mechanical adhesion of sol–gel titanium dioxide coatings containing calcium phosphate for metal implant application, Biomaterials 21 (21) (2000) 2193–2201.
[49] I. Nishimura, Y. Huang, F. Butz, T. Ogawa, L. Lin, W.C. Jake, Discrete deposition of hydroxyapatite nanoparticles on a titanium implant with predisposing substrate microtopography accelerated osseointegration, Nanotechnology 18 (2007).
[50] J.L. Ricci, J. Charvet, S.R. Frenkel, R. Chang, P. Nadkarni, J. Turner, J.E. Davies, Bone response to laser microtextured surfaces, in: J.E. Davies (Ed.), Bone Engineering, Em2 Inc, Toronto, 2000.
[51] D. Buser, R.K. Schenk, S. Steinemann, J.P. Fiorellini, C.H. Fox, H. Stich, Influence of surface characteristics on bone integration of titanium implants. A histomorphometric study in miniature pigs, J. Biomed. Mater. Res. 25 (7) (1991) 889–902.
[52] K. Gotfredsen, E. Hjorting-Hansen, E. Budtz-Jorgensen, Clinical and radiographic evaluation of submerged and nonsubmerged implants in monkeys, Int. J. Prosthodont. 3 (5) (1990) 463–469.
[53] Y. Germanier, S. Tosatti, N. Broggini, M. Textor, D. Buser, Enhanced bone apposition around biofunctionalized sandblasted and acid-etched titanium implant surfaces. A histomorphometric study in miniature pigs, Clin. Oral Implants Res. 17 (3) (2006) 251–257.
[54] E. De Giglio, S. Cometa, C.D. Calvano, L. Sabbatini, P.G. Zambonin, S. Colucci, A new titanium biofunctionalized interface based on poly(pyrrole-3-acetic acid)

coating: proliferation of osteoblast-like cells and future perspectives, J. Mater. Sci. 18 (9) (2007) 1781–1789.
[55] E. De Giglio, S. Cometa, N. Cioffi, L. Torsi, L. Sabbatini, Analytical investigations of poly(acrylic acid) coatings electrodeposited on titanium-based implants: a versatile approach to biocompatibility enhancement, Anal. Bioanal. Chem. 389 (7–8) (2007) 2055–2063.
[56] E. De Giglio, M.R. Guascito, L. Sabbatini, G. Zambonin, Electropolymerization of pyrrole on titanium substrates for the future development of new biocompatible surfaces, Biomaterials 22 (19) (2001) 2609–2616.
[57] E. De Giglio, L. Sabbatini, S. Colucci, G. Zambonin, Synthesis, analytical characterization, and osteoblast adhesion properties on RGD-grafted polypyrrole coatings on titanium substrates, J. Biomater. Sci., Polym. Ed. 11 (10) (2000) 1073–1083.
[58] G. Helary, F. Noirclere, J. Mayingi, V. Migonney, A new approach to graft bioactive polymer on titanium implants: improvement of MG 63 cell differentiation onto this coating, Acta Biomater. 5 (1) (2009) 124–133.
[59] J.E. Ho, T.A. Barber, A.S. Virdi, D.R. Sumner, K.E. Healy, The effect of enzymatically degradable IPN coatings on peri-implant bone formation and implant fixation, J. Biomed. Mater. Res. 81 (3) (2007) 720–727.
[60] F. Zeifang, M. Grunze, G. Delling, H. Lorenz, C. Heisel, G. Tosounidis, Improved osseointegration of PTFEP-coated titanium implants, Med. Sci. Monit. 14 (2) (2008) BR35–BR40.
[61] E. De Giglio, S. Cometa, L. Sabbatini, P.G. Zambonin, G. Spoto, Electrosynthesis and analytical characterization of PMMA coatings on titanium substrates as barriers against ion release, Anal. Bioanal. Chem. 381 (3) (2005) 626–633.
[62] R. Muller, J. Abke, E. Schnell, D. Scharnweber, R. Kujat, C. Englert, Influence of surface pretreatment of titanium- and cobalt-based biomaterials on covalent immobilization of fibrillar collagen, Biomaterials 27 (22) (2006) 4059–4068.
[63] Z. Shi, K.G. Neoh, E.T. Kang, C. Poh, W. Wang, Bacterial adhesion and osteoblast function on titanium with surface-grafted chitosan and immobilized RGD peptide, J. Biomed. Mater. Res. 86 (4) (2008) 865–872.
[64] M.M. Deckers, R.L. van Bezooijen, G. van der Horst, J. Hoogendam, C. van Der Bent, S.E. Papapoulos, Bone morphogenetic proteins stimulate angiogenesis through osteoblast-derived vascular endothelial growth factor A, Endocrinology. 143 (4) (2002) 1545–1553.
[65] R.A. Carano, E.H. Filvaroff, Angiogenesis and bone repair, Drug Discov. Today 8 (21) (2003) 980–989.
[66] T. Furumatsu, Z.N. Shen, A. Kawai, K. Nishida, H. Manabe, T. Oohashi, Vascular endothelial growth factor principally acts as the main angiogenic factor in the early stage of human osteoblastogenesis, J. Biochem. 133 (5) (2003) 633–639.
[67] H.P. Gerber, T.H. Vu, A.M. Ryan, J. Kowalski, Z. Werb, N. Ferrara, VEGF couples hypertrophic cartilage remodeling, ossification and angiogenesis during endochondral bone formation, Nat. Med. 5 (6) (1999) 623–628.
[68] J.M. Kanczler, R.O. Oreffo, Osteogenesis and angiogenesis: the potential for engineering bone, Eur. Cell Mater. 15 (2008) 100–114.
[69] C. Wolf-Brandstetter, A. Lode, T. Hanke, D. Scharnweber, H. Worch, Influence of modified extracellular matrices on Ti6Al4V implants on binding and release of VEGF, J. Biomed. Mater. Res. 79 (4) (2006) 882–894.

[70] C.K. Poh, Z. Shi, T.Y. Lim, K.G. Neoh, W. Wang, The effect of VEGF functionalization of titanium on endothelial cells *in-vitro*, Biomaterials (2009) 4.
[71] L. Tang, J.W. Eaton, Natural responses to unnatural materials: a molecular mechanism for foreign body reactions, Mol. Med. 5 (6) (1999) 351–358.
[72] L. Chenglong, Y. Dazhi, L. Guoqiang, Q. Min, Corrosion resistance and hemocompatibility of multilayered Ti/TiN-coated surgical AISI 316L stainless steel, Mater. Lett. 59 (29–30) (2005) 3813–3819.
[73] M. Terada, S. Abe, T. Akasaka, M. Uo, Y. Kitagawa, F. Watari, Multiwalled carbon nanotube coating on titanium, Biomed. Mater. Eng. 19 (1) (2009) 45–52.
[74] P. Roach, D. Farrar, C.C. Perry, Surface tailoring for controlled protein adsorption: effect of topography at the nanometer scale and chemistry, J. Am. Chem. Soc. 128 (12) (2006) 3939–3945.
[75] K.S. Brammer, S. Oh, C.J. Cobb, L.M. Bjursten, H. van der Heyde, S. Jin, Improved bone-forming functionality on diameter-controlled TiO(2) nanotube surface, Acta Biomater. 5 (8) (2009) 3215–3223.
[76] K.S. Brammer, S. Oh, J.O. Gallagher, S. Jin, Enhanced cellular mobility guided by TiO_2 nanotube surfaces, Nano Lett. 8 (3) (2008) 786–793.
[77] B. Chi, E.S. Victorio, T. Jin, Synthesis of TiO_2-based nanotube on Ti substrate by hydrothermal treatment, J. Nanosci. Nanotechnol. 7 (2) (2007) 668–672.
[78] D. Ding, C. Ning, L. Huang, F. Jin, Y. Hao, S. Bai, Anodic fabrication and bioactivity of Nb-doped TiO_2 nanotubes, Nanotechnology 20 (30) (2009) 305103.
[79] S. Oh, K.S. Brammer, Y.S. Li, D. Teng, A.J. Engler, S. Chien, Stem cell fate dictated solely by altered nanotube dimension, Proc. Natl. Acad. Sci. U.S.A. 106 (7) (2009) 2130–2135.
[80] K.C. Popat, L. Leoni, C.A. Grimes, T.A. Desai, Influence of engineered titania nanotubular surfaces on bone cells, Biomaterials 28 (21) (2007) 3188–3197.
[81] T. Sjostrom, M.J. Dalby, A. Hart, R. Tare, R.O. Oreffo, B. Su, Fabrication of pillar-like titania nanostructures on titanium and their interactions with human skeletal stem cells, Acta Biomater. 5 (5) (2009) 1433–1441.
[82] T. Sjostrom, N. Fox, B. Su, Through-mask anodization of titania dot- and pillar-like nanostructures on bulk Ti substrates using a nanoporous anodic alumina mask, Nanotechnology 20 (13) (2009) 135305.
[83] Z. Huang, R.H. Daniels, R.J. Enzerink, V. Hardev, V. Sahi, S.B. Goodman, Effect of nanofiber-coated surfaces on the proliferation and differentiation of osteoprogenitors *in-vitro*, Tissue Eng. 14 (11) (2008) 1853–1859.
[84] J. He, W. Zhou, X. Zhou, X. Zhong, X. Zhang, P. Wan, The anatase phase of nanotopography titania plays an important role on osteoblast cell morphology and proliferation, J. Mater. Sci. 19 (11) (2008) 3465–3472.
[85] R. Karpagavalli, A. Zhou, P. Chellamuthu, K. Nguyen, Corrosion behavior and biocompatibility of nanostructured TiO_2 film on Ti6Al4V, J. Biomed. Mater. Res. 83 (4) (2007) 1087–1095.
[86] J.M. Rice, J.A. Hunt, J.A. Gallagher, P. Hanarp, D.S. Sutherland, J. Gold, Quantitative assessment of the response of primary derived human osteoblasts and macrophages to a range of nanotopography surfaces in a single culture model *in-vitro*, Biomaterials 24 (26) (2003) 4799–4818.
[87] P.T. de Oliveira, S.F. Zalzal, M.M. Beloti, A.L. Rosa, A. Nanci, Enhancement of *in-vitro* osteogenesis on titanium by chemically produced nanotopography, J. Biomed. Mater. Res. 80 (3) (2007) 554–564.

[88] S. Valencia, C. Gretzer, L.F. Cooper, Surface nanofeature effects on titanium-adherent human mesenchymal stem cells, Int. J. Oral Maxillofac. Implants 24 (1) (2009) 38–46.

[89] J. Guo, R.J. Padilla, W. Ambrose, I.J. De Kok, L.F. Cooper, The effect of hydrofluoric acid treatment of TiO_2 grit blasted titanium implants on adherent osteoblast gene expression *in-vitro* and *in-vivo*, Biomaterials 28 (36) (2007) 5418–5425.

[90] O. Zinger, K. Anselme, A. Denzer, P. Habersetzer, M. Wieland, J. Jeanfils, Time-dependent morphology and adhesion of osteoblastic cells on titanium model surfaces featuring scale-resolved topography, Biomaterials 25 (14) (2004) 2695–2711.

[91] B.C. Ward, T.J. Webster, The effect of nanotopography on calcium and phosphorus deposition on metallic materials *in-vitro*, Biomaterials 27 (16) (2006) 3064–3074.

[92] I. Wall, N. Donos, K. Carlqvist, F. Jones, P. Brett, Modified titanium surfaces promote accelerated osteogenic differentiation of mesenchymal stromal cells *in-vitro*, Bone 45 (1) (2009) 17–26.

[93] Z. Qu, X. Rausch-Fan, M. Wieland, M. Matejka, A. Schedle, The initial attachment and subsequent behavior regulation of osteoblasts by dental implant surface modification, J. Biomed. Mater. Res. 82 (3) (2007) 658–668.

[94] X. Rausch-fan, Z. Qu, M. Wieland, M. Matejka, A. Schedle, Differentiation and cytokine synthesis of human alveolar osteoblasts compared to osteoblast-like cells (MG63) in response to titanium surfaces, Dent. Mater. 24 (1) (2008) 102–110.

[95] S.F. Lamolle, M. Monjo, M. Rubert, H.J. Haugen, S.P. Lyngstadaas, J.E. Ellingsen, The effect of hydrofluoric acid treatment of titanium surface on nanostructural and chemical changes and the growth of MC3T3-E1 cells, Biomaterials 30 (5) (2009) 736–742.

[96] C.Y. Wang, B.H. Zhao, H.J. Ai, Y.W. Wang, Comparison of biological characteristics of mesenchymal stem cells grown on two different titanium implant surfaces, Biomed. Mater. (Bristol, England) 3 (1) (2008) 15004.

[97] E. Eisenbarth, D. Velten, J. Breme, Biomimetic implant coatings, Biomol. Eng. 24 (1) (2007) 27–32.

[98] H. Zreiqat, S.M. Valenzuela, B.B. Nissan, R. Roest, C. Knabe, R.J. Radlanski, The effect of surface chemistry modification of titanium alloy on signalling pathways in human osteoblasts, Biomaterials 26 (36) (2005) 7579–7586.

[99] T.K. Monsees, K. Barth, S. Tippelt, K. Heidel, A. Gorbunov, W. Pompe, Effects of different titanium alloys and nanosize surface patterning on adhesion, differentiation, and orientation of osteoblast-like cells, Cells Tissues Organ. 180 (2) (2005) 81–95.

[100] T.J. Webster, C. Ergun, R.H. Doremus, W.A. Lanford, Increased osteoblast adhesion on titanium-coated hydroxyapatite that forms $CaTiO_3$, J. Biomed. Mater. Res. 67 (3) (2003) 975–980.

[101] C. Ergun, H. Liu, J.W. Halloran, T.J. Webster, Increased osteoblast adhesion on nanograined hydroxyapatite and tricalcium phosphate containing calcium titanate, J. Biomed. Mater. Res. 80 (4) (2007) 990–997.

[102] T.J. Webster, C. Ergun, R.H. Doremus, R.W. Siegel, R. Bizios, Specific proteins mediate enhanced osteoblast adhesion on nanophase ceramics, J. Biomed. Mater. Res. 51 (3) (2000) 475–483.

[103] M.C. Advincula, F.G. Rahemtulla, R.C. Advincula, E.T. Ada, J.E. Lemons, S.L. Bellis, Osteoblast adhesion and matrix mineralization on sol–gel-derived titanium oxide, Biomaterials 27 (10) (2006) 2201–2212.
[104] H. Liu, E.B. Slamovich, T.J. Webster, Increased osteoblast functions among nanophase titania/poly(lactide-*co*-glycolide) composites of the highest nanometer surface roughness, J. Biomed. Mater. Res. 78 (4) (2006) 798–807.
[105] G. Mendonca, D.B. Mendonca, L.G. Simoes, A.L. Araujo, E.R. Leite, W.R. Duarte, Nanostructured alumina-coated implant surface: effect on osteoblast-related gene expression and bone-to-implant contact *in-vivo*, Int. J. Oral Maxillofac. Implants 24 (2) (2009) 205–215.
[106] G. Mendonca, D.B. Mendonca, L.G. Simoes, A.L. Araujo, E.R. Leite, W.R. Duarte, The effects of implant surface nanoscale features on osteoblast-specific gene expression, Biomaterials 30 (25) (2009) 4053–4062.
[107] C.R. Choi, H.S. Yu, C.H. Kim, J.H. Lee, C.H. Oh, H.W. Kim, Bone cell responses of titanium blasted with bioactive glass particles, J. Biomater. Appl. 25 (2) (2010) 99–117.
[108] L. Meirelles, F. Currie, M. Jacobsson, T. Albrektsson, A. Wennerberg, The effect of chemical and nanotopographical modifications on the early stages of osseointegration, Int. J. Oral Maxillofac. Implants 23 (4) (2008) 641–647.
[109] Y.C. Jung, C.H. Han, I.S. Lee, H.E. Kim, Effects of ion beam-assisted deposition of hydroxyapatite on the osseointegration of endosseous implants in rabbit tibiae, Int. J. Oral Maxillofac. Implants 16 (6) (2001) 809–818.
[110] Y.S. Park, K.Y. Yi, I.S. Lee, C.H. Han, Y.C. Jung, The effects of ion beam-assisted deposition of hydroxyapatite on the grit-blasted surface of endosseous implants in rabbit tibiae, Int. J. Oral Maxillofac. Implants 20 (1) (2005) 31–38.
[111] P.G. Coelho, M. Suzuki, Evaluation of an IBAD thin-film process as an alternative method for surface incorporation of bioceramics on dental implants: a study in dogs, J. Appl. Oral Sci. 13 (2005) 87–92.
[112] L. Meirelles, T. Albrektsson, P. Kjellin, A. Arvidsson, V. Franke-Stenport, M. Andersson, Bone reaction to nano hydroxyapatite modified titanium implants placed in a gap-healing model, J. Biomed. Mater. Res. 87 (3) (2008) 624–631.
[113] L. Meirelles, A. Arvidsson, M. Andersson, P. Kjellin, T. Albrektsson, A. Wennerberg, Nano hydroxyapatite structures influence early bone formation, J. Biomed. Mater. Res. 87 (2) (2008) 299–307.
[114] R.S. Faeda, H.S. Tavares, R. Sartori, A.C. Guastaldi, E. Marcantonio, Biological performance of chemical hydroxyapatite coating associated with implant surface modification by laser beam: biomechanical study in rabbit tibias, J. Oral Maxillofac. Surg. 67 (8) (2009) 1706–1715.
[115] L. Meirelles, L. Melin, T. Peltola, P. Kjellin, I. Kangasniemi, F. Currie, Effect of hydroxyapatite and titania nanostructures on early *in-vivo* bone response, Clin. Implant Dent. Relat. Res. 10 (4) (2008) 245–254.
[116] I.S. Yeo, J.S. Han, J.H. Yang, Biomechanical and histomorphometric study of dental implants with different surface characteristics, J. Biomed. Mater. Res., B Appl. Biomater. 87 (2) (2008) 303–311.
[117] P.M. Burgos, L. Rasmusson, L. Meirelles, L. Sennerby, Early bone tissue responses to turned and oxidized implants in the rabbit tibia, Clin. Implant Dent. Relat. Res 10 (3) (2008) 181–190.

[118] S.J. Ferguson, J.D. Langhoff, K. Voelter, B. von Rechenberg, D. Scharnweber, S. Bierbaum, Biomechanical comparison of different surface modifications for dental implants, Int. J. Oral Maxillofac. Implants 23 (6) (2008) 1037–1046.

[119] V.C. Mendes, R. Moineddin, J.E. Davies, The effect of discrete calcium phosphate nanocrystals on bone-bonding to titanium surfaces, Biomaterials 28 (32) (2007) 4748–4755.

[120] V.C. Mendes, R. Moineddin, J.E. Davies, Discrete calcium phosphate nanocrystalline deposition enhances osteoconduction on titanium-based implant surfaces, J. Biomed. Mater. Res. 90 (2) (2009) 577–585.

[121] M.G. Tavares, P.T. de Oliveira, A. Nanci, A.C. Hawthorne, A.L. Rosa, S.P. Xavier, Treatment of a commercial, machined surface titanium implant with H_2SO_4/H_2O_2 enhances contact osteogenesis, Clin. Oral Implants Res. 18 (4) (2007) 452–458.

[122] L.F. Cooper, Y. Zhou, J. Takebe, J. Guo, A. Abron, A. Holmen, Fluoride modification effects on osteoblast behavior and bone formation at TiO_2 grit-blasted c.p. titanium endosseous implants, Biomaterials 27 (6) (2006) 926–936.

[123] J.E. Ellingsen, C.B. Johansson, A. Wennerberg, A. Holmen, Improved retention and bone-tolmplant contact with fluoride-modified titanium implants, Int. J. Oral Maxillofac. Implants. 19 (5) (2004) 659–666.

[124] T. Berglundh, I. Abrahamsson, J.P. Albouy, J. Lindhe, Bone healing at implants with a fluoride-modified surface: an experimental study in dogs, Clin. Oral Implants Res. 18 (2) (2007) 147–152.

[125] P.T. de Oliveira, A. Nanci, Nanotexturing of titanium-based surfaces upregulates expression of bone sialoprotein and osteopontin by cultured osteogenic cells, Biomaterials 25 (3) (2004) 403–413.

[126] C. Yao, V. Perla, J.L. McKenzie, E.B. Slamovich, T.J. Webster, Anodized Ti and Ti6Al4V possessing nanometer surface features enhances osteoblast adhesion, J. Biomed. Nanotechnol. 1 (2005) 68–73.

[127] D. Khang, J. Lu, C. Yao, K.M. Haberstroh, T.J. Webster, The role of nanometer and sub-micron surface features on vascular and bone cell adhesion on titanium, Biomaterials 29 (8) (2008) 970–983.

[128] N.R. Washburn, K.M. Yamada, C.G. Simon, S.B. Kennedy, E.J. Amis, High-throughput investigation of osteoblast response to polymer crystallinity: influence of nanometer-scale roughness on proliferation, Biomaterials 25 (7–8) (2004) 1215–1224.

[129] R. Chiesa, G. Giavaresi, M. Fini, E. Sandrini, C. Giordano, A. Bianchi, *In-vitro* and *in-vivo* performance of a novel surface treatment to enhance osseointegration of endosseous implants, Oral Surg. Oral Med. Oral Pathol. Oral Radiol. Endod. 103 (6) (2007) 745–756.

[130] C.H. Lohmann, R. Sagun, V.L. Sylvia, D.L. Cochran, D.D. Dean, B.D. Boyan, Surface roughness modulates the response of MG63 osteoblast-like cells to 1,25-(OH)(2)D(3) through regulation of phospholipase A(2) activity and activation of protein kinase A, J. Biomed. Mater. Res. 47 (2) (1999) 139–151.

[131] C. Yao, E.B. Slamovich, T.J. Webster, Enhanced osteoblast functions on anodized titanium with nanotube-like structures, J. Biomed. Mater. Res. A 85 (1) (2008) 157–166.

[132] D.T. Scadden, The stem-cell niche as an entity of action, Nature 441 (7097) (2006) 1075–1079.

[133] A. Curtis, C. Wilkinson, Nanotechniques and approaches in biotechnology, Trends Biotechnol. 19 (3) (2001) 97–101.

[134] Y. Dzenis, Material science. Spinning continuous fibers for nanotechnology, Science 304 (5679) (2004) 1917–1919.

[135] Y. Geissler MaX, Patterning: principles and some new developments, Adv. Mater. 16 (2004) 1249–1269.

[136] J.J. Norman, T.A. Desai, Methods for fabrication of nanoscale topography for tissue engineering scaffolds, Ann. Biomed. Eng. 34 (1) (2006) 89–101.

[137] N.J. Sniadecki, R.A. Desai, S.A. Ruiz, C.S. Chen, Nanotechnology for cell–substrate interactions, Ann. Biomed. Eng. 34 (1) (2006) 59–74.

[138] J. Park, S. Bauer, K. von der Mark, P. Schmuki, Nanosize and vitality: TiO_2 nanotube diameter directs cell fate, Nano Lett. 7 (6) (2007) 1686–1691.

[139] K.N. Chua, C. Chai, P.C. Lee, Y.N. Tang, S. Ramakrishna, K.W. Leong, Surface-aminated electrospun nanofibers enhance adhesion and expansion of human umbilical cord blood hematopoietic stem/progenitor cells, Biomaterials. 27 (36) (2006) 6043–6051.

[140] M.J. Dalby, N. Gadegaard, R. Tare, A. Andar, M.O. Riehle, P. Herzyk, The control of human mesenchymal cell differentiation using nanoscale symmetry and disorder, Nat. Mater. 6 (12) (2007) 997–1003.

[141] C.J. Bettinger, Z. Zhang, S. Gerecht, J.T. Borenstein, R. Langer, Enhancement of *in-vitro* capillary tube formation by substrate nanotopography, Adv. Mater. Deerfield. 20 (1) (2008) 99–103.

[142] C.J. Bettinger, B. Orrick, A. Misra, R. Langer, J.T. Borenstein, Microfabrication of poly (glycerol-sebacate) for contact guidance applications, Biomaterials 27 (12) (2006) 2558–2565.

[143] J.D. Foley, E.W. Grunwald, P.F. Nealey, C.J. Murphy, Cooperative modulation of neuritogenesis by PC12 cells by topography and nerve growth factor, Biomaterials. 26 (17) (2005) 3639–3644.

[144] S. Gerecht, C.J. Bettinger, Z. Zhang, J.T. Borenstein, G. Vunjak-Novakovic, R. Langer, The effect of actin disrupting agents on contact guidance of human embryonic stem cells, Biomaterials 28 (28) (2007) 4068–4077.

[145] S. Jungbauer, R. Kemkemer, H. Gruler, D. Kaufmann, J.P. Spatz, Cell shape normalization, dendrite orientation, and melanin production of normal and genetically altered (haploinsufficient NF1)-melanocytes by microstructured substrate interactions, ChemPhysChem. 5 (1) (2004) 85–92.

[146] A.I. Teixeira, G.A. Abrams, P.J. Bertics, C.J. Murphy, P.F. Nealey, Epithelial contact guidance on well-defined micro- and nanostructured substrates, J. Cell Sci. 116 (Pt 10) (2003) 1881–1892.

[147] X.F. Walboomers, H.J. Croes, L.A. Ginsel, J.A. Jansen, Contact guidance of rat fibroblasts on various implant materials, J. Biomed. Mater. Res. 47 (2) (1999) 204–212.

[148] E.K. Yim, R.M. Reano, S.W. Pang, A.F. Yee, C.S. Chen, K.W. Leong, Nanopattern-induced changes in morphology and motility of smooth muscle cells, Biomaterials 26 (26) (2005) 5405–5413.

[149] K.A. Diehl, J.D. Foley, P.F. Nealey, C.J. Murphy, Nanoscale topography modulates corneal epithelial cell migration, J. Biomed. Mater. Res. 75 (3) (2005) 603–611.

[150] E.K. Yim, S.W. Pang, K.W. Leong, Synthetic nanostructures inducing differentiation of human mesenchymal stem cells into neuronal lineage, Exp. Cell Res. 313 (9) (2007) 1820–1829.

[151] C.C. Berry, G. Campbell, A. Spadiccino, M. Robertson, A.S. Curtis, The influence of microscale topography on fibroblast attachment and motility, Biomaterials 25 (26) (2004) 5781–5788.

[152] C.S. Chen, M. Mrksich, S. Huang, G.M. Whitesides, D.E. Ingber, Geometric control of cell life and death, Science 276 (5317) (1997) 1425–1428.

[153] P. Clark, P. Connolly, A.S. Curtis, J.A. Dow, C.D. Wilkinson, Topographical control of cell behaviour. I. Simple step cues, Development 99 (3) (1987) 439–448.

[154] P. Clark, P. Connolly, A.S. Curtis, J.A. Dow, C.D. Wilkinson, Topographical control of cell behaviour: II. Multiple grooved substrata, Development 108 (4) (1990) 635–644.

[155] M.J. Dalby, M.O. Riehle, D.S. Sutherland, H. Agheli, A.S. Curtis, Use of nanotopography to study mechanotransduction in fibroblasts—methods and perspectives, Eur. J. Cell Biol. 83 (4) (2004) 159–169.

[156] M.J. Dalby, M.O. Riehle, S.J. Yarwood, C.D. Wilkinson, A.S. Curtis, Nucleus alignment and cell signaling in fibroblasts: response to a micro-grooved topography, Exp. Cell Res. 284 (2) (2003) 274–282.

[157] B. Wojciak-Stothard, Z. Madeja, W. Korohoda, A. Curtis, C. Wilkinson, Activation of macrophage-like cells by multiple grooved substrata. Topographical control of cell behaviour, Cell Biol. Int. 19 (6) (1995) 485–490.

[158] E.K. Yim, E.M. Darling, K. Kulangara, F. Guilak, K.W. Leong, Nanotopography-induced changes in focal adhesions, cytoskeletal organization, and mechanical properties of human mesenchymal stem cells, Biomaterials 31 (6) (2010) 1299–1306.

[159] C.C. Berry, M.J. Dalby, R.O. Oreffo, D. McCloy, S. Affrossman, The interaction of human bone marrow cells with nanotopographical features in three dimensional constructs, J. Biomed. Mater. Res. A 79 (2) (2006) 431–439.

[160] T. Gustafson, L. Wolpert, Studies on the cellular basis of morphogenesis in the sea urchin embryo. Directed movements of primary mesenchyme cells in normal and vegetalized larvae, Exp. Cell Res. 24 (1961) 64–79.

[161] M.J. Dalby, M.O. Riehle, H. Johnstone, S. Affrossman, A.S. Curtis, *In-vitro* reaction of endothelial cells to polymer demixed nanotopography, Biomaterials 23 (14) (2002) 2945–2954.

[162] M.J. Dalby, M.O. Riehle, H.J. Johnstone, S. Affrossman, A.S. Curtis, Polymer-demixed nanotopography: control of fibroblast spreading and proliferation, Tissue Eng. 8 (6) (2002) 1099–1108.

[163] B. Wojciak-Stothard, A. Curtis, W. Monaghan, K. MacDonald, C. Wilkinson, Guidance and activation of murine macrophages by nanometric scale topography, Exp. Cell Res. 223 (2) (1996) 426–435.

[164] A.S. Andersson, F. Backhed, A. von Euler, A. Richter-Dahlfors, D. Sutherland, B. Kasemo, Nanoscale features influence epithelial cell morphology and cytokine production, Biomaterials 24 (20) (2003) 3427–3436.

CHAPTER

9

Corrosion resistance of Ti–6Al–4V with nanostructured TiO_2 coatings

Qian Chen, Gerald D. McEwen, Nikita Zaveri, Ramji Karpagavalli and Anhong Zhou

Utah State University, Logan, UT, United States

CHAPTER OUTLINE

9.1 Introduction 165
9.1.1 SiO_2–CaO Coatings on Ti–6Al–4V Alloys 166
9.1.2 SiO_2 and SiO_2–TiO_2 Intermediate Coatings on Titanium and Ti–6Al–4V Alloy 167
9.1.3 Coated Hydroxyapatite on Ti–6Al–4V by Electrophoretic Deposition 167
9.1.4 Double-Layer Glass–Ceramic Coatings on Ti–6Al–4V 168
9.2 Nanostructured TiO_2 Deposited on Ti–6Al–4V 168
9.2.1 Preparation of the Ti–6Al–4V Electrode 168
9.2.2 TiO_2 Nanoparticles Coating 169
9.3 Characterization Techniques 169
9.3.1 Scanning Electron Microscopy 169
9.3.2 Raman Microscopy 169
9.4 Corrosion Test With Electrochemical Techniques 173
9.4.1 Open-Circuit Voltage (OCV) and Tafel Analysis 174
9.4.2 Electrochemical Impedance Spectroscopy 175
9.5 Conclusion 176
References 177

9.1 INTRODUCTION

Mechanical properties are important factors for implant materials because they are exposed to the corrosive biological environment chronically, which can only be achieved by the use of metallic materials. However, the biocompatibility of the metals is limited by their reactivity and degradation in the corrosive biological environment. Titanium and its alloys, such as Ti–6A1–4V [1,2] has excellent

Emerging Nanotechnologies in Dentistry. DOI: http://dx.doi.org/10.1016/B978-0-12-812291-4.00009-1

properties, e.g., biocompatibility, corrosion resistance, low density, mechanical strength and relatively low cost [3–5], all of these properties make titanium and its alloys to a potential dental implant materials. Among these features, the corrosion resistance is of great importance, not only because it determines the device service life, but also because of the harmfulness of corrosion processes taking place in the living organism. However, an important issue also needs to be addressed, namely, how to obtain suitable bonding between dental implant and metal substrate [4–7]. A few factors may lead to decrease the bonding strength, e.g., high affinity of titanium for oxygen [3,8–10] and mismatch of the coefficients of thermal expansion between titanium and porcelain [10–13]. Insufficient bonding leads to failure, reducing durability and functionality of the system, posing clinical and esthetic problems. Recently, it is considered that suitable bonding of dental implant and metal substrate could be obtained using various methods of modification of the metal substrate surface [3,4,6,14–19].

The modified metal substrate surface should address problems such as (1) incompatibility between the artificial implant and the natural bone; (2) metallic dental implants materials release metallic ions which could be toxic or detrimental to the host. Moreover, metallic surfaces are in general not adequately bioactive, and surface treatment is usually needed to enhance the bioactivity. To overcome these problems, coating the metallic implant with other materials were explored in previous studies, such as glass–ceramics [20], SiO_2–CaO [21], hydroxyapatite (HA) [22,23], TiO_2 nanoparticles [24], SiO_2 and SiO_2–TiO_2 [25]. In this way, the mechanical properties of a metallic implant are not only complemented with biocompatible and bioactive features using different coatings, but such coatings also reduce the release of metallic ions to the biological environment.

In this chapter, some different coatings on titanium and Ti–6Al–4V alloy that have potential in improving bioactivity when applied as dental materials are reviewed at the beginning. Then, some conventional characterization techniques which are employed to evaluate the behaviors of titanium and Ti–6Al–4V alloy with different coatings will be also addressed.

9.1.1 SiO_2–CaO COATINGS ON Ti–6Al–4V ALLOYS

Sol–gel technology is a common method to create surface layers; the required equipment is quite simple and the thermal treatments are not too aggressive. Most importantly, the sol–gel process allows control of the microstructure, thickness, and composition of the coating, ensuring high levels of purity and homogeneity [26]. However, the coating process of a metal by a bioactive ceramic or glass is somewhat complex. The expected clinical success depends greatly on this process, since the quality and durability of the interfacial bond between prosthesis and host tissue is closely related to several parameters of the coating—chemical composition, purity, particle size, thickness—and also to the surface quality of the metallic substrate [27]. Recently, in vitro assays of binary glasses in the

SiO_2–CaO system have proved their bioactivity; glass pieces soaked in fluids mimicking the composition of human plasma grew a hydroxycarbonate apatite surface layer similar to the mineral component of natural bone tissue [28,29]. Besides, vitreous coatings of SiO_2–CaO on Ti–6Al–4V alloy substrates by sol–gel method revealed that the porosity and roughness of the coatings were determined by the precursor solutions used in the coating procedures [30,31].

9.1.2 SiO_2 AND SiO_2–TiO_2 INTERMEDIATE COATINGS ON TITANIUM AND Ti–6Al–4V ALLOY

The bond strength of titanium–porcelain systems can be increased by intermediate coatings of SiO_2 and SiO_2–TiO_2, produced by sol–gel between metal substrate and ceramics. The coatings obtained using the sol–gel process are characterized by low thickness, high homogeneity, satisfactory mechanical and chemical stability, good bonding to the substrate, high biocompatibility and corrosion resistance [32–35]. Moreover, the process of manufacturing sol–gel coatings runs at low temperatures, and there is also the possibility of obtaining both mono- and multicomponents [25], which can be multilayered, and the possibility of producing coatings for implants with complex geometries [25,32–37]. These sol–gel coatings seem to be very attractive in the field of dentistry.

9.1.3 COATED HYDROXYAPATITE ON Ti–6Al–4V BY ELECTROPHORETIC DEPOSITION

HA, a bioceramic which has similar mineral constituents of human bones and teeth, is a common material used for metallic coating. The HA coating not only would reduce the release of metallic ions by acting a barrier but also enhance the bone bioactivity by virtue of its chemical constituents. Many techniques, such as dip coating, ion sputtering, sol–gel coating, pulsed-laser deposition, electrophoretic deposition (EPD), etc., have been proposed to fabricate HA coatings [38]. Among the various fabrication methods, EPD is a promising technique, with advantages including less formation time, simplicity in instrumentation and capability of coating complex-shaped implants [39].

EPD is a colloidal process which offers easy control of thickness and quality of coating through simple adjustment of deposition time, applied potential and powder morphology [39]. The corrosion behavior of EPD HA coatings on Ti alloys has not been fully studied with respect to morphology and size of the starting HA particles. Furthermore, HA coatings reinforced with carbon nanotubes (CNTs) was reported to be a viable route for forming surface composite with higher hardness, elastic modulus, and interlaminar shear strength than that of the monolithic HA layers [40]. Addition of CNTs also helps to prevent peeling-off of the coating layers by acting as reinforcement network with the deposit [40].

9.1.4 DOUBLE-LAYER GLASS–CERAMIC COATINGS ON Ti–6Al–4V

Many different methods such as enameling, sputtering techniques, and vacuum Plasma Spray [41–43] were applied to coat Ti–6Al–4V. However, the sputtering and Plasma Spray techniques could present some problems related to difficulty of coating complex shapes, to eventual compositional modifications of the coating, and to their high cost. On the contrary, enameling technique is a very inexpensive way to coat substrates, even of complex shape, and it is easily applicable to glasses and glass–ceramics because of the peculiar softening properties of these materials. For this reason, bioactive glasses and glass–ceramics do represent a good alternative to HA, commonly used as bioactive coating on metallic prostheses in order to improve their adhesion to the bone.

Double-layer bioactive glass–ceramic coatings were prepared on Ti–6Al–4V substrates by dipping and firing. A SiO_2–CaO–Na_2O–MgO–P_2O_5–K_2O (SCP)-based glass was used as first layer in direct contact with the metallic substrate, and a SiO_2–Al_2O_3–P_2O_5–K_2O–CaO–F^--based glass–ceramic was used as outer bioactive layer. The deposition of the intermediate SCP layer was necessary in order to obtain a good adhesion of the coating to the substrate, to minimize the reactivity between the substrate and the outer glass–ceramic coating, and thus to preserve the nature of its crystalline phases.

9.2 NANOSTRUCTURED TiO_2 DEPOSITED ON Ti–6Al–4V

Regardless their vast biomedical applications [44], titanium and its alloy form an external superficial passive oxide film that can reach nanometer (nm) thickness and protect against corrosion [45]. It is well known that nanostructured materials exhibited a greater surface area than the conventional materials [24,46], which may significantly influence the corrosion behavior of nanocoatings. In the current study, researchers began to use the nanostructured materials to modify the implant surface. For example, Karpagavalli et al. [46] studied the corrosion behavior and biocompatibility of nanostructured TiO_2 film on Ti–6Al–4V. They found that nano-TiO_2 coated Ti–6Al–4V showed a better corrosion resistance in simulated biofluid than uncoated Ti–6Al–4V. Furthermore, Zaveri et al. [24] also addressed that the TiO_2 nanoparticle coatings increased the thickness of the pre-existing oxide layer on the Ti–6Al–4V surface, serving to improve the bioimplant corrosion resistance. Thus, it is indicated that nanoimplants prepared by various technologies will exhibit different corrosion resistance.

In this section, the behaviors of TiO_2 nanoparticles-coated Ti–6Al–4V implant in simulated biofluids will be mainly discussed.

9.2.1 PREPARATION OF THE Ti–6Al–4V ELECTRODE

The Ti–6Al–4V alloys were used to prepare the samples, which was polished with sand paper and polish pad, respectively. A titanium wire was spot welded at one end of the Ti electrode. Then, the total surface of the electrode was coated with an insulating material (epoxy) leaving only the electrode surface exposed. The coated electrode was thoroughly rinsed by distilled water and left for 24 h to completely dry.

9.2.2 TiO_2 NANOPARTICLES COATING

A compound was formed by mixing together 1.2 g of TiO_2, 0.4 mL of distilled water, 0.02 mL of Triton-X100, and 0.04 mL of acetyl acetone. Then, in order to obtain a 10% (mass) TiO_2 aqueous colloidal suspension, the compound was diluted in deionized water. The suspension was sonicated for 30 min. The spin coating was repeated six times on the Ti–6Al–4V substrate by a spin coater. After each spin coating, the sample was allowed to air dry for 30 s. Finally, the sample was heat-treated in air at 550°C.

9.3 CHARACTERIZATION TECHNIQUES

Presently, a variety of instrumental analysis techniques were applied to access the corrosion phenomena. For example, microscopy techniques include scanning electron microscopy (SEM) [47], atomic force microscope [48], confocal laser scanning microscopy, scanning Auger microprobe, and X-ray diffraction [49].

9.3.1 SCANNING ELECTRON MICROSCOPY

The SEM images of TiO_2 nanoparticles were taken with Hitachi S-4700 at 15 kV.

Typical SEM micrographs of TiO_2 deposited on Ti–6Al–4V substrates under different deposition currents, before and after annealing at 550°C, are shown in Fig. 9.1. The amorphous particles and nonuniform clumps were observed before heat treatment (Fig. 9.1A and C). The agglomeration of the particles resulted in the formation of clusters. However, after annealing at 550°C, bricklike crystallites were observed (Fig. 9.1B and D). The SEM micrograph of as-deposited TiO_2 (Fig. 9.1A) revealed that the mean particle size of TiO_2 is in the nanometer level ($\sim$30–50 nm). It is found that the TiO_2 crystallite size has been greatly increased during crystallization, a higher crystallite size after thermal annealing than before heat treatment. The particle size was found to be grown in size due to thermal annealing. Moreover, the applied deposition current density can also affect the particle size. Fig. 9.1A, C, E, and F shows that, when the applied deposition current density was increased, the agglomeration phenomenon increased.

9.3.2 RAMAN MICROSCOPY

Raman microspectroscopy is a nondestructive vibrational/structural characterization method that can be utilized to identify naturally occurring crystalline forms. Here, Raman spectroscopic characterization technique was applied to analyze

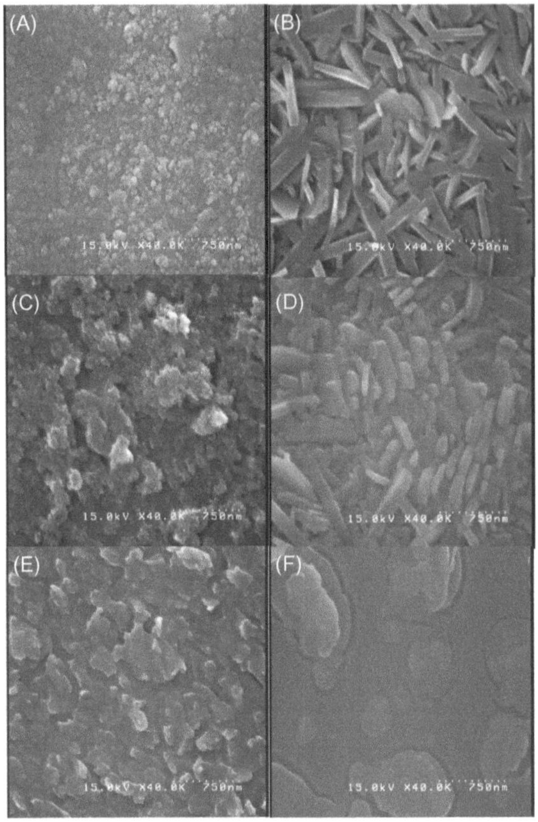

FIGURE 9.1

SEM micrographs of nanostructured TiO_2 deposited on Ti–6Al–4V: (A), (C), (E), (F) as-deposited; (B) and (D) annealed at 550°C. Applied deposition current densities: (A) 30 mA/cm^2, (C) 40 mA/cm^2, (E) 50 mA/cm^2, (F) 60 mA/cm^2.

crystalline distribution based on thermal induced properties of TiO_2 nanoparticles. An inVia confocal Raman microspectroscopy system (Renishaw, United Kingdom) equipped with a 785-nm laser, 1200 line/mm grating, and a 20× objective (Leica). Spectra and image data were acquired at 100% laser power, 3-s integration time under WiRE 3.0 software with streamline mapping capability.

Fig. 9.2 shows Raman spectra of four TiO_2 nanoparticle samples at different temperatures; room temperature (RT) and calcine temperatures of 400, 700, and 1000°C. In the case of the calcined samples through 700°C, a series of peaks are visible, which correspond to both TiO_2 crystalline anatase (A) and rutile (R). From analysis of these Raman spectra, there are three distinctive Raman modes: A1g, B1g, and Eg. They were mode Eg for rutile at c. 236 cm^{-1}, B1g for anatase and rutile at 395 and 445 cm^{-1}, B1g/A1g for anatase and rutile at 515 and

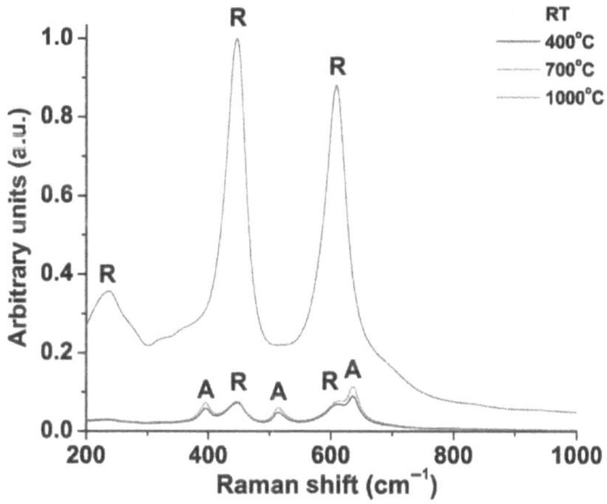

FIGURE 9.2

Raman spectra collected from three calcined TiO_2 samples and processed at RT, 400, 700, and 1000°C. Each individual spectrum is an average of 15 randomly collected spectra per sample.

609 cm^{-1}, and Eg for rutile at 637 cm^{-1}. However, in the case of the 1000°C sample, there are only modes associated with rutile observed at peaks 236, 445, and 609 cm^{-1}. It is well established that the crystalline transition of amorphous TiO_2 powders can be viewed as a two-step process; (1) amorphous to anatase and (2) anatase to rutile.

TiO_2 Raman spectra have been studied extensively, but Raman mapping investigations of the anatase/rutile ratio distribution of nanocrystalline TiO_2 is limited. Fig. 9.3 shows the bright field image (A) of RT TiO_2 sample along with selected Raman map area (B), where each 2.6 × 2.6 mm square within that area is an individual Raman spectrum over a 410-wave-number range beginning at 324 and ending at 734 cm^{-1}. Direct classical least squares (DCLS) was used to process the map into the individual components (1–4) and a combined overlay (C) based on Fig. 9.2 reference spectra. Each individual component was present in the RT TiO_2 sample (C), and associated spectra (i–iv, solid line) were processed according to Fig. 9.2 standard spectra (i–iv, dashed line). The RT component (C, colored yellow) along with the actual spectrum (i, yellow line) at the arrow point compared to the standard RT spectrum (i, black dashed line). The 400°C component (C, colored blue) along with the actual spectrum (ii, blue line) at the arrow point compared with the 400°C standard spectrum (ii, black dashed line). Next, the 700°C component (C, colored green) along with actual spectrum (iii, green line) compared with the 700°C standard spectrum (iii, black dashed line). Finally, the 1000°C component (C, colored red) along with actual spectrum (iv, red line) compared with the 1000°C standard spectrum (iv, black dashed line). In the actual spectra for the RT (i), 400°C (ii), and 700°C (iii) components, the peaks for anatase (395, 515, and

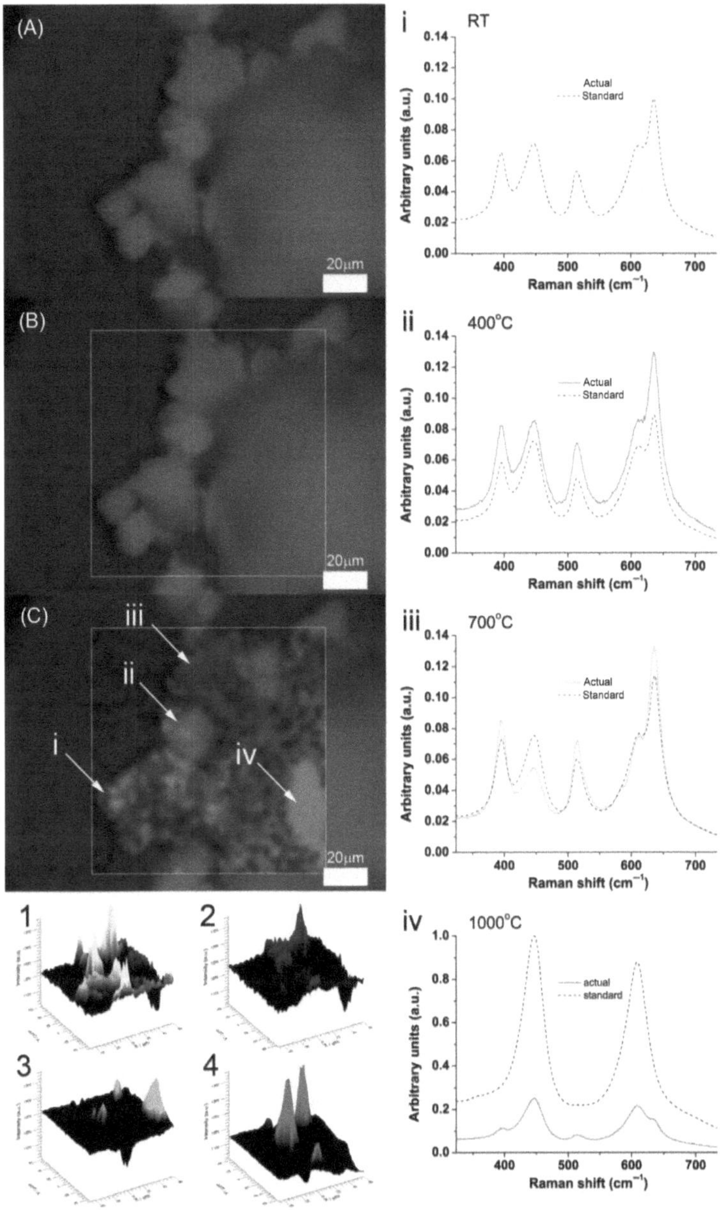

FIGURE 9.3

Images from the RT sample; bright field view (A) and Raman map area (B). Arrow points indicate various components within the sample; RT component (i), 400°C component (ii), 700°C component (iii), and 1000°C component (iv). Actual spectra for each arrow are compared with each Fig. 9.2 TiO_2 standard (the colors of standard spectra match those in Fig. 9.3). 3-D surface figures for each component (1–4), corresponding to components i–iv, are shown to give a phase distribution (C).

637 cm^{-1}) are more dominant than rutile (445, and 609 cm^{-1}). On the other hand, in the actual spectra for the 1000°C (iv), it is apparent that the peaks for rutile are more prominent than anatase. The combination of Raman mapping (C) along with the 3D surface plot for individual components (1 4) illustrate the overall TiO_2 crystalline distribution, whereas the separate crystalline forms in the bright field image are indistinguishable (Fig. 9.3A and B).

The temperature effect of anatase to rutile transformation was verified spectroscopically. Raman microspectroscopy is a powerful technique to distinguish different crystalline phases by using Raman mapping at selected characteristic peaks related to individual crystalline form.

9.4 CORROSION TEST WITH ELECTROCHEMICAL TECHNIQUES

Various electrochemical techniques have been applied to assess the behavior of corrosion, for example, open-circuit voltage (OCV), potentiodynamic polarization (Tafel analysis), and electrochemical impedance spectroscopy (EIS). Tafel analysis is a well-established electrochemical technique, the current is recorded when the open-circuit potential is imposed on a metal sample. Eq. (9.1) shows that the current and voltage have the linear relationship, and the slope is the polarization resistance (R_p). Polarization resistance is the resistance of the metal to oxidation during the application of an external potential. Combined Eqs. (9.1) and (9.2), it shows that the corrosion rate is inversely related to Rp and can be calculated if the anodic and cathodic Tafel slope were known, respectively. The anodic and cathodic Tafel slope can be obtained from linear polarization measurements (Eq. (9.1)). The equation used for calculating the corrosion rate (mmpy) is as indicated below in Eq. (9.2).

$$I_{corr} = \frac{\beta_a \times \beta_c}{2.3R_P(\beta_a + \beta_c)} \tag{9.1}$$

where I_{corr} is the corrosion current (in A), R_P is the polarization resistance (in Ω), β_a is the anodic Tafel slope, and β_c is the cathodic Tafel slope.

$$\text{Corrosion Rate (CR)} = \frac{I_{corr} \times K \times EW}{d \times A} \tag{9.2}$$

K is the constant that defines the units of the corrosion rate = 3272 mm/(A cm year), EW is the equivalent weight (in g/equivalent) = 11.768 g/eq, d is the density(in g/cm^3) = 4.420 g/cm^3, A = sample area (in cm^2) = 0.151 cm^2, for all current Ti alloy samples.

The electrochemical techniques, including OCV, Tafel analysis, and EIS, were conducted using a VMP2/Z multichannel potentiostat (PAR, TN), controlled by EC-lab Win 9.01. In Tafel experiment, scan rate at 1 mV/s and the scan range of ±25 mV of open-circuit potential were used after the specimens have reached stable E_{corr}. All potentials were measured with respect to Ag/AgCl reference electrode (CH Instruments, TX). The EIS measurements were obtained in the potentiodynamic mode with voltage perturbation amplitude of 10 mV in the frequency range 100 kHz to 10 mHz. The equivalent electrical circuit models were developed by ZSimpwin 3.21 (EChem Software).

9.4.1 OPEN-CIRCUIT VOLTAGE (OCV) AND TAFEL ANALYSIS

OCV is the other typical technique to evaluate the corrosion behaviors. Fig. 9.4A shows the OCV of the bare and the TiO_2 nanoparticle-coated Ti–6Al–4V in simulated biological NaCl's solution. When the uncoated Ti–6Al–4V were exposed to NaCl's solution, the initial open-circuit potential is about −0.56 mV, and then it gradually increases to a steady value of −0.32 V.

While the initial potential of the TiO_2 nanoparticles-coated Ti–6Al–4V was 0.134 V and then it decreases to −0.064 V.

Fig. 9.4B shows the polarization curves of the bare and the TiO_2 nanoparticle-coated Ti–6Al–4V in simulated biological NaCl's solution.

It is found that the average corrosion potential estimated of the uncoated Ti–6Al–4V is −0.316 V, while TiO_2 nanoparticle-coated Ti–6Al–4V alloy in NaCl's solution is −0.07 V. Moreover, the corrosion current was obtained from the polarization curves by extrapolation of the anodic and the cathodic branches of the polarization curve to the corrosion potential. The bare Ti electrode has a higher I_{corr} compared to TiO_2 nanoparticle-coated Ti–6Al–4V substrate, 0.004 and 0.002 μA, respectively.

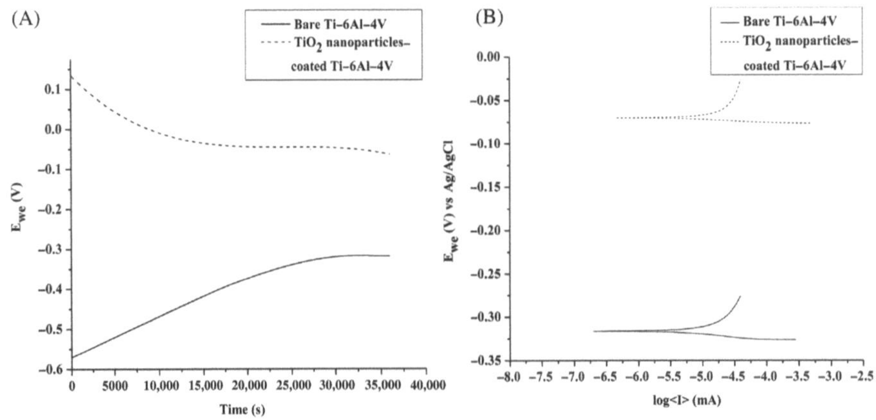

FIGURE 9.4

Open-circuit voltage (OCV) (A) Tafel analysis results (B) for the bare and the TiO_2 nanoparticle-coated Ti–6Al–4V in simulated biological NaCl's solution.

9.4.2 ELECTROCHEMICAL IMPEDANCE SPECTROSCOPY

Fig. 9.5A and B shows the Nyquist plots for the bare Ti–6Al–4V and the TiO_2 nanoparticle-coated Ti–6Al–4V alloy in NaCl's solution, respectively. The Nyquist plots for the bare Ti–6Al–4V shows arc-shape, which indicated that the electrochemical corrosion process of bare Ti–6Al–4V in NaCl's solution is

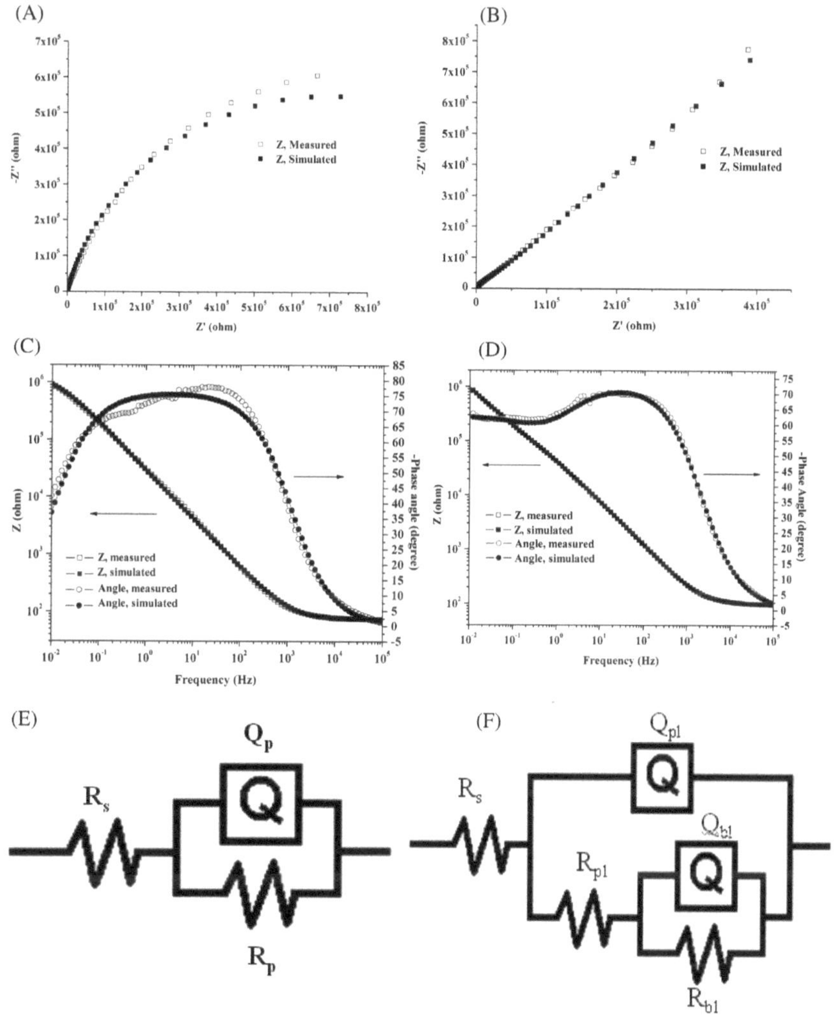

FIGURE 9.5

Nyquist plot (A, B) and Bode plot (C, D) for the bare (A, C) and TiO_2 nanoparticle-coated Ti–6Al–4V alloy (B, D) in NaCl's solution. Electrical equivalent circuit used to simulate the measured data for bare (E) and TiO_2 nanoparticle-coated Ti–6Al–4V (F).

electron-transfer controlled. Compared to the bare Ti–6Al–4V, straight-lines were observed in the Nyquist plots for TiO_2 nanoparticle-coated Ti–6Al–4V (Fig. 9.5B) indicate that the corrosion process is mass transport (diffusion) controlled. This is different from the bare Ti–6Al–4V which is electron-transfer controlled.

Bode plots for bare Ti–6Al–4V and TiO_2 nanoparticle-coated Ti–6Al–4V are shown in Fig. 9.5C and D. It is found that the maximum phase angles of both bare Ti–6Al–4V and TiO_2 nanoparticle-coated Ti–6Al–4V are less than 90° and n values are less than 1. However, the shapes of the phase angle plots are quite different. According to the proposed EIS model, the electrochemical interface of bare Ti–6Al–4V in solutions can be described as the equivalent circuit, as shown in Fig. 9.5E. The impedance data obtained for bare Ti–6Al–4V are simulated by a simple R (QR) circuit wherein R_s is the solution resistance, in series with the constant phase element (CPE or Q) and in parallel with the polarization resistance (R_p), i.e., a simple Randles equivalent circuit. After the Ti–6Al–4V, substrates were coated with TiO_2 nanoparticles, the modified electrical equivalent circuit (Fig. 9.5F) used to simulate the measured data were somewhat different from that used for bare Ti–6Al–4V (Fig. 9.5E). Using this oxide-layer-coating approach, R_s corresponds to the solution resistance, R_{pl} to the resistance of the outer porous layer, Q_{pl} to the capacitance of the outer porous layer, R_{bl} to the resistance of the barrier layer, and Q_{bl} to the capacitance of the barrier layer. It is reasonable to introduce an "outer porous layer" to this model due to the nanostructured TiO_2 particle deposition as well as the formation of oxide layer during the electrochemical corrosion process.

9.5 CONCLUSION

Nanoscale modification can alter the chemistry and/or topography of the implant surface. Different methods have been described to modify or to embellish titanium substrates with nanoscale features. Such changes alter the implant surface interaction with ions, biomolecules, and cells. These interactions can favorably influence molecular and cellular activities and alter the process of osseointegration. In this chapter, a nanoscale modification of TiO_2 nanoparticle-coated Ti–6Al–4V was introduced in detail. From the results mentioned above, it can be concluded that the TiO_2 nanoparticle-coated Ti–6Al–4V is more corrosion resistant than the bare Ti–6Al–4V in the simulated biofluids. Furthermore, Raman microspectroscopy, as a powerful technique to distinguish different crystalline phases by using Raman mapping spectroscopically, was also discussed. The results show that the temperature affects anatase to rutile transformation.

REFERENCES

[1] E.P. Lautenschlager, P. Monaghan, Titanium and titanium alloys as dental materials, Int. Dent. J. 43 (3) (1993) 245.

[2] R.R. Wang, A. Fenton, Titanium for prosthodontic applications: a review of the literature, Quintessence Int. 27 (6) (1996) 401.

[3] Y. Oshida, A. Hashem, Titanium–porcelain system. Part I. Oxidation kinetics of nitrided pure titanium, simulated to porcelain firing process, Biomed. Mater. Eng. 3 (4) (1993) 185.

[4] N. Suansuwan, M.V. Swain, Adhesion of porcelain to titanium and titanium alloy, J. Dent. 31 (7) (2003) 509.

[5] M. Yoda, T. Konno, Y. Takada, K. Iijima, J. Griggs, O. Okuno, Bond strength of binary titanium alloys to porcelain, Biomaterials 22 (2001) 167.

[6] J. Fischer, Ceramic bonding to a dental gold–titanium alloy, Biomaterials 23 (2002) 1303.

[7] L. Pröbster, U. Maiwald, H. Weber, Three-point bending strength of ceramics fused to cast titanium, Eur. J. Oral. Sci 104 (3) (1996) 313.

[8] R.R. Wang, K.K. Fung, Oxidation behaviour of surface-modified titanium for titanium–ceramic restorations, J. Prosthet. Dent. 77 (4) (1997) 423.

[9] J.A. Hautaniemi, H. Hero, J.T. Juhanoja, On the bonding of porcelain on titanium, J. Mater. Sci 3 (3) (1992) 186.

[10] M. Adachi, J.R. Mackert, E.E. Parry, C.W. Fairhurst, Oxide adherence and porcelain bonding to titanium and Ti–6Al–4V alloy, J. Dent. Res. 69 (6) (1990) 1230.

[11] H. Yilmaz, C. Dincer, Comparison of the bond compatibility of titanium and an NiCr alloy to dental porcelain, J. Dent. 27 (1999) 215.

[12] A. Sadeq, Z. Cai, R.D. Woody, A.W. Miller, Effect of interfacial variables on ceramic adherence to cast and machined commercially pure titanium, J. Prosthet. Dent. 90 (1) (2003) 10.

[13] S. Atsü, S. Berksun, Bond strength of three porcelains to two forms of titanium using two firing atmospheres, J. Prosthet. Dent. 84 (5) (2000) 567.

[14] I. Nergiz, P. Schmage, W. Herrmann, M. Özcan, Effect of alloy type and surface conditioning on roughness and bond strength of metal brackets, Am. J. Orthod. Dentofac. Orthop. 125 (1) (2004) 42.

[15] J.L. Gilbert, D.A. Covey, E.P. Lautenschlager, Bond characteristic of porcelain fused to milled titanium, Dent. Mater. 10 (2) (1994) 134.

[16] Z. Cai, N. Bunce, M.E. Nunn, T. Okabe, Porcelain adherence to dental cast CP titanium: effects of surface modifications, Biomaterials 22 (9) (2001) 979.

[17] I. Hussaini, K. Wazzan, Effect of surface treatment on bond strength of low-fusing porcelain to commercially pure titanium, J. Prosthet. Dent. 94 (4) (2005) 350.

[18] R.R. Wang, G.E. Welsch, O. Monterio, Silicon nitride coating on titanium to enable titanium–ceramic bonding, J. Biomed. Mater. Res. 46 (2) (1999) 262.

[19] M. Kõnönen, J. Kivilahti, Bonding of low-fusing dental porcelain to commercially pure titanium, J. Biomed. Mater. Res. 28 (9) (1994) 1027.

[20] E. Verne, C. Fernandez Valles, C. Vitale Brovarone, S. Spriano, C. Moisescu, Double-layer glass–ceramic coatings on Ti6Al4V for dental implants, J. Eur. Ceram. Soc. 24 (9) (2004) 2699.

[21] I. Izquierdo-Barba, F. Conde, N. Olmo, M.A. Lizarbe, M.A. García, M. Vallet-Regı, Vitreous SiO_2–CaO coatings on Ti6Al4V alloys: reactivity in simulated body fiuid versus osteoblast cell culture, Acta Biomaterialia 2 (2006) 445.
[22] C.T. Kwok, P.K. Wong, F.T. Cheng, H.C. Man, Characterization and corrosion behavior of hydroxyapatite coatings on Ti6Al4V fabricated by electrophoretic deposition, Appl. Surf. Sci. 255 (13–14) (2009) 6736.
[23] E.C.S. Rigoa, A.O. Boschi, M. Yoshimotob, S. Allegrini Jr., B. Konig Jr., M.J. Carbonari, Evaluation in vitro and in vivo of biomimetic hydroxyapatite coated on titanium dental implants, Mater. Sci. Eng. C 24 (5) (2004) 647.
[24] N. Zaveri, G.D. McEwen, R. Karpagavalli, A.H. Zhou, Biocorrosion studies of TiO_2 nanoparticle-coated Ti–6Al–4V implant in simulated biofluids, J. Nanopart. Res. 12 (5) (2010) 1609.
[25] J. Bieniaś, B. Surowska, A. Stoch, H. Matraszek, M. Walczak, The influence of SiO_2 and SiO_2–TiO_2 intermediate coatings on bond strength of titanium and Ti6Al4V alloy to dental porcelain, Dent. Mater. 25 (9) (2009) 1128.
[26] C.J. Brinker, G.W. Scherer, Sol–gel science, The Physics and Chemistry of Sol–Gel Processing, Academic Press, New York (NY), 1990.
[27] C. Gabbi, A. Cacchioli, B. Locardi, F. Guadagninno, Bioactive glass coating: physicochemical aspect and biological finding, Biomaterials 16 (7) (1995) 515.
[28] J.P. Borrajo, J. Serra, P. González, B. León, F.M. Muñoz, M. López, In vivo evaluation of titanium implants coated with bioactive glass by pulsed laser deposition, J. Mater. Sci.: Mater. Med. 18 (12) (2007) 2371.
[29] A. Martínez, I. Izquierdo-Barba, M. Vallet-Regí, Bioactivity of a CaO–SiO_2 binary glass system, Chem. Mater. 12 (10) (2000) 3080.
[30] I. Izquierdo-Barba, A. Asenjo, L. Esquivias, M. Vallet-Regí, SiO_2–CaO vitreous films deposited onto Ti6Al4V substrates, Eur. J. Inorg. Chem. 8 (2003) 1608.
[31] I. Izquierdo-Barba, M. Vallet-Regí, J.M. Rojo, E. Blanco, L. Esquivias, The role of precursor concentration on the characteristics of SiO_2–CaO films, J. Sol–Gel Sci. Tech. 26 (1–3) (2003) 1179.
[32] J. Livage, Sol–gel processing of metal oxides, Chem. Scripta 28 (1988) 9.
[33] B. Surowska, J. Bieniás, M. Walczak, K. Sangwal, A. Stoch, Microstructure and mechanical properties of ceramic coatings on Ti and Ti-based alloy, Appl. Surf. Sci. 238 (2004) 288.
[34] M.E. Pätsi, J.A. Hautaniemi, H.M. Rahiala, T.O. Peltola, I.M.O. Kangasniemi, Bonding strengths of titania sol–gel derived coatings on titanium, J. Sol–Gel Sci. Technol. 11 (1) (1998) 55.
[35] H. Matraszek, A. Stoch, C.Z. Paluszkiewicz, A. Brożek, E. Długoń, The application of sol–gel technique in dentistry, Eng. Biomater. 23 (25) (2002) 81.
[36] D.B. Haddow, S. Kothari, P.F. James, R.D. Short, P.V. Hatton, R. Noort, Synthetic implant surfaces: 1. The formation and characterization of sol–gel titania films, Biomaterials 17 (5) (1996) 501–507.
[37] J. Breme, Y. Zhou, L. Groh, Development of titanium alloy suitable for an optimized coating with hydroxyapatite, Biomaterials 16 (3) (1995) 239.
[38] F. Lusquinos, J. Pou, J.L. Arias, M. Boutinguiza, B. Leon, M. Perez-Amor, Alloying of hydroxyapatite onto Ti6Al4V by high power laser irradiation, J. Mater. Sci.: Mater. Med. 13 (6) (2002) 601.
[39] L. Besra, M. Liu, A review on fundamentals and applications of electrophoretic depositions (EPD), Prog. Mater. Sci. 52 (1) (2007) 1.

[40] C. Kaya, Electrophoretic deposition of carbon nanotube-reinforced hydroxyapatite bioactive layers on Ti−6Al−4V alloys for biomedical applications, Ceram. Inter. 34 (2008) 1843.
[41] M. Ferraris, P. Rabajoli, F. Brossa, L. Paracchini, Vacuum plasma spray deposition of titanium particle/glass−ceramic matrix biocomposites, J. Am. Ceram. Soc. 79 (6) (1996) 1515.
[42] E. Verne, M. Ferraris, A. Ventrella, L. Paracchini, A. Krajewski, A. Ravaglioli, Sintering and plasma spray deposition of bioactive glass−matrix composites for biomedical applications, J. Eur. Ceram. Soc. 18 (4) (1998) 363.
[43] J.M. Gomez Vega, E. Saiz, A.P. Tomsia, G.W. Marshall, S.J. Marshall, Bioactive glass coatings with hydroxyapatite and Bioglass particles on Ti-based implants. 1. Processing, Biomaterials 21 (2000) 105.
[44] B. Grosgogeat, M. Boinet, F. Dalard, M. Lissac, Electrochemical studies of the corrosion behaviour of titanium and the Ti−6Al−4V alloy using electrochemical impedance spectroscopy, Bio-Med. Mater. Eng. 14 (2004) 323.
[45] R. Narayanan, P. Mukherjee, S.K. Seshadri, Corrosion properties of anodic oxide coatings on Ti−6Al−4V in simulated body solution, Trans. Inst. Metal. Finish. 84 (2006) 134.
[46] R. Karpagavalli, A. Zhou, P. Chellamuthu, K. Nguyen, Corrosion behavior and biocompatibility of nanostructured TiO_2 film on Ti6Al4V, J. Biomed. Mater. Res. A 83 (2007) 1087.
[47] G. Carbajal, A. Martinez-Villafane, J.G. Gonzalez-Rodriguez, V.M. Castano, Corrosion-resistant coatings: a nanotechnology approach, Anti-Corros Methods Mater. 48 (2001) 241.
[48] M.L. Zheludkevich, R. Serra, M.F. Montemor, I.M.M. Salvado, M.G.S. Ferreira, Corrosion protective properties of nanostructured sol−gel hybrid coatings to AA2024−T3, Surf. Coat. Technol. 200 (2006) 3084.
[49] K.M.S. Youssef, C.C. Koch, P.S. Fedkiw, Improved corrosion behavior of nanocrystalline zinc produced by pulsecurrent electrodeposition, Corros. Sci. 46 (2004) 51.

CHAPTER

Multiwalled Carbon nanotubes/hydroxyapatite nanoparticles incorporated GTR membranes

10

Qing Cai and Xiaoping Yang
Beijing University of Chemical Technology, Beijing, China

CHAPTER OUTLINE

10.1 Introduction 182
10.2 Periodontal Defects and GTR 183
10.2.1 Studies Using Nonresorbable Membranes 183
10.2.2 Studies Using Bioresorbable Membranes 184
10.2.3 Layer-Designed Membranes for GTR 185
10.2.4 Cell-Sheet-Based Technology for GTR 186
10.3 Use of Electrospinning for Preparation of Nanocomposites 187
10.3.1 Electrospinning 187
10.3.2 Carbon Nanotubes Incorporated Into Nanofibers 188
10.3.3 Organic–Inorganic Composite Nanofibers 190
10.4 GTR Membranes Based on Electrospun CNT/HA Nanoparticles Incorporated Composite Nanofibers 192
10.4.1 Fabrication of MWCNTs/HA Hybrids 192
10.4.2 Electrospun Nanofibers With Different Fiber Arrangements 194
10.4.3 Fabrication of PLLA/MWCNTs/HA Composite Nanofibers 195
10.4.4 Characterization of PLLA/MWCNTs/HA Composite Nanofibers 195
10.4.5 Cell Culture on PLLA/MWCNTs/HA Composite Nanofibers Membranes 197
10.4.6 In Vivo Implantation of PLLA/MWCNTs/HA Membranes 199
10.5 Conclusions 200
References 200

Emerging Nanotechnologies in Dentistry. DOI: http://dx.doi.org/10.1016/B978-0-12-812291-4.00010-8

10.1 INTRODUCTION

The periodontium is a topographically complex organ consisting of epithelial tissue, soft and mineralized connective tissue. The structures comprise the gingiva, periodontal ligament (PDL), cementum and alveolar bone. Several diseases affect the composition and integrity of periodontal structures, which gradually causes the destruction of the connective tissue matrix and cells, the loss of fibrous attachment and the resorption of alveolar bone. Bone defects associated with periodontal diseases may be regarded as risk sites for periodontal destruction. These changes often lead to tooth loss eventually [1].

The goals of periodontal therapy have been defined in many ways over the years. The key concept is to improve periodontal health, to prevent further attachment loss and thereby to satisfy a patient's aesthetic and functional needs or demands [2]. To achieve this goal, periodontal treatments including rotational and advanced gingival flaps, free gingival or connective tissue grafts, have been applied to reduce probing depths and maintain or improve attachment levels [3,4]. The characteristic of such treatments is that clinical improvements are associated with periodontal healing by repair. That is, the methods may present an inherent disadvantage of tissue loss as well as the formation of a long epithelial attachment. The traditional techniques for managing periodontal and endodontic defects are effective but often require the removal of bone, necessitating healing by repair. Despite this limitation, the conventional treatments can be successful in achieving the goals of periodontal therapy.

Over the last two decades in particular, there has been a tremendous interest in attempting to regenerate the periodontal tissues and thereby reversing some of the damage to the periodontium caused by the disease process [5,6]. The rationale for adopting such an approach in practice would result in greater patient benefits than conventional approaches. To achieve successful periodontal regeneration, issues including the formation of a functional epithelial seal, the insertion of new connective tissue fibers into the root, the reformation of a new acellular cementum on the root surface and the restoration of alveolar bone height are required [7]. However, the obstacle to achieve periodontal regeneration after being treated with conventional modalities is the proliferation of epithelial tissues into the defect at a faster rate than that of mesenchymal tissues. Epithelial cells migrate about 10 times faster than other periodontal cell types [8]. This leads to the formation of a long junctional epithelium, preventing the selective migration, proliferation, and differentiation of cells derived from the PDL. The complete regeneration of lost periodontal tissues, is now close to being achieved as a result of many recent clinical studies, one of the best strategies is identified guided tissue regeneration (GTR). It is based on the principle to guide the proliferation of periodontal tissue components by means of membranes, which has been an area evoking active research interests in the past decade [1,9,10].

10.2 PERIODONTAL DEFECTS AND GTR

Actually, the GTR technique has been developed and used in the periodontal therapy for almost three decades, which was identified efficient in promoting the regeneration of periodontal tissues, as well as the bone tissue around natural teeth and dental implants. Since the concept of GTR was initially introduced by Nyman et al. in the early 1980s [11,12], this technique has been recognized as a predictable and effective method in enhancing periodontal defect healing, and has become a standard procedure for periodontal regeneration therapy [9].

The potential advantage of applying GTR principle in the treatment of periodontal defects is the possibility to achieve periodontal regeneration rather than connective tissue repair onto exposed root surfaces [5]. For this goal, GTR is in fact a barrier technique that membranes are employed in the hope to exclude the ingressive proliferation of extraneous tissue-forming cells into the specific space to be reconstructed. Treatment with GTR involves the placement of a barrier membrane covering the periodontal defect and the denuded root surface, thus allowing PDL and bone cells to selectively repopulate the isolated space. It is believed that exclusion of gingival cells and epithelial cells will lead to success.

The barrier membranes used in GTR process should meet several prerequisites. General requirements are biocompatible, nonimmunogenic, and nontoxic, and the membranes are supposed to have suitable mechanical stability and optimal porosity. Other properties such as tissue integration, cell occlusivity, nutrient transfer, space making ability and ease of use in the clinic are also of interest [13]. Mechanical stability is believed closely relating to membrane barrier function for selective cell growth during the implantation period.

A variety of synthetic and naturally derived GTR barriers have been used to facilitate periodontal tissue regeneration. Clinical data and numerous reports indicated that the use of membranes with different compositions ad structures, including both nonresorbable and resorbable barrier membranes, allowed the regenerative potential cells (such as PDL cells, bone cells, and cementoblasts) to proliferate and migrate into the protected wound area. It has been shown that GTR treatment of periodontal lesions resulted in significantly greater clinical improvements than only surgical debridement [14].

10.2.1 STUDIES USING NONRESORBABLE MEMBRANES

Nonresorbable membranes, including cellulose acetate membrane (Millipore filter), expanded polytetrafluoroethylene (ePTFE) membranes (Gore-Tex), and even titanium mesh (FriosBoneShields), were initially used for GTR treatment [15,16]. In addition, it was demonstrated that periodontal tissues regenerated by means of nonbioresorbable membranes were not more susceptible to periodontitis than either the pristine periodontium. The purpose-designed ePTFE membranes had been proven biocompatibility by numerous animal and human tests, and were predominantly used in reconstruction of periodontal defects. Defects treated with this membrane have healed faster and with greater quality and quantity of regenerative bone than defects without this treatment [13,17]. Experimentally, a porous ePTFE membrane is inserted over exposed tooth surface, and sutured with the soft tissues. As the newly formed tissues fill the space, the nonresorbable ePTFE membrane should be removed by a second surgical procedure. This additional surgical trauma is obviously inconvenient to both the patient and the clinician. More importantly, it may increase the risk of patient infection and other undesirable effects like disruption of the newly regenerated tissues [18–20].

On the other hand, some authors had argued that rat PDL-derived cells (PDLCs) could not proliferate on ePTFE membrane. Their study showed that PDLCs attached and proliferated well on collagen or copolymer of lactic acid and glycolic acid (PLGA) membranes [21].

To avoid the removal of the membrane after healing and to achieve more efficient periodontal regeneration, thus many researchers have assessed the value of biodegradable membranes, which received high interest for clinical trials in recent years [22].

10.2.2 STUDIES USING BIORESORBABLE MEMBRANES

The development and manufacture of bioresorbable GTR devices have been spurred by the concerns as aforementioned, by using natural derivatives and synthetic organic polymer technologies. A variety of biomaterials have been suggested and used, most commonly type I collagen, polylactide (PLA), and PLGA, which mimic the chemical and physical properties of natural extracellular matrix (ECM) [23–28]. Collagen membranes have excellent cell affinity and biocompatibility to regenerate tissues [29,30]. In addition, negatively charged collagen membranes were reported able to affect cell behavior and the distribution of ECM components to allow for fast defect healing without delay [31]. Immobilization of alkaline phosphatase onto collagen membrane was effective in further enhancing the osteogenic potential of the membrane [32]. Nevertheless, there might exhibit disadvantages in use of collagen membranes. The mechanical strength of collage membrane is normally weak and it is therefore a little difficult to manipulate, moreover, it might cause localized chronic inflammatory response and rapid degradation behavior [27,33]. Enzymes, bacterial proteases, and even the immobilized alkaline phosphatase [32,34,35] could accelerate the degradation of collagen membranes

Membranes of PLA and its copolymer with glycolide have provided an alternative choice for GTR technique. PLA-based polymers indicate sufficient mechanical properties and adjustable biodegradation rates. By controlling their chemical compositions, the membranes can maintain for a period long enough to achieve tissue regeneration before membrane disintegration [14,36]. However, the PLA-based membranes are inherently hydrophobic and exhibit poor cell affinity.

In addition, surrounding soft tissue inflammation took place in clinical trials from time to time when this kind of GTR membranes were used, which was due to their acidic degradation products [22,37,38]. Considering these problems, to find an ideal barrier membrane material is still the key to promote GTR technique.

Composite GTR membranes developed from osteoinductive calcium phosphate ceramics and biodegradable polymers have been highlighted in recent years. As early as in 1995, Jansen had mixed hydroxyapatite (HA) grains with a segmented copolymer consisting of poly(ethylene glycol terephthalate) and poly (butylene terephthalate) (polyactive) to prepare biodegradable GTR membranes, and demonstrated their biocompatibility with tissues [39]. Their histological findings in rabbit demonstrated that this kind of HA-polymer composite membrane could maintain its integrity for 10-month period. The polymer appeared to be completely resorbed in about 4–6 months, while the inorganic component appeared to begin to be resorbed after the 10th month [40]. Lee et al. [41] had used molded porous PLA and PLA/β-tricalcium phosphate (β-TCP) composites with platelet-derived growth factor to guide bone regeneration (GBR). Kikuchi et al. [42,43] developed membranes composed of β-TCP and PLA, PLGA or poly (L-lactide-*co*-glycolide-*co*-e-caprolactone) and reported their applications for periodontal tissue regeneration and GBR. More recently, novel GTR membranes, which contained mineralized nano-HA collagen/polyester (PLA or PLGA), have been biomimetically fabricated by precipitation and casting procedures [44–46]. β-TCP or HA/chitosan composite membranes were also investigated as candidates for potential bioresorbable barrier membranes [47,48]. These composite membranes exhibited excellent biocompatibility, biodegradability, and mechanical properties. Besides, these biodegradable membranes can have extra function to deliver antiinfective agents in a sustained manner. GTR/GBR procedures have been shown to fail when the sites became infected. To avoid contamination, drugs such as tetracycline hydroxide and metronidazole were incorporated into the membranes. The release of these antibacterial drugs demonstrated obvious inhibition of related bacteria, but displayed insignificant cytotoxicity to osteoblasts and PDLCs, giving an aid in the regeneration of a new periodontium [49–52].

10.2.3 LAYER-DESIGNED MEMBRANES FOR GTR

The design of GTR membranes attempts to meet the criteria acting as a barrier to deflect the gingival tissue away from the root surface and creates a protected space over the defect. In treating bone defects and alveolar ridge augmentation, the GTR membranes should allow the migration of regenerative bone cells into the defect, in addition to exclude the epithelium and connective tissue growing into the defect. It is hard to balance all these needs including biocompatibility, cell-occlusiveness, space making, tissue integration, osteoconductivity and clinical manageability by a single layer without gradual changes in structure and composition. Functional graded materials provide us one new concept for GTR membrane design with graded component and graded structure. Commercially available

GTR membranes such as Bio-Gide (collagen types I and III; Geistlich Biomaterials, Wolhusen, Switzerland) and Guidor (PLA; Guidor AB, Huddinge, Sweden) employ a two-layer design [53–56]. Hong et al. [57] prepared an asymmetric gradationally changed porous chitosan membrane with similar graded structures, including a dense skin layer, a transition region and a sponge-like porous layer, by using an immersion–precipitation phase inversion technique. In other literatures, asymmetric chitosan membranes were developed for GTR by importing small pores on the top layer via liquid-liquid phase separation and interconnected cellular pores ($\sim 100\ \mu m$) on the bottom layer via granular porogens [58,59]. Milella et al. [60] prepared a bilamellar membrane constituted of an asymmetric poly (L-lactide) (PLLA) membrane with an alginate film, in which, the alginate film contained β-TCP was in contact with the bone defect to accelerate osteogenesis. The PLLA membrane functioned to both support the alginate film and separate the soft tissue and was less rougher than the alginate film. Park et al. [61] had synthesized a kind of PLGA grafted hyaluronic acid copolymer and used it with PLGA to fabricate by-layered films with a nonporous PLGA top layer and a porous hyaluronic acid-riched bottom layer. Reis et al. [62] developed a kind of bilayer PLGA/calcium phosphate construct to promote periodontal regeneration by connecting a continuous outer PLGA barrier membrane on the inner PLGA/calcium phosphate foam. And Liao et al. [44] proposed a kind of three-layered membrane for GTR. They developed a composite membrane with one face of 8% nanocarbonated HA/collagen/PLGA (nCHAC/PLGA) porous membrane, the opposite face of pure PLGA nonporous membrane, the middle layer of 4% nCHAC/PLGA as the transition through layer-by-layer casting method. In these approaches, it can be seen the porous face of the membrane allows cell growth and sufficient nutrient permeation, while the dense face of the membrane is essential for preventing apical migration of gingival epithelial cells (GECs).

10.2.4 CELL-SHEET-BASED TECHNOLOGY FOR GTR

Besides GTR therapy using nonresorbable or resorbable membranes, PDL cell sheet had been fabricated and replanted into artificially created periodontal defects in animals (immuno-compromised or in autologous manner) to induce periodontal healing [63–66]. The results displayed cell sheet transplantation was able to regenerate both new bone and cementum connecting with well-oriented collagen, as well as PDL. Cell sheet technology is a concept of tissue engineering that provides a mass of cells to the site of regeneration by the transplantation of an in vitro cultured multilayered cell sheet. It enables transplanted cells to be engrafted for a long time, while fully maintaining their viability [67]. This technology has already been applied to various diseases in the clinical setting, including the cornea, esophagus, heart, PDL, and cartilage using autologous cells. Transplanted cell sheets not only replace the injured tissue and compensate for impaired function, but also lead to promotion of tissue repair via delivering bioactive components in situ. For periodontal regeneration, PDL cell sheet transplantation is believed useful in clinical settings.

10.3 USE OF ELECTROSPINNING FOR PREPARATION OF NANOCOMPOSITES

Until so far, most GTR membranes were made in the shape of porous foam with pore size in the range of micron, which were created by traditional methods such as particulate leaching, solvent casting or gas foaming. While cells are inherently sensitive to local mesoscale, microscale, and nanoscale patterns of chemistry and topography [68]. In these days, cell attachment and proliferation are found being significantly enhanced on nanostructure surface than on micronstructure surface [69]. Especially, synthetic nanofibers were deemed exhibiting certain, properties comparable to the natural collagen fibers, and thus nanofibrous architecture may be superior versus solid-walled architecture for promoting cell behaviors [70–75]. With this regard, several purpose-designed nanofibers had been fabricated by electrospinning and examined for GTR application. Kim et al. [76] evaluated the electrospun silk fibrin nanofiber membrane for clinical applications as a device in bone and periodontal regenerative therapy using a rabbit calvarial model. Kim et al. [77] reported the preparation of biomimetic gelatin-HA nanofibers. Fujihara et al. [78] and Yang et al. [79] had electrospun calcium carbonate or apatite embedded polycaprolactone (PCL) nanofibers. These nanofibers can ultimately be equipped into layered 3D biomedical membrane with desired composition and porous structure gradient for GTR by layer-by-layer electrospinning, to guide periodontal regeneration more effectively [77,80].

10.3.1 ELECTROSPINNING

Electrospinning has been recognized as an efficient technique for the fabrication of polymer nanofibers. It is evident from the vast literatures that electrospinning of any biodegradable and biocompatible polymer (synthetic and natural) is no more a problem [81,82]. Electrospinning relies on the application of an electrostatic force to drive fiber formation. There are three components to fulfill the electrospinning process: a high voltage supplier, a capillary tube with a pipette or needle of small diameter, and a metal collector. A high electric field is applied to the droplet of a fluid (polymer solution or melt) coming out from the tip of a needle, which acts as one of the electrodes. The other electrode is attached to the collector. This induces a charge on the surface of the liquid. With the intensity of the electric field increasing, the fluid is ejected from the tip of the Taylor cone, when the electric field strength has exceeded the surface tension of the liquid. The discharged jet is driven electrostatically, elongated, and polymer fibers formed as the solvent evaporating or solidification, which are deposited on the metal collector [83]. Typical electrospinning processes create very long fibers having diameters in the order of a few micrometers down to the tens of nanometers with large surface areas, superior mechanical properties, and ease of fictionalization for various purposes. Biomedical field is one of the important application areas among others like filtration, protective material and nanofiber reinforced composites, etc., because electrospinning can generate loosely connected 3D porous matrixes to physically resemble the nanofibrous features of the natural ECM structure [84–87].

While electrospinning has been proven to be a relatively simple and versatile method for forming fibrous mats, a number of parameters like molecular weight and architecture of polymer, polymer solution properties (viscosity, conductivity, surface tension, etc.), and processing variables (electric field, flow rate, capillary—collector distance, etc.), as well as environmental temperature and humidity, all greatly influence the electrospinning process and the properties of the generated fibers [88].

In addition to pure polymer or polymer mixtures, the electrospinning technique also has the capacity to tailor the electrospun nanofiber network with a variety of types of additives. Ceramic nanoparticles (HA and β-TCP) and carbon nanotubes (CNTs) were both encapsulated into polymer nanofibers by being dispersed in polymer solution and electrospun [89,90]. Control of the process, fibers with various surface morphology, cross-sectional shapes, beads, branches and buckling coils or zigzags can also be produced [91].

Thus, by selecting a combination of proper components and by adjusting the component ratio, properties of electrospun fiber network can be tailored with desired new functions. The adjustment of several electrospinning parameters allows for further control and refinement of fibrous matrix characteristics.

10.3.2 CARBON NANOTUBES INCORPORATED INTO NANOFIBERS

CNTs are a macromolecular form of well-ordered, high aspect ratio allotropes of carbon. Both single-walled carbon nanotubes (SWCNTs) and multiwalled carbon nanotubes (MWCNTs) are constructed of graphite sheet, possessing excellent thermal and chemical stability, high aspect ratio, high tensile strengths, electrical and magnetic properties [92]. With these remarkable features, there has been growing interest in using CNTs for biomedical applications [93–95]. CNTs can be used as nanofillers in polymeric materials to both dramatically improve mechanical properties and create highly anisotropic nanocomposites [96]. The idea of dispersing or aligning CNTs in a nanofiber matrix to form composite materials seems to be very promising from a mechanical and electrical perspective. One thing is noted that the morphology of resulting composite nanofibers is closely related to the dispersion of CNTs in the fibers. At low loading levels, the CNTs can be normally well embedded in the polymeric matrix and oriented along the fiber axis, forming bead-free nanofibrous structure (Fig.10.1). But at high loading levels, the CNTs are liable to aggregate as entangled bundles and as a result, the obtained fibers likely

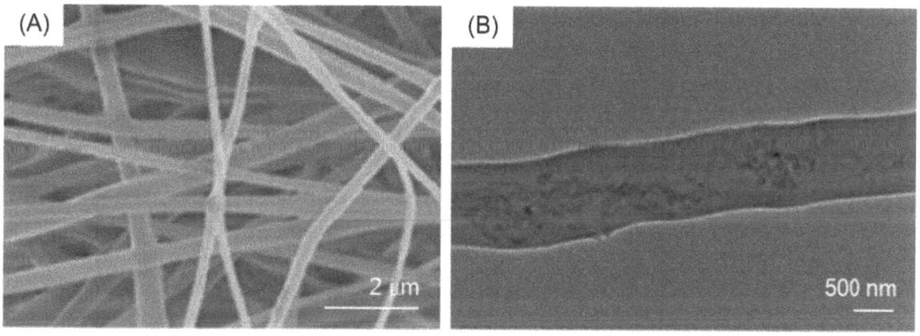

FIGURE 10.1

SEM (A) and TEM (B) observations of electrospun PLLA/MWCNT composite nanofibers.

present misshaped morphology [97]. Surface modification of CNTs before application is effective to ameliorate this problem [98].

Sen et al. [99] electrospun CNTs reinforced polystyrene and polyurethane nanofibers with diameters in the range of 50–100 nm. The incorporation of ester functionalized SWCNTs into polymer matrix increased the tensile strength more effectively than as-prepared SWCNTs being incorporated. In addition, some literatures investigated the reinforcement of electrospun CNT-polyacrylonitrile (PAN) composite nanofibers, the results demonstrated the composite nanofiber sheets possessed enhanced electrical conductivity, mechanical properties, thermal deformation temperature, thermal stability, and dimensional stability [100,101]. Ji et al. [102] had obtained continuous macroscopic aligned PAN composite nanofiber sheets embedded with PAN-grafted MWCNTs by electrospinning followed by hot stretching. The MWCNTs dispersed homogeneous and aligned highly along the fiber, which led to a significant enhancement in the mechanical performance of the resulting composite nanofiber sheets.

CNTs can also be electrospun with biopolymers to form aligned matrices [103]. SWCNTs had been embedded into electrospun poly (ethylene oxide) by Salalha et al.; the CNTs within the nanofibers were in a straight and aligned form [104]. Ko et al. [105] reported the preparation of CNT reinforced spider silk fiber by electrospinning process. Initial tensile testing of the aligned silk composites showed a 10-fold increase in modulus, 5-fold increase in strength, and 3-fold increase in toughness with only 1 wt% of SWCNT in a silk matrix. Kang et al. [106] also prepared the MWCNTs incorporated silk fibroin nanofibers by electrospinning, and identified that the MWCNTs were well incorporated along the nanofibers and the tensile properties of the nanofibers being enhanced by the incorporation of a small amount of MWCNTs.

CNT/PCL nanofibers were fabricated by electrospinning PCL-grafted CNT nanocomposite, and the graft of PCL onto CNTs was conducted by in situ polymerization [107]. The nanocomposite CNT/PCL nanofibers showed a relatively broader diameter than the pure PCL nanofibers. The MWCNTs were embedded within the nanofibers and were well oriented along the axis of the electrospun nanofibers, as confirmed by transmission electron microscopy. In another research, the uniform and highly smooth nanofibers of biodegradable PCL composited with different MWCNTs contents were electrospun [108] Compared with pure PCL nanofibers, on the one hand, the tensile strength of PCL/CNT nanofibers increased, and on the other, PCL/CNT nanofibers showed accelerating degradation behavior. The cytotoxicity results revealed that the electrospun PCL/CNT nanofibrous membranes possessed good in vitro biocompatibility, and the hemolysis and kinetic clotting tests indicated that the PCL nanofibrous membranes did not enhance blood coagulation after the addition of MWCNTs. More recently, electrically conductive PLA/CNT nanofibers were prepared and characterized in terms of morphology, degradation and osteoblast function to study the synergistic effect of topographic cues and electrical stimulation on osteoblasts outgrowth. The results showed that the cellular elongation and proliferation were dependent on the electrical stimulation [109].

All these data demonstrated that the electrospinning of CNT-embedded biodegradable polymer composite nanofibrous network could potentially supply useful options for the fabrication of biomaterial scaffolds, such as wound dressings and GTR membranes. However, there are contradictory reports in regards to the biocompatible character of CNTs [110,111]. On the one hand, some studies reported that CNT are cytotoxic [112–114], which decreased the viability of certain cells significantly, such as A549, mesothelial cells, keratinocyte, glial and HEK293 cell survival [113,115–118]. On the other, chemically functionalized CNTs have been used successfully as substrates for neuronal [119–121], and mesenchymal stem cell growth [122], especially for osteoblasts [123–125].

Overall, for biomedical applications, the toxicity and biocompatibility of CNTs need to be thoroughly investigated. Detailed understanding of the balanced evaluation of risk/benefit ratios is crucial and required before CNTs being incorporated into biomedical devices and applied in vivo.

10.3.3 ORGANIC–INORGANIC COMPOSITE NANOFIBERS

Naturally, bones are generally strengthened by the nucleation of HA into nano-sized gaps between collagen molecules. In order to mimic chemical compositions and fibrous structure of natural bone ECM, the concept of combining a biodegradable polymer and calcium phosphates (HA and β-TCP), most preferably in a nanoparticulate form, has already resulted in numerous investigations [126–129]. The feasibility of incorporation of inorganic nanoparticles into electrospun fibers has made it more attractive for the preparation of inorganic–organic composite nanofibers. Many researchers have investigated a number of characteristics of electrospun polymer/HA or β-TCP composites nanofibers, using biodegradable polymers as PLA, PCL, PLGA, polyhydroxyalkanoate, collagen and gelatin, etc. [77,130–138].

In the conventional method, inorganic powders were directly dispersed in polymeric solutions via ultrasonic. The agglomeration between the nanoparticles tended to take place irreversibly due to their high surface energy; thus, it was a challenge to get a stable suspension. Taking advantage of this disadvantage, Paik et al. [139] developed a method to prepare PCL/TCP fibrous matrix with a gradient mineral content via electrospinning. They centrifuged the PCL solution containing TCP powders to induce gradient content in TCP, and then electrospun the solution to produce TCP-gradient fibrous matrix with high content of TCP at the bottom and nearly zero content of TCP at the top of the matrix. This material could be potentially used in bone-to-tendon regeneration or as GTR/GBR membrane.

To produce electrospun inorganic–organic composite fibers with homogeneous structure, homogeneous dispersion of ceramic nanoparticles in polymer solutions is crucial. Several methods have been proposed to effectively disperse HA or β-TCP nanoparticles into electrospun polymeric fibers. Kim et al. [133] dispersed HA powder in a surfactant hydroxysteric acid mediated chloroform solution, followed by dissolving PLA into the solution. Continuous and uniform fibers were electrospun successfully with diameters of 1–2 μm, and featured a well-developed nanocomposite structure of HA nanopowder-dispersed PLA fiber. The amphiphilic surfactant existed between hydrophilic HA powder and hydrophobic PLA helpful to obtain a homogeneous mixture. Different from most literatures, in which dry ceramic powders are used to prepare composite fibers, Teng et al. [140] explored to mix gel-like calcium phosphate precipitates directly with collagen/hexafluoropropane (HFP) solution. The wet HA precipitates made a good blending with collagen/HFP solution, which could stand up for several days. The uniform composite solution with 30 wt% HA had been successfully obtained by this modified method.

Grafting PLA onto HA nanoparticles (HA–PLA) can also help the inorganic powders dispersing easily in PLA or PLGA matrix [141,142]. The PLA grafting could be readily conducted by dehydration the mixture of HA and PLA oligomers, or by lactide polymerization initiated by hydroxyls on HA surface. To control the length of the grafted PLA oligomers more precisely, in our lab, brominated HA was prepared to initiate the atom transfer radiation polymerization of PLA oligomers with methylacrylate end group. The obtained HA–PLA demonstrated narrow particle size distribution and stable dispersion in chloroform, and resulted in quite uniform dispersion in electrospun PLA nanofibers without hint of HA aggregations even at 30 wt% addition (Fig. 10.2) [143]. Thus, the electrospun composite fibers showed improved tensile strength and tensile modulus due to the uniform distribution of PLA-*g*-HAP in the composite fibers and the relative good interaction and adhesion between the fillers and PLA matrix. These calcium phosphate nanoparticle loaded polymeric fibers showed good biomineralization behavior by being immersed in simulated body fluid (SBF) and have potential use as bone scaffolds and GTR/GBR applications.

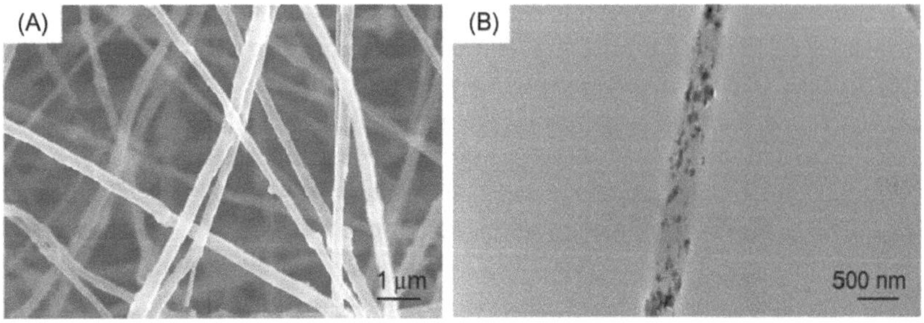

FIGURE 10.2

TEM images of electrospun PLLA/HA composite nanofibers containing 30 wt% of PLA grafted HA.

10.4 GTR MEMBRANES BASED ON ELECTROSPUN CNT/HA NANOPARTICLES INCORPORATED COMPOSITE NANOFIBERS

New types of nanofibrous GTR membrane were developed in our lab, and were described in details in the following section. Electrospinning had been applied to fabricate this novel composite fibrous network consisting of PLLA, MWCNTs, and HA (PLLA/MWCNTs/HA). Briefly, the MWCNTs/HA hybrids were first prepared by depositing HA particles onto CNTs [144–146], followed by being dispersed into PLLA solution and electrospun. The fiber arrangement was regulated by using aluminum foil, rotating metal roller or metal net as the receiver. Cell culture revealed that bone-related cells as MG63, MC3T3-E1 and BMSCs proliferated and osteogenic differentiated well on the substrates, but showing dependence on fiber arrangement and composition [74,75,147,148]. The PLLA/MWCNT/HA composite nanofibrous membrane showed a selective enhancement to different cell lines. It was found that PDLCs functioned well on the membrane both in vitro and in vivo, while the growth of GECs was prohibited on the membrane. This new type of membrane showed excellent dual biological functions and satisfied the requirement of the GTR technique.

10.4.1 FABRICATION OF MWCNTS/HA HYBRIDS

To incorporate both CNTs and HA into electrospun polymeric nanofibers, it is not suggested to mix CNTs and HA nanoparticles into polymeric solution at the same time. Due to the large surface specific surface areas of both CNTs and nanoscaled HA, poor dispersion of them in the polymeric matrix can be envisioned easily. Therefore, MWCNTs/HA hybrids were prepared before the electrospinning by depositing HA onto MWCNTs via coprecipitation or biomineralization in SBF.

In the coprecipitation method, MWCNTs were firstly modified by anodic oxidation (sulfonic acid/nitric acid, v/v = 3/1, 80°C), and then mixed into the solution containing $Ca(NO_3)_2 \cdot 4H_2O$ and $(NH_4)_2HPO_4$ (Ca/P = 1.67). Under continuous ultrasonication, HA precipitated from the solution and embedded MWCNTs inside to get the MWCNTs/HA complexes, in which, the content of MWCNTs being ~3 wt%. As shown in Fig. 10.3, the needle-like HA nanoparticles were about 40–70 nm in length and 15 nm in diameter, and well combined on the surfaces of anodic oxidized MWCNTs [144].

In the biomineralization method, functionalization of MWCNTs were carried out to introduce different chemical groups including phosphate, carbonxyl, hydroxyl, and amine groups to induce apatite nucleation and growth on MWCNT surface [145,146]. By soaking the functionalized MWCNTs in SBF, amorphous calcium phosphate was initially observed on the surface of MWCNT (Fig. 10.4). Gradually, the deposited mineralites transformed into HA with longer soaking time, but with average Ca/P molar ratios slightly below 1.67 due to the fast nucleation and other ions doping from supersaturated SBF [149]. It was found the amine group and the phosphate group showed the fasted nucleation and crystal growth rate, and the addition of amino acid into the SBF could regulate the biomineralization behavior [150]. Particularly, the resulting HA displayed crystalline structure biomimetic to natural bone minerals if carbon dioxide gas bubbling was used to buffer the SBF instead of the conventional Tris–HCl [151].

After the preparation, the obtained MWCNTs/HA powders were washed with 1,4-dioxane (or trifluoroethanol) repeatedly to remove the water and dispersed again in 1,4-dioxane (or trifluoroethanol) to form a suspension for further use.

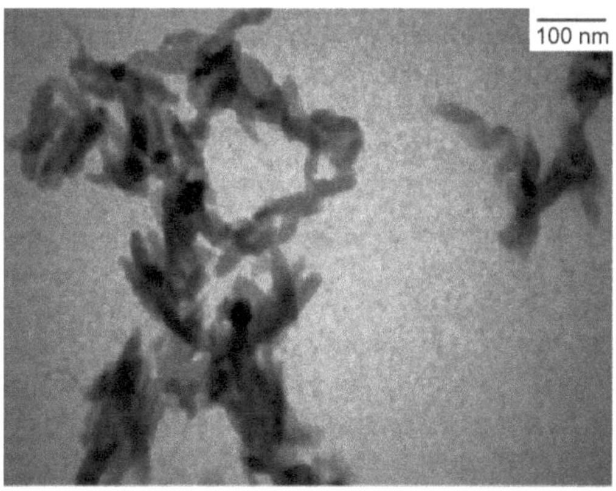

FIGURE 10.3

TEM observation of coprecipitated MWCNT/HA complexes.

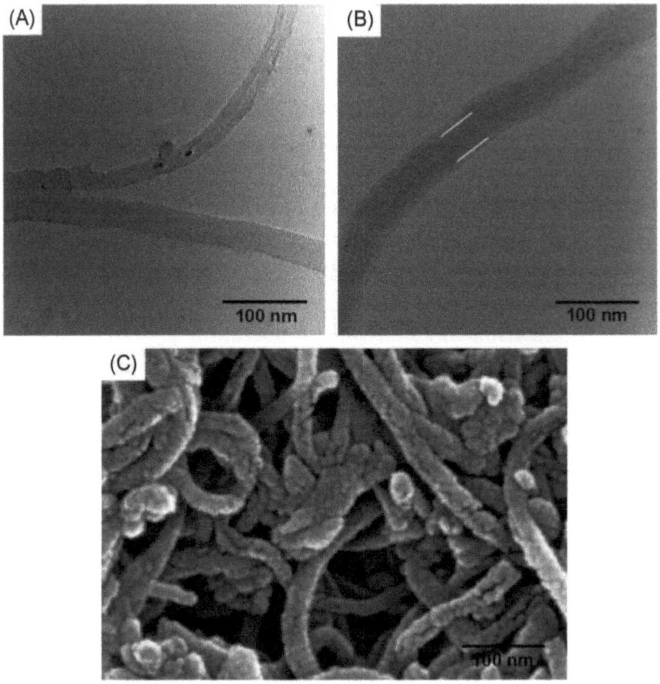

FIGURE 10.4

SEM and TEM images of MWCNTs with surface apatite deposition (B and C) prepared by soaking phosphorylated MWCNTs (A) in SBF for 12 h.

10.4.2 ELECTROSPUN NANOFIBERS WITH DIFFERENT FIBER ARRANGEMENTS

Topographical feature of substrate is a crucial factor to influence cell biological behaviors due to the contact guide effect. In electrospinning, the fiber arrangement can be regulated by selecting different receivers (Fig. 10.5). Nonwoven nanofibrous membranes are prepared when a flat metallic plate is used to collect electrospun nanofibers. While parallel-aligned nanofibers are resulted as a rotating roller is applied. Interestingly, lattice-like nanofibrous sheets are obtained if a kind of metallic network is used as the receiver. These nanofibrous membranes displayed different responses to evaluations on mechanical strength, volume shrinkage, and structural stability with hydrolysis [152]. The lattice-like nanofibrous membranes, which consisted of both paralleled ridge area and nonwoven area, demonstrated the smallest volume shrinkage in contacting with aqueous solution, and the best structure-remaining ability during degradation. It provided a better choice than those nonwoven nanofibers for constructing GTR membranes.

In the studies of their biological properties by coculturing bone-related cells on the nanofibrous membranes, it was found the cell proliferation and osteogenic

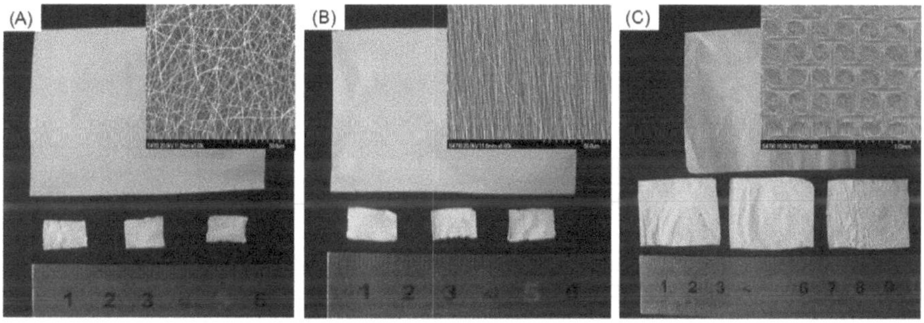

FIGURE 10.5

Electrospun PLLA nanofibrous membranes with nonwoven (A), parallel-aligned (B) and lattice-like (C) fiber arrangements, and their volume shrinkage after being soaked in PBS (37°C) for 24 h.

differentiation were enhanced the most on lattice-like membranes, followed by nonwoven and parallel-aligned membranes [74,75,147]. The explanation came to the effect of nanotopographical features of nanofibrous membranes on cell attachment and spreading, and thus affecting their proliferation and differentiation through the Ras homolog gene family member A (RhoA) and ERK signaling pathways. A combined nanotopographical structure such as patterned nanofibrous meshes provided a more instructive microenvironment for bone-related cells to differentiate into osteolineage than those simple fiber arrangements.

10.4.3 FABRICATION OF PLLA/MWCNTs/HA COMPOSITE NANOFIBERS

Into the aforementioned MWCNT/HA dispersion in 1,4-dioxane (or trifluoroethanol), PLLA was added until a weight ratio of MWCNT/HA to PLLA of 1:9 was achieved. The suspension was stirred continuously overnight thermoset at 30–50°C (if necessary) to get a well-mixed viscous solution. The solution was then electrospun at following parameters: voltage = 10–15 kV, flow rate = 0.8–1.0 mL/h, receiving distance = 100–150 mm. The resulted fibrous membranes were placed in a vacuum oven at 50°C over 2 days to remove the residual solvent. Before the electrospinning, ultrasonication might be required to ensure the fine dispersion MWCNTs/HA complexes in PLLA solution. PLLA and PLLA/HA nanofibers were electrospun in a similar way as references.

10.4.4 CHARACTERIZATION OF PLLA/MWCNTS/HA COMPOSITE NANOFIBERS

The diameters of electrospun PLLA/MWCNTs/HA fibers were about several hundred nanometers to ~1 μm, and the incorporation of MWCNTs/HA particles resulted in fibers slightly irregular in diameter and showing some beaded morphology along the fibers. EDX analysis confirmed the presence of P and Ca elements distributing uniformly in the fibrous membrane, while the presence of MWCNTs was identified by Raman spectrum (Fig. 10.6).

The results of the in vitro degradation tests (PBS, pH 7.4, 37°C) demonstrated the incorporation of HA or MWCNTs/HA slowed down the rate of mass loss, especially displaying a lag phase in the early stage (Fig. 10.7). The presence of alkaline HA component could neutralize the acidic degradation products of hydrolyzed PLLA, and thus the acid autocatalytical degradation of PLLA might be eliminated. It was indicated that the incorporation of HA or MWCNTs/HA had prevented PLLA nanofibrous membranes from rapid decomposition by avoiding the rapid decrease of pH in the surrounding environment along with the hydrolysis.

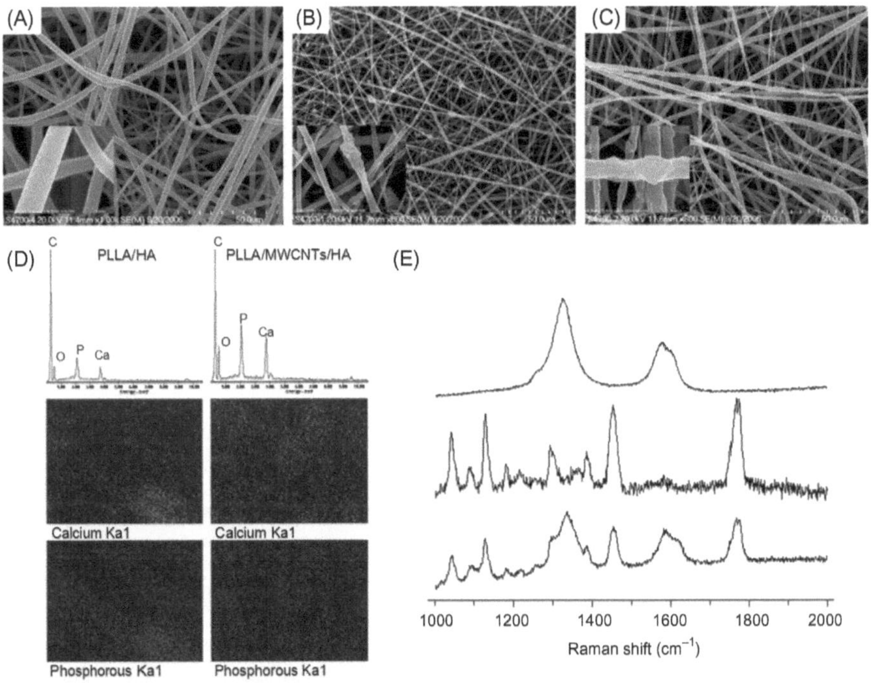

FIGURE 10.6

Characteristics of three kinds of membranes: (A–C) represent typical scanning electron microscope (SEM) images of PLLA, PLLA/HA, and PLLA/MWCNTs/HA membranes, respectively; (D) represents the EDX mapping of PLLA/HA (left) and PLLA/MWCNTs/HA (right) membranes for Ca (middle) and P (lower) elements; (E) represents Raman spectra of MWCNTs (upper), PLLA/HA membrane (middle), and PLLA/MWCNTs/HA membranes (lower).

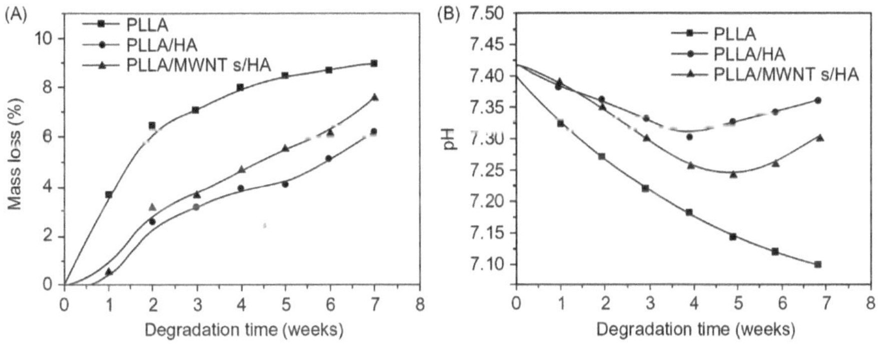

FIGURE 10.7

Characteristics of three kinds of membranes during in vitro degradation: (A) and (B) represent the changes of mass loss and pH, respectively.

10.4.5 CELL CULTURE ON PLLA/MWCNTs/HA COMPOSITE NANOFIBERS MEMBRANES

PDLCs were obtained from healthy teeth extracted for orthodontic reasons. GECs were obtained from the gingival tissue of systemic healthy individuals removed during periodontal surgery. Both the cells between the third and the fifth passages were used in the following studies. PDLCs and GECs were harvested with 0.25% trypsin/0.02% EDTA, and PDLCs were transferred to an osteogenic differentiation medium, α-MEM containing 10% fetal bovine serum and antibiotics supplemented with 10 nM dexamethasone, 10 mM β-glycero-phosphate, and 50 mg/L of ascorbic acid. PDLCs or GECs were seeded onto PLLA, PLLA/HA and PLLA/MWCNTs/HA nanofibrous membranes, as well as tissue culture polystyrene (control group) at a density of 5000 cells/well (24-well culture plate).

The density of PDLCs on PLLA/MWCNTs/HA membranes was the highest among those of the three membranes. The cells spread over the membrane fibers, linked with fibers by cytoplasmic extensions. PDLCs extended more actively on the PLLA/MWCNTs/HA membrane than on the PLLA and PLLA/HA membranes within the same culture time. More granulates, which was an implication of mineralization, were observed on the surface of attached cells on PLLA/MWCNTs/HA and PLLA/HA membranes than on PLLA membrane (Fig. 10.8).

By methyl thiazolyl tetrazolium (MTT) assay (Fig. 10.9), the number of PDLCs was similar at 1 day for the three tests and control group, while the most active proliferation was observed on the PLLA/MWCNTs/HA composite membrane. It was almost three times that of the initial seeding cells and 30% larger than that of the PLLA/HA membrane or control group after 7 days of culturing ($P<0.05$). However, the PLLA/MWCNTs/HA membrane prohibited the adhesion and proliferation of GECs compared with the control group.

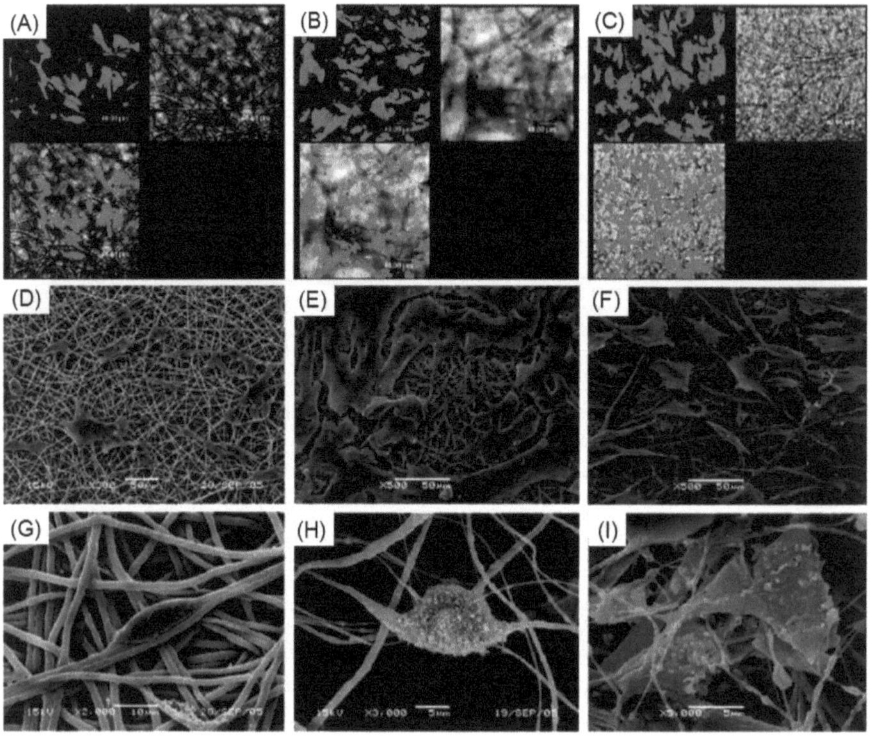

FIGURE 10.8

Morphology observation of PDLCs cultured on three kinds of membranes: (A–C) represent confocal laser microscope images of PDLCs on PLLA, PLLA/HA, and PLLA/MWCNTs/HA membranes, respectively: (D–F) represent SEM images of PDLCs on PLLA, PLLA/HA, and PLLA/MWCNTs/HA membranes, respectively, at low magnification, and (G–I) represent them at high magnification.

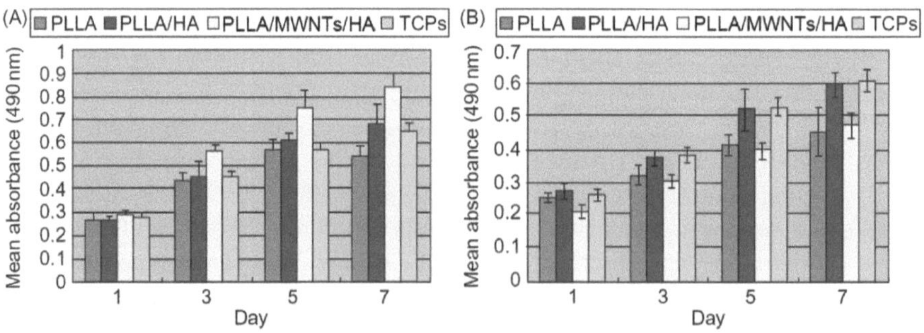

FIGURE 10.9

Effect of three kinds of membranes on the adhesion and proliferation of PDLCs (A) and GECs (B) determined by MTT assay. Standard deviations are shown as bars.

The results suggested that the PLLA/MWCNTs/HA membrane would selectively increase the adhesion of osteoblast cells and decrease the adhesion of osteoblast competitive cell lines, which was a valuable feature for GTR application.

10.4.6 IN VIVO IMPLANTATION OF PLLA/MWCNTs/HA MEMBRANES

PDLCs were cultured in an osteogenic differentiation medium for 7 days and then seeded onto PLLA/MWCNTs/HA membranes and cultured in the osteogenic differentiation medium for another 48 h. Under general anesthesia, cell/membrane complexes were implanted into muscle pouches on one side leg of 10 immuno-deficient mice. Four weeks postsurgery, all mice were sacrificed. The block sections were rinsed in sterile saline and fixed in 10% buffered formalin and prepared for paraffin sections (of 5 μm) to conduct histological examinations.

It was observed that residual electrospun fibers could be clearly identified; no obvious inflammation was found in the implant areas. As shown in Fig. 10.10, bone-like tissues were formed with a round or irregular shape (areas pointed by

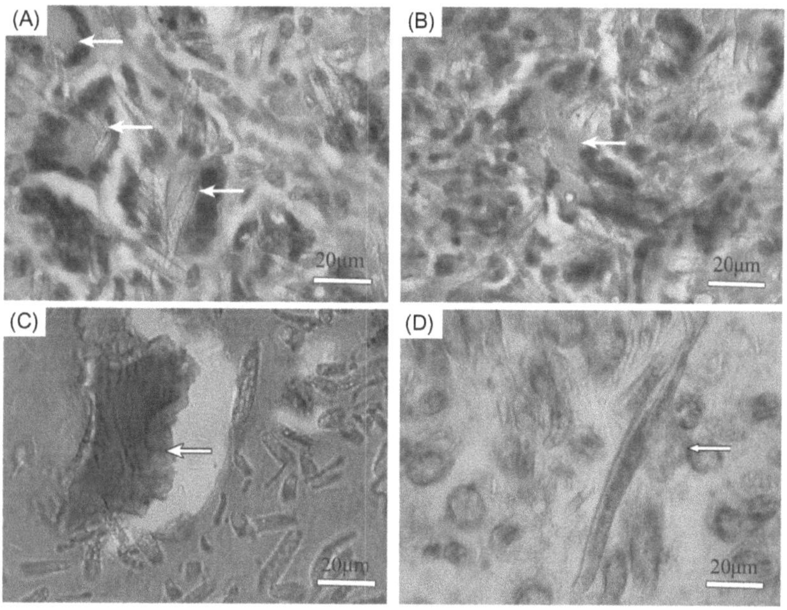

FIGURE 10.10

Histological examination of cell/membrane composites implanted into immuno-deficient mice: (A–C) show new-formed bone-like tissues in round or irregular shape (white arrow), and osteoblast-like cells were well arranged around bone-like tissues. Abundant blood vessels were found in the implanted area. In (C), alizarin red staining confirmed calcium deposits (white arrow) in newly formed bone-like tissues. In (D), osteocalcin, which was stained in brown, was detected in the cytoplasm and outside the cells.

white arrow) and were stained into homogeneous pink by hematoxylin/eosin, and osteoblast-like cells were well arranged around the bone-like tissues. Calcium deposits were confirmed in new-formed bone-like tissue by alizarin red staining. It was notable that abundant blood vessels were grown into the newly formed tissues. Osteocalcin, which was stained in brown, was detected in the cytoplasm and outside the cells (pointed by white arrow). The results indicated that PLLA/MWCNTs/HA membranes were of good biocompatibility in vivo during the 4-week period of implantation, and human PDLCs seemed to function well on the membrane.

10.5 CONCLUSIONS

With respect to PLLA/MWCNTs/HA membranes, PLLA is absorbable, while HA is a major inorganic component of bone and has shown good osteoconductivity and bone-binding ability. CNTs can be used not only to stimulate bone regeneration but also to serve a permanent mechanical role. Therefore, PLLA/MWCNTs/HA membranes need not be taken out after healing, which lightens the burden of patients. Although more extensive work is required to investigate the exact mechanism, the unique biologic properties of the PLLA/MWCNTs/HA membrane have shown great potential for application in GTR. Compared with commercial or in research GTR membranes of similar function, this new type of material fabricated via electrospinning simplifies the manufacturing process, lowers the fabrication cost, and avoids possible mistakes in clinical application. Moreover, electrospinning is a strong tool to fabricate nanofibers of different topographical and compositional features. Therefore, in addition to electrospun PLLA/MWCNTs/HA membrane, more flexible GTR membranes with different designs are readily developed from electrospun nanofibers via layer-by-layer combinations of various functional nanofibers to meet diverse clinic needs.

REFERENCES

[1] B.B. Benatti, K.G. Silvério, M.Z. Casati, E.A. Sallum, F.H. Nociti Jr., Physiological features of periodontal regeneration and approaches for periodontal tissue engineering utilizing periodontal ligament cells, J. Biosci. Bioeng. 103 (2007) 1.

[2] P.M. Bartold, C.A. McCulloch, A.S. Narayanan, S. Pitaru, Tissue engineering: a new paradigm for periodontal regeneration based on molecular and cell biology, Periodontology 24 (2000) 253.

[3] H.E. Grupe, R.F. Warren, Repair of gingival defects by a sliding flap operation, J. Periodontol. 27 (1956) 92.

[4] S.W. Nelson, The subpedicle connective tissue graft. A bilaminar reconstructive procedure for the coverage of denuded root surfaces, J. Periodontol. 58 (1987) 95.

[5] S. Jepsen, J. Eberhard, D. Herrera, I. Needleman, A systematic review of guided tissue regeneration for periodontal furcation defects. What is the effect of guided tissue regeneration compared with surgical debridement in the treatment of furcation defects? J. Clin. Periodontol. 29 (2002) 103.
[6] I. Needleman, R. Tucker, E. Giedrys-Leeper, H. Worthington, A systematic review of guided tissue regeneration for periodontal infrabony defects, J. Periodontal. Res. 37 (2002) 380.
[7] M. Shimono, T. Ishikawa, H. Ishikawa, H. Matsuzaki, S. Hashimoto, T. Muramatsu, et al., Regulatory mechanisms of periodontal regeneration, Microsc. Res. Tech. 60 (2003) 491.
[8] S. Amar, Implications of cellular and molecular biology advances in periodontal regeneration, Anat. Rec. 245 (1996) 361.
[9] M.J. Danesh-Meyer, U.M.E. Wikesjö, Gingival recession defects and guided tissue regeneration: a review, J. Periodontal Res. 36 (2001) 341.
[10] C. Stoecklin-Wasmer, A.W.S. Rutjes, B.R. da Costa, G.E. Salvi, P. Jüni, A. Sculean, Absorbable collagen membranes for periodontal regeneration: a systematic review, J. Dent. Res. 92 (2013) 773.
[11] T. Karring, S. Nyman, J. Lindhe, Healing following implantation of periodontitis affected roots into bone tissue, J. Clin. Periodontal. 7 (1980) 96.
[12] S. Nyman, J. Gottlow, T. Karring, J. Lindhe, The regenerative potential of the periodontal ligament. An experimental study in the monkey, J. Clin. Periodontol. 9 (1982) 257.
[13] R.S. Bhumbra, A.B. Berman, P.S. Walker, D.S. Barrett, G.W. Blunn, Enhanced bone regeneration and formation around implants using guided bone regeneration, J. Biomed. Mater. Res. (Appl. Biomater.) 43 (1998) 162.
[14] A. Stavropoulos, T. Karring, Long-term stability of periodontal conditions achieved following guided tissue regeneration with bioresorbable membranes: case series results after 6–7 years, J. Clin. Periodontol. 31 (2004) 939.
[15] G. Zellin, A. Gritli-Linde, A. Linde, Healing of mandibular defects with different biodegradable and non-biodegradable membranes: an experimental study in rats, Biomaterials 16 (1995) 601.
[16] Y.D. Rakhmatia, Y. Ayukawa, A. Furuhashi, K. Koyano, Current barrier membranes: titanium mesh and other membranes for guided bone regeneration in dental applications, J. Prosthodont. Res. 57 (2013) 3.
[17] A. Piattelli, A. Scarano, M. Paolantonio, Bone formation inside the material interstices of e-PTFE membranes: a light microscopical and histochemical study in man, Biomaterials 17 (1996) 1725.
[18] L. Trombelli, L. Minenna, R. Farina, A. Scabbia, Guided tissue regeneration in human gingival recessions: a 10-year follow-up study, J. Clin. Periodontol. 32 (2005) 16.
[19] G. Polimeni, K.-T. Koo, M. Qahash, A.V. Xiropaidis, J.M. Albandar, U.M.E. Wikesjö, Prognostic factors for alveolar regeneration: effect of a space-providing biomaterial on guided tissue regeneration, J. Clin. Periodontol. 31 (2004) 725.
[20] A.L. Nastri, A.C. Smith, Guided osteogenesis using synthetic membranes: an experimental pilot study, J. Craniomaxillofac. Surg. 24 (1996) 163.
[21] T. Takata, H.-L. Wang, M. Miyauchi, Attachment, proliferation and differentiation of periodontal ligament cells on various guided tissue regeneration membranes, J. Periodontal Res. 36 (2001) 322.

[22] L. Dupoirieux, D. Pourquier, M.C. Picot, M. Neves, Comparative study of three different membranes for guided bone regeneration of rat cranial defects, Int. J. Oral Maxillofac. Surg. 30 (2001) 58.
[23] K. Kubo, N. Tsukasa, K. Iki, M. Uehara, A. Shimotsu, Y. Seto, et al., Occlusive effects of lactic acid–glycolic acid copolymer membrane on gingival fibroblasts in-vitro, J. Biomed. Mater. Res. 39A (1998) 554.
[24] G.Rh Owen, J. Jackson, B. Chehroudi, H. Burt, D.M. Brunette, A PLGA membrane controlling cell behaviour for promoting tissue regeneration, Biomaterials 26 (2005) 7447.
[25] Y. Taguchi, N. Amizuka, M. Nakadate, H. Ohnishi, N. Fujii, K. Oda, et al., A histological evaluation for guided bone regeneration induced by a collagenous membrane, Biomaterials 26 (2005) 6158.
[26] W.-Y. Ip, S. Gogolewski, Clinical application of resorbable polymers in guided bone regeneration, Macromol. Symp. 253 (2007) 139.
[27] E. Milella, P.A. Ramires, E. Brescia, G. La Sala, L. Di Paola, V. Bruno, Physicochemical, mechanical, and biological properties of commercial membranes for GTR, J. Biomed. Mater. Res. (Appl. Biomater.) 58 (2001) 427.
[28] B.S. Kim, I.K. Park, T. Hoshiba, H.L. Jiang, Y.J. Choi, T. Akaikef, et al., Design of artificial extracellular matrices for tissue engineering, Prog. Polym. Sci. 36 (2011) 238.
[29] E.A.A. Neel, L. Bozec, J.C. Knowles, O. Syed, V. Mudera, R. Day, et al., Collagen—emerging collagen based therapies hit the patient, Adv. Drug Delivery Rev. 65 (2013) 429.
[30] C. Stoecklin-Wasmer, A.W.S. Rutjes, B.R. da Costa, G.E. Salvi, P. Jüni, A. Sculean, Absorbable collagen membranes for periodontal regeneration: a systematic review, J. Dent. Res. 92 (2013) 773.
[31] D.M.L.R.F.C. Veríssimo, R.A. Ribeiro, S.D. Figueiró, A.S.B. Sombra, J.C. Góes, G. A.C. Brito, Polyanionic collagen membranes for guided tissue regeneration: effect of progressive glutaraldehyde cross-linking on biocompatibility and degradation, Acta Biomater. 6 (2010) 4011.
[32] D.A.W. Oortgiesen, A.S. Plachokova, C. Geenen, G.J. Meijer, X.F. Walboomers, J.J.J.P. van den Beucken, et al., Alkaline phosphatase immobilization onto Bio-Gide® and Bio-Oss® for periodontal and bone regeneration, J. Clin. Periodontol. 39 (2012) 546.
[33] E. Picfte, P. Alberius, N. Samman, A. Linde, Experience with e-PTFE membrane application to bone grafting of cleft maxilla, Int. J. Oral Maxillofac. Surg. 24 (1995) 327.
[34] A. Kozlovsky, G. Aboodi, O. Moses, H. Tal, Z. Artzi, M. Weinreb, et al., Bio-degradation of a resorbable collagen membrane (Bio-Gide®) applied in a double-layer technique in rats, Clin. Oral Implants Res 20 (2009) 1116–1123.
[35] M.N. Sela, E. Babitski, D. Steinberg, D. Kohavi, G. Rosen, Degradation of collagen-guided tissue regeneration membranes by proteolytic enzymes of *Porphyromonas gingivalis* and its inhibition by antibacterial agents, Clin. Oral Implants Res. 20 (2009) 496–502.
[36] T. Nieminen, I. Kallela, J. Keränen, I. Hiidenheimo, H. Kainulainen, E. Wuolijoki, et al., in-vivo and in-vitro degradation of a novel bioactive guided tissue regeneration membrane, Int. J. Oral Maxillofac. Surg. 35 (2006) 727.
[37] M. Simion, A. Scarano, L. Gionso, A. Piattelli, Guided bone regeneration using resorbable and nonresorbable membranes: a comparative histologic study in humans, Int. J. Oral Maxillofac. Implants 11 (1996) 735.

[38] R.G. Caffesse, C.E. Nasjleti, E.C. Morrison, R. Sanchez, Guided tissue regeneration: comparison of bioabsorbable and non-bioabsorbable membranes: histologic and histometric study in dogs, J. Periodontol. 65 (1994) 583.
[39] J.A. Jansen, J.E. de Ruijter, P.T.M. Janssen, Y.G.C.J. Paquay, Histological evaluation of a biodegradable polyactive®/hydroxyapatite membrane, Biomaterials 16 (1995) 819.
[40] A. Piattelli, M. France, G. Ferronato, M.T. Santello, R. Martinetlit, A. Scarano, Resorption of composite polymer-hydroxyapatite membranes: a time-course study in rabbit, Biomaterials 18 (1997) 629.
[41] S.J. Lee, Y.J. Park, S.N. Park, Y.M. Lee, Y.J. Seol, Y. Ku, et al., Molded porous poly(L-lactide) membranes for guided bone regeneration with enhanced effects by controlled growth factor release, J. Biomed. Mater. Res. 55A (2001) 295.
[42] M. Kikuchi, Y. Koyama, K. Takakuda, H. Miyairi, N. Shirahama, J. Tanaka, in-vitro change in mechanical strength of β-tricalcium phosphate/copolymerized poly(L-lactide) composites and their application for guided bone regeneration, J. Biomed. Mater. Res. 62A (2002) 265.
[43] M. Kikuchi, Y. Koyama, T. Yamada, Y. Imamura, T. Okada, N. Shirahama, et al., Development of guided bone regeneration membrane composed of β-tricalcium phosphate and poly (L-lactide-*co*-glycolide-*co*-ε-caprolactone) composites, Biomaterials 25 (2004) 5979.
[44] S. Liao, W. Wang, M. Uo, S. Ohkawa, T. Akasaka, K. Tamura, et al., A three-layered nano-carbonated hydroxyapatite/collagen/PLGA composite membrane for guided tissue regeneration, Biomaterials 26 (2005) 7564.
[45] S. Liao, F. Watari, Y. Zhu, M. Uo, T. Akasaka, W. Wang, et al., The degradation of the three layered nano-carbonated hydroxyapatite/collagen/PLGA composite membrane in-vitro, Dent. Mater. 23 (2007) 1120.
[46] S.X. Pan, Y. Li, H.L. Feng, W. Bai, Y.Y. Gu, in-vitro aging of mineralized collagen-based composite as guided tissue regeneration membrane, Mater. Sci. Eng. C 26 (2006) 724.
[47] X. Cheng, Y. Li, Y. Zuo, L. Zhang, J. Li, H. Wang, Properties and in-vitro biological evaluation of nano-hydroxyapatite/chitosan membranes for bone guided regeneration, Mater. Sci. Eng. C 29 (2009) 29.
[48] S.M. Kuo, S.J. Chang, G.C.-C. Niu, C.-W. Lan, W.T. Cheng, C.Z. Yang, Guided tissue regeneration with use of β-TCP/chitosan composite membrane, J. Appl. Polym. Sci. 112 (2009) 3127.
[49] E.R. Kenawy, G.L. Bowlin, K. Mansfield, J. Layman, D.G. Simpson, E.H. Sanders, et al., Release of tetracycline hydrochloride from electrospun poly(ethylene-*co*-vinylacetate), poly(lactic acid), and a blend, J. Controlled Release 81 (2002) 57.
[50] M. Zamani, M. Morshed, J. Varshosaz, M. Jannesari, Controlled release of metronidazole benzoate from polyepsilon-caprolactone electrospun nanofibers for periodontal diseases, Eur. J. Pharm. Biopharm. 75 (2010) 179.
[51] N. Yan, X.H. Zhang, Q. Cai, X.P. Yang, X.G. Zhou, B. Wang, et al., The effects of lactidyl/glycolidyl ratio and molecular weight of poly(lactide-*co*-glycolide) on the tetracycline entrapment and release kinetics of hybrid nanofibers, J. Biomater. Sci. 23 (2012) 1005.
[52] M. Ranjbar-Mohammadi, M. Zamani, M.P. Prabhakaran, S. Hajir Bahrami, S. Ramakrishna, Electrospinning of PLGA/gum tragacanth nanofibers containing tetracycline hydrochloride for periodontal regeneration, Mater. Sci. Eng. C 58 (2016) 521.

[53] G. Hillmann, A. Steinkamp-Zucht, W. Geurtsen, G. Gross, A. Hoffmann, Culture of primary human gingival fibroblasts on biodegradable membranes, Biomaterials 23 (2002) 1461.
[54] B. Alpar, G. Leyhausen, H. Gunay, W. Geurtsen, Compatibility of resorbable and nonresorbable guided tissue regeneration membranes in cultures of primary human periodontal ligament fibroblasts and human osteoblast-like cells, Clin. Oral Investig. 4 (2000) 219.
[55] H.L. Wang, M. Miyauchi, T. Takata, Initial attachment of osteoblasts to various guided bone regeneration membranes: an in-vitro study, J. Periodontal. Res. 37 (2002) 340.
[56] J.M. Payne, C.M. Cobb, J.W. Rapley, W.J. Killoy, P. Spencer, Migration of human gingival fibroblasts over guided tissue regeneration barrier materials, J. Periodontol. 67 (1996) 236.
[57] H. Hong, J. Wei, C. Liu, Development of asymmetric gradational-changed porous chitosan membrane for guided periodontal tissue regeneration, Compos. Part B – Eng. 38 (2007) 311.
[58] M.-H. Ho, C.-C. Hsieh, S.-W. Hsiao, H. Thien, Fabrication of asymmetric chitosan GTR membranes for the treatment of periodontal disease, Carbohydr. Polym. 79 (2010) 955.
[59] S.Q. Ma, Z. Chen, F. Qiao, Y.C. Sun, X.P. Yang, X.L. Deng, et al., Guided bone regeneration with tripolyphosphate cross-linked asymmetric chitosan membrane, J. Dent. 42 (2014) 1603.
[60] E. Milella, G. Barra, P.A. Ramires, G. Leo, P. Aversa, A. Romito, Poly(L-lactide) acid/alginate composite membranes for guided tissue regeneration, J. Biomed. Mater. Res. 57A (2001) 248.
[61] J.K. Park, J. Yeom, E.J. Oh, M. Reddy, J.Y. Kim, D.-W. Cho, et al., Guided bone regeneration by poly(lactic-*co*-glycolic acid) grafted hyaluronic acid bi-layer films for periodontal barrier applications, Acta Biomater. 5 (2009) 3394.
[62] E.C.C. Reis, A.P.B. Borges, M.V.F. Araújo, V.C. Mendes, L. Guan, J.E. Davies, Periodontal regeneration using a bilayered PLGA/calcium phosphate construct, Biomaterials 32 (2011) 9244.
[63] A. Dogan, A. Ozdemir, A. Kubar, T. Oygur, Healing of artificial fenestration defects by seeding of fibroblast-like cells derived from regenerated periodontal ligament in a dog: a preliminary study, Tissue Eng. 9 (2003) 1189.
[64] M. Hasegawa, M. Yamato, A. Kikuchi, T. Okano, I. Ishikawa, Human periodontal ligament cell sheets can regenerate periodontal ligament tissue in an athymic rat model, Tissue Eng. 11 (2005) 469.
[65] T. Iwata, M. Yamato, H. Tsuchioka, R. Takagi, S. Mukobata, K. Washio, et al., Periodontal regeneration with multi-layered periodontal ligament-derived cell sheets in a canine model, Biomaterials 30 (2009) 2716.
[66] Y. Zhou, Y. Li, L. Mao, H. Peng, Periodontal healing by periodontal ligament cell sheets in a teeth replantation model, Arch. Oral Biol. 57 (2012) 169.
[67] K. Matsuura, R. Utoh, K. Nagase, T. Okano, Cell sheet approach for tissue engineering and regenerative medicine, J. Controlled Release 190 (2014) 228.
[68] M.M. Stevens, J.H. George, Exploring and engineering the cell surface interface, Science 310 (2005) 1135.
[69] S. Park, G.-I. Im, Stem cell responses to nanotopography, J. Biomed. Mater. Res. 103A (2015) 1238.

[70] E.K.F. Yim, K.W. Leong, Significance of synthetic nanostructures in dictating cellular response, Nanomed. Nanotechnol. Biol. Med. 1 (2005) 10.
[71] L. Zhang, T.J. Webster, Nanotechnology and nanomaterials: promises for improved tissue regeneration, Nano Today 4 (2009) 66.
[72] J. Hu, X. Liu, P.X. Ma, Induction of osteoblast differentiation phenotype on poly(L-lactic acid) nanofibrous matrix, Biomaterials 29 (2008) 3815.
[73] K.M. Woo, J.-H. Jun, V.J. Chen, J. Seo, J.-H. Baek, H.-M. Ryoo, et al., Nano-fibrous scaffolding promotes osteoblast differentiation and biomineralization, Biomaterials 28 (2007) 335.
[74] J.X. Zhu, Q. Cai, X. Zhang, X.Q. Hu, L. Li, W.P. Wang, et al., Biological characteristics of mesenchymal stem cells grown on different topographical nanofibrous poly-L-lactide meshes, J. Biomed. Nanotechnol. 9 (2013) 1757.
[75] W. Liu, Y. Wei, X. Zhang, M. Xu, X. Yang, X. Deng, Lower extent but similar rhythm of osteogenic behavior in hBMSCs cultured on nanofibrous scaffolds versus induced with osteogenic supplement, ACS Nano 7 (2013) 6928.
[76] K.-H. Kim, L. Jeong, H.-N. Park, S.-Y. Shin, W.-H. Park, S.-C. Lee, et al., Biological efficacy of silk fibroin nanofiber membranes for guided bone regeneration, J. Biotechnol. 120 (2005) 327.
[77] H.-W. Kim, J.-H. Song, H.-E. Kim, Nanofiber generation of gelatin-hydroxyapatite biomimetics for guided tissue regeneration, Adv. Funct. Mater. 15 (2005) 1988.
[78] K. Fujihara, M. Kotaki, S. Ramakrishna, Guided bone regeneration membrane made of polycaprolactone/calcium carbonate composite nanofibers, Biomaterials 26 (2005) 4139.
[79] F. Yang, S.K. Both, X. Yang, X.F. Walboomers, J.A. Jansen, Development of an electrospun nano-apatite/PCL composite membrane for GTR/GBR application, Acta Biomater. 5 (2009) 3295.
[80] M.C. Bottino, V. Thomas, G. Schmidt, Y.K. Vohra, T.-M.G. Chu, M.J. Kowolik, et al., Recent advances in the development of GTR/GBR membranes for periodontal regeneration—a materials perspective, Dent. Mater. 28 (2012) 703.
[81] D. Liang, B.S. Hsiao, B. Chu, Functional electrospun nanofibrous scaffolds for biomedical applications, Adv. Drug Delivery Rev. 59 (2007) 1392.
[82] C.P. Barnes, S.A. Sell, E.D. Boland, D.G. Simpson, G.L. Bowlin, Nanofiber technology: designing the next generation of tissue engineering scaffolds, Adv. Drug Delivery Rev. 59 (2007) 1413.
[83] G.C. Rutledge, S.V. Fridrikh, Formation of fibers by electrospinning, Adv. Drug Delivery Rev. 59 (2007) 1384.
[84] S. Agarwal, J.H. Wendorff, A. Greiner, Use of electrospinning technique for biomedical applications, Polymer 49 (2008) 5603.
[85] Z.-M. Huang, Y.-Z. Zhang, M. Kotaki, S. Ramakrishna, A review on polymer nanofibers by electrospinning and their applications in nanocomposites, Compos Sci. Technol. 63 (2003) 2223.
[86] A. Greiner, J.H. Wendorff, Electrospinning: a fascinating method for the preparation of ultrathin fibers, Angew. Chem. Int. Ed. 46 (2007) 5670.
[87] V. Beachley, X. Wen, Polymer nanofibrous structures: fabrication, biofunctionalization, and cell interactions, Prog. Polym. Sci. 35 (2010) 868.
[88] T.J. Sill, H.A. von Recum, Electrospinning: applications in drug delivery and tissue engineering, Biomaterials 29 (2008) 1989.
[89] I.S. Chronakis, Novel nanocomposites and nanoceramics based on polymer nanofibers using electrospinning process—a review, J. Mater. Process. Technol. 167 (2005) 283.

[90] W.-E. Teo, S. Ramakrishna, Electrospun nanofibers as a platform for multifunctional, hierarchically organized nanocomposite, Compos. Sci. Technol. 69 (2009) 1804.
[91] D.H. Reneker, A.L. Yarin, Electrospinning jets and polymer nanofibers, Polymer 49 (2008) 2387.
[92] I. Kang, Y.Y. Heung, J.H. Kim, J.W. Lee, R. Gollapudi, S. Subramaniam, et al., Introduction to carbon nanotube and nanofiber smart materials, Compos. Part B – Eng. 37 (2006) 382.
[93] B.S. Harrison, A. Atala, Carbon nanotube applications for tissue engineering, Biomaterials 28 (2007) 344.
[94] A. Bianco, K. Kostarelos, M. Prato, Applications of carbon nanotubes in drug delivery, Curr. Opin. Chem. Biol. 9 (2005) 674.
[95] A. Abarrategi, M.C. Gutierrez, C. Moreno-Vicente, M.J. Hortiguela, V. Ramos, J.L. Lopez-Lacomba, et al., Multiwall carbon nanotube scaffolds for tissue engineering purposes, Biomaterials 29 (2008) 94.
[96] X.-L. Xie, Y.-W. Mai, X.-P. Zhou, Dispersion and alignment of carbon nanotubes in polymer matrix: a review, Mater. Sci. Eng. R 49 (2005) 89.
[97] T. Yang, D. Wu, L. Lu, W. Zhou, M. Zhang, Electrospinning of polylactide and its composites with carbon nanotubes, Polym. Compos. 32 (2011) 1280.
[98] S. Mazinani, A. Ajji, C. Dubois, Morphology, structure and properties of conductive PS/CNT nanocomposite electrospun mat, Polymer 50 (2009) 3329.
[99] R. Sen, B. Zhao, D. Perea, M.E. Itkis, H. Hu, J. Love, et al., Preparation of single-walled carbon nanotube reinforced polystyrene and polyurethane nanofibers and membranes by electrospinning, Nano Lett. 4 (2004) 459.
[100] J.J. Mack, L.M. Viculis, A. Ali, R. Luoh, G. Yang, H.T. Hahn, et al., Graphite nanoplatelet reinforcement of electrospun polyacrylonitrile nanofibers, Adv. Mater. 17 (2005) 77.
[101] J.J. Ge, H. Hou, Q. Li, M.J. Graham, A. Greiner, D.H. Reneker, et al., Assembly of well-aligned multiwalled carbon nanotubes in confined polyacrylonitrile environments: electrospun composite nanofiber sheets, J. Am. Chem. Soc. 126 (2004) 15754.
[102] J. Ji, G. Sui, Y. Yu, Y. Liu, Y. Lin, Z. Du, et al., Significant improvement of mechanical properties observed in highly aligned carbon-nanotube-reinforced nanofibers, J. Phys. Chem. C 113 (2009) 4779.
[103] J. Ayutsede, M. Gandhi, S. Sukigara, H. Ye, C. Hsu, Y. Gogotsi, et al., Carbon nanotube reinforced bombyx mori silk nanofibers by the electrospinning process, Biomacromolecules 7 (2006) 208.
[104] W. Salalha, Y. Dror, R. Khalfin, Y. Cohen, A. Yarin, E. Zussman, CNTs embedded in oriented polymer nanofibers by electrospinning, Langmuir 19 (2003) 7012.
[105] F.K. Ko, M. Gandhi, C. Karatzas, Carbon nanotube reinforced spider silk—a model for the next generation of super strong and tough fibers, in: Proceedings of the 19th American Society for Composites Annual Technical Conference, October 17–20, 2004, Atlanta, GA, 2004.
[106] M. Kang, P. Chen, H.-J. Jin, Preparation of multiwalled carbon nanotubes incorporated silk fibroin nanofibers by electrospinning, Curr. Appl. Phys. 9 (2009) S95.
[107] K. Saeed, S.-Y. Park, H.-J. Lee, J.-B. Baek, W.-S. Huh, Preparation of electrospun nanofibers of carbon nanotube/polycaprolactone nanocomposite, Polymer 47 (2006) 8019.

[108] Z.X. Meng, W. Zheng, L. Li, Y.F. Zheng, Fabrication and characterization of three-dimensional nanofiber membrane of PCL-MWCNTs by electrospinning, Mater. Sci. Eng., C 30 (2010) 1014.
[109] S. Shao, S. Zhou, L. Li, J. Li, C. Luo, J. Wang, et al., Osteoblast function on electrically conductive electrospun PLA/MWCNTs nanofibers, Biomaterials 32 (2011) 2821.
[110] S.K. Smart, A.I. Cassady, G.Q. Lu, D.J. Martin, The biocompatibility of carbon nanotubes, Carbon 44 (2006) 1034.
[111] A.A. Shvedova, E.R. Kisin, D. Porter, P. Schulte, V.E. Kagan, B. Fadeel, et al., Mechanisms of pulmonary toxicity and medical applications of carbon nanotubes: two faces of Janus? Pharmacol. Ther. 121 (2009) 192.
[112] G. Jia, H. Wang, L. Yan, X. Wang, R. Pei, T. Yan, Cytotoxicity of carbon nanomaterials: single-wall nanotube, multi-wall nanotube and fullerene, Environ. Sci. Technol. 39 (2005) 1378.
[113] D.X. Cui, F.R. Tian, C.S. Ozkan, M. Wang, H.J. Gao, Effect of single wall carbon nanotubes on human HEK293 cells, Toxicol. Lett. 55 (2005) 73.
[114] E. Herzog, A. Casey, F.M. Lyng, G. Chambers, H.J. Byrne, M. Davoren, A new approach to the toxicity testing of carbon-based nanomaterials, The clonogenic assay, Toxicol. Lett. 174 (2007) 49.
[115] P. Wick, P. Manser, L.K. Limbach, U. Dettlaff-Weglikowska, F. Krumeich, S. Roth, et al., The degree and kind of agglomeration affect carbon nanotube toxicity, Toxicol. Lett. 168 (2007) 121.
[116] P. Wick, P. Manser, P. Spohn, A. Bruinink, In-vitro evaluation of possible adverse effects of nanosized materials, Phys. Status Solidi. 243 (2006) 3556.
[117] A.A. Shvedova, V. Castranova, E.R. Kisin, D. Schwegler-Berry, A.R. Murray, V.Z. Gandelsman, et al., Exposure to carbon nanotube material: assessment of nanotube cytotoxicity using human keratinocyte cells, Toxicol. Environ. Health, A 66 (2003) 1909.
[118] J.L. McKenzie, M.C. Waid, R. Shi, T.J. Webster, Decreased functions of astrocytes on carbon nanofiber materials, Biomaterials 25 (2004) 1309.
[119] H. Hu, Y.C. Ni, V. Montana, R.C. Haddon, V. Parpura, Chemically functionalized carbon nanotubes as substrates for neuronal growth, Nano Lett. 4 (2004) 507.
[120] T. Gabay, E. Jakobs, E. Ben-Jacob, Y. Hanein, Engineered selforganization of neural networks using carbon nanotube clusters, Physica A 250 (2005) 611.
[121] H. Hu, Y. Ni, S.K. Mandal, V. Montana, B. Zhao, R.C. Haddon, et al., Polyethyleneimine functionalized single-walled carbon nanotubes as a substrate for neuronal growth, J. Phys. Chem. B 109 (2005) 4285.
[122] S.Y. Park, S.Y. Park, S. Namgung, B. Kim, J. Im, J.Y. Kim, et al., Carbon nanotube monolayer patterns for directed growth of mesenchymal stem cells, Adv. Mater. 19 (2007) 2530.
[123] L.P. Zanello, B. Zhao, H. Hu, R.C. Haddon, Bone cell proliferation on carbon nanotubes, Nano Lett. 6 (2006) 562.
[124] K.L. Elias, R.L. Price, T.J. Webster, Enhanced functions of osteoblasts on nanometer diameter carbon fibers, Biomaterials 23 (2002) 3279.
[125] P.R. Supronowicz, P.M. Ajayan, K.R. Ullmann, B.P. Arulanandam, D.W. Metzger, R. Bizios, Novel current-conducting composite substrates for exposing osteoblasts to alternating current stimulation, J. Biomed. Mater. Res. 59A (2002) 499.

[126] M. Wang, Developing bioactive composite materials for tissue replacement, Biomaterials 24 (2003) 2133.
[127] R. Murugan, S. Ramakrishna, Development of nanocomposites for bone grafting, Compos. Sci. Technol. 65 (2005) 2385.
[128] E.M. Christenson, K.S. Anseth, J.J.J.P. van den Beucken, C.K. Chan, B. Ercan, J.A. Jansen, et al., Nanobiomaterial applications in orthopedics, J. Orthop. Res. 25 (2007) 11.
[129] V. Uskoković, When $1+1>2$: nanostructured composites for hard tissue engineering applications, Mater. Sci. Eng., C 57 (2015) 434.
[130] X.-L. Deng, G. Sui, M.-L. Zhao, G.-Q. Chen, X.-P. Yang, Poly(L-lactic acid)/hydroxyapatite hybrid nanofirous scaffolds prepared by electrospinning, J. Biomater. Sci. Polym. Ed. 18 (2007) 117.
[131] G. Sui, X. Yang, F. Mei, X. Hu, G. Chen, X. Deng, et al., Poly-L-lactic acid/hydroxyapatite hybrid membrane for bone tissue regeneration, J. Biomed. Mater. Res. 82A (2007) 445.
[132] H.J. Gong, X.P. Yang, X.L. Deng, X.Y. Hu, Study on PLA/MWNT/HA hybrid nanofibers prepared via electrospinning technology, Acta Polym. Sin. 2 (2005) 297.
[133] H.-W. Kim, H.-H. Lee, J.C. Knowles, Electrospinning biomedical nanocomposite fibers of hydroxyapatite/poly(lactic acid) for bone regeneration, J. Biomed. Mater. Res. 79A (2006) 643.
[134] V. Thomas, S. Jagani, K. Johnson, M.V. Jose, D.R. Dean, Y.K. Vohra, et al., Electrospun bioactive nanocomposite scaffolds of polycaprolactone and nanohydroxyapatite for bone tissue engineering, J. Nanosci. Nanotechnol. 6 (2006) 487.
[135] Y. Ito, H. Hasuda, M. Kamitakahara, C. Ohtsuki, M. Tanihara, I.K. Kang, et al., A composite of hydroxyapatite with electrospun biodegradable nanofibers as a tissue engineering material, J. Biosci. Bioeng. 100 (2005) 43.
[136] M.V. Jose, V. Thomas, K.T. Johnson, D.R. Dean, E. Nyairo, Aligned PLGA/HA nanofibrous nanocomposite scaffolds for bone tissue engineering, Acta Biomater. 5 (2009) 305.
[137] V. Thomas, D.R. Dean, M.V. Jose, B. Mathew, S. Chowdhury, Y.K. Vohra, Nanostructured biocomposite scaffolds based on collagen coelectrospun with nanohydroxyapatite, Biomacromolecules 8 (2007) 631.
[138] A. Bishop, C. Balazsi, J.H.C. Yang, P.-I. Gouma, Biopolymer-hydroxyapatite composite coatings prepared by electrospinning, Polym. Adv. Technol. 17 (2006) 902.
[139] D.-H. Paik, K.-Y. Jeong, S.-K. Moon, M.-J. Oh, T.-K. Ryu, S.-E. Kim, et al., A facile method for preparation of polycaprolactone/tricalcium phosphate fibrous matrix with a gradient mineral content, Colloids Surf. A 429 (2013) 134.
[140] S.-H. Teng, E.-J. Lee, P. Wang, H.-E. Kim, Collagen/hydroxyapatite composite nanofibers by electrospinning, Mater. Lett. 62 (2008) 3055.
[141] X. Xu, X. Chen, A. Liu, Z. Hong, X. Jing, Electrospun poly(L-lactide)-grafted hydroxyapatite/poly(L-lactide) nanocomposite fibers, Eur. Polym. J. 43 (2007) 3187.
[142] Y. Cui, Y. Liu, Y. Cui, X. Jing, P. Zhang, X. Chen, The nanocomposite scaffold of poly(lactide-*co*-glycolide) and hydroxyapatite surface-grafted with L-lactic acid oligomer for bone repair, Acta Biomater. 5 (2009) 2680.
[143] J.Q. He, X.P. Yang, J.F. Mao, F.J. Xu, Q. Cai, Hydroxyapatite-poly (L-lactide) nanohybrids via surface-initiated ATRP for improved bone-like apatite-formation ability, Appl. Surf. Sci. 258 (2012) 6823.

[144] F. Mei, J. Zhong, X. Yang, X. Ouyang, S. Zhang, X. Hu, et al., Improved biological characteristics of poly(L-lactic acid) electrospun membrane by incorporation of multiwalled carbon nanotubes/hydroxyapatite nanoparticles, Biomacromolecules 8 (2007) 3729.

[145] Q. Cai, J.F. Mao, X.L. Li, X.P. Yang, Macroporous and nanofibrous PLLA scaffold reinforced with biomimetic CaP-coated multiwalled carbon nanotubes, Mater. Lett. 128 (2014) 238.

[146] X.L. Li, J.L. Lan, M. Ai, Y.G. Guo, Q. Cai, X.P. Yang, Biomineralization on polymer-coated carbon nanotubes with different surface functional groups, Colloids Surf. B: Biointerfaces 123 (2014) 753.

[147] B. Wang, Q. Cai, S. Zhang, X.P. Yang, X.L. Deng, The effect of poly (L-lactic acid) nanofiber orientation on osteogenic responses of human osteoblast-like MG63 cells, J. Mech. Behav. Biomed. Mater. 4 (2011) 600.

[148] X.H. Zhang, Q. Cai, S. Zhang, H.Y. Liu, X.P. Yang, X.L. Deng, et al., Calcium ion release and osteoblastic behavior of gelatin_beta-tricalcium phosphate composite nanofibers fabricated by electrospinning, Mater. Lett. 73 (2012) 172.

[149] Q. Cai, Q.Q. Xu, Q.F. Feng, X.Y. Cao, X.P. Yang, X.L. Deng, Biomineralization of electrospun poly(L-lactic acid)/gelatin composite fibrous scaffold by using a supersaturated simulated body fluid with continuous CO_2 bubbling, Appl. Surf. Sci. 57 (2011) 10109.

[150] Y. Guo, J. Lan, C. Zhang, M. Cao, Q. Cai, X. Yang, Effects of amino acids on biomimetic mineralization of electrospun polylactide/gelatin composite nanofibers, Appl. Surf. Sci. 349 (2015) 538.

[151] Q. Cai, Q.F. Feng, H.Y. Liu, X.P. Yang, Preparation of biomimetic hydroxyapatite by biomineralization and calcination using poly(L-lactide)/gelatin composite fibrous mat as template, Mater. Lett. 91 (2013) 275.

[152] X.G. Zhou, Q. Cai, N. Yan, X.L. Deng, X.P. Yang, In-vitro hydrolytic and enzymatic degradation of nest-like patterned electrospun poly(D,L-lactide-*co*-glycolide) scaffolds, J. Biomed. Mater. Res. 95A (2010) 755.

CHAPTER

11 Nanoapatitic composite scaffolds for stem cell delivery and bone tissue engineering

Hockin H.K. Xu[1], Mike D. Weir[1], Liang Zhao[1], Jennifer L. Moreau[1], Dwayne D. Arola[1] and Carl G. Simon[2]

[1]*University of Maryland, Baltimore, MD, United States* [2]*National Institute of Standards and Technology, Gaithersburg, MD, United States*

CHAPTER OUTLINE

11.1 Introduction ... 211
11.2 Development of Nanoapatitic and Macroporous Scaffolds ... 213
11.3 Cell Infiltration into Scaffold ... 215
11.4 Biomimetic Nanoapatite–Collagen Fiber Scaffold ... 215
11.5 Fast Fracture of Nanoapatite Scaffold ... 217
11.6 Fatigue of Nanoapatite Scaffold ... 219
11.7 Nanoapatite Scaffold–Human Umbilical Cord Stem Cell Interactions ... 220
11.8 Seeding Bone Marrow Stem Cells on Nanoapatite Scaffolds ... 223
11.9 Conclusions ... 226
Acknowledgments ... 226
References ... 227

11.1 INTRODUCTION

Seven million people have bone fractures each year in the United States [1], and musculoskeletal conditions cost $215 billion annually [1,2]. These numbers are predicted to increase dramatically because of an aging population [3]. The introduction of stem cells into the clinical setting opens new therapeutic horizons [4–8]. Embryonic stem cells are pluripotent, able to become over 200 types of cells in the body. Adult bone-marrow-derived mesenchymal (or stromal) stem cells (BMSCs) are multipotent, able to differentiate into bone tissue, neural tissue, cartilage, muscle, and fat [9,10]. BMSCs can be harvested from the patient, expanded in culture, induced to differentiate, and combined with a scaffold to repair bone defects.

Emerging Nanotechnologies in Dentistry. DOI: http://dx.doi.org/10.1016/B978-0-12-812291-4.00011-X

The development of suitable scaffolds will constitute a centerpiece for bone tissue engineering. The structure needs to be maintained to define the shape of the regenerated tissue. Mechanical properties are of crucial importance for the regeneration of load-bearing tissues such as bone, to withstand stresses to avoid scaffold fracture. Bio-inert implants can induce undesirable fibrous capsules in vivo, while bioactive implants with bone-like calcium phosphate minerals beneficially bind to native bone. This is because calcium phosphate minerals provide a preferred substrate for cell attachment and support the proliferation and expression of osteoblast phenotype [11,12]. Hence, hydroxyapatite and other bioactive calcium phosphate scaffolds are important for bone repair [13–19]. However, for sintered hydroxyapatite and other bioactive ceramics to fit into a bone cavity, the surgeon needs to machine the graft to the desired shape or carve the surgical site around the implant. This leads to an increase in bone loss, trauma, and surgical time [3].

By contrast, calcium phosphate cements can be molded and set in situ to provide intimate adaptation to bone defects [20–27]. The first calcium phosphate cement was comprised of a mixture of tetracalcium phosphate (TTCP: $Ca_4(PO_4)_2O$) and dicalcium phosphate anhydrous (DCPA: $CaHPO_4$), and was referred to as calicium phosphate cement (CPC) [28]. The CPC powder can be mixed with an aqueous liquid to form a paste that can be sculpted during surgery to conform to the defects in hard tissues. The paste self-hardens to form a resorbable hydroxyapatite implant [29–31]. Due to its excellent bioactivity and ability to be replaced by new bone, CPC was approved in 1996 by the US Food and Drug Administration (FDA) for repairing craniofacial defects, thus becoming the first CPC for clinical use [30]. However, because it is brittle and weak, the use of CPC was "limited to the reconstruction of nonstress-bearing bone" [29], and "none of the indications include significant stress-bearing applications" [30]. Recent studies used resorbable fibers to provide the needed early strength to CPC and then to create macropores after fiber degradation [32–34]. These previous studies measured the mechanical properties using single-load, fast fracture methods.

However, implants in vivo are subjected to repeated loadings. In periodontal repair, for example, tooth mobility resulted in the fracture and eventual exfoliation of the brittle CPC [35]. Other potential uses of CPC include mandibular and maxillary ridge augmentation, since CPC could be molded to the desired shape and set to form a scaffold for bone ingrowth. However, these implants would be subject to cyclic loading by provisional dentures and need to be resistant to flexure. Major reconstructions of the maxilla or mandible after trauma or tumor resection would also require a moldable implant with improved fracture resistance and rapid osteoconduction, as would the support of metal dental implants or augmentation of deficient implant sites. All these dental and craniofacial applications and many other orthopedic repairs would be better served with an improved CPC having higher fracture resistance and more rapid bone regeneration via macroporosity and stem cell delivery. This chapter describes recent studies on tetracalcium phosphate–dicalcium phosphate cement scaffolds, focusing on biomimetic nanoapatite–fiber composite scaffolds and stem cell delivery for bone tissue engineering.

11.2 DEVELOPMENT OF NANOAPATITIC AND MACROPOROUS SCAFFOLDS

Natural bone consists of an extracellular matrix with nanosized apatitic minerals and collagen fibers that support bone cell functions [36]. It is advantageous for a synthetic scaffold to contain nanoapatite crystals similar to those in bone along with fibers to form a matrix that supports cell attachment.

In addition, it is desirable for the scaffold to possess mechanical properties similar to those of bone, and to have the capability to encapsulate and deliver bone cells for osteogenic differentiation and bone regeneration. During CPC setting, it undergoes a dissolution and reprecipitation process to form nanosized hydroxyapatite [31–34]. Scanning electron microscopy (SEM) shows a micrograph of elongated nanosized hydroxyapatite crystals that make up the CPC matrix (Fig. 11.1A). Fig. 11.1B plots the crystal length distribution. Based on 300 crystals measured with SEM, the apatite crystal length ranged between 75 and 550 nm (median length = 299 nm), and the crystal width ranged between 50 and 150 nm (median width = 87 nm).

Macropores have been built into biomaterials to facilitate bony ingrowth and implant fixation [14–19]. One advantage of a macroporous CPC is that it can form macroporous hydroxyapatite implants in situ without high-temperature sintering and machining. Particulate and fibrous porogens can be mixed in the CPC paste to create interconnected macropores [32,33,37]. One approach was to combine slow-dissolving fibers with fast-dissolving porogens to develop strong scaffolds with controlled strength histories and tailored macropore formation rates [33]. The rationale was that CPC containing fast-dissolving porogen and slow-dissolving fibers would have two stages of macropore formation in vivo. The fast porogen would dissolve quickly upon contact with the physiological solution to form macropores to start tissue ingrowth, while the fibers would provide the needed early strength to the graft. After significant new bone ingrowth into the macropores thus increasing the CPC strength, the fibers would then dissolve and create the second group of macropores for further tissue ingrowth. The fiber degradation rate can be controlled to match the new bone formation rate for specific applications. In one study, an absorbable polyglactin fiber (Vicryl suture, Ethicon, Somerville, NJ) was cut to 8-mm-long filaments [32]. This suture consisted of fibers braided into a bundle with a diameter of approximately 322 μm, suitable for producing macropores after fiber degradation [32].

Fig. 11.1C shows a macropore in CPC formed by the dissolution of a fast-dissolving porogen, mannitol crystals ($CH_2OH[CHOH]_4CH_2OH$, Sigma, St. Louis, MO) [38]. The letter "O" refers to the preosteoblastic cells (MC3T3-E1) that infiltrated into the macropore in CPC. Fig. 11.1D shows examples of macropore channels in CPC created from the dissolution of absorbable polyglactin fibers. These pores were approximately 300 μm in diameter and 8 mm in length. A high magnification in Fig. 11.1E shows a human umbilical cord mesenchymal stem cell (hUCMSC)

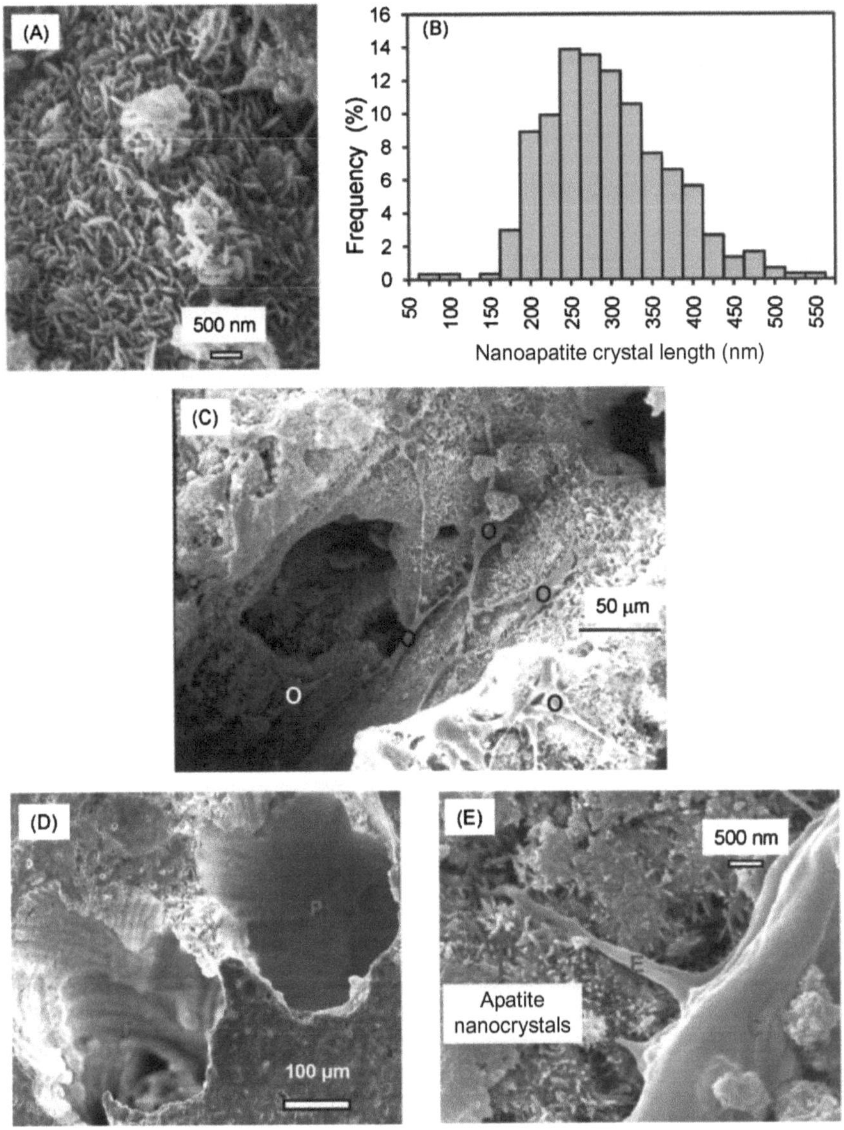

FIGURE 11.1

CPC-based nanoapatite scaffold. (A) SEM of the nanosized hydroxyapatite crystals that make up CPC. (B) Size distribution of apatite crystals in CPC. The crystal length ranged from 75 to 550 nm. The width ranged from 50 to 150 nm. (C) Macroporosity was created in CPC via the dissolution of mannitol porogen, and preosteoblastic cells (mouse MC3T3-E1 cells) (indicated by "O") in the pore. (D) Macropore channels "P" in CPC formed via fibers after fiber dissolution. (E) Human umbilical cord stem cells (hUCMSC) attaching to the nanoapatite in CPC via cytoplasmic extensions. "C" refers to the hUCMSC. "E" designates the cytoplasmic extensions of the cell.

attaching to the nanoapatite in CPC via the cytoplasmic extensions "E." These studies demonstrate the feasibility of developing macroporous, CPC-based nanoapatite scaffolds which are promising for cell delivery and tissue engineering.

11.3 CELL INFILTRATION INTO SCAFFOLD

To investigate cell infiltration into the macropores of the CPC scaffold, in vitro cell culture was performed on the new CPC scaffolds. MC3T3-E1 mouse preosteoblastic cells were cultured following established protocols [39]. Traditional CPC control and the new, mechanically stronger CPC scaffold composites were tested. Fifty thousand cells diluted into 2 mL of media were added to each well containing a specimen or to an empty well of tissue culture polystyrene (TCPS, as a biocompatible control), and incubated for 1 or 14 days.

As shown in Fig. 11.2, the pore in the CPC scaffold was large enough for the osteoblast cell "O," and the cell had developed long cytoplasmic extensions "E" anchoring to the pore bottom. Cells had established cell–cell junction (arrow) in the pore (B). In (C), secondary extensions sprouted from the primary extension "E." The cell extensions were shown (D) to have anchored to the nanoapatite crystals that make up the CPC matrix. Hence, cells were able to infiltrate into the macroporous CPC and attach to the nanoapatite crystals.

11.4 BIOMIMETIC NANOAPATITE–COLLAGEN FIBER SCAFFOLD

Human bone is comprised of collagen fibers (mainly type I) and nanosized hydroxyapatite crystals. Hence, synthetic hydroxyapatite/collagen fiber composites are promising in mimicking and replacing bone. Only a few studies have incorporated collagen into moldable, self-setting calcium phosphate cements [41,42]. In one study, collagen decreased the strength of the cement [41], whilst in another study, collagen decreased the cell activities in vitro [42]. Therefore, there is a need to develop an in situ-setting collagen–calcium phosphate cement composite with increased cell attachment and increased toughness.

In a recent study, a type I bovine collagen fiber (Sigma, St. Louis, MO) was added to CPC to develop an in situ-setting, bone-mimicking, nanoapatite–collagen composite [43]. Bovine Achilles tendon was cleaned of all noncollagenous tissue and extracted at a temperature of 0°C with a 3% Na_2HPO_4 solution to remove soluble proteins. To remove mucopolysaccharides, the tendon was extracted with a 25% KCl solution. The resultant collagen was washed with water, dehydrated with absolute alcohol, and air dried. The collagen was in the form of elongated, flexible bundles, with a bundle diameter ranging from approximately 0.1 to 3 μm. The length of the bundles ranged from about 20 to 100 μm [43].

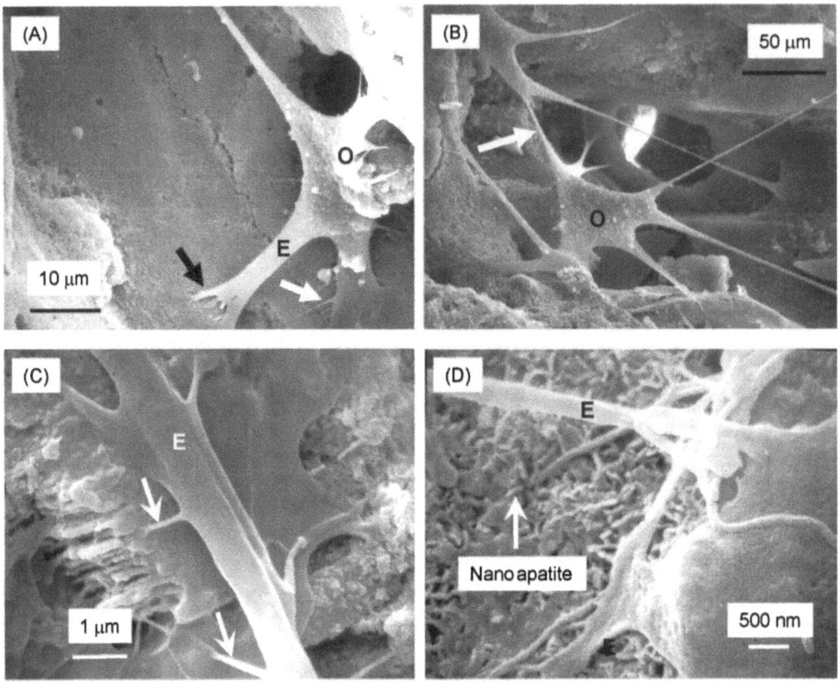

FIGURE 11.2

SEM showing cell infiltration into the macropores of CPC. (A) The pore was large enough for the osteoblastic cell "O" (mouse MC3T3-E1 cells), and the cell had developed cytoplasmic extensions "E." (B) Two cells had established cell–cell junction (arrow) in the pore. (C) Secondary extensions sprouting from the primary extension "E." (D) Cell extensions anchored to the nanoapatite crystals of CPC, similar to that of the stem cells in Fig. 11.1.

Adapted from H.H.K. Xu, M.D. Weir, C.G. Simon, Injectable and strong nano-apatite scaffolds for cell/growth factor delivery and bone regeneration. Dent. Mater. 24 (2008) 1212–1222. Ref. [40] with permission.

The CPC powder was mixed with distilled water to form a paste to which the collagen fibers were added and incorporated by mixing. The following collagen mass fractions of collagen/(collagen + CPC) were used: 0%, 2.5%, and 5.0%. Fig. 11.3A shows collagen fibers covered with small hydroxyapatite crystals that make up the CPC matrix [43]. This indicates an intimate contact between collagen fibers and hydroxyapatite crystals. The work-of-fracture (or toughness, which measures the energy required to fracture the specimen) (Fig. 11.3B) was increased from $9.4 \pm 3.1\ J/m^2$ with 0% collagen to $430 \pm 103\ J/m^2$ with 5% collagen ($P \leq 0.05$) at a powder:liquid ratio (P/L) of 2.5. At a P/L of 3, the work-of-fracture was increased from $22.2 \pm 3.8\ J/m^2$ without collagen to $381 \pm 119\ J/m^2$ with 5% collagen ($P < 0.05$). Fig. 11.3C shows cell attachment for 1 day on CPC–collagen composite. Increasing the collagen content increased the cell number per specimen area ($P < 0.05$). The number of cells/area was 382 ± 99 cells/mm^2 at 5% collagen,

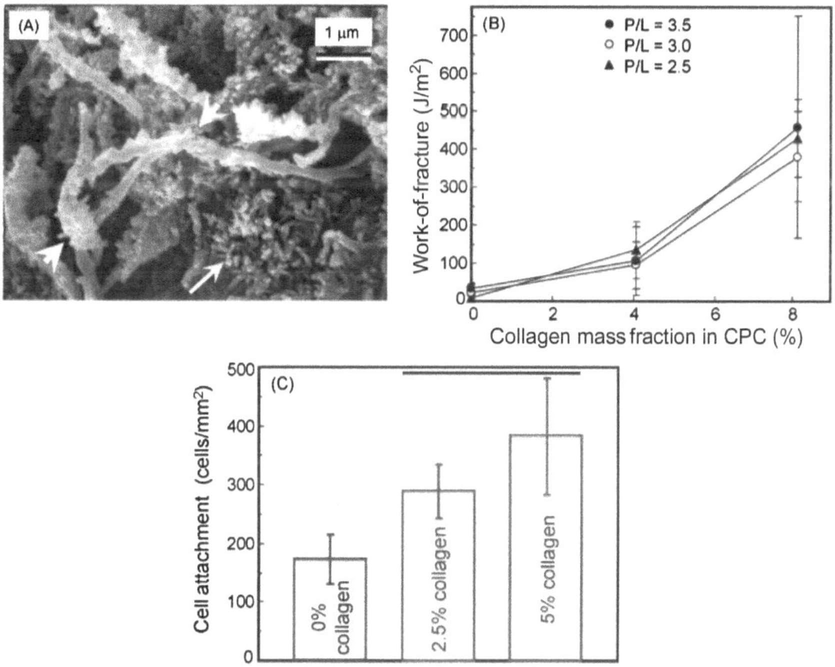

FIGURE 11.3

Nanoapatitic collagen–CPC composite. (A) SEM showing collagen fibers (short arrows) covered with nanoapatite crystals (long arrow). The nanoapatite had intimate contact with collagen fibers. (B) Work-of-fracture, or toughness (mean ± SD; $n = 6$) of CPC was increased with collagen content. P/L is the CPC powder-to-liquid ratio. (C) MC3T3-E1 preosteoblastic cell attachment = number of live cells/area of CPC (mean ± SD; $n = 5$). It was increased with collagen content. Horizontal lines indicate values that are not significantly different ($P > 0.1$).

Adapted from J.L. Moreau, M.D. Weir, H.H.K. Xu, Self-setting collagen–calcium phosphate bone cement: mechanical and cellular properties. J. Biomed. Mater. Res. A. 91 (2009) 605–613 with permission.

which was higher than ($P \leq 0.05$) that at 0% collagen (173 ± 42 cells/mm^2) [43]. Hence, the incorporation of collagen fibers into CPC not only improved its toughness but also increased the osteoblastic cell attachment.

11.5 FAST FRACTURE OF NANOAPATITE SCAFFOLD

To improve the mechanical properties for load-bearing dental and orthopedic applications, chitosan and absorbable fibers were used to reinforce CPC. Chitosan and its derivatives are natural polymers that are biocompatible, biodegradable, and osteoconductive [44]. Chitosan has been shown to strengthen and increase toughness of CPC [45] and resist the washout of CPC paste in physiological solution as well as accelerate CPC setting [33]. Chitosan lactate (referred to as chitosan; VANSON, Redmond, WA) was dissolved in water at concentrations of 0%, 5%, 10%, or 15% to form CPC liquids.

Polyglactin fibers (Ethicon) were mixed into the CPC paste to form a composite paste, which was placed into a rectangular mold of 3 × 4 × 25 mm [32]. A fiber volume fraction of 20% was used to obtain a CPC–chitosan–fiber paste that was readily mixed and not dry. The fiber volume fraction was equal to the volume of fibers divided by the volume of the entire specimen. The paste in the mold was set in a humidor with 100% relative humidity at 37°C.

Fig. 11.4 plots the results from the single-load, three-point flexure (mean ± SD; $n = 6$) [46]. CPC with 15% chitosan reached strength of 16.2 ± 3.0 MPa, higher

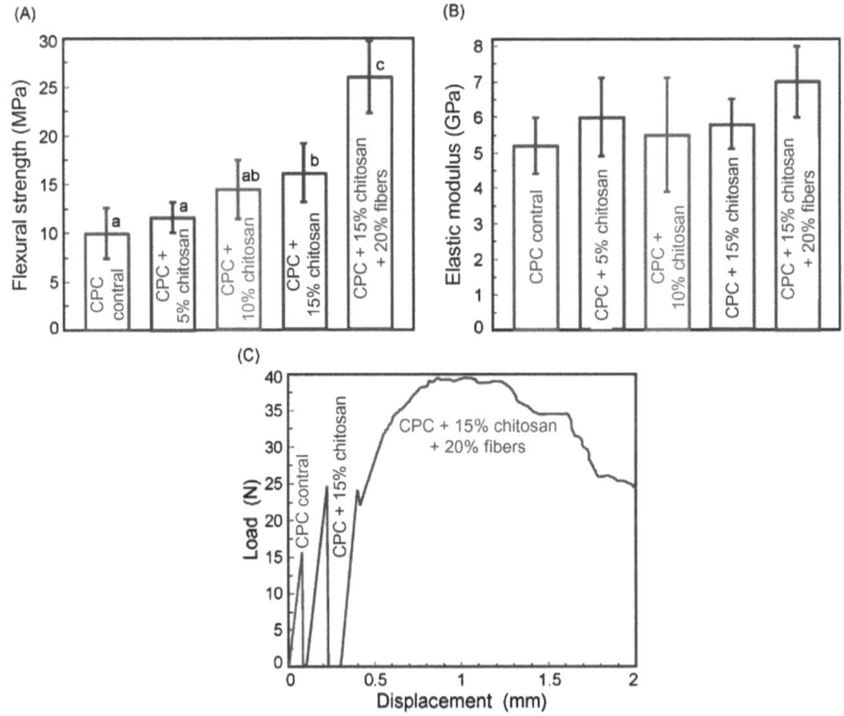

FIGURE 11.4

Single-load fast fracture of nanoapatitic CPC scaffolds. (A) Flexural strength (mean ± SD; $n = 6$), (B) elastic modulus, and (C) load–displacement curves. In (A), values indicated by dissimilar letters are significantly different ($P < 0.05$). In (B), all the values are statistically similar ($P > 0.1$). In (C), CPC with 15% chitosan and 20% fibers showed a tough, noncatastrophic failure mode.

Adapted from R. Zhao, E.F. Burguera, H.H.K. Xu, N. Amin, H. Ryou, D.D. Arola, Fiber-reinforced calcium phosphate scaffold: fatigue and human umbilical cord stem cell seeding. Biomaterials 93 (1) (2010) 93–105 with permission.

($P < 0.05$) than MPa for CPC control (10.2 ± 2.1) (Fig. 11.4A). By contrast, CPC with 15% chitosan and 20% fibers had a strength of 26.0 ± 3.7 MPa, which is higher ($P \leq 0.05$) than all of the other materials investigated in this study. Elastic modulus ranged from about 5.5 to 7 GPa, which was similar for all materials investigated (Fig. 11.4B). In Fig. 11.4C, the load–displacement curves of CPC control and CPC with chitosan showed a brittle, catastrophic failure mode. By contrast, the CPC with 15% chitosan and 20% fibers showed a tough, noncatastrophic failure mode. Hence, the incorporation of chitosan and fibers progressively strengthened and toughened CPC scaffold.

11.6 FATIGUE OF NANOAPATITE SCAFFOLD

Most studies have measured the mechanical properties of CPC using single-load, fast fracture methods. However, implants in vivo are subjected to repeated loadings. Hence, a recent study investigated the fatigue behavior of fiber-reinforced CPC [46]. Two materials were tested: CPC–chitosan–fiber (15% chitosan and 20% fibers) and CPC control. Fatigue testing was conducted in cyclic four-point flexure. Cyclic loading was performed at room temperature (22°C) in a physiological solution using load-control actuation, at a frequency of 5 Hz and a ratio of minimum to maximum stress of 0.1. Successive specimens were then tested using a maximum cyclic load decreased in increment of 1 MPa, according to a staircase fatigue method [46–48].

The fatigue results are shown in Fig. 11.5A and B for CPC control, and CPC with 15% chitosan and 20% fibers (designated as CPC–chitosan–fiber), respectively [46]. CPC control failed in a single cycle at stress values ≥ 6 MPa. However, at slightly lower stress values (e.g., ≤ 5 MPa), none of the specimens failed after 2×10^6 cycles. CPC–chitosan–fiber failed at higher stresses. CPC–chitosan–fiber specimens that survived 2×10^6 cycles reached the highest stress of 10 MPa. This was 2 times higher than the highest stress of 5 MPa for the CPC control. Fig. 11.6 plots the mean and standard deviation for specimens failed after one cycle (A), or after multiple cycles (B). When a tough material is loaded cyclically, microdamage such as microcracks would be created in the material. As the number of cycles increases, microcracks may accumulate and coalesce, leading to failure of the specimen. For CPC control, at stresses ≤ 5 MPa, no specimens failed after 2 million cycles; however, with a slight increase in the stress (≥ 6 MPa), all the specimens failed even when a single cycle was performed. There was little room in the stress level for gradual damage accumulation without specimen failure, indicating that the CPC control has a very low damage tolerance. This is likely to be attributed to CPC being extremely brittle, and as soon as a microcrack is formed, it propagates catastrophically through the entire specimen. Incorporation of chitosan and polyglactin fibers reinforced the CPC. In fatigue, the highest stress reached 10 MPa for the CPC–chitosan–fiber to survive 2 million cycles, compared to the highest stress of 5 MPa for the CPC control.

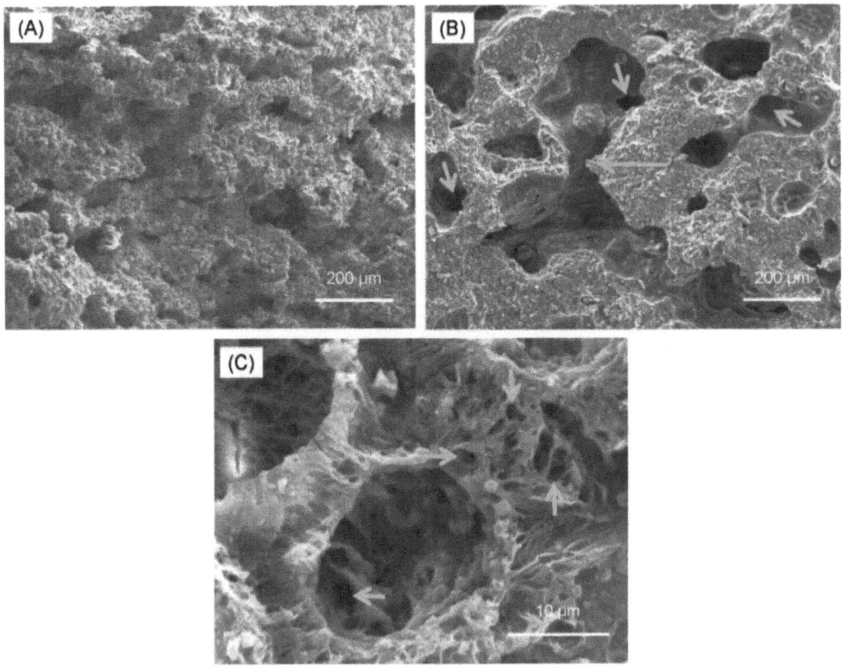

FIGURE 11.5

Fatigue of (A) CPC control and (B) CPC with 15% chitosan and 20% fibers (designated as CPC–chitosan–fiber). Data points near the left axis are for specimens that failed at 1 cycle. Arrows near the right axis indicate specimens that survived 2×10^6 cycles without fracture. CPC–chitosan–fiber specimens failed at higher stresses than CPC control.

Adapted from R. Zhao, E.F. Burguera, H.H.K. Xu, N. Amin, H. Ryou, D.D. Arola, Fiber-reinforced calcium phosphate scaffold: fatigue and human umbilical cord stem cell seeding. Biomaterials 93 (1) (2010) 93–105 with permission.

After one cycle, the stress-to-failure for CPC–chitosan–fiber was nearly doubled compared to the CPC control. After multiple cycles, the stress-to-failure for CPC–chitosan–fiber was 1.5 times that of the CPC control.

11.7 NANOAPATITE SCAFFOLD–HUMAN UMBILICAL CORD STEM CELL INTERACTIONS

hUCMSCs were purchased from ScienCell laboratories (hUCMSCs #7530, Carlsbad, CA). They were obtained from the umbilical cord of a healthy baby born by normal term delivery. The hUCMSCs were harvested as described in the literature [49–53]. Cells were cultured in a low glucose Dulbecco's modified

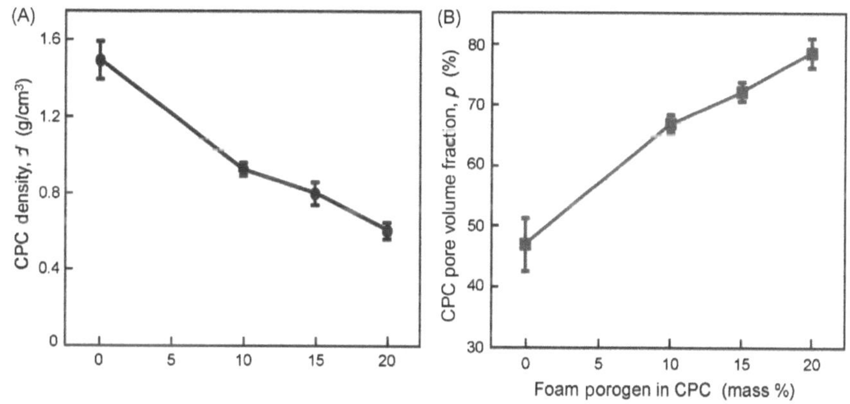

FIGURE 11.6

Stress-to-failure values for CPC control and CPC–chitosan–fiber specimens failed after (A) one cycle and (B) multiple cycles. In each plot, values with dissimilar letters are significant different ($P < 0.05$).

Adapted from R. Zhao, E.F. Burguera, H.H.K. Xu, N. Amin, H. Ryou, D.D. Arola, Fiber-reinforced calcium phosphate scaffold: fatigue and human umbilical cord stem cell seeding. Biomaterials 93 (1) (2010) 93–105 with permission.

Eagle's medium (DMEM) with 10% fetal bovine serum (FBS) and 1% penicillin/streptomycin (PS) (Invitrogen, Carlsbad, CA). Two materials were seeded with hUCMSCs: CPC control and CPC–chitosan–fiber; 150,000 cells [54] were diluted into 2 mL of media and added to each well containing a CPC disk. Fig. 11.7 shows SEM micrographs of hUCMSCs seeded on: (A) CPC control and (B) CPC–chitosan–fiber (with 20% by volume of fibers). In (A), hUCMSCs (designated as "C") had polygonal shapes and were anchored on CPC. In (B), cells attached to the fibers of CPC–chitosan–fiber scaffold. The cells had developed cytoplasmic extensions "E," which are visible in (A) and (B), and are shown in (C) attaching to the fiber "F."

A Wst-1 assay [37] quantified the metabolic activity of the hUCMSCs cultured for 8 days on CPC control and CPC–chitosan–fiber scaffold. The cell viability, measured using the absorbance at 450 nm, was proportional to the amount of dehydrogenase activity in the cells. This absorbance was measured to be 1.3 ± 0.2 for the CPC control and 1.2 ± 0.1 for the CPC–chitosan–fiber scaffold. Hence the stronger and tougher CPC–chitosan–fiber scaffold yielded hUCMSC viability that matched that of the CPC control.

Quantitative real-time reverse transcription polymerase chain reaction (qRT-PCR) was used to determine osteogenic differentiation of hUCMSCs on CPC. hUCMSCs were seeded on three materials which are CPC control (containing no chitosan or fibers), CPC–chitosan, and CPC–chitosan–fiber scaffold (20% fibers). CPC control was referred to as the FDA-approved CPC because

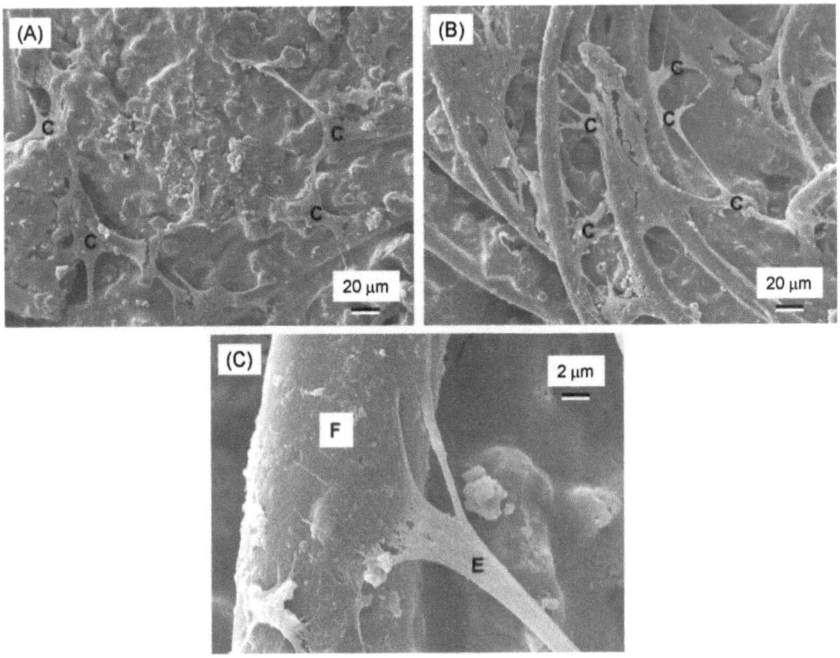

FIGURE 11.7

SEM of hUCMSC attachment on: (A) CPC control and (B) CPC–chitosan–fiber scaffold. Cells are designated as "C," which anchored to CPC in (A), and to the fibers in the scaffold in (B). Cells developed long, cytoplasmic extensions "E," shown in (C) at a higher magnification, attaching firmly to the fiber in the CPC–chitosan–fiber scaffold.

Adapted from R. Zhao, E.F. Burguera, H.H.K. Xu, N. Amin, H. Ryou, D.D. Arola, Fiber-reinforced calcium phosphate scaffold: fatigue and human umbilical cord stem cell seeding. Biomaterials 93 (1) (2010) 93–105 with permission.

that material consisted of the same TTCP–DCPA mixture with no chitosan or fibers [30]. A total 150,000 cells were diluted into 2 mL of osteogenic media and added to each well containing a CPC disk of 2 mm thickness and 12 mm in diameter. For qRT-PCR (7900HT, Applied Biosystems, Foster City, CA), cells were incubated for 1, 4, and 8 days. The total cellular RNA on the scaffolds was extracted with TRIzol reagent (Invitrogen) and reverse transcribed into cDNA. Relative expression level for each target gene was evaluated using the $2^{-\Delta\Delta Ct}$ method [54]. The results are plotted in Fig. 11.8. Alkaline phosphatase (ALP) gene expression was minimal at 1 day; it peaked at 4 days, and then slightly decreased at 8 days. Osteocalcin (OC) gene expression was at highest after 8 days. In both cases, incorporating chitosan and fibers, which greatly improved the mechanical properties, did not adversely affect ALP and OC. Instead, the ALP and OC on the CPC–chitosan–fiber scaffold were higher ($P < 0.05$) than those on the FDA-approved CPC control.

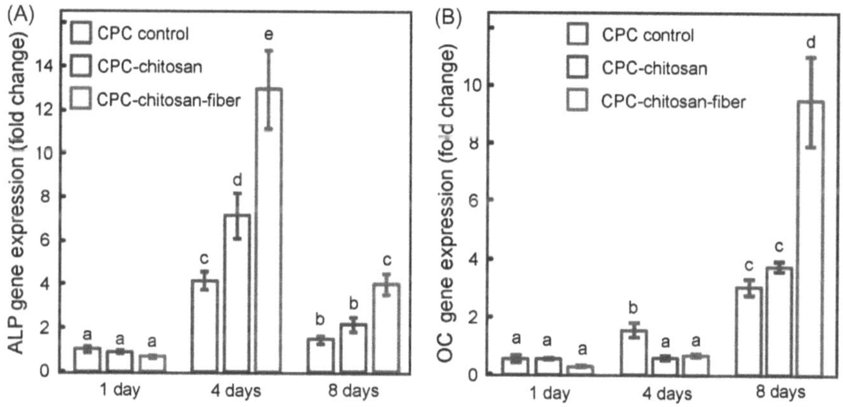

FIGURE 11.8

hUCMSC osteogenic differentiation. (A) qRT-PCR for hUCMSCs yielded the ALP gene expression. (B) qRT-PCR yielded OC gene expression. Each value is mean ± SD; $n = 5$. In each plot, bars with dissimilar letters indicate significantly different values ($P < 0.05$).

11.8 SEEDING BONE MARROW STEM CELLS ON NANOAPATITE SCAFFOLDS

Besides hUCMSCs, human bone marrow mesenchymal stem cells (hBMSC) were also seeded on CPC-based nanoapatite scaffolds. Adult mesenchymal (or stromal) stem cells derived from the bone marrow are multipotent, able to differentiate into neural tissue, cartilage, bone, or fat [5–10]. hBMSCs are emerging as an important tool to engineer bone tissues. hBMSCs can be harvested from patient's bone marrow, expanded in culture, and combined with a scaffold carrier to deliver the cells to defective bone.

hBMSCs were cultured at 37°C with 5% CO_2 in low glucose DMEM (Gibco, Carlsbad, CA). This media was supplemented with 10% FBS (Hyclone, Minneapolis, MN), 1% PS, 0.25% gentamicin, and 0.25% fungizone. This media is referred to as the control media. The osteogenic media consisted of the control media supplemented with 100 nM dexamethasone, 0.05 mM ascorbic acid, and 10 mM β-glycerophosphate. At 90% confluence, cells were harvested by rinsing with 0.25% trypsin, 0.03% EDTA solution, and incubated at room temperature until the cells detached. A live/dead cell viability assay was performed to assess hBMSC viability. Cell viability and proliferation were investigated at 1, 4, and 8 days [54]. The cells were observed using epifluorescence microscopy. The percentage of live cells was defined as $P_{LIVE} = N_{LIVE}/(N_{LIVE} + N_{DEAD})$, where N_{LIVE} is the number of live cells and N_{DEAD} is the number of dead cells, in the same image [55]. Fig. 11.9 shows hBMSCs on CPC-based nanoapatite scaffolds [56]. Live cells formed an almost confluent monolayer and attained a polygonal morphology consistent with osteogenic differentiation. There were few dead cells on

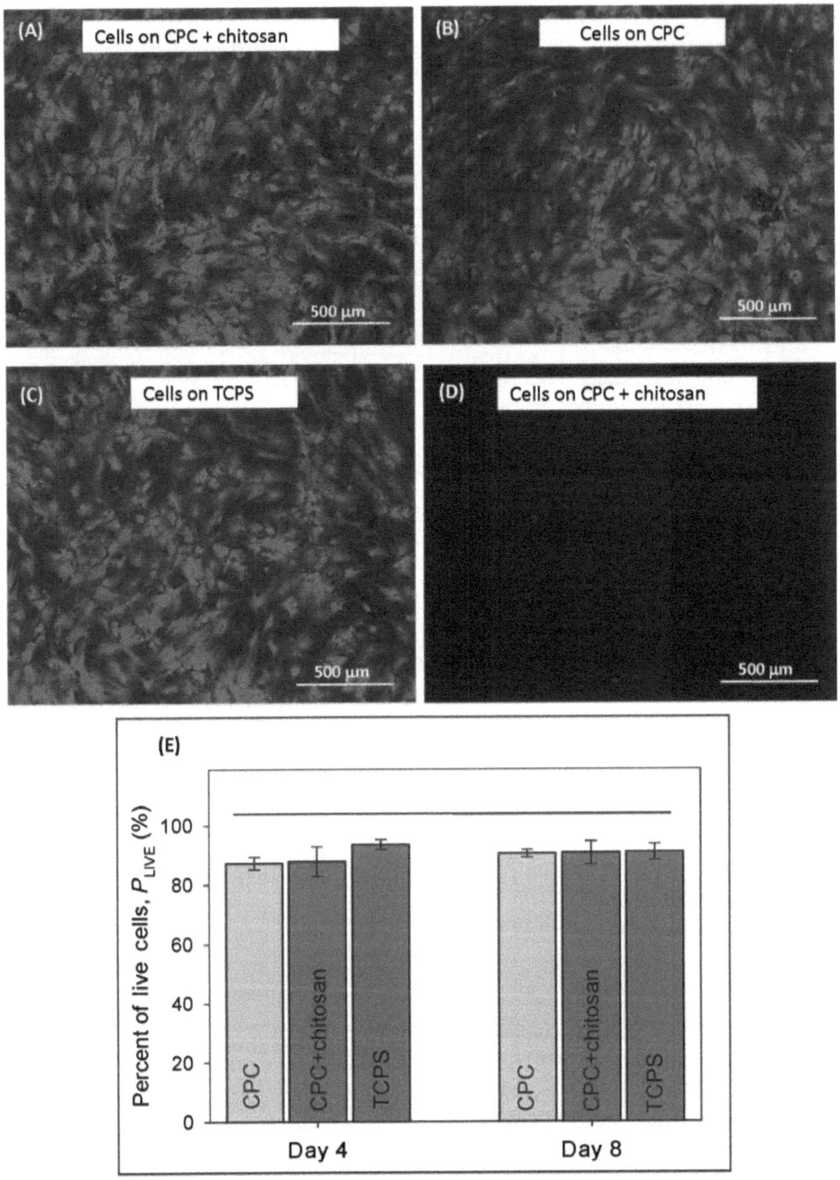

FIGURE 11.9

hBMSCs cultured for at 8 days in osteogenic media. (A) Live cells (stained green) forming a spread, polygonal morphology on CPC–chitosan, (B) live cells on CPC, (C) live cells on TCPS, (D) dead cells on CPC–chitosan, (E) percent of live cells at days 4 and 8. Horizontal line indicates values that are not significantly different ($P > 0.1$). Each value is mean ± SD; $n = 5$.

Adapted from M.D. Weir, H.H.K. Xu, Culture human mesenchymal stem cells with calcium phosphate cement scaffolds for bone repair. J. Biomed. Mater. Res. B. 93 (1) (2010) 93–105 with permission.

all materials at day 8. In (E), the percent of live cells at 4 days was similar for CPC at 87.8% ± 5.1% and CPC–chitosan at 87.3% ± 2.1% ($P > 0.1$). At 8 days, the viability of cells increased to 90.5% ± 1.3% for CPC–chitosan and 90.7% ± 3.8% for CPC control. At 8 days, there was no significant difference in the percent of live cells between CPC, CPC–chitosan, and TCPS ($P > 0.1$).

Fig. 11.10 shows hBMSC attachment and mineral formation on CPC–chitosan at (A and B) 4 days and (C and D) 8 days [56]. The hBMSCs had a polygonal morphology, typical for stem cells undergoing osteogenic differentiation. Arrows in Fig. 11.10A indicate globular mineral formation and collagen bundles. In Fig. 11.10B, mineral formation is more clearly seen as granules. Similar features and morphologies have been shown to be mineralization by cells in a previous study [57]. The cell in Fig. 11.10C is indicated by the letter "C" and the cytoplasmic extensions are indicated by "E." Mineralization on the cell surface at 8 days is shown in Fig. 11.10D.

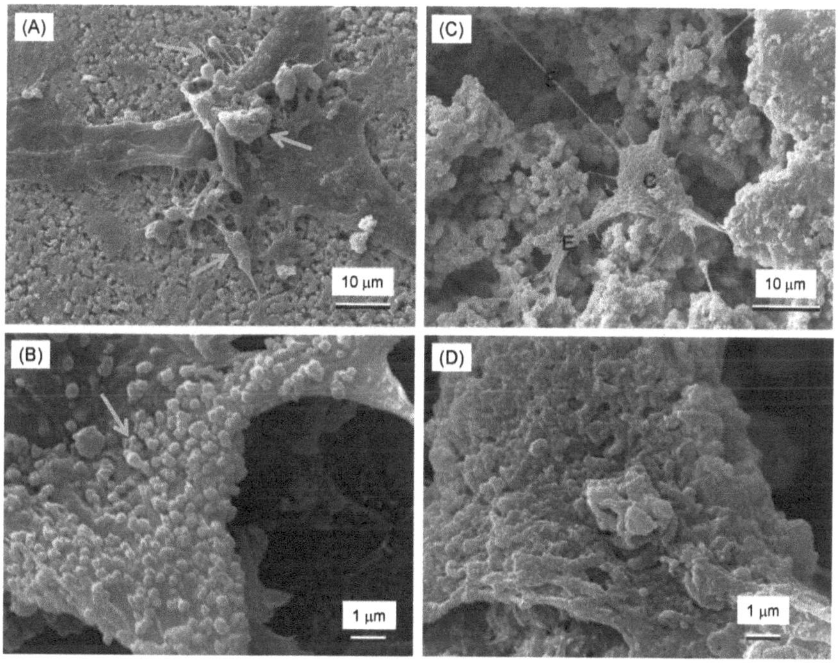

FIGURE 11.10

SEM of hBMSCs on CPC–chitosan specimen: (A and B) 4 days; (C and D) 8 days. Arrows in (A) indicate globular mineral formation and collagen bundles. At higher magnification in (B), mineral formation is more clearly seen as individual mineral granules. The cell body in (C) is indicated by the letter C, and the cytoplasmic extensions of the cell are indicated by E. (D) Mineralization on cell surface at 8 days, similar to those at 4 days in (B).

Adapted from M.D. Weir, H.H.K. Xu, Culture human mesenchymal stem cells with calcium phosphate cement scaffolds for bone repair. J. Biomed. Mater. Res. B. 93 (1) (2010) 93–105 with permission.

Potential dental and craniofacial applications of the high-strength, nanoapatitic CPC-based scaffold with the delivery of stem cells include major reconstructions of the maxilla or mandible after trauma or tumor resection. These treatments would greatly benefit from a moldable implant with shaping and aesthetic capabilities, possessing improved fracture resistance and rapid bone regeneration. Other potential uses of the nanoapatitic composite scaffold include direct filling of bulk bone defects, and minimally invasive surgeries such as in situ fracture fixation and percutaneous vertebroplasty to fill and strengthen osteoporotic bone lesions at risk for fracture. The strong and tough CPC-based nanoapatitic composite scaffolds may be advantageous vehicles to deliver stem cells to facilitate bone regeneration in a wide range of dental, craniofacial, and orthopedic applications.

11.9 CONCLUSIONS

A new class of biomimetic nanoapatite scaffolds was developed using calcium phosphate cement, biopolymer chitosan, collagen fibers, other degradable fibers, and porogens. This family of nanoapatitic composite scaffolds showed relatively high strength and toughness, and the feasibility for stem cell delivery and bone tissue engineering. The macroporous scaffolds were suitable for cell infiltration, and the apatite crystals and the collagen fibers enhanced cell attachment. Resistance to fracture and cyclic fatigue were greatly increased, rendering the new scaffolds promising for a wide range of moderate load-bearing dental, craniofacial, and orthopedic repairs. The biomimetic scaffolds showed excellent biocompatibility with stem cells. hUCMSCs and hBMSCs attached to these scaffolds and successfully differentiated down the osteogenic lineage. The nanoapatite–fiber scaffold morphologically mimicked the extracellular matrix of natural bone and supported stem cell attachment and function. Potential applications include the major reconstructions of the maxilla, mandible, and other craniofacial restorations, bone regeneration after trauma or tumor resection, in situ fracture fixation, and the filling and strengthening of osteoporotic bone lesions at risk for fracture. All these applications could benefit from a self-setting, biomimetic nanoapatite scaffold/stem cell construct with enhanced fracture and fatigue resistance, and rapid bone regeneration capability. Further studies are needed to investigate the in vivo bone regeneration efficacy of this new class of nanoapatitic composite scaffold/stem cell constructs.

ACKNOWLEDGMENTS

We thank Drs. L.C. Chow, S. Takagi, and A.A. Giusseppetti at the Paffenbarger Research Center, and Dr. John Fisher at the University of Maryland for discussions and help. This study was supported by NIH R01 grants DE14190 and DE17974 (HX), Maryland Stem Cell Research Fund (HX), and the University of Maryland Dental School.

REFERENCES

[1] A. Praemer, S. Furner, D.P. Rice, Musculoskeletal Conditions in the United States, American Academy of Orthopaedic Surgeons, Rosemont, IL, 1999.

[2] A.M.A. Ambrosio, J.S. Sahota, Y. Khan, C.T. Laurencin, A novel amorphous calcium phosphate polymer ceramic for bone repair. I. Synthesis and characterization, J. Biomed. Mater. Res. 58B (2001) 295–301.

[3] C.T. Laurencin, A.M.A. Ambrosio, M.D. Borden, J.A. Cooper, Tissue engineering: orthopedic applications, Ann. Rev. Biomed. Eng. 1 (1999) 19–46.

[4] K.Y. Lee, E. Alsberg, D.J. Mooney, Degradable and injectable poly(aldehyde guluronate) hydrogels for bone tissue engineering, J. Biomed. Mater. Res. 56 (2001) 228–233.

[5] J. Yao, S. Radin, G. Reilly, P.S. Leboy, P. Ducheyne, Solution-mediated effect of bioactive glass in poly(lactic-*co*-glycolic acid)-bioactive glass composites on osteogenesis of marrow stromal cells, J. Biomed. Mater. Res. 75A (2005) 794–801.

[6] N. Datta, H.L. Holtorf, V.I. Sikavitsas, J.A. Jansen, A.G. Mikos, Effect of bone extracellular matrix synthesized *in-vitro* on the osteoblastic differentiation of marrow stromal cells, Biomaterials 26 (2005) 971–977.

[7] J.J. Mao, W.V. Giannobile, J.A. Helms, S.J. Hollister, P.H. Krebsbach, M.T. Longaker, Craniofacial tissue engineering by stem cells, J. Dent. Res. 85 (2006) 966–979.

[8] Y. Wang, H.J. Kim, G. Vunjak-Novakovic, D.L. Kaplan, Stem cell based tissue engineering with silk biomaterials, Biomaterials 27 (2006) 6064–6082.

[9] D.S.W. Benoit, A.R. Durney, K.S. Anseth, The effect of heparin-functionalized PEG hydrogels on three-dimensional human mesenchymal stem cell osteogeneic differentiation, Biomaterials 28 (2007) 66–77.

[10] J.J. Mao, G. Vunjak-Novakovic, A.G. Mikos, A. Atala, Regenerative Medicine: Translational Approaches and Tissue Engineering, Artech House, Boston and London, 2007.

[11] P. Ducheyne, Q. Qiu, Bioactive ceramics: the effect of surface reactivity on bone formation and bone cell function, Biomaterials 20 (1999) 2287–2303.

[12] W.L. Murphy, S. Hsiong, T.P. Richardson, G.A. Simmons, D.J. Mooney, Effects of a bone-like mineral film on phenotype of adult human mesenchymal stem cells *in-vitro*, Biomaterials 26 (2005) 303–310.

[13] R.Z. LeGeros, Biodegradation and bioresorption of calcium phosphate ceramics, Clin. Mater. 14 (1993) 65–88.

[14] K.A. Hing, S.M. Best, W. Bonfield, Characterization of porous hydroxyapatite, J. Mater. Sci. Mater. Med. 10 (1999) 135–145.

[15] R.M. Pilliar, M.J. Filiaggi, J.D. Wells, M.D. Grynpas, R.A. Kandel, Porous calcium polyphosphate scaffolds for bone substitute applications—*in-vitro* characterization, Biomaterials 22 (2001) 963–972.

[16] T.M.G. Chu, D.G. Orton, S.J. Hollister, S.E. Feinberg, J.W. Halloran, Mechanical and *in-vivo* performance of hydroxyapatite implants with controlled architectures, Biomaterials 23 (2002) 1283–1293.

[17] S. Radin, G. Reilly, G. Bhargave, P.S. Leboy, P. Ducheyne, Osteogenic effects of bioactive glass on bone marrow stromal cells, J. Biomed. Mater. Res. 73A (2005) 21–29.

[18] J. Russias, E. Saiz, S. Deville, K. Gryn, G. Liu, R.K. Nalla, Fabrication and *in-vitro* characterization of three-dimensional organic/inorganic scaffolds by robocasting, J. Biomed. Mater. Res 83A (2007) 434–445.
[19] P. Miranda, A. Pajares, E. Saiz, A.P. Tomsia, F. Guiberteau, Mechanical properties of calcium phosphate scaffolds fabricated by robocasting, J. Biomed. Mater. Res. 85A (2008) 218–227.
[20] C. Durucan, P.W. Brown, Low temperature formation of calcium-deficient hydroxyapatite-PLA/PLGA composites, J. Biomed. Mater. Res. 51 (2000) 717–725.
[21] M.P. Ginebra, A. Rilliard, E. Fernández, C. Elvira, J.S. Román, J.A. Planell, Mechanical and rheological improvement of a calcium phosphate cement by the addition of a polymeric drug, J. Biomed. Mater. Res. 57 (2001) 113–118.
[22] J.E. Barralet, T. Gaunt, A.J. Wright, I.R. Gibson, J.C. Knowles, Effect of porosity reduction by compaction on compressive strength and microstructure of calcium phosphate cement, J. Biomed. Mater. Res. 63B (2002) 1–9.
[23] A. Yokoyama, S. Yamamoto, T. Kawasaki, T. Kohgo, M. Nakasu, Development of calcium phosphate cement using chitosan and citric acid for bone substitute materials, Biomaterials 23 (2002) 1091–1101.
[24] M. Bohner, U. Gbureck, J.E. Barralet, Technological issues for the development of more efficient calcium phosphate bone cements: a critical assessment, Biomaterials 26 (2005) 6423–6429.
[25] M. Bohner, G. Baroud, Injectability of calcium phosphate pastes, Biomaterials 26 (2005) 1553–1563.
[26] D.P. Link, J. van den Dolder, J.J. van den Beucken, J.G. Wolke, A.G. Mikos, J.A. Jansen, Bone response and mechanical strength of rabbit femoral defects filled with injectable CaP cements containing TGF-β1 loaded gelatin microspheres, Biomaterials 29 (2008) 675–682.
[27] D.P. Link, J. van den Dolder, J.J. van den Beucken, V.M. Cuijpers, J.G. Wolke, A.G. Mikos, Evaluation of the biocompatibility of calcium phosphate cement/PLGA microparticle composites, J. Biomed. Mater. Res. 87A (2008) 760–769.
[28] W.E. Brown, L.C. Chow, P.W. Brown, A new calcium phosphate water setting cement, in: P.W. Brown (Ed.), Cements Research Progress, The American Ceramic Society, Westerville, OH, 1986, pp. 352–379.
[29] M.L. Shindo, P.D. Costantino, C.D. Friedman, L.C. Chow, Facial skeletal augmentation using hydroxyapatite cement, Arch. Otolaryngol. Head Neck Surg. 119 (1993) 185–190.
[30] C.D. Friedman, P.D. Costantino, S. Takagi, L.C. Chow, Bone source hydroxyapatite cement: a novel biomaterial for craniofacial skeletal tissue engineering and reconstruction, J. Biomed. Mater. Res. (Appl. Biomater.) 43 (1998) 428–432.
[31] L.C. Chow, Calcium phosphate cements: chemistry, properties, and applications, Mater. Res. Symp. Proc. 599 (2000) 27–37.
[32] H.H.K. Xu, J.B. Quinn, Calcium phosphate cement containing resorbable fibers for short-term reinforcement and macroporosity, Biomaterials 23 (2002) 193–202.
[33] H.H.K. Xu, S. Takagi, J.B. Quinn, L.C. Chow, Fast-setting and anti-washout calcium phosphate scaffolds with high strength and controlled macropore formation rates, J. Biomed. Mater. Res. 68A (2004) 725–734.
[34] E.F. Burguera, H.H.K. Xu, S. Takagi, L.C. Chow, High early-strength calcium phosphate bone cement: effects of dicalcium phosphate dehydrate and absorbable fibers, J. Biomed. Mater. Res. 75A (2005) 966–975.

[35] G.D. Brown, B.L. Mealey, P.V. Nummikoski, S.L. Bifano, T.C. Waldrop, Hydroxyapatite cement implant for regeneration of periodontal osseous defects in humans, J. Periodontol. 69 (1998) 146–157.
[36] M.H. Ross, G.I. Kaye, W. Pawlina, Histology, fourth ed, Wolters Kluwer Co, Baltimore, MD, 2003 (Chapter 8).
[37] H.H.K. Xu, C.G. Simon, Fast setting calcium phosphate–chitosan scaffold: mechanical properties and biocompatibility, Biomaterials 26 (2005) 1337–1348.
[38] H.H.K. Xu, L.E. Carey, C.G. Simon, Macroporous scaffold of premixed calcium phosphate bone cement: mechanical properties and cell response, J. Mater. Sci. Mater. Med. 18 (2007) 1345–1353.
[39] International Standards Organization. ISO 10993-5, Biological Evaluation of Medical Devices. Part 5. Tests for *in-vitro* Cytotoxicity. International Standards Organization, Geneva, Switzerland, 1999.
[40] H.H.K. Xu, M.D. Weir, C.G. Simon, Injectable and strong nano-apatite scaffolds for cell/growth factor delivery and bone regeneration, Dent. Mater. 24 (2008) 1212–1222.
[41] Y. Miyamoto, K. Ishikawa, M. Takechi, T. Toh, T. Yuasa, M. Nagayama, Basic properties of calcium phosphate cement containing atelocollagen in its liquid or powder phases, Biomaterials 19 (1998) 707–715.
[42] U. Hempel, A. Reinstorf, M. Poppe, U. Fischer, M. Gelinsky, W. Pompe, Proliferation and differentiation of osteoblasts on biocement D modified with collagen type I and citric acid, J. Biomed. Mater. Res. 71B (2004) 130–143.
[43] J.L. Moreau, M.D. Weir, H.H.K. Xu, Self-setting collagen–calcium phosphate bone cement: mechanical and cellular properties, J. Biomed. Mater. Res. A 91 (2009) 605–613.
[44] R.A.A. Muzzarelli, G. Biagini, M. Bellardini, L. Simonelli, C. Castaldini, G. Fraatto, Osteoconduction exerted by methylpyrolidinone chitosan in dental surgery, Biomaterials 14 (1993) 39–43.
[45] H.H.K. Xu, J.B. Quinn, S. Takagi, L.C. Chow, Processing and properties of strong and non-rigid calcium phosphate cement, J. Dent. Res. 81 (2002) 219–224.
[46] R. Zhao, E.F. Burguera, H.H.K. Xu, N. Amin, H. Ryou, D.D. Arola, Fiber-reinforced calcium phosphate scaffold: fatigue and human umbilical cord stem cell seeding, Biomaterials 93 (1) (2010) 93–105.
[47] D. Arola, R. Reprogel, Tubule orientation and the fatigue strength of human dentin, Biomaterials 27 (2006) 2131–2140.
[48] D. Arola, R. Reprogel, Effects of aging on the mechanical behavior of human dentin, Biomaterials 26 (2005) 4051–4061.
[49] H.S. Wang, S.C. Hung, S.T. Peng, Mesenchymal stem cells in the Wharton's jelly of the human umbilical cord, Stem Cells 22 (2004) 1330–1337.
[50] A. Can, S. Karahuseyinoglu, Concise review: human umbilical cord stroma with regard to the source of fetus-derived stem cells, Stem Cells 25 (2007) 2886–2895.
[51] D. Baksh, R. Yao, R.S. Tuan, Comparison of proliferative and multilineage differentiation potential of human mesenchymal stem cells derived from umbilical cord and bone marrow, Stem Cells 25 (2007) 1384–1392.
[52] M.M. Bailey, L. Wang, C.J. Bode, K.E. Mitchell, M.S. Detamore, A comparison of human umbilical cord matrix stem cells and temporomandibular joint condylar chondrocytes for tissue engineering temporomandibular joint condylar cartilage, Tissue Eng. 13 (2007) 2003–2010.

[53] L. Wang, M. Singh, L.F. Bonewald, M.S. Detamore, Signalling strategies for osteogenic differentiation of human umbilical cord mesenchymal stromal cells for 3D bone tissue engineering, J. Tissue Eng. Regen. Med. 3 (2009) 398–404.
[54] K. Kim, D. Dean, A.G. Mikos, J.P. Fisher, Effect of initial cell seeding density on early osteogenic signal expression of rat bone marrow stromal cells cultured on cross-linked poly(propylene fumarate) disks, Biomacromolecules 10 (2009) 1810–1817.
[55] J.L. Moreau, H.H.K. Xu, Mesenchymal stem cell proliferation and differentiation on an injectable calcium phosphate–chitosan composite scaffold, Biomaterials 30 (2009) 2675–2682.
[56] M.D. Weir, H.H.K. Xu, Culture human mesenchymal stem cells with calcium phosphate cement scaffolds for bone repair, J. Biomed. Mater. Res. B 93 (1) (2010) 93–105.
[57] Y. Zhang, J.R. Venugopal, A. El Turki, S. Ramakrishna, B. Su, C.T. Lim, Electrospun biomimetic nanocomposite nanofibers of hydroxyapatite/chitosan for bone tissue engineering, Biomaterials 29 (2008) 4314–4322.

CHAPTER

Self-assembly of proteins and peptides and their applications in bionanotechnology and dentistry

12

Karthikeyan Subramani[1] and Waqar Ahmed[2]

[1]*Roseman University of Health Sciences, Henderson, NV, United States*

[2]*University of Lincoln, Lincoln, United Kingdom*

CHAPTER OUTLINE

12.1 Introduction ... 231
12.2 Mechanism of Molecular Self-Assembly ... 232
12.3 Classification of Self-Assembly ... 232
12.4 Self-Assembly of Proteins and Peptides ... 233
12.5 Bionanotechnology Applications ... 236
12.6 Peptide Nanofibers, Nanotubes, and Nanowires ... 236
12.7 Three-Dimensional Peptide Matrix Scaffolds ... 242
12.8 Advantages and Limitations of Self-Assembling Peptide Matrix Scaffolds ... 244
12.9 Self-Assembly in Regenerative Biology and Dentistry ... 244
12.10 Conclusions ... 246
References ... 246

12.1 INTRODUCTION

Self-assembly is the fundamental principle which generates and governs structural organization on all scales from molecules to the galaxies. It is the autonomous organization of individual components into patterns or structures without human intervention. Self-assembly has been defined as "the noncovalent interaction of two or more molecular subunits to form an aggregate whose novel structure and properties are determined by the nature and positioning of the individual components" [1]. In other words, self-assembly is defined as "the spontaneous assembly of molecules into structured, stable, noncovalently joined aggregates" [2,3]. Self-assembly is the inherent ability of numerous multimeric biological structures to

Emerging Nanotechnologies in Dentistry. DOI: http://dx.doi.org/10.1016/B978-0-12-812291-4.00012-1

assemble from their component parts through random movements of molecules and formation of weak chemical bonds between surfaces with complementary shapes. A few examples of self-assembled structures that can be found in biological systems are the phospholipid bilayer of human cell membranes [4], RNA [5], and DNA complexes [6]. The detergent surfactant molecules exhibit self-assembly phenomenon due to their amphiphilic properties. The molecular building mechanisms underlying the formation of bacteriophage and viral particles are all based on self-assembly.

12.2 MECHANISM OF MOLECULAR SELF-ASSEMBLY

Molecular self-assembly is a powerful phenomenon borrowed from nature by scientists for fabricating novel supramolecular architectures. Molecular self-assembly is the assembly of molecules without guidance from an outside source. Molecular self-assembly is, by definition, the spontaneous organization of molecules under near-thermodynamic equilibrium conditions into structurally well-defined and stable arrangements through noncovalent interactions [7]. This is mainly governed by weak noncovalent bonds like electrostatic interactions (ionic bonds), hydrogen bonds, hydrophobic and hydrophilic interactions, water-mediated hydrogen bonds, and van der Waals interactions [7,8]. Although these forces are weak, their collective interactions can produce structurally and chemically stable structures. Frequently, molecular self-assembly relies on chemical complementarity and structural compatibility [7,9]. The molecular components need complementary properties such as specific surface characteristics, surface charge, polarizability, mass, and surface functionalities to self-assemble into different physiological forms [9]. Biological molecules like proteins, peptides, nucleic acids, lipids, and other cellular components with complementary properties self-assemble to form the basic biological unit, the cell. Cellular events like amyloid fibril formation, antigen–antibody recognition, chromatin assembly, and phospholipid membrane self-assembly are excellent examples of molecular self-assembly.

12.3 CLASSIFICATION OF SELF-ASSEMBLY

Self-assembly is a native process. It can be classified into two types: static and dynamic [10]. Most research studies done in self-assembly have been focused on static type, while the study of dynamic self-assembly is still in its infancy. Static self-assembly involves systems that are at global or local equilibrium and do not dissipate energy [10]. In static self-assembly, formation of the ordered structure may require energy, but once it is formed, it is stable. A few examples of static self-assembly phenomenon tailored by nature are lipid molecules forming oil droplets in water, four hemoglobin polypeptides forming a functional hemoglobin protein, and the combination of RNA and ribosomal proteins to form a functional ribosome. The other common examples of static self-assembled structures have been illustrated in Fig. 12.1. Dynamic self-assembly occurs when the formation of an ordered state of equilibrium requires dissipation of energy. In other words, the interactions responsible for the formation of structures or patterns between components occur only if the system dissipates energy [10]. The examples of dynamic self-assembled structures have been illustrated in Fig. 12.2.

Self-assembly takes place at molecular, mesoscopic, and macroscopic scales. Based on this criteria, self-assembly has been classified as molecular and nanoscale self-assembly (classical form of self-assembly in chemistry involving atoms, molecules, and crystal formation) and meso- and macroscopic self-assembly (e.g., engineered microparts). Molecular and nanoscale self-assembly can be further classified as intramolecular and intermolecular self-assembly [8]. In intramolecular self-assembly, molecules are often complex polymers which can assemble from the random coil into a well-defined stable structure (secondary and tertiary structure). Protein folding to form secondary and tertiary amine structures is a good example of intramolecular self-assembly. Intermolecular self-assembly is the ability of molecules to form supramolecular assemblies (quaternary structure). Supramolecular micelle formation by surfactant molecules in solution and self-assembled monolayers (SAMs) on a substrate are classic examples of intermolecular self-assembly. Magnetic, capillary, electrostatic, and gravitational forces play a vital role in self-assembly of structures on the meso- and macroscale level [8].

12.4 SELF-ASSEMBLY OF PROTEINS AND PEPTIDES

Proteins are fundamental components and building blocks of all living cells. Proteins have been classified as a group of complex organic macromolecules containing carbon, hydrogen, nitrogen, oxygen, and sulfur. They are composed of one or more chains of amino acids. Amino acids contain an amino ($-NH_2$) and a carboxyl ($-COOH$) group. Two or more amino acids linked by a peptide bond form a peptide molecule. Large numbers of peptide molecules arrange themselves in different fashions to make up different kinds of proteins. Enzymes, hormones, and antibodies are few examples of biological substances that are made up of proteins and are required for the proper functioning of a living organism. Structural analysis of protein molecules has revealed that they take up various shapes to form a stable macroscopic structure. Nature has used self-assembly of proteins and peptides to build a vast array of structures like keratin, collagen, coral, pearl, shell, etc. In the past few decades, numerous researches have been done to understand the structural characteristics that influence the self-assembly of protein and peptide molecules. The self-assembly process is highly dependent on the peptide sequence, concentration, pH, presence of salts, and time or kinetics.

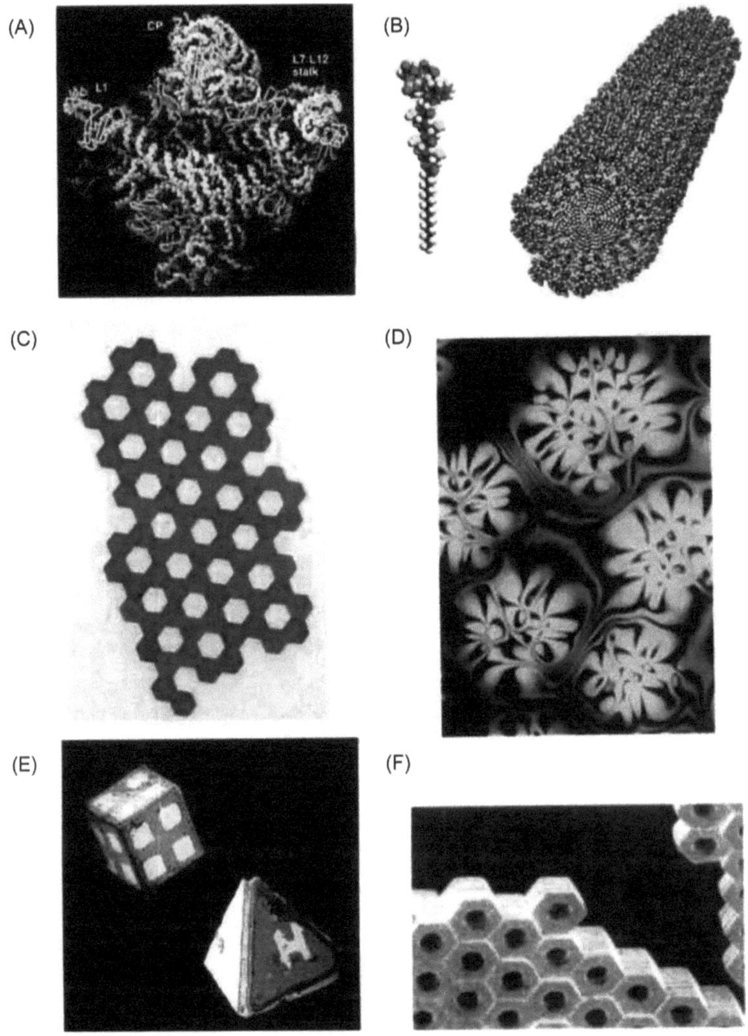

FIGURE 12.1

Examples of static self-assembly. (A) Crystal structure of a ribosome. (B) Self-assembled peptide–amphiphile nanofibers. (C) An array of millimeter-sized polymeric plates assembled at a water–perfluorodecalin interface by capillary interactions. (D) Thin film of a nematic liquid crystal on an isotropic substrate. (E) Micrometer-sized metallic polyhedra folded from planar substrates. (F) A three-dimensional aggregate of micrometer plates assembled by capillary forces.

Image credits: G.M. Whitesides, B. Grzybowski, Self-Assembly at all scales. Science 295 (2002) 2418.

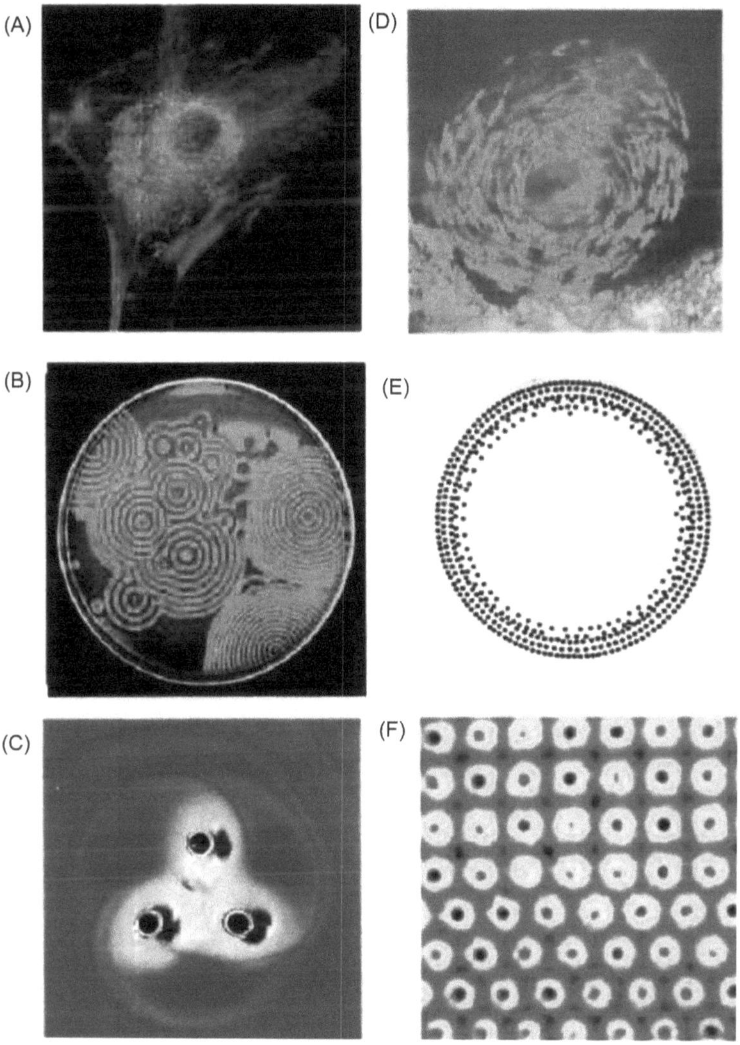

FIGURE 12.2

Examples of dynamic self-assembly. (A) An optical micrograph of a cell with fluorescently labeled cytoskeleton and nucleus. (B) Reaction–diffusion waves in a Belousov–Zabatinski reaction in a 3.5-in petri dish. (C) A simple aggregate of 3-mm sized, rotating, magnetized disks interacting with one another via vortex–vortex interactions. (D) A school of fish. (E) Concentric rings formed by charged metallic beads 1 mm in diameter rolling in circular paths on a dielectric support. (F) Convection cells formed above a micropatterned metallic support. The distance between the centers of the cells is ~2 mm.

Image credits: G.M. Whitesides, B. Grzybowski, Self-Assembly at all scales. Science 295 (2002) 2418.

Self-assembling proteins and peptides have been effectively utilized to design novel biomimetic nanomaterials which have tremendous applications in the fields of bionanotechnology and biomedicine [9–20].

12.5 BIONANOTECHNOLOGY APPLICATIONS

As understood from the word "bionanotechnology," it is the emerging field of science which utilizes biological molecules for nanotechnological applications. Bionanotechnology takes advantage of the unique properties of biological molecules like amphiphilic peptides by utilizing their self-assembling property for the nanoengineering of molecular templates and supramolecular structures [9]. Amphiphiles are molecules containing a nonpolar hydrophobic region and a polar hydrophilic region will self-assemble in aqueous solution to form distinct structures such as micelles, vesicles, and tubules. When suspended in an aqueous solution, the nonpolar hydrophobic regions of amphiphilic molecules are attracted toward each other and away from water (hydrophobic effect). The shape and dimensions of supramolecular structures formed from such assemblies will then depend on different factors, such as the structure of the polar head group and the shape of each amphiphile [11]. Several self-assembling amphiphilic peptide and protein systems that self-assemble to form various nanostructures like nanofibers, nanotubes, vesicles, helical ribbons, and fibrous scaffolds have been described extensively for their potential applications in the field of bionanotechnology [12].

12.6 PEPTIDE NANOFIBERS, NANOTUBES, AND NANOWIRES

Stupp et al. [13] utilized the change in pH to induce self-assembly of amphiphilic peptides to nanoscale fibers which formed nanostructured fibrous scaffold reminiscent of extracellular matrix (ECM). In a later study, Stupp et al. [14] fabricated nanofibers by exploiting the electrostatic attraction between two bioactive peptide–amphiphile molecules. Artificial proteins that self-assemble to form hydrogels in response to pH and environmental changes have been reported by Tirrell et al. [15]. The self-assembling proteins were made up of ionic self-complementary peptide group that had an alternating polar and nonpolar fashion of arrangement of peptide molecules. These peptides formed stable β-strand and β-sheet structures which self-assembled to form nanofibers. These nanofibers formed interwoven matrices that self-assembled to form a scaffold hydrogel with high water content (Fig. 12.3). Hydrogel has water as its dispersion medium and responds to changes in pH and other environmental factors. These protein hydrogels can be used for advanced wound closure and tissue repair in regenerative medicine [16,17] and tissue engineering [18,19]. Biodegradable self-assembled protein and peptide nanofibers can also act as target-specific drug delivery systems delivering drug

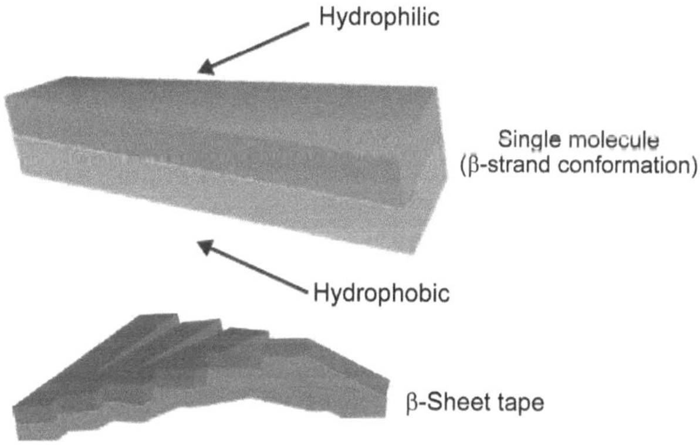

FIGURE 12.3

Amphiphilic peptides in β-strand and β-sheet conformation self-assemble into interwoven matrices that further form a scaffold hydrogel.

Image credits: S. Zhang, D.M. Marini, W. Hwang, S. Santoso, Design of nano biological materials through self-assembly of peptides and proteins. Curr. Opin. Chem. Biol. 6 (2002) 865.

molecules [20]. Immunohistochemical studies of self-assembling peptides injected in vivo did not show obvious inflammation or immune response, thus demonstrating that self-assembled peptides can act as effective drug delivery systems with better biocompatibility [21–23].

The self-assembling amphiphilic peptide molecules have also been utilized as scaffolds to fabricate nanowires. A nanowire is a solid metallic cylinder-like structure with a diameter ranging from 10 nm to a several 100 nm [12]. The amphiphilic peptide nanotubes can be used as templates for metallization [13]. One such example of the peptide sequence used was the histidine-rich peptide nanotubes (Fig. 12.4). This structure was metallized with gold nanocrystals, and the organic peptide scaffold was removed to make a conducting gold nanowire [14–16]. Specific peptide sequences recognize specific metal ions and bind to them. Thus, by varying the peptide sequence, much efficient metal coatings can be coated on peptide nanotubes. The examples of other efficient metals that can be coated on peptide nanotubes are silver, platinum, copper, and nickel. Zhang et al. [12] have designed peptide nanotubes using surfactant-like peptides. These peptide nanotubes have been used as templates for growing metal nanocrystals, and thus nanowires can be fabricated (Fig. 12.5). Surface-binding peptides can bind covalently with metal surfaces like gold (Fig. 12.6). These peptides can also form complexes with DNA which in turn can be bound to a metal surface.

Supramolecular self-assembly of hydrophobic dipeptides into nanotubes was later described by Görbitz [25]. These peptide nanotubes can also serve as ion channels when incorporated in phospholipid bilayer of the cell membrane. Biomimetic protein structures and peptide systems that can form complexes with metals and semiconducting elements have also been reported earlier [26].

FIGURE 12.4

(A) Scanning electron microscopy (SEM) and (B) TEM image of a peptide nanotube.

Image credits: M. Yang, J. Jiang, Y. Lu, Y. He, G. Shen, R. Yu, Functional histidine/nickel hexacyanoferrate nanotube assembly for biosensor applications. Biomaterials 28 (2007) 3408 [24].

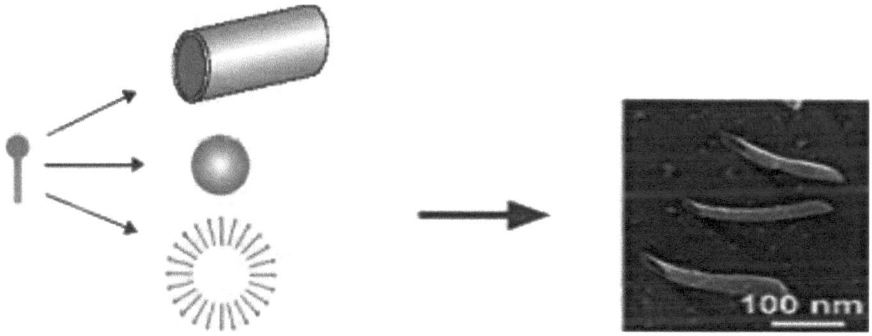

FIGURE 12.5

Surfactant-like peptides self-assemble to form nanotubes.

Image credits: S. Zhang, D.M. Marini, W. Hwang, S. Santoso, Design of nano biological materials through self-assembly of peptides and proteins. Curr. Opin. Chem. Biol. 6 (2002) 865.

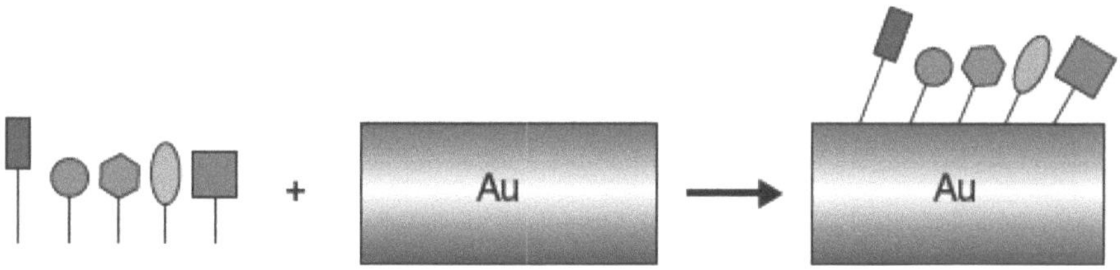

FIGURE 12.6

Surface-binding peptides binding with gold surface.

Image credits: S. Zhang, D.M. Marini, W. Hwang, S. Santoso, Design of nano biological materials through self-assembly of peptides and proteins. Curr. Opin. Chem. Biol. 6 (2002) 865.

Certain peptides with strong dipoles undergo drastic conformational changes between the α-helical structure and the β-sheet [27]. These are called molecular switch peptides (Fig. 12.7). Gold nanoparticles can be attached to these dipolar peptides to fabricate tiny molecular switches. As described earlier, amphiphilic peptide molecules have a hydrophilic (polar) and a hydrophobic (nonpolar) component. In aqueous solution, they self-assemble into distinct structures whose shape is largely determined by the size and shape of the hydrophilic polar head. These molecules have one or two amino acids in the polar region and four or more consecutive hydrophobic amino acids in their nonpolar end. One such example described is the V6D amino acid complex [17]. The V6D amino acid sequence (VVVVVVD) has six valine (V) residues that are hydrophobic and an aspartic acid (D) residue that is negatively charged. Valine (α-aminoisovaleric acid) is an essential amino acid, and aspartic acid a nonessential amino acid. In aqueous solution, the V6D peptide complex self-assembled to form various nanostructures like nanotubes and nanovesicles (Fig. 12.8). The samples of the

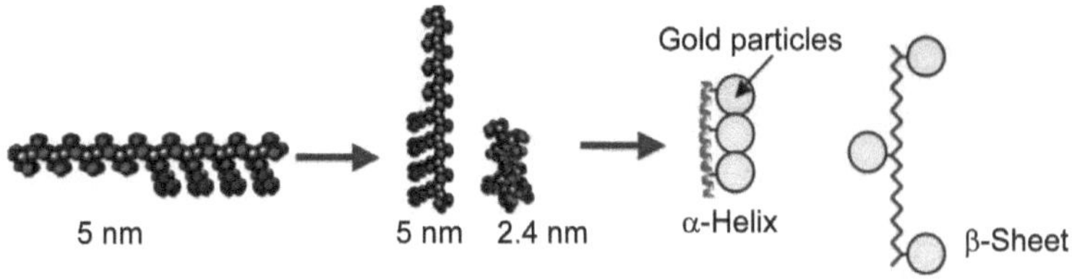

FIGURE 12.7

Helical dipolar peptides undergoing conformational changes and can be used as tiny molecular switches with gold nanoparticles attached.

Images adapted from S. Zhang, D.M. Marini, W. Hwang, S. Santoso, Design of nano biological materials through self-assembly of peptides and proteins. Curr. Opin. Chem. Biol. 6 (2002) 865; M. Altman, P. Lee, A. Rich, S. Zhang, Conformational behavior of ionic self-complementary peptides. Protein Sci. 9 (2000) 1095.

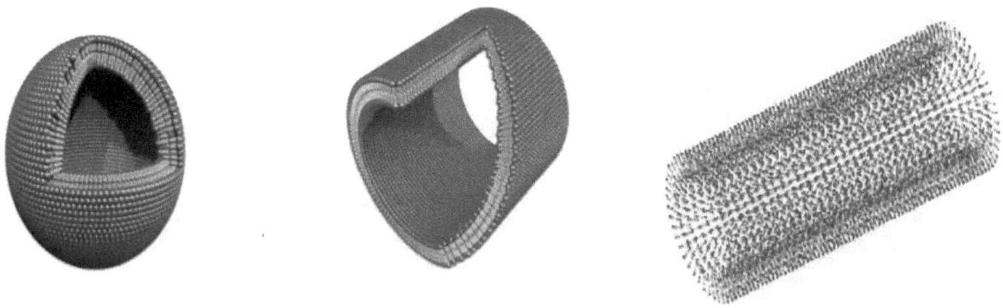

FIGURE 12.8

Molecular models of V6D peptide nanovesicle and nanotube.

Image credits: S. Zhang, D.M. Marini, W. Hwang, S. Santoso, Design of nano biological materials through self-assembly of peptides and proteins. Curr. Opin. Chem. Biol. 6 (2002) 865.

aqueous solution were frozen in liquid propane (−180°C) and surface coated with a thin layer of platinum and carbon to preserve the structures formed, and when investigated using transmission electron microscopy, Zhang et al. [12,17] observed nanotubes and nanovesicles (Fig. 12.9). The nanotubes measured about 30−50 nm. The researchers concluded that the self-assembled structure can be modified by changing the sequence of the amino acids in the peptide chain and the environmental factors. These peptide nanotubes can be incorporated into self-assembled membranes for application in nanobiosensor devices [17].

The researchers also suggest that these surfactant peptides can be engineered for improved functionality by using a technique called biotinylation [12]. Biotinylation is a process of incorporation of biotin groups into molecules to visualize specific substrates by incubating them with biotin-labeled probes and avidin or streptavidin [28]. It is a rapid method of detecting nucleic acids for use in Western blot technique [29]. When these surfactant peptides nanostructures are made to undergo the process of biotinylation, they can be bound to streptavidin-coated

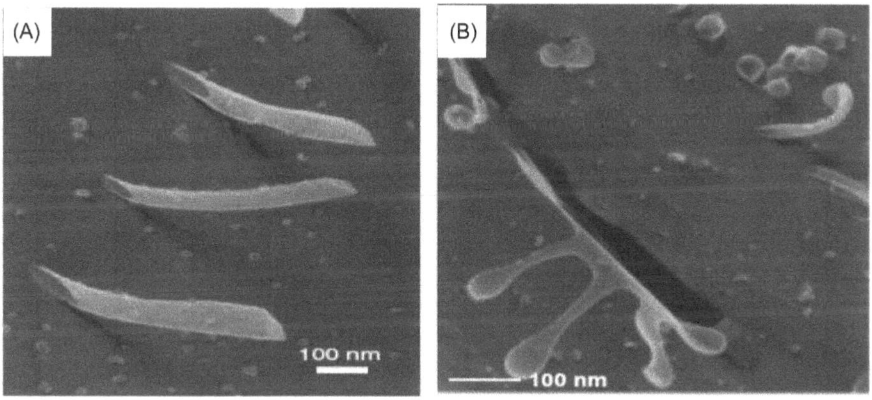

FIGURE 12.9

Quick-freeze/deep-etch transmission electron micrographs of V6D peptide complex. (A) Nanotubes with a diameter of 30–50 nm; (B) nanovesicles and nanotubes.

Image credits: S. Zhang, D.M. Marini, W. Hwang, S. Santoso, Design of nano biological materials through self-assembly of peptides and proteins. Curr. Opin. Chem. Biol. 6 (2002) 865; S. Vauthey, S. Santoso, H. Gong, N. Watson, S. Zhang, Molecular self-assembly of surfactant-like peptides to form nanotubes and nanovesicles. Proc. Natl. Acad. Sci. USA 99 (2002) 5355.

inorganic metal surface. Histidine-tagged peptides and proteins can be bound to nickel surfaces. Thus, by utilizing the standard techniques in peptide chemistry, these nanostructures can be bound to metallic surfaces.

Various other methods have been developed to attach peptides and gold nanocrystals [30]. Crystalline glycylglycine bolaamphiphile peptide tubules and studied the pH-sensitive structural transformation of the peptide tubules. These bolaamphiphiles undergo self-assembly in solution and form tubular structures with diameters ranging from 20 nm to 1 μm. The peptide nanotube was coated with a metalloporphyrin compound. Metalloporphyrins are a group of chemical compounds which have a metallic group attached to a porphyrin ring [31]. Porphyrins are a group of closely related tetrapyrollic pigments occurring widely in nature and play a vital role in various biological processes. The heme component of the hemoglobin is a good example of this group of compounds. The researchers utilized the electron-transferring property of the metalloporphyrins to bind to the peptide nanotubes and were able to grow and immobilize the peptide nanotubes on a gold substrate. They also demonstrated that the peptide nanotubes can be coated with avidin that enabled them to bind to gold surfaces treated with biotinylated SAMs [32] which can be applied to fabricate nanobiosensors. A wider range of applications and the design flexibility of self-assembling protein and peptide complexes together with knowledge of cell biology, biochemistry, and molecular biology have enabled scientists to identify potential newer applications in the field of material science and nanotechnology [33–35].

12.7 THREE-DIMENSIONAL PEPTIDE MATRIX SCAFFOLDS

A wider variety of self-assembling peptide nanofibers has been utilized to fabricate three-dimensional peptide matrix scaffolds [13,14,36,37]. One of the main factors influencing the entire process was the chirality of the peptide molecules. Two molecules are said to be chiral if their mirror images do not superimpose on each other. One such peptide sequence is the KFE8 peptide complex which is a natural amphiphilic eight-residue peptide complex [34]. The KFE8 has the peptide chain sequence of FKFEFKFE (Figs. 12.10 and 12.11). This structure represents a group of self-assembling peptides that spontaneously self-assemble under certain physiological conditions. This peptide molecule self-assembles in aqueous solution into left-handed helical ribbons when the peptide backbone is twisted in the opposite direction. The researchers replaced certain amino acids in the hydrophobic side chains with another amino acid sequence and observed the changes [38]. When a positively charged lysine was replaced by a positively charged arginine (Arg) or when a negatively charged glutamate was replaced by a negatively charged aspartate, they observed very little changes in the nanofibers that were formed. But when the positively charged residue was replaced by a negatively charged residue or vice versa, the peptides did not self-assemble. When the alanines were replaced by hydrophobic residues, there was a greater tendency to self-assemble and form peptide matrices with enhanced strength.

This behavior of the peptide molecules has led the researchers to concentrate their attention toward understanding the basis of protein conformational diseases. Protein conformational diseases are a group of disorders characterized by accumulation of malformed protein structures in cells. Proteins must fold into a proper three-dimensional structure to perform their normal physiological functions. But when they do not fold properly, they form malfolded protein structures that accumulate in cells leading to pathological conditions. Alzheimer's disease, Prion disease, and Parkinson's disease are a few examples of protein conformational diseases. Thus, by understanding the mechanism of formation of peptide

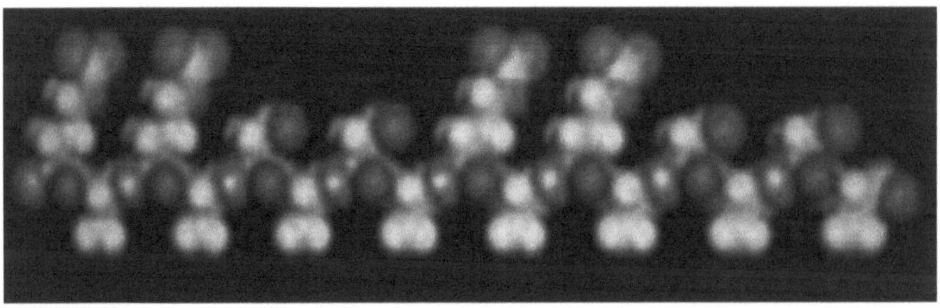

FIGURE 12.10

Molecular model of KFE8 peptide.

Image credits: S. Zhang, Building from bottom-up. Mater. Today 5 (2003) 20 [38].

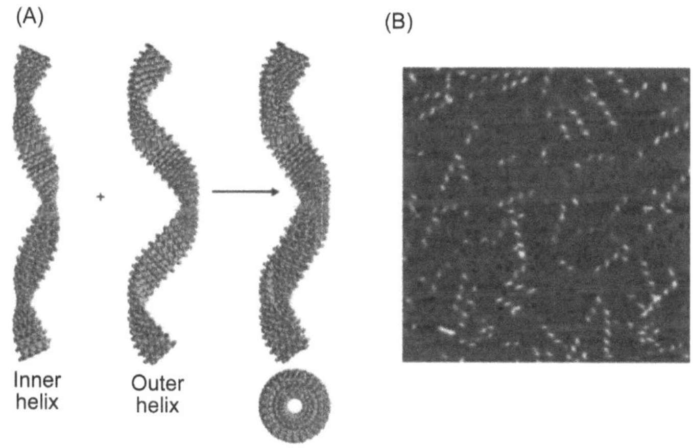

FIGURE 12.11

(A) The inner and outer β-helices of KFE8 peptide form a double-sheet helix with hydrophobic side chains sandwiched between the two layers. (B) Atomic force microscopy (AFM) image (500 nm × 500 nm) of peptide solution deposited over mica.

Image credits: S. Zhang, Building from bottom-up. Mater. Today 5 (2003) 20.

nanofibers and the factors controlling their self-assembly, future research work can be aimed in formulating remedies for protein conformational diseases. Earlier researches done on three-dimensional peptide matrix scaffolds were not only limited to natural amphiphilic peptides, and synthetic complex amphiphilic peptides were also produced by joining 12 hydrophilic peptides into long alkyl chains by pH control, divalent ion induction, and change in concentration [36]. The peptide end of the molecule was designed to regulate biomineralization. Bone is produced as a result of deposition of calcium and phosphate ions to form hydroxyapatite crystals. This process is known as mineralization. Serine is a nonessential amino acid. When a phosphorylated serine was incorporated with the synthetic amphiphilic peptide complex, it served to attract and organize calcium and phosphate ions to form hydroxyapatite crystals [37]. The researchers further functionalized the synthetic amphiphilic peptide complex by adding a cell-adhesion motif [37]. It was the Arg–glycine–aspartic acid (RGD) complex that was attached to the C-terminus of the peptide. This can be used to study the ability of the bone cells to differentiate, proliferate, and adhere to a biomaterial surface like titanium. Titanium alloys offer several benefits, including lower elastic modulus, excellent corrosion resistance, and enhanced biocompatibility [39]. Due to these characteristics, titanium is the most widely used biomaterial surface to produce orthopedic implants, dental implants, and hip replacements [40–42]. In spite of its excellent biocompatibility, titanium implants still fail [43,44]. Most orthopedic implants have a lifetime of 15 years to the maximum [45]. In order to fabricate titanium implants with lesser failure rate, its surface has to be modified with nanosized surface patterns so that bone cells (osteoblasts) can attach, differentiate, and migrate into these patterns resulting in enhanced bone-implant adhesion. For such a purpose, poly(ethylene glycol) (PEG) hydrogel micropatterns with osteoinductive growth factors have been reported [46]. These osteoinductive PEG micropatterns were shown to stimulate rat osteoblast migration. Self-assembling synthetic amphiphiles and peptide hydrogels can be fabricated on titanium surfaces to regulate and control the osteoblasts [47]. These protein and peptide sequences can be efficiently functionalized with osteoinductive growth factors and cell-adhesion motifs like the RGD peptide sequence [47].

12.8 ADVANTAGES AND LIMITATIONS OF SELF-ASSEMBLING PEPTIDE MATRIX SCAFFOLDS

As understood from the earlier studies, macroscopic three-dimensional peptide matrices can be engineered from various self-assembling peptides. The advantages of using designer peptides for the fabrication of three-dimensional matrices for tissue engineering include (1) easy design using known biologically active peptide sequences, (2) these scaffolds provide the opportunity to incorporate a number of different functional motifs and their combinations to study cell behavior in a well-defined ECM-analog microenvironment, not only without any chemical cross-link reactions but also with fully bioresorbable scaffolds, (3) biodegradable by a variety of proteases in the body with superior biocompatibility with the tissues, (4) designer peptides can be readily modified at the single amino acid level at will, inexpensively and quickly using conventional, commercially chemical peptide synthesis methods, and (5) these designer peptide scaffolds are pure with known motifs and can be used to study controlled gene expression or cell-signaling processes [48,49]. Thus, these new designer nanofiber scaffolds have proven to be promising tools for bionanotechnological applications. In enthusing about the many good properties of self-assembling proteins and peptides, it is important not to lose sight of the limitations. If a material is copied from nature, then one cannot assume that the material will perform well in environments or applications that are very different from those found in nature. Self-assembly of proteins and peptides have unique capabilities as a chemical processing method and the biologic characteristics. Thus far, it has penetrated little from the laboratory to manufacturing and practical applications. The limitations on its employment of these self-assembled matrix scaffolds appear to be practical rather than fundamental [50,51].

12.9 SELF-ASSEMBLY IN REGENERATIVE BIOLOGY AND DENTISTRY

As self-assembled peptide matrix scaffolds encouraged cell proliferation and differentiation and were also able to support various types of cell attachments

[52,53], the ability of the peptides to support attachment of mouse neuronal cells was evaluated using self-assembling ionic self-complementary β-sheet oligopeptides [52]. The primary mouse neuron cells formed active connections with the peptide scaffolds. These three-dimensional scaffolds are being tested for their applications in neuron regeneration and treatment of related neuropathology [54,55]. In regenerative medicine, these peptide matrices can be used to cultivate chondrocyte ECM which can further be used to repair cartilage tissue. Cartilage tissue engineering has been done by placing the primary chondrocytes and mesenchymal stem cells into these self-assembled peptide hydrogels to produce collagen and glycosaminoglycans [53,56]. These peptide matrices can also be used in regeneration of bone by incorporating a phosphorylated serine which can attract and organize calcium ions to form hydroxyapatite crystals and functionalizing them with a cell-adhesion motif like RGD complex, and recent research developments have proved the feasibility of this concept [57–60].

The production of bone, dentine, and enamel-like biomaterials for the engineering of mineralized (hard) tissues is a high priority in regenerative medicine and dentistry. Amelogenin is a protein found in developing tooth enamel, and it belongs to the family of ECM proteins. Developing enamel contains about 30% protein, and 90% of this is amelogenins. Other significant proteins in enamel are ameloblastins, enamelins, and tuftelins. The function of amelogenins is believed to be in organizing enamel rods during tooth development. The latest research indicates that this protein regulates the initiation and growth of hydroxyapatite crystals during the mineralization of enamel. In addition, amelogenins appear to aid in the development of cementum by directing cells that form cementum on the root surface of teeth. Under near-physiological pH, temperature, and ionic strength, amelogenin accelerates hydroxyapatite nucleation kinetics, in vitro [61]. Hierarchically organized apatite microstructures have been achieved by self-assembly involving nucleated nanocrystallites and amelogenin oligomers and nanospheres at low supersaturation and protein concentrations in a slow and well-controlled constant composition system. This method has allowed the capture of an intermediate structure, the nanorod, following the formation of the critical nuclei at the earliest nucleation stages of calcium phosphate crystallization. The nanorod building blocks are formed spontaneously by synergistic interactions between flexible amelogenin protein assemblies and rigid calcium phosphate nanocrystallites. These intermediate structures further assembled by a self-epitaxial growth mechanism to form the final hierarchically organized microstructures that are compositionally and morphologically similar to natural enamel. This in vitro observation showed that amelogenin promoted apatite crystallization and organization. Another recent study on amelogenin supramolecular self-assembly showed that recombinant full-length amelogenin (rP172) self-assembled into nanochain structures through electrolytic deposition process and had a significant effect on the induction of the parallel bundles of calcium phosphate nanocrystals grown on semiconductive silicon wafer surface [62]. This study suggests that the formation of organized bundles in amelogenin–apatite composites is mainly driven by amelogenin nanochain self-assembly and highlights the potential of such composite for future application as dental restorative materials. Such self-assembly systems of amelogenin may maximally exert an influence on the structural control of developing enamel crystals at the earliest nucleation stages in vivo for enamel repair or caries treatment.

The β-sheet-forming peptides that spontaneously self-assembled to form three-dimensional fibrillar scaffolds in response to specific environmental triggers have been suggested to be of use in skeletal tissue engineering and the treatment/prevention of dental caries, via bioactive surface groups. In a recent study, it was hypothesized that infiltration of caries and lesions with monomeric low-viscosity peptide solutions would be followed by in situ polymerization triggered by conditions of pH and ionic strength, providing a biomimetic scaffold capable of hydroxyapatite nucleation, promoting enamel repair [63]. This study explored the effect of an anionic peptide applied to caries-like lesions in human dental enamel under simulated intraoral conditions of pH cycling. The results showed that peptide treatment significantly increased net mineral gain by the lesions, due to both increased remineralization and inhibition of demineralization over a 5-day period. The assembled peptide was also capable of inducing hydroxyapatite nucleation de novo. The study suggested that self-assembling peptides may be useful in the modulation of mineral behavior during in situ dental tissue engineering.

12.10 CONCLUSIONS

As understood from the various research studies, proteins and peptides can self-assemble into various structures like nanotubes, nanovesicles, and three-dimensional peptide matrices with interwoven nanofibers. The increasing interest in bionanotechnology has stimulated numerous researchers to scrutinize biological elements and to learn from nature. These nanoscale structures and their versatile applications have inspired numerous researchers to apply them in various fields of bionanotechnology, tissue engineering, regenerative medicine, and dentistry. The formation of self-assembled structure is more energy efficient than directed self-assembly. Further investigation of self-assembly of peptides will contribute in a significant way to the development of novel nanobiomaterials and play an important role in the progress of emerging field of nanobiotechnology, regenerative medicine, and dentistry.

REFERENCES

[1] P. Tecilla, R.P. Dixon, G. Slobodkin, D.S. Alavi, D.H. Waldeck, A.D. Hamilton, Hydrogen-bonding self-assembly of multichromophore structures, J. Am. Chem. Soc. 112 (1990) 9408.

[2] G.M. Whitesides, J.P. Mathias, C.T. Seto, Molecular self-assembly and nanochemistry: a chemical strategy for the synthesis of nanostructures, Science 254 (1991) 1312.

[3] G.M. Whitesides, M. Boncheva, Beyond molecules: self-assembly of mesoscopic and macroscopic components, Proc. Natl. Acad. Sci. USA 99 (2002) 4769.
[4] C.C. Ng, Y.L. Cheng, P.S. Pennefather, Properties of a self-assembled phospholipid membrane supported on lipobeads, Biophys. J. 87 (2004) 323.
[5] L. Nasalean, S. Baudrey, N.B. Leontis, L. Jaeger, Controlling RNA self-assembly to form filaments, Nucleic Acids Res. 34 (2006) 1381.
[6] J. Won, S.K. Chae, J.H. Kim, H.H. Park, Y.S. Kang, H.S. Kim, Self-assembled DNA composite membranes, J. Membr. Sci. 249 (2005) 113.
[7] S. Zhang, Molecular Self-Assembly, Encyclopedia of Materials: Science & Technology, Elsevier Science, Oxford, UK, 2001.
[8] J.A. Pelesko, Self Assembly: The Science of Things That Put Themselves Together, Chapman & Hall/CRC, London, 2007.
[9] S. Zhang, Fabrication of novel biomaterials through molecular self-assembly, Nat. Biotechnol. 21 (2003) 1171.
[10] G.M. Whitesides, B. Grzybowski, Self-assembly at all scales, Science 295 (2002) 2418.
[11] J. Lee, C.W. Macosko, D.W. Urry, Elastomeric polypentapeptides crosslinked into matrixes and fibers, Biomacromolecules 2 (2001) 170.
[12] S. Zhang, D.M. Marini, W. Hwang, S. Santoso, Design of nano biological materials through self-assembly of peptides and proteins, Curr. Opin. Chem. Biol. 6 (2002) 865.
[13] J.D. Hartgerink, E. Beniash, S.I. Stupp, Self-assembly and mineralization of peptide−amphiphile nanofibers, Science 294 (2001) 1684.
[14] K.L. Niece, J.D. Hartgerink, J.J. Donners, S.I. Stupp, Self-assembly combining two bioactive peptide−amphiphile molecules into nanofibers by electrostatic attraction, J. Am. Chem. Soc. 125 (24) (2003) 7146.
[15] W.A. Petka, J.L. Harden, K.P. McGrath, D. Wirtz, D.A. Tirrell, Reversible hydrogels from self-assembling artificial proteins, Science 281 (1998) 389.
[16] S. Halstenberg, A. Panitch, S. Rizzi, H. Hall, J.A. Hubbell, Biologically engineered protein-*graft*-poly(ethylene glycol) hydrogels: a cell adhesive and plasmin-degradable biosynthetic material for tissue repair, Biomacromolecules 3 (2002) 710.
[17] S. Vauthey, S. Santoso, H. Gong, N. Watson, S. Zhang, Molecular self-assembly of surfactant-like peptides to form nanotubes and nanovesicles, Proc. Natl. Acad. Sci. USA 99 (2002) 5355.
[18] J. Kisiday, M. Jin, B. Kurz, H. Hung, C. Semino, S. Zhang, Self-assembling peptide hydrogel fosters chondrocyte extracellular matrix production and cell division: implications for cartilage tissue repair, Proc. Natl. Acad. Sci. USA 99 (2002) 9996.
[19] S. Zhang, F. Gelain, X. Zhao, Designer self-assembling peptide nanofiber scaffolds for 3D tissue cell cultures, Semin. Cancer Biol. 15 (2005) 413.
[20] V.F.M. Segers, R.T. Lee, Local delivery of proteins and the use of self-assembling peptides, Drug Discov. Today 12 (2007) 561.
[21] T.C. Holmes, S. de Lacalle, X. Su, G. Liu, A. Rich, S. Zhang, Extensive neurite outgrowth and active synapse formation on self-assembling peptide scaffolds, Proc. Natl. Acad. Sci. USA 97 (2000) 6728.
[22] M.E. Davis, J.P. Motion, D.A. Narmoneva, T. Takahashi, D. Hakuno, R.D. Kamm, Injectable self-assembling peptide nanofibers create intramyocardial microenvironments for endothelial cells, Circulation 111 (2005) 442.
[23] M. Antonietti, Surfactants for novel templating applications, Curr. Opin. Coll. Int. Sci. 6 (2001) 244.

[24] M. Yang, J. Jiang, Y. Lu, Y. He, G. Shen, R. Yu, Functional histidine/nickel hexacyanoferrate nanotube assembly for biosensor applications, Biomaterials 28 (2007) 3408.
[25] C.H. Görbitz, An exceptionally stable peptide nanotube system with flexible pores, Acta Crystallogr. B. 58 (2002) 849.
[26] S.R. Whaley, D.S. English, E.L. Hu, P.F. Barbara, A.M. Belcher, Selection of peptides with semiconductor binding specificity for directed nanocrystal assembly, Nature 405 (2000) 665.
[27] M. Altman, P. Lee, A. Rich, S. Zhang, Conformational behavior of ionic self-complementary peptides, Protein Sci. 9 (2000) 1095.
[28] J.W. Jarvik, C.A. Telmer, Epitope tagging, Annu. Rev. Genet. 32 (1998) 601.
[29] G. Rigaut, A. Shevchenko, B. Rutz, M. Wilm, M. Mann, B. Seraphin, A generic protein purification method for protein complex characterization and proteome exploration, Nat. Biotechnol. 17 (1999) 1030.
[30] H. Matsui, B. Gologan, Crystalline glycylglycine bolaamphiphile tubules and their pH-sensitive structural transformation, J. Phys. Chem. B 104 (2000) 3383.
[31] H. Matsui, R. MacCuspie, Metalloporphyrin nanotube fabrication using peptide nanotubes as templates, Nano Lett. 1 (2001) 671.
[32] H. Matsui, P. Porrata, G.E. Douberly, Protein tubule immobilization on self-assembled monolayers on Au substrates, Nano Lett. 1 (2001) 461.
[33] M.R. Caplan, P.N. Moore, S. Zhang, R.D. Kamm, D.A. Lauffenburger, Self-assembly of a β-sheet protein is governed by relief of electrostatic repulsion relative to van der Waals attraction, Biomacromolecules 1 (2000) 627.
[34] D.M. Marini, W. Hwang, D.A. Lauffenburger, S. Zhang, R.D. Kamm, Left-handed helical ribbon intermediates in the self-assembly of a beta-sheet peptide, Nano Lett. 2 (2002) 295.
[35] C.A. Hauser, S. Zhang, Designer self-assembling peptide nanofiber biological materials, Chem. Soc. Rev. (2010) 2780.
[36] J.D. Hartgerink, E. Beniash, S.I. Stupp, Peptide−amphiphile nanofibers: a versatile scaffold for the preparation of self-assembling materials, Proc. Natl. Acad. Sci. USA 99 (2002) 5133.
[37] E. Beniash, J.D. Hartgerink, H. Storrie, J.C. Stendahl, S.I. Stupp, Self-assembling peptide amphiphile nanofiber matrices for cell entrapment, Acta Biomater. 1 (2005) 387.
[38] S. Zhang, Building from bottom-up, Mater. Today 5 (2003) 20.
[39] M. Long, H.J. Rack, Titanium alloys in total joint replacement—a materials ... materials science perspective, Biomaterials 19 (1998) 1621.
[40] E.B. Taddei, V.A.R. Henriques, C.R.M. Silva, C.A.A. Cairo, Production of new titanium alloy for orthopedic implants, Mater. Sci. Eng. C 24 (2004) 683.
[41] G.R. Parr, L.K. Gardner, R.W. Toth, Titanium: the mystery metal of implant dentistry, J. Prosthet. Dent. 54 (1995) 410.
[42] M. Semlitsch, Titanium alloys for hip joint replacements, Clin. Mater. 2 (1987) 1.
[43] R.L. Buly, M.H. Huo, E. Salvati, W. Brien, M. Bansal, Titanium wear debris in failed cemented total hip arthroplasty. An analysis of 71 cases, J. Arthroplasty 7 (1992) 315.
[44] A. Piattelli, A. Scarano, M. Piattelli, Microscopical aspects of failure in osseointegrated dental implants: a report of five cases, Biomaterials 17 (1996) 1235.
[45] H.J. Rack, J.I. Qazi, Titanium alloys for biomedical applications, Mater. Sci. Eng. C 26 (2006) 1269.
[46] K. Subramani, M.A. Birch, Fabrication of poly(ethylene glycol) hydrogel micropatterns with osteoinductive growth factors and evaluation of the effects on osteoblast activity and function, Biomed. Mater. 1 (2006) 144.

[47] T.D. Sargeant, M.O. Guler, S.M. Oppenheimer, A. Mata, R.L. Satcher, D.C. Dunand, Self-assembly of peptide amphiphile nanofibers within porous titanium, Biomaterials 29 (2008) 161.
[48] F. Gelain, D. Bottai, A. Vescovi, S. Zhang, Designer self-assembling peptide nanofiber scaffolds for adult mouse neural stem cell 3-dimensional cultures, PLoS One 1 (2007) e119.
[49] A. Horii, X. Wang, F. Gelain, S. Zhang, Biological designer self-assembling peptide nanofiber scaffolds significantly enhance osteoblast proliferation, differentiation and 3-D migration, PLoS One 2 (2007) e190.
[50] F. Gelain, A. Horii, S. Zhang, Designer self-assembling peptide scaffolds for 3-D tissue cell cultures and regenerative medicine, Macromol. Biosci. 7 (2005) 544.
[51] C. Viney, Self-assembly as a route to fibrous materials: concepts, opportunities and challenges, Curr. Opin. Solid State Mater. Sci. 8 (2004) 95.
[52] S. Zhang, T.C. Holmes, C.M. DiPersio, R.O. Hynes, X. Su, A. Rich, Self-complementary oligopeptide matrices support mammalian cell attachment, Biomaterials 16 (1995) 1385.
[53] J. Kisiday, M. Jin, B. Kurz, H. Hung, C. Semino, S. Zhang, Self-assembling peptide hydrogel fosters chondrocyte extracellular matrix production and cell division: implications for cartilage tissue repair, Proc. Natl. Acad. Sci. USA 99 (2002) 9996.
[54] R.G. Ellis-Behnke, Y.X. Liang, S.W. You, D.K. Tay, S. Zhang, K.F. So, Nano neuro knitting: peptide nanofiber scaffold for brain repair and axon regeneration with functional return of vision, Proc. Natl. Acad. Sci. USA 103 (2006) 5054.
[55] J. Guo, H. Su, Y. Zeng, Y. Liang, W.M. Wong, R.G. Ellis-Behnke, Reknitting the injured spinal cord by self-assembling peptide nanofiber scaffold, Nanomed. Nanotechnol. Biol. Med. 3 (2007) 311.
[56] J. Kisiday, M. Jin, M.A. DiMicco, B. Kurz, A.J. Grodzinsky, Effects of dynamic compressive loading on chondrocyte biosynthesis in self-assembling peptide scaffolds, J. Biomech. 37 (2004) 595.
[57] H. Hosseinkhani, M. Hosseinkhani, F. Tian, H. Kobayashi, Y. Tabata, Osteogenic differentiation of mesenchymal stem cells in self-assembled-peptide amphiphile nanofibers, Biomaterials 27 (2006) 4079.
[58] H. Hosseinkhani, M. Hosseinkhani, F. Tian, H. Kobayashi, Y. Tabata, Ectopic bone formation in collagen sponge-self assembled peptide amphiphile nanofibers hybrid scaffold in a perfusion culture bioreactor, Biomaterials 27 (2006) 5089.
[59] H. Hosseinkhani, M. Hosseinkhani, A. Khademhosseini, H. Kobayashi, Bone regeneration through controlled release of bone morphogenetic protein-2 from 3-D tissue engineered nano-scaffold, J. Control. Release 117 (2007) 380.
[60] H. Hosseinkhani, M. Hosseinkhani, F. Tian, H. Kobayashi, Y. Tabata, Impregnation of plasmid DNA into 3D scaffold and medium perfusion enhance in vitro DNA expression of mesenchymal stem cells, Tissue Eng. 13 (2007) 11.
[61] L. Wang, X. Guan, H. Yin, J. Moradian-Oldak, G.H. Nancollas, Mimicking the self-organized microstructure of tooth enamel, J. Phys. Chem. C 112 (15) (2008) 5892.
[62] Y. Fan, Z. Sun, R. Wang, C. Abbott, J. Moradian-Oldak, Enamel inspired nanocomposite fabrication through amelogenin supramolecular assembly, Biomaterials 28 (19) (2007) 3034.
[63] J. Kirkham, A. Firth, D. Vernals, N. Boden, C. Robinson, R.C. Shore, Self-assembling peptide scaffolds promote enamel remineralization, J. Dent. Res. 86 (5) (2007) 426–430.

CHAPTER

Surface engineering of dental tools with diamond for enhanced life and performance

13

Waqar Ahmed[1], Htet Sein[2], Mark Jackson[3,4], Christopher Rego[2], Israr-Ul Hassan[5] and Karthikeyan Subramani[6]

[1]*University of Lincoln, Lincoln, United Kingdom* [2]*Manchester Metropolitan University, Manchester, United Kingdom* [3]*Purdue University, West Lafayette, IN, United States* [4]*Kansas State University, Salina, KS, United States* [5]*Dhofar University, Salalah, Oman* [6]*Roseman University of Health Sciences, Henderson, NV, United States*

CHAPTER OUTLINE

13.1 Tooth Materials 251
13.2 Dental Burs 253
13.3 Chemical Vapor Deposition of Diamond Films Onto Dental Burs 256
13.3.1 Plasma-Enhanced CVD 257
13.3.2 Hot Filament CVD 259
13.3.3 Controlling Structure and Morphology 263
13.4 Bur Performance Investigations 273
13.4.1 Tool Preparation 273
13.4.2 CVD Diamond Deposition on the Dental Burs 273
13.4.3 Dental Bur Machining: Drilling Experiments 273
13.4.4 Dental Bur Machining: Machining Experiments on Human Teeth 274
13.4.5 Performance Testing 276
13.4.6 Drilling Experiments 277
13.4.7 Performance Results 278
13.5 Conclusions 284
References 284
Further Reading 288

13.1 TOOTH MATERIALS

The human mouth provides a highly complex environment for the teeth [1]. There are variations in a number of factors. For example, the temperature, which

Emerging Nanotechnologies in Dentistry. DOI: http://dx.doi.org/10.1016/B978-0-12-812291-4.00013-3

can be between below 0 and 70°C, the pH, which can vary from acid to alkaline, the chemical composition, which is highly variable, humidity, the mechanical impact, and applied pressure have a wide range. The conditions are highly dependent on what is being consumed and chemicals released in the mouth. The human teeth need to be able to withstand such a wide range of chemical, thermal, and mechanical conditions. The human teeth are hard calcified structures attached to the upper and lower jaw for cutting, grinding, and chewing actions [2,3]. A typical tooth consists of a crown sticking out of the jaw, the exposed neck area, and root extending down into the socket in the jaw.

The surface of the tooth is covered with enamel with dentine underneath and pulp cavity in the center. There are numerous sensory nerve endings. A typical tooth structure is shown in Fig. 13.1. Over years of repeated use and often lack of proper care, the teeth get damaged due to acid erosion and wear. Cavities develop which require fillings and some teeth to be removed during a lifetime. These can be replaced by artificial teeth, which also need to withstand the severe environment that they will encounter [4,5]. They need to be tough and strong to resist fracture and abrasive forces encountered during brushing. Acrylic resin and porcelain are commonly used materials for artificial teeth. Acrylic resins are crosslinked and made using molding methods such as injection molding. It is a good thermal insulator but susceptible to abrasive forces and has poor resistance to fatigue fracture and impact strength. Porcelain is made from an inner opaque

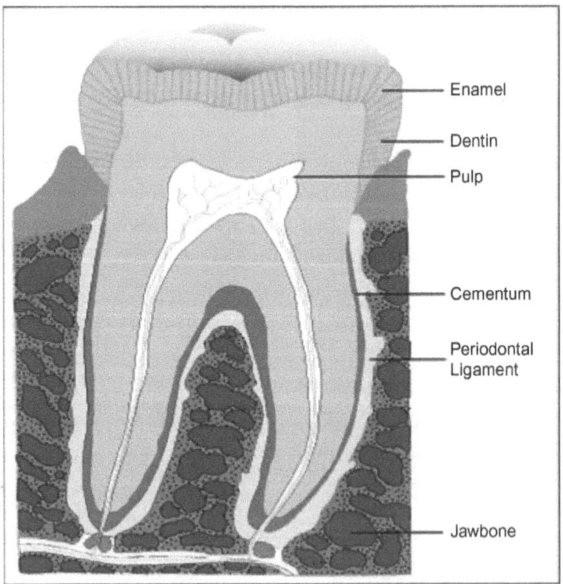

FIGURE 13.1

Structure of the human tooth [2].

core material overlaid with translucent dentine material and a final layer of porcelain on the outside. It is very rigid, hard, and brittle. One of the solutions to improving the properties is to use nanosized powders of alumina to the porcelain. Titanium dental implants are also becoming increasingly popular for replacing teeth that have been removed from the mouth due to cheaper/improved manufacturing technologies becoming available.

Natural and artificial human teeth need regular treatment from qualified dentists. Due to the complex environment, dental technologist and dentists have developed numerous tools of various sizes and shapes to manipulate, drill, and remove unwanted materials. The correct use of the dental tool is dictated by the function required and preference of the dentist. One of the most common dental tools is the dental bur. These also come in various shapes and sizes to enable the dentist to reach complex and hard-to-access regions in the mouth.

13.2 DENTAL BURS

Dental burs have revolutionized the dentistry field and are used for cutting hard tissues such as bone or tooth [1,4,6–9]. They are normally made of stainless steel, diamond grit or particles, and tungsten carbide and fitted to a dental drill incorporating an air turbine. The dental bur was developed <300 years ago [8] and still widely used.

The dental bur [4,8,9] has three parts: the head, the neck, and the shank. The head contains the blades, which produce cutting action by rotary motion. The blades are positioned at various degree angles in order to change the property of the bur. Dentists use this to bore through tooth enamel and to clean and remove plaque from the tooth's surface, and dental technicians use this to prepare dental materials in laboratory. Decayed tooth material is also removed before applying a filling. Dental burs come in many shapes and sizes designed for specific applications and can rotate at speeds of up to 500,000 revolutions per minute (rpm). They can be made of steel and then coated with a hard coating, such as tungsten carbide coating, or they can be entirely tungsten carbide [8,10–13]. Numerous shapes of burs are manufactured for various applications, cutting, and drilling abilities.

The basic design of an eight blades fissure bur [14] is shown in Fig. 13.2.

Generally, geometrical features of the bur are manufactured to have a negative rake angle (Fig. 13.3). However, those with a positive rank angle are designed mainly for cutting soft materials (e.g., acrylics) to remove material during cutting to prevent the tool from clogging with the chips.

When the rake angle of a bur blade is too steep, it can damage the subsurface of the tooth. These weak areas become sites for subsequent bacterial infection. When the rake angle is decreased, a gentler action is achieved. The life of the bur is decreased due to the acute angle of the cutting tips, but less residual subsurface damage occurs to the tooth.

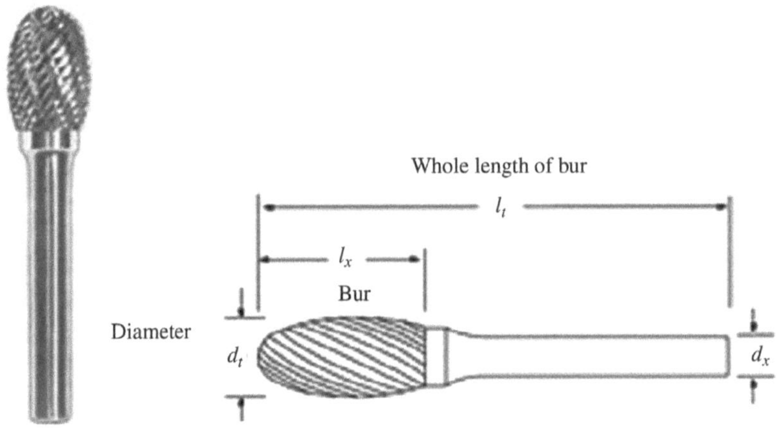

FIGURE 13.2

Dental burs with dimensional parameters showing the bur head [14].

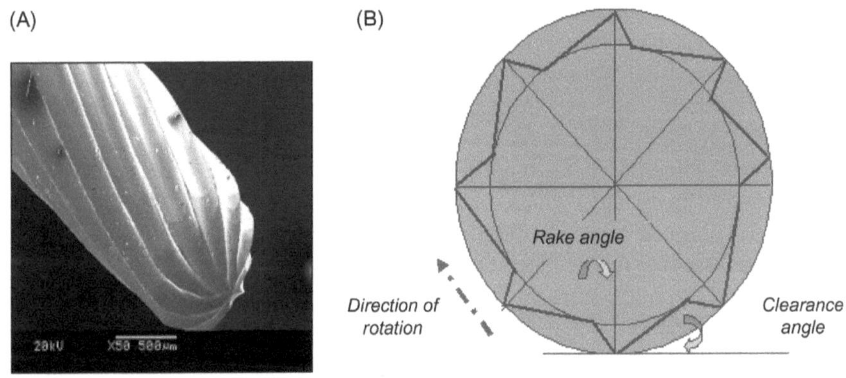

FIGURE 13.3

(A) Geometry of a typical tungsten carbide bur. (B) Schematic diagram showing the rake angle relative to the cutting direction [15].

As mentioned earlier, the two most common dental bur materials used are stainless steel and tungsten carbide [7–9,11,16,17]. Each material has its own specific advantages and disadvantages. Steel wears rapidly and corrodes easily and thus appears to be a poor choice of material for burs. However, at lower speeds, the wear rate associated with the cutting process is reduced significantly, and steel may be the preferred material. The main corrosive environment experienced by the bur is during the sterilization process, although the conditions in the oral cavity will also contribute. Stainless steel provides a less efficient cutting edge than carbon steel but has a greater corrosion resistance. Stainless steel is a cheap material which makes it suitable for burs provided that they are considered single-use and classed as disposable burs.

The high hardness makes WC extremely wear resistant, but it is brittle compared to stainless steel. Therefore, only the blades of a bur should be made of tungsten carbide, and the shank can be made of steel. Sintering is used to join the carbide blades onto the steel shank. Tungsten carbide is also suitable for burs designated for use at lower speeds, when they are to be used many times. Instruments made from tungsten carbide are much more expensive than their steel equivalents, but they compensate for this by their increased working life. The WC burs contains cobalt as about 6% binder, which provides additional toughness to the tool. Two parameters, the Co/WC ratio and the WC particle size, control the bulk properties of the material. In general, course-grained WC combined with a high %Co gives better shock resistance and impact strength. However, the finer-grained WC and thus larger surface area and lower %Co are harder and impart greater wear resistance. Therefore, to achieve optimum performance premature, failure needs to be avoided whilst achieving a higher wear resistance.

Conventional diamond dental bur [15] uses small diamond particles bonded onto the substrate using a binder matrix material (Fig. 13.4). It has inherent limitations, due to the heterogeneity of the grain sizes and shapes, difficulty in automation during fabrication, and short tool life times. Additional problems are also encountered due to repeated sterilization of the instruments which decreases their cutting effectiveness and resulting in diamond particle loss causing oral contamination.

The hand piece is a slender, lightweight tube-shaped device connected to the drill bit and has a driving motor. The hand piece needs to withstand high-pressure steam sterilization. The most important part of the dental drill is the drill bit, or bur, which needs to be robust, durable, and be able to rotate at very high speeds. High speeds generate a considerable amount of heat, which needs to be removed

FIGURE 13.4

SEM micrograph of a conventional sintered diamond bur showing one diamond particles embedded in a binder matrix material [15].

using water-cooling. Often, they are fitted with illumination devices to aid the dentist in performing cutting, drilling, and grinding operations.

Each bur is designed to operate at an optimum speed and intended for use in a specific hand piece motor combination [7]. The optimum speed is also dependent upon the nature of the material being cut. Cutting speed can be separated into two categories, each using different burs and designed for different types of operations.

A miniature compressed air or electric motor is used to drive low-speed hand pieces. They operated with rotational speeds of up to 4000 rpm. The burs used in low-speed operations are used to trim dentures and remove decayed tissue and can be made of stainless steel.

Air turbine hand pieces can run up to 500,000 rpm, although they are rarely used at such high speeds. General operating speeds are 20,000–50,000 rpm varying with the diameter of the bur. The burs used at high speeds are mainly diamond coated or made of tungsten carbide. High speeds are used in the removal of enamel, dentine, and old fillings. When operating at such high speeds, a built in water spray is used as a coolant, protecting both the bur and tooth.

Conditions in the surgery are different from the laboratory. The purpose of coating the bur with diamond is to lengthen the working life of a bur so that it can be used for multiple patients and be able to withstand repeated sterilization [18,19]. This process can increase the risk of instrument fracture and also has the potential to be corrosive and may affect the cutting surface or coating of the bur.

13.3 CHEMICAL VAPOR DEPOSITION OF DIAMOND FILMS ONTO DENTAL BURS

In order to increase the life and performance of dental burs, they have been coated with diamond films. Diamond is an attractive material [12,15,19–44] for this type of application because it has a high hardness value, high thermal conductivity, chemical inertness, and high wear resistance (Table 13.1). Normally, natural diamond is obtained from mines, and it has formed over many years under

Table 13.1 Physical Properties of Diamond [19,21,25]

Property	Value
Hardness (kg/mm^2)	10,000
Strength, tensile (GPa)	>1.2
Strength, compressive (GPa)	>110
Density (g/cm^3)	3.52
Young's modulus (GPa)	1.22
Thermal expansion coefficient (K^{-1})	0.0000011
Thermal conductivity ($W/cm^{-1}/K^{-1}$)	20.0

extreme conditions of high-temperature and -pressure underground. Diamond needs to be formed under conditions that are thermodynamically favorable. From the carbon phase diagram, it is evident that heating carbon under pressure can form diamond. Hence, the high-pressure, high-temperature (HPHT) growth technique [21–23,25] was developed to produce "industrial" diamond. However, this is not suitable for dental burs and therefore a new chemical vapor deposition (CVD) process has been developed to grow diamond onto various substrate materials in thin film form [19,21–30]. This process is conceptually very simple and requires thermal decomposition of carbon containing precursors, such as acetylene, at low pressure. The films deposited have many of the properties such as high hardness and thermal conductivity associated with natural diamond. In addition, polycrystalline diamond (PCD) deposited by CVD has a rough surface making it highly suitable for abrasion.

Diamond films have extreme properties and have been deposited onto a range of substrates. Their structure ranges from single crystals, polycrystalline to amorphous materials depending on the type of growth process used, and the specific process conditions applied. Crystal sizes can be micro- or nanocrystalline depending on the temperature, pressure, and the feed gases used. The success of the CVD methods expands the potential applications of diamond compared to natural or HPHT-synthetic diamond into automotive, consumer, defense, and space applications [19,25,28].

CVD from energetically activated hydrocarbon/hydrogen gas mixtures is the most successful process for depositing diamond. The process is defined as the formation of a thin solid from the gaseous phase via a chemical reaction onto a substrate at an elevated temperature [19,26–28].

A CVD reactor provides several functions including:

- Transport the reactant and diluents gases to the substrate
- Provide sufficient activation energy to reactants
- Maintain specific system pressure and temperature
- Allow film deposition to proceed
- Remove the gaseous by-products.

There are several different approaches for activating precursor gases for diamond [26–31], which include the following.

13.3.1 PLASMA-ENHANCED CVD

Plasmas generated by various forms of electrical discharges or inductions heating have been employed for many years to deposit diamond films. The plasma contains atomic hydrogen and other species containing carbon necessary for diamond growth. These highly energetic species enable diamond films to be deposited at lower substrate temperatures compared to thermal CVD processes. The efficiencies of the different plasma processes vary from method to method. Three plasma frequency regimes are commonly employed. These are *microwave* plasma-enhanced CVD, which typically uses excitation frequencies of 2.45 GHz, *radio-frequency* (RF) plasma excitation, which employs frequencies of usually 13.56 MHz, and *direct-current* (DC) plasmas, which can be run at low electric powers (a "cold" plasma) or at high electric powers (which create an *arc* or a *thermal* plasma).

13.3.1.1 Microwave plasma-enhanced CVD

Microwave plasma-enhanced CVD has been used more extensively and successfully than any other plasma-based method for the growth of diamond films. Microwave plasmas are different from other plasmas in which the microwave frequency can oscillate electrons. Collision of electrons with gaseous atoms and molecules generates a high degree of ionization. This method of diamond film growth has a number of advantages over other plasma methods of diamond film growth. Microwave deposition, being an electrodeless process, avoids contamination of the films due to electrode erosion. Furthermore, the microwave discharge at 2.45 GHz, being a higher-frequency process than the RF, produces a higher plasma density with higher-energy electrons resulting in higher concentrations of atomic hydrogen and other hydrocarbon radicals producing efficient high-quality diamond films. In addition, the plasma is confined to the center of the deposition chamber as a ball, preventing carbon deposition onto the walls of the chamber.

13.3.1.2 RF plasma-enhanced CVD

In this method, RF power is applied to two electrodes for creating a plasma in an inductively coupled or a capacitively coupled parallel plate arrangement. The growth of diamond crystals and thin films using inductively coupled RF plasma methods [34,35] and capacitively coupled methods [36,37] has been widely reported. A high power in the discharge leading to greater electron densities is necessary for efficient diamond growth. However, the use of higher power results in physical and chemical sputtering from the reactor walls, leading to contamination of the diamond films. The advantage of RF plasmas is that they can easily be generated over much larger areas compared to microwave plasmas, but the method is not routinely used for the deposition of diamond films because the quality of the diamond films is inferior.

13.3.1.3 DC plasma-enhanced CVD

In this method, plasma in an H_2–hydrocarbon mixture is excited by applying a DC bias across two parallel plates, one of which is the substrate [38–40]. DC plasma-enhanced CVD has the advantage of being able to coat large areas as the diamond deposition area is limited by the electrodes and the DC power supply. In addition, the technique has the potential for very high growth rates. However, diamond films produced by DC plasmas were reported to be under high stress and contain high concentrations of hydrogen as well as impurities resulting from plasma erosion of the electrodes.

13.3.2 HOT FILAMENT CVD

In this chapter, we consider the hot filament CVD (HFCVD) method in detail because it has particular advantages over the other methods mentioned earlier. Atomic hydrogen can be produced by the passage of H_2 over a refractory metal filament, such as tungsten, molybdenum, and tantalum, heated to a temperature between 2000 and 2500K. When atomic hydrogen was added to the hydrocarbon typically with a C/H ratio of <0.01, it was observed that diamond could be deposited while graphite formation was suppressed [40–45]. The generation of atomic hydrogen during diamond HFCVD enables the following to occur:

- A dramatic increase in the diamond deposition rate is <1 μm/h
- The nucleation and growth of diamond on nondiamond substrates.

Due to its inherent simplicity and comparatively low operating cost, HFCVD has become widely used in industry. Table 13.2 outlines typical deposition parameters used in the growth of diamond films by HFCVD.

A wide variety of refractory materials have been used as filaments including tungsten, tantalum, and rhenium due to their high melting point and high electron emissivity.

13.3.2.1 Growth mechanisms

Various mechanisms for diamond growth [42–46] have been postulated. Thermodynamic near-equilibrium is established in the gas phase at the filament surface. At temperatures of around 2300K, molecular hydrogen dissociates into atomic hydrogen and the H atoms are able to convert methane into methyl radicals, which are generally regarded as the active diamond precursor species, and thereafter into acetylenic species and other hydrocarbons, which are stable at these elevated temperatures. Atomic hydrogen and the neutral and radical hydrocarbon species then diffuse to the substrate surface. Although the gaseous species generated at the filament are in equilibrium at the filament temperature, the species are at a superequilibrium concentration when they arrive at the much cooler substrate. The reactions, which generate these high-temperature species close to the filament where there is a high concentration of hydrogen atoms, proceed faster than any other reactions which decompose these species during the

Table 13.2 Typical Deposition Parameters Used in the Growth of Diamond Films by HFCVD

Gas Mixture	Total Pressure (Torr)	Temperature (K) Substrate	Filament
CH_4 (0.5%–2.0%)/H_2	10–50	1000–1400	2200–2500

transit time from the filament to the substrate. Consider the equilibrium between methane and acetylene:

$$CH^4 + H^* = CH_3^* + H_2 = \frac{1}{2}C_2H_2 + 2H_2$$

At the filament surface, the reaction is immediately driven to the right, creating methyl radicals and thereafter acetylene. Subsequently, the CH_3 radicals and other hydrocarbon species diffuse to the substrate. Thermodynamic equilibrium at the substrate temperature of approximately 1100K calls for the formation of methane, but the reverse reaction proceeds much slower. Solid carbon precipitates on the substrate to reduce the superequilibrium concentration of species such as CH_3 in the gas phase. The allotrope diamond is "stabilized" relative to graphite by a concurrent superequilibrium concentration of atomic hydrogen. This simple explanation emphasizes the importance of reaction kinetics in diamond synthesis by HFCVD. The mechanism of diamond CVD is summarized in Fig. 13.5. Not only do the H atoms play a crucial role in diamond CVD by driving the formation of active hydrocarbon species, but they also prevent both the reconstruction of the growing diamond surface and the formation of graphitic nuclei. The deposition of high quality diamond at appreciable growth rates depends on the ratio of C:H in the precursor gas mixture. Increasing the proportion of carbon results in higher concentrations of CH_3 at the growing diamond surface and hence increased growth rates. However, higher concentrations of carbon also lead to poorer quality films, as there is insufficient H to etch away nondiamond carbon deposits. If the concentration of H at the growing surface is too high, then diamond will be etched, resulting in very low growth rates or no diamond growth at all.

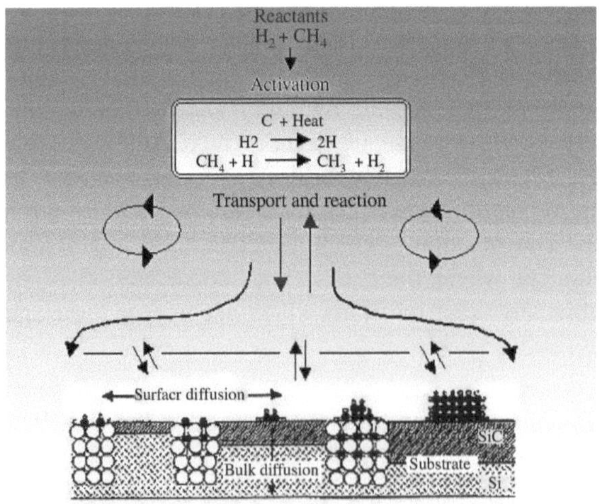

FIGURE 13.5

Schematic of the CVD diamond growth mechanism.

In HFCVD, the transport of the active species and their mixing with the surrounding gas takes place as a result of the interaction of free and forced convection as well as diffusion. A combination of concentration and thermal gradients is the major driving force. In order to achieve uniform film coating and uniform thickness, the temperature of the substrate and the local specie concentration close to the substrate surface need to be uniform.

In diamond HFCVD, the mechanism of heat transfer also plays a major role. In addition to conduction, convection and radiation, and heat transfer is also achieved via atomic hydrogen. This is as expected since the formation of atomic hydrogen at or near the filament surface is highly endothermic. Atomic hydrogen readily recombines on solid surfaces to form molecular hydrogen with the recombination reaction being highly exothermic. Thus, atomic hydrogen acts as a carrier of heat from the filament to the growth surface. Since the quality, morphology, and defect density of diamond films are sensitive to temperature, a uniform substrate temperature distribution is crucial for the deposition of uniform diamond films with consistent properties. An important feature of HFCVD is that the growth step is separate from the film activation step, which ultimately allows independent control of both growth and activation temperatures.

13.3.2.2 Filament characteristics

It is obvious that in a process such as HFCVD, the filament plays a critical role. The key features of the filament characteristics and assembly have been summarized [44,45]. The most commonly used filament materials are tantalum and tungsten due to their high melting point and high electron emissivity. Refractory metals, which form carbides (e.g., tungsten and tantalum), typically must carburize their surface before supporting the deposition of diamond films. The process of filament carburization results in carbon consumption from the hydrocarbon precursor gas. Hence, there is a specific incubation time for the nucleation process which produces diamond films. Therefore, this process may affect the early stages of film growth, although it is insignificant over longer growth periods. Furthermore, the volume expansion produced by carbon incorporation results in cracks along the length of the wire. The development of these cracks is undesirable as it reduces the lifetime of the filament but does not adversely affect the quality of the resulting films.

As well as having a high melting point, suitable filament materials must also have an appreciable electron emissivity to cause dissociation of molecular hydrogen and ultimately initiate the growth process. For these reasons, tungsten (W, melting point 3695K) and tantalum (Ta, melting point 3293K) filaments are typically used in HFCVD processes. The physical data for tungsten and tantalum is shown in Table 13.3. In this research work, Ta filaments are used because there is less contamination of the diamond films than when W filaments are used. As the differences in the chemical natures and physical properties of the filament materials have a direct influence on gas activation and decomposition, diamond deposition may vary slightly given otherwise constant deposition parameters.

Table 13.3 Selected Physical Data for Prospective Filament Materials [47]

Material	Melting Point (K)	Resistivity (Ω m)	Density (kg/m^3)
Tungsten	3695	39×10^{-8}	19,254
Tantalum	3293	63×10^{-8}	16,670

A wide range of literature on HFCVD-based diamond deposition places the low filament temperature limit at 2200K and the upper limit at 2700K for successful diamond growth.

The filament carburization process [44,48] proceeds according to the following sequence of steps:

- Transport of CH_4 from the gas phase to the metal surface and subsequent physical adsorption
- Decomposition of CH_4 on the metal surface where the resulting carbon and hydrogen atoms are chemisorbed
- Liberation of gaseous H, H_2, and CH_X
- Transformation of the adsorbed C atoms into the dissolved state
- Diffusion of carbon atoms into the metal lattice to form subcarbides (M_2C) and carbides (MC).

If the filament temperature falls below a critical value, depending on the carbon concentration, a carbon film forms on the filament. This film deactivates the filament surface (i.e., the filament temperature decreases), and the production of atomic hydrogen is rapidly reduced. If the filament power is not increased immediately to dissolve or vaporize the carbon film, then it will grow thicker leading to a further decrease in the filament temperature. Spectral emissivity measurements aim at measuring the spectral emissivity of the filaments. The filament activity is highest if the filament surface is totally free of carbon. The changes in filament emissivity, resistance, and power consumption have been attributed to the deposition and etching of carbon occurring at the surface.

13.3.2.3 Diamond nucleation process

For obvious reasons, the initial stage of diamond nucleation on a nondiamond substrate is critical particularly when considering the application of the process to dental burs [48–54]. It affects the adhesion, growth, uniformity, and morphology of the diamond films deposited by CVD.

Seeding or abrading with diamond powder or immersing in diamond paste containing small crystallites processed in an ultrasonic bath has been used for many years for diamond film deposition [49–52]. The method promotes nucleation of diamond crystals onto the substrate surface by creating a high density of nucleation sites, which reduces the induction time. This process can be carried out mechanically. The substrate is polished with abrasive grit. Typically, diamond powder of 0.1–10 μm particle sizes is employed.

Bias-enhanced nucleation is another method for the growth of heteroepitaxial diamond films and is much more controllable than mechanical abrasion [44,53]. In this method, the substrate is negatively biased relative to the filament from 0 to −500 V. Intense plasma is created because diamond has a negative electron affinity and emits electron. At the same time, the positively charged ion, generated as a result of electron bombardment, bombard the substrate resulting in the creation of nucleation sites for subsequent growth. The higher density of diamond nucleation sites initiate diamond growth. As the bias voltage is increased, the crystal size is decreased due to increase in ion bombardment of the surface prior to and during initial stages of growth at higher bias voltages.

13.3.3 CONTROLLING STRUCTURE AND MORPHOLOGY

The structure and morphology of the diamond films deposited are highly dependent on a number of factors, which include the CVD growth and nucleation conditions, substrate type, and substrate preparation. Numerous studies have been carried out onto flat silicon substrates, and the general trends are well established [54–74]. However, deposition of adherent high-quality diamond onto substrates such as cemented carbides, stainless steel, and various metal alloys containing transition elements present a considerable challenge due to poor adhesion and low nucleation density [44,45,57–60].

As mentioned earlier, high purity diamond films can be grown using HFCVD with a mixture containing hydrogen (H_2) and methane (CH_4). Whilst these highly crystalline diamond films are exceptionally hard, they are typically very rough and very brittle. For applications requiring smooth surfaces (for low friction), high hardness (for high wear resistance), and high toughness (for preventing catastrophic brittle fracture), a more viable solution is needed. This is particularly true for deposition on metallic substrates where large thermally induced residual stresses can be present in the film, providing a driving force for delamination. Ideally, the film must be well bonded to the substrate and be able to undergo some plastic deformation without hard elastic-to-brittle fracture occurring. One promising material is that of nanostructured diamond, generally defined as containing small diamond grains (from about 5 to 100 nm) imbedded in an amorphous carbon phase that will typically include both sp^2 (graphite-like) and sp^3 (diamond-like) carbon bonds. These films can still be quite hard (up to 80% that of natural diamond) but can also offer the advantage of much lower surface roughness and much higher fracture toughness. Several processing routes for these nanostructured films exist, depending on deposition technique and/or feed gas mixture. Nucleation of diamond is an important step in the growth of diamond thin films, because it strongly influences the diamond growth process, film quality, and morphology. The most promising in situ method for diamond nucleation enhancement in this application is negative substrate biasing during the initial stage of deposition. Complex 3D-shaped dental burs can be biased readily using this method [64]. A modified vertical HFCVD [61–64,66–79] was employed and the key process parameters, which affect the structure and morphology, have also been investigated and summarized.

13.3.3.1 Effects of temperature

The substrate temperature has a profound effect on the crystal size and morphology of the diamond films. Using the vertical arrangement in this new process, the dental burs were placed concentrically within the coils of the filament. With this configuration, the temperature varies significantly between the center and the ends of the filament resulting in the variation of the substrate temperature, which in turn affects the morphology and structure of the resulting diamond films. Therefore, an analysis of temperature distribution along the coiled filament is required. Filament temperature measurements were carried out using a two-color optical pyrometer at various positions along the filament (Fig. 13.6). The bur substrate temperature was also measured parallel to the positions A, B, and C using a K-type thermocouple. These measurements show that the substrate temperature varies according to the bur position within the filament from 950°C at position A to 840°C at position C (Fig. 13.7). The diamond growth is as a result also nonuniform with the highest growth rates being at the position on the burs which was located at the center of the filament during deposition. The films morphology at the positions was examined using a scanning electron microscope (SEM). Fig. 13.8 shows SEM micrographs of the as-grown diamond film at the tip, middle, and base of the WC–Co dental bur (cf. positions A, B, and C, respectively, in Fig. 13.7).

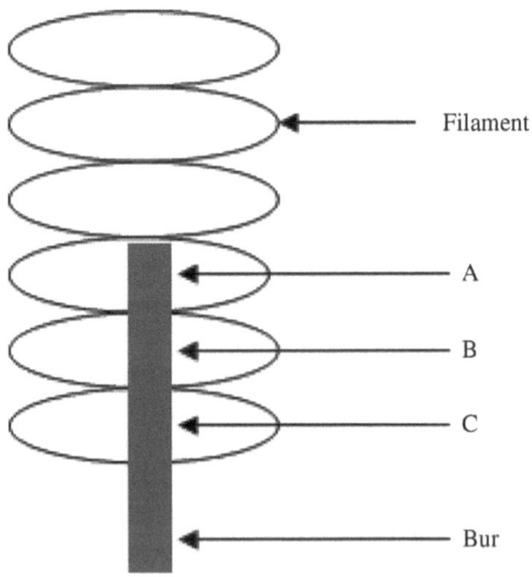

FIGURE 13.6

Schematic diagram of the bur situated within the filament defining positions A–C on the bur.

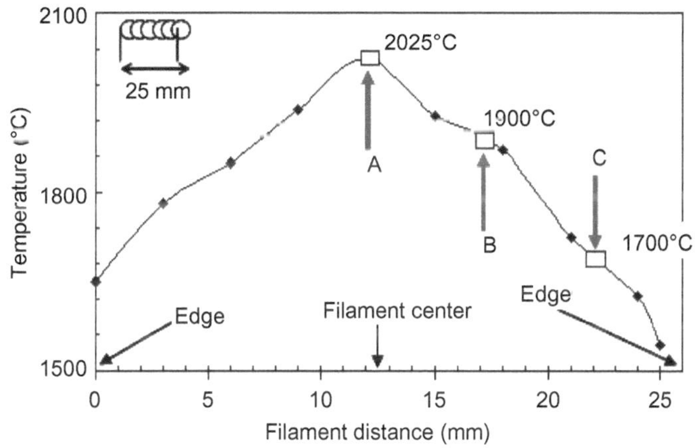

FIGURE 13.7

Substrate temperatures as a function of filament dimensions.

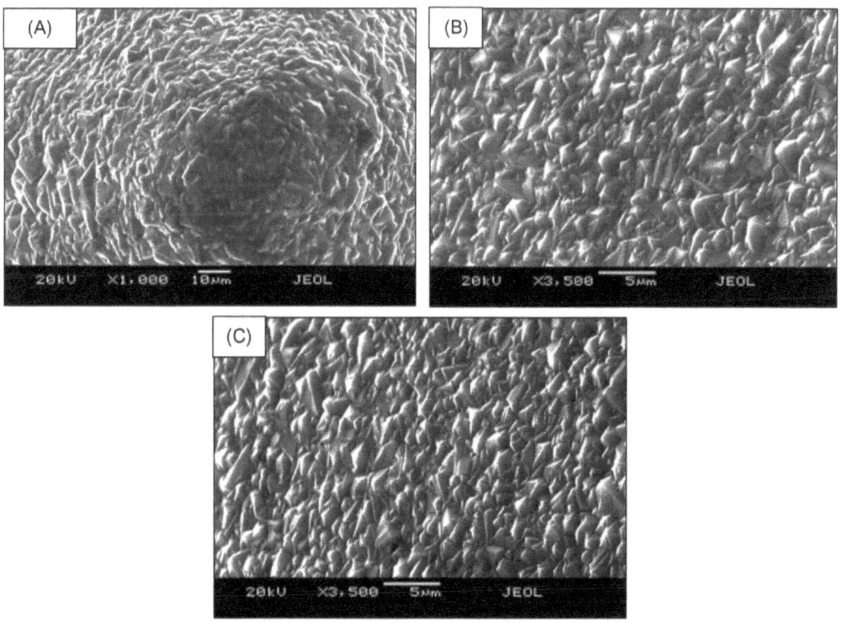

FIGURE 13.8

The effect of bur position in the filament coil on the diamond crystallite size (position within the filament) is indicated on the micrograph.

It can be seen that regardless of the position within the filament, the films are polycrystalline with predominantly (1 1 1) faceted octahedral diamond. It is evident that the films are continuous, uniform, and with good coverage. In addition, the grooves and the cutting edges were successfully covered with the diamond coating. It is significant that the diamond morphology does not change significantly between positions A, B, and C on the coated bur. However, the crystal size does vary with position. The grain sizes are 8–10, 3–5, and 2–4 μm, for positions A, B, and C, respectively. The variation in the average crystallite size along the length of the bur is most likely caused by the variation in substrate temperature, with larger crystal sizes growing at higher substrate temperatures. Raman spectra were measured to investigate further the differences between the different regions on the bur. A shift in the diamond peak centered normally at 1332 cm^{-1} indicates stress in the films.

From the Raman spectra of the as-grown diamond film coating at positions A, B, and C (Fig. 13.9), it is evident from the spectra that all the films display a rather broad band centered at <1500 cm^{-1}. This broad band indicates the presence of an amorphous carbon phase in the coating. The intensity of the Raman diamond peak relative to the broad band is almost identical in all three positions, which indicates that the quality of the diamond is similar at the three positions on the bur. However, the peak energies are slightly different from one another. The peaks at positions A, B, and C on the bur are at 1338, 1336, and 1335 cm^{-1}, respectively.

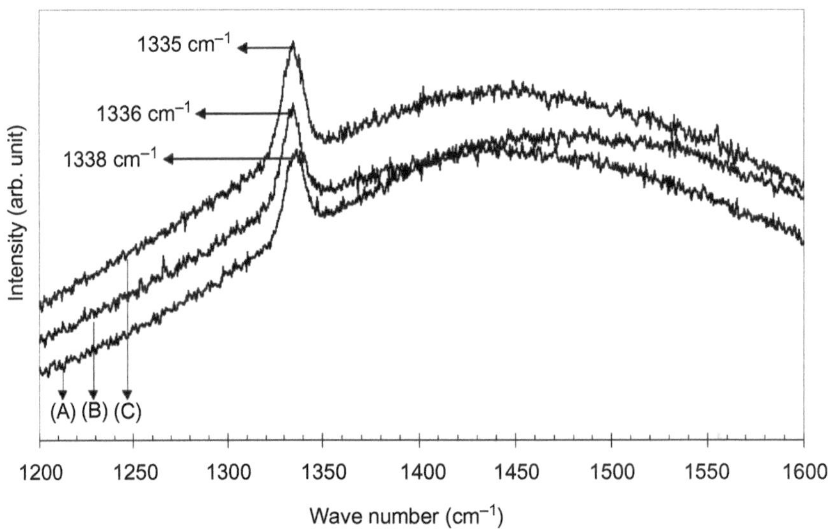

FIGURE 13.9

Raman spectra at positions A, B, and C corresponding to SEMs.

The variation in the peak position can be used to calculate the stress in the films. Sein et al [63] discussed the residual biaxial stress in diamond films grown on a titanium alloy by Raman spectroscopy and described quantitatively the relations between singlet and doublet phonon scattering and the biaxial stress σ as follows:

$\sigma = -0.567(\nu_m - \nu_0)$ (GPa)where ν_0 is 1332 cm^{-1}, and ν_s is the observed maximum of the singlet phonon in the spectrum. Film stress at positions A, B, and C was calculated to be −3.4, −2.3, and −1.7 GPa, in compression, respectively. Fig. 13.10 shows the residual stress in the diamond film grown on the cemented WC–Co dental bur as a function of the substrate temperature. It is entirely reasonable that the highest stress occurs where the deposition temperature is greatest.

Work by Wang et al. [56] has shown Raman shift of diamond coating can be related to the cobalt content in the WC substrate. Higher cobalt (10%) concentration of WC–Co dental bur shows compressive stress in the diamond coating. However, in this work, it is clear that variations in the film thickness and crystal sizes are due to variation in the deposition temperature at various positions on the bur.

13.3.3.2 Effect of negative BEN on the dental bur

A negative bias voltage up to − 300 V was applied to the substrate relative to the filament using a gas mixture of 3% CH_4 in H_2 at 20 Torr. This produced emission currents up to 200 mA. The nucleation times used were between 10 and 30 min. The nucleation density of diamond was determined from SEMs. Figs. 13.11 and 13.12 show the effects of the substrate bias time on the diamond nucleation density. It is evident that as the bias time is increased, the nucleation density also

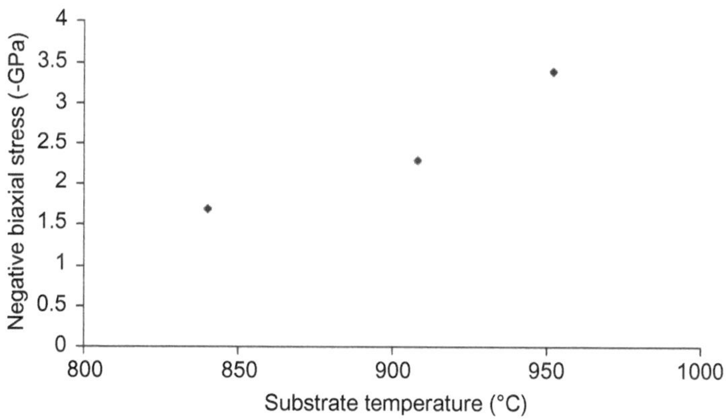

FIGURE 13.10

Effects of substrate temperature on the biaxial stress at positions A, B, and C on the dental bur. Positions A, B, and C relate to those in Fig. 13.6.

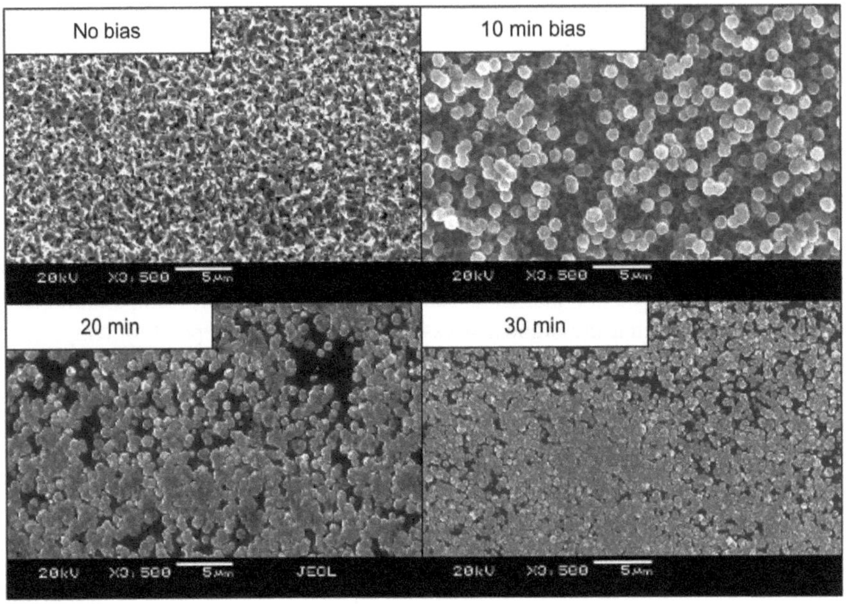

FIGURE 13.11

SEMs showing the effect of bias time on nucleation density on the WC bur.

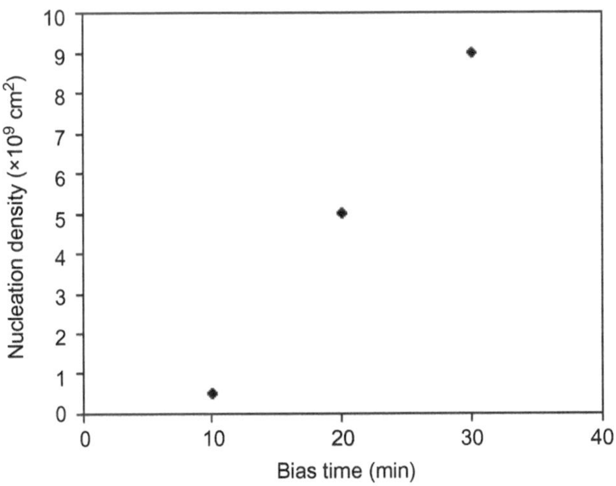

FIGURE 13.12

The effect of biasing time on the nucleation density.

increases. The highest nucleation density was calculated to be $9 \times 10^9\ cm^{-2}$ for a bias time of 30 min. At a bias time of 10 min, the nucleation density obtained was $5 \times 10^8\ cm^{-2}$.

Fig. 13.13 shows the variation of the emission current as a function of the negative bias applied to the substrate holder where the bias emission current increased rapidly after −180 V. Wang et al. [56] also reported that an increase in the emission current produced higher nucleation densities. Since the bias voltage and emission current are related, the enhancement of the nucleation density cannot be attributed to solely ion bombardment or electron emission of the diamond-coated molybdenum substrate holder but may be a combination of these mechanisms. Our result was purely based on negatively BEN related to the grounded filament. However, Polo et al. [63] reported that very low electric biasing current values (μA) were detected for applied substrate biases voltages, either positive or negative. Furthermore, increasing negative biases of up to −200 V resulted in value of nucleation density, similar to that obtained with positively BEN. In contrast, an application of negative bias applied to the substrate at −250 V resulted in ($10^{10}\ cm^{-2}$) maximum values of nucleation density. The enhancement in the nucleation density could be attributed to the electron current from the filament by increasing the decomposition of H_2 and CH_4 [61,68]. The increase in the nucleation density is as expected since negatively biasing the substrate increases the rate of ion bombardment into the surface creating greater numbers and density of nucleation sites.

Therefore, the greater the density of nucleation sites, the higher the expected nucleation density. Kamiya et al. [65] reported that reproducibility of the

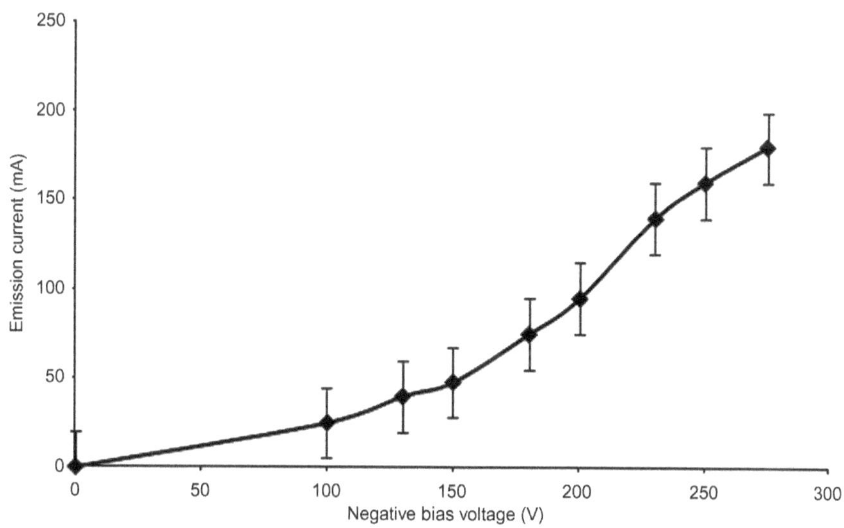

FIGURE 13.13

Effect of negative bias voltage on the emission current.

experiment was poor and that no definite trend in the nucleation density could be found with respect to different bias conditions.

Fig. 13.14 shows SEMs of diamond on the cutting tool edges with negative biasing (A) and (B) and without any bias treatment (C) and (D). It is evident that the cutting edges are uniformly coated in both cases. However, when biasing is employed, there is a considerable reduction of average crystal size; therefore, negative BEN creates more nucleation sites for diamond growth, resulting in smaller average crystallite sizes. The negatively bias-assisted and diamond-coated WC–Co dental burs were tested with human teeth in order to observe their adhesive strength of diamond particles on the surface.

13.3.3.3 Effects of substrate preparation on diamond deposition

The effects of the Murakami/acid chemical etching can clearly be seen on the WC–Co substrate surfaces. Fig. 13.15 shows that cobalt etching was successfully achieved because no cobalt peaks could be detected in the Energy dispersive X-ray (EDX) spectrum. In addition, the etching process produced a roughened surface.

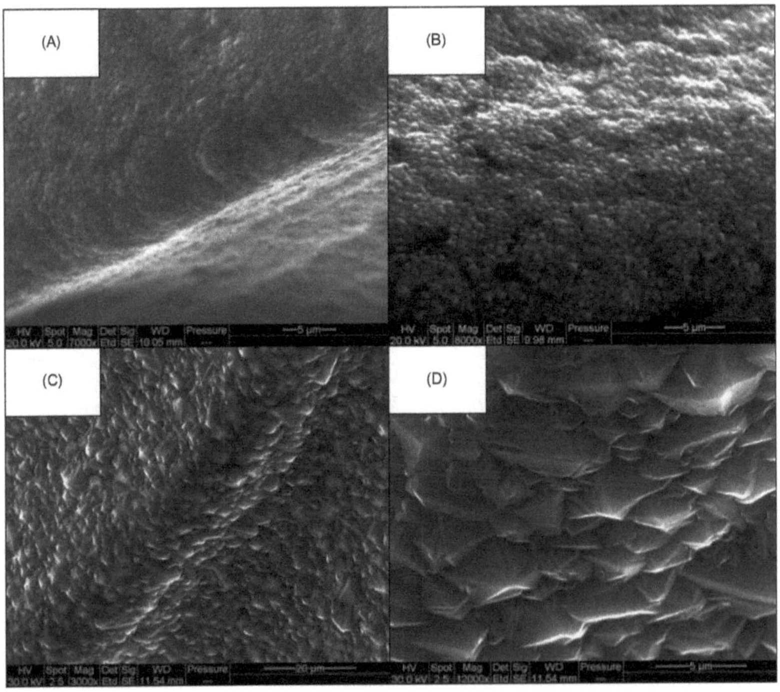

FIGURE 13.14

SEMs showing (A) and (B) with −180 V negative substrate and (C) and (D) without bias of the bur cutting edges.

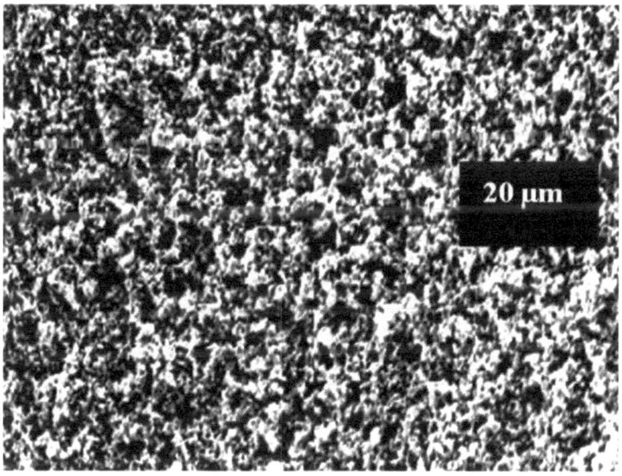

FIGURE 13.15

The cutting edge of a WC–Co dental bur after etching with acid and Murakami's solution.

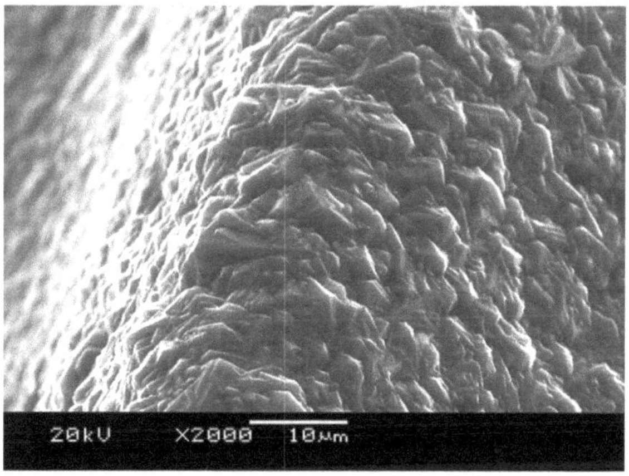

FIGURE 13.16

(1 1 1) Faceted CVD diamond-coated dental bur.

Diamond layers were deposited on pretreated WC–Co dental burs. Adherent diamonds on WC–Co dental burs consisting of mainly (1 1 1) faceted diamond crystals were deposited as shown in Fig. 13.16. Micrographs of the burs show that the diamond film is uniformly coated onto the complex 3D substrate surface. The modified filament arrangement produced a uniform and dense diamond coating even though the substrate is nonplanar with a complex geometry. The morphology of the diamond surface of the dental bur is rough making the bur extremely desirable for dental machining applications.

The coated dental bur was cut in order to study the cross section of the tool as can be seen in Fig. 13.17. It was found that the coating is thicker at the cutting teeth with average thickness of about 23 μm (growth rate of ~1.5 μm/h) due to the slightly higher temperature at the bur tip because cutting teeth is closer to the filament coil. At the base of the bur, the heat is carried away faster, and therefore, it is at a lower temperature, thus giving rise to lower growth rates and hence thinner films, at about 8 μm in thickness. The thicker coating at the tip is expected to give the tool longer life.

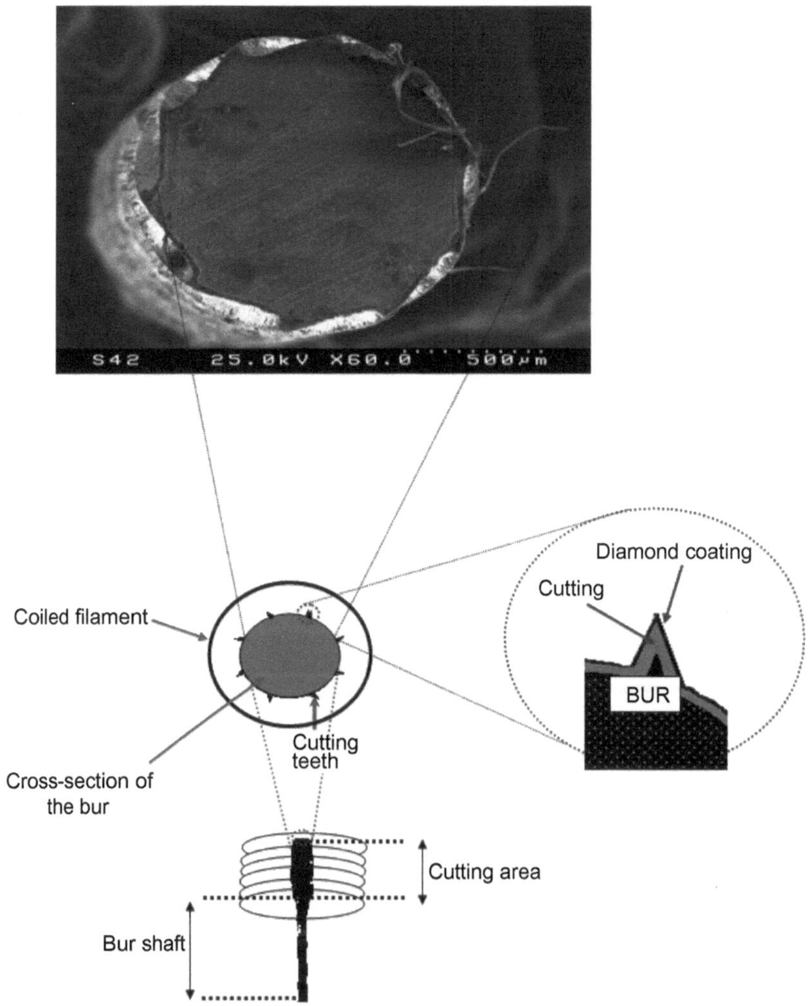

FIGURE 13.17

Cross section of dental burs coated with CVD diamond.

13.4 BUR PERFORMANCE INVESTIGATIONS

CVD diamond films have attracted considerable interest for cutting tool applications, including rotary tools and inserts due to their excellent physical and chemical properties.

In this chapter, we report an investigation of diamond film deposited on etched and unetched WC–Co dental burs, and subsequent machining results on human teeth, borosilicate glass, and acrylic teeth are presented [63,64].

13.4.1 TOOL PREPARATION

Two sets of laboratory tungsten carbide (WC–Co) dental burs (AT23 LR) with fine WC grain sizes (1 μm) 20–30 mm in length and 1.0–1.5 mm in diameter (supplied by Metrodent Ltd, United Kingdom) were used as substrates. SEM and EDX were used to characterize the etched surfaces of substrates.

13.4.2 CVD DIAMOND DEPOSITION ON THE DENTAL BURS

Diamond films were deposited onto the cutting edge of the burs. The coiled tantalum filament (0.5 mm diameter) was held vertically within the vacuum deposition chamber, and the dental burs were positioned centrally and coaxially within the coils of the filament such that the cutting edges were 5 mm from the filament coils. Prior to deposition, the tantalum filament was carburized for 30 min with 3% CH_4 with excess hydrogen. Standard deposition conditions described in Section 13.3 were used. The deposition time and pressure in the vacuum chamber were 5–15 h and 20 Torr, respectively.

13.4.3 DENTAL BUR MACHINING: DRILLING EXPERIMENTS

In order to examine the cutting performance of the diamond-coated dental burs, a variety of clinical and laboratory materials such as borosilicate glass, acrylic teeth, and natural human teeth were drilled. The drilling unit (Fig. 13.18) was constructed. It was constructed with a water-cooling system so that maximum spindle speeds of 250,000 rpm feed rates of between 5 and 20 μm per revolution and cutting speeds in the range 100–200 m/min for drilling with dental burs could be achieved.

After the dental burs were coated and examined for adhesion, the coated burs were used to machine a number of dental materials. The coated burs were compared with uncoated burs to distinguish them in terms of their drilling behavior. The machine shown in Fig. 13.18 was constructed using three principal axes each controlled using a DC motor connected to a Motionmaster controller. A laser light source was focused on the rotating spindle in order to measure the speed of the dental bur during drilling.

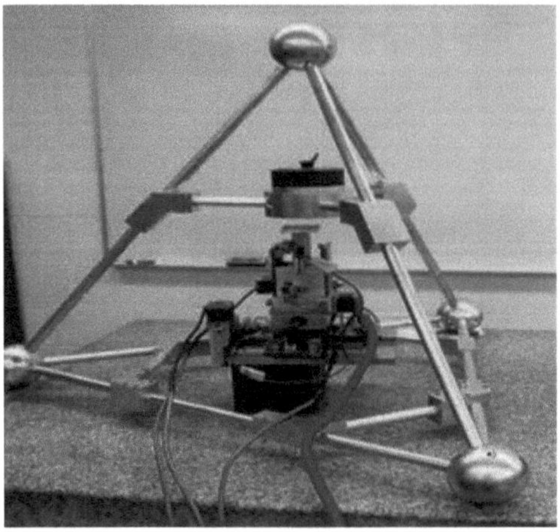

FIGURE 13.18

Dental bur drilling machine.

The flank wear of the burs was analyzed by SEM at selected machining time intervals of between 1 and 3 min. Prior to SEM analysis, the diamond-coated burs were ultrasonically washed with 6 M H_2SO_4 solutions to remove any unwanted machining material, which eroded the surface of the CVD diamond-coated bur. For comparison, conventional diamond sintered burs with different geometry were also tested on the same substrate materials.

13.4.4 DENTAL BUR MACHINING: MACHINING EXPERIMENTS ON HUMAN TEETH

To examine the machining characteristics of coated and uncoated dental burs, a specially constructed clamp was developed to locate over the tooth to prepare it for the location of a crown. Fig. 13.19 shows the construction of the clamping device.

The clamping device is located onto the tooth to be machined and allows the tooth to be machined by incorporating a wire-driven driving mechanism that attaches itself onto the clamp so that the dental bur can rotate at the appropriate cutting speed. Fig. 13.20 shows the driving mechanism attached to the clamping device and a tooth. The driving mechanism is attached to the air-operated hand piece that provides the power to drive the mechanism, clamp, and dental bur. Fig. 13.21 shows the construction of the full assembly that allows machining of the tooth materials to take place. The bur was rotated at 20,000–30,000 rpm, with a feed rate of between 0.2 and 0.5 mm/rev without water-cooling. The uncoated and coated dental burs were compared with sintered diamond burs machining acrylic teeth, and borosilicate glass was used to simulate the

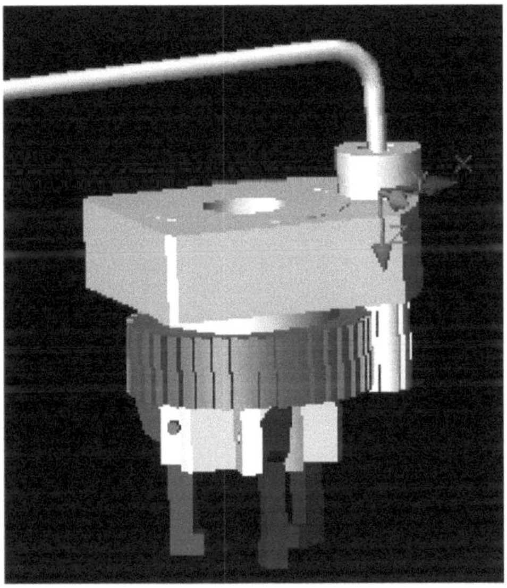

FIGURE 13.19

Human tooth clamping device.

FIGURE 13.20

Tooth clamping device and driving stage used to locate clamping device to the surface of the tooth.

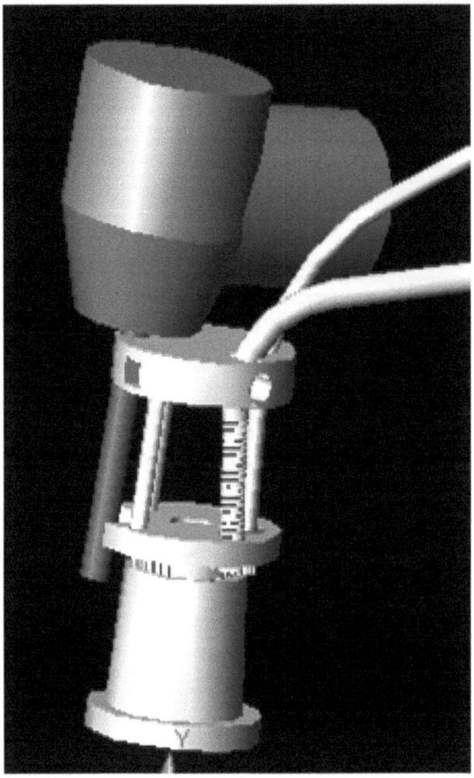

FIGURE 13.21

Air-operated spindle unit attached to the clamping device and driving unit attached firmly to the tooth.

machining of porcelain teeth. The flank wear of the dental burs was estimated using SEM.

13.4.5 PERFORMANCE TESTING

Fig. 13.22 shows a SEM of a conventional PCD sintered bur. The diamond particles are imbedded to surface with a suitable binder matrix material, in this case nickel. Typically, the surface is inhomogeneous, and sizes of particles range from 50 to 200 μm causing considerable variation in the cutting performance of the tool.

An important factor that could affect the final performance of the dental bur is the adhesive toughness of the diamond on the substrates. Endler et al. [49] and Polini et al. [70] have developed a new method for the quantitative evaluation of the adhesive toughness of diamond films onto Cocemented WC

FIGURE 13.22

Inhomogeneous surface of PCD diamond sintered bur.

substrates. They found that the adhesive toughness of diamond on WC to be in the range of 20–37 J/m^2. Commercial burs exhibited much higher adhesive toughness than flat substrates due to the large surface roughness and the absence of interfacial voids. This factor needs to be investigated in detail for nonplanar dental burs.

The effectiveness of using HFCVD-coated dental burs was measured by comparing uncoated burs, HFCVD dental burs, and PCD sintered diamond burs when drilling and machining extracted human teeth, acrylic teeth, and borosilicate glass.

13.4.6 DRILLING EXPERIMENTS

A sequence of 50 drillings was employed in each drilling experiment. The sharpness and initial condition of the burs were inspected visually after the burs had drilled 10 holes in sequence. An abrading coefficient of drilling, C_a, has been defined as a quality criterion for small drilling tools. It is defined as the ratio between the bur's total abraded area, S, and the effective coated area of the bur used during the drilling process. The effective coated area is given by the difference of the nominal coated bur area, D_b, and the area of the bur consumed during drilling, W_b. C_a is defined by the following equation:

$$C_a = \frac{S}{D_b - W_b} \tag{13.1}$$

A high-quality-coated dental bur is one that produces accurate drilling that has an area S close to D_b and does not lose its coating during the machining process. The cutting will therefore have an abrading coefficient close to unity. One must remember that the quality of machining is dependent on the cutting speed, V_c, defined by the ratio of the drilling depth by the drilling duration. A comparative factor of merit for the dental bur is defined as

$$F = \frac{C_a}{V_c} \tag{13.2}$$

where F, the figure of merit, is directly related to the lifetime of the dental bur for a specific drilling process.

13.4.7 PERFORMANCE RESULTS

Figs. 13.23–13.25 show the results of drilling the dental materials with the three types of burs on borosilicate glass, acrylic tooth materials, and human teeth, respectively.

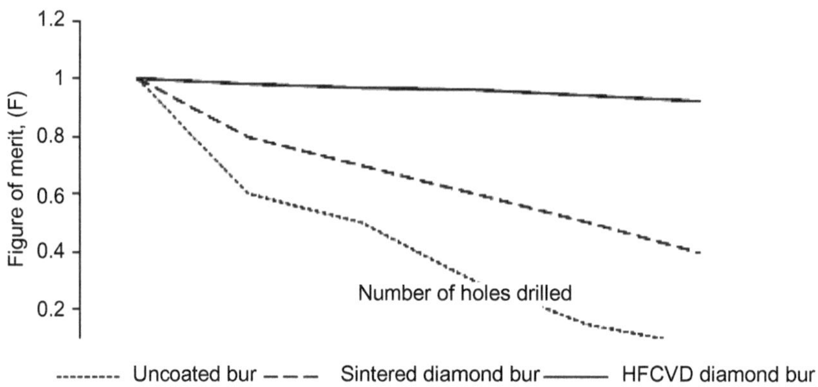

FIGURE 13.23

Figure of merit as a function of number of holes drilled for dental burs drilling borosilicate glass.

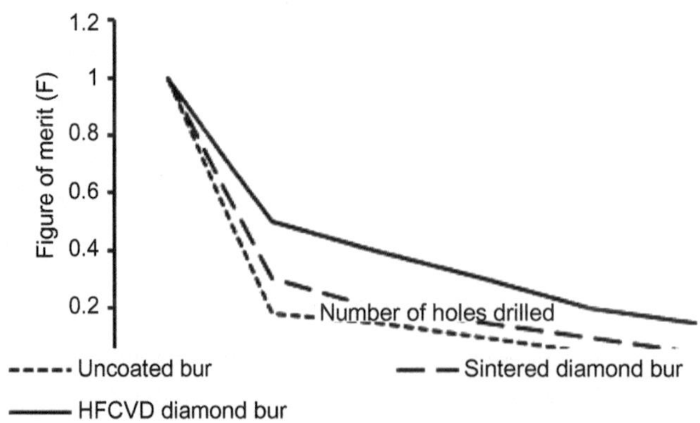

FIGURE 13.24

Figure of merit as a function of number of holes drilled for dental burs drilling acrylic material.

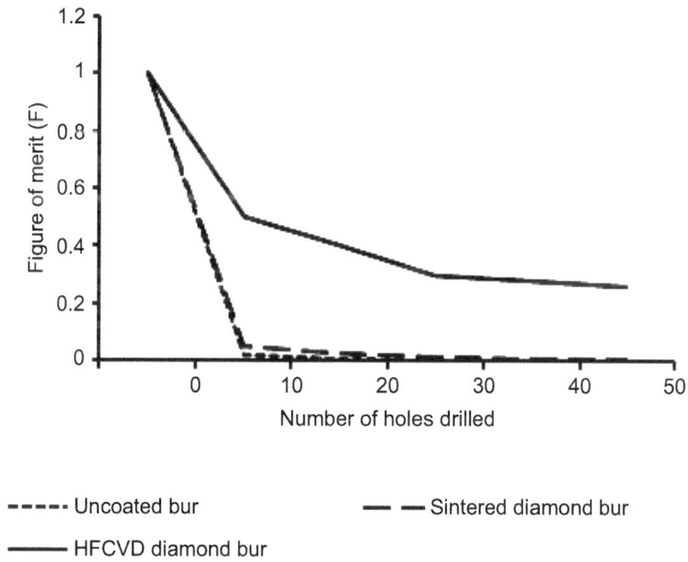

FIGURE 13.25

Figure of merit as a function of number of holes drilled for dental burs drilling human tooth material.

For all the materials investigated including acrylic, borosilicate glass, and human tooth material, it is evident that the best-performing bur in terms of figure of merit were the HFCVD diamond-coated burs on the WC–Co substrates and TiN.

Fig. 13.26 shows a micrograph of uncoated WC–Co dental bur tested on the borosilicate glass using the same machining conditions. The uncoated WC–Co burs displayed flank wear along the cutting edge of the bur. The areas of flank wear were investigated at the cutting edge of the dental bur. Fig. 13.31 shows flank wear as a function of cutting time when drilling borosilicate glass. It is evident that the action of machining causes high rates of flank wear on the cutting edge of dental bur. Therefore, the cutting edges of WC–Co dental burs should have a minimum thickness of CVD diamond, which will enhance not only quality of cutting but also prolong the life of the bur.

Fig. 13.27 shows the surface of a sintered diamond bur after being tested on borosilicate glass at a cutting speed of 30,000 rpm for 5 min with an interval at every 30 s. It is clear that there is significant removal of diamond particles from the surface of the tool after 50 holes. As expected, there is deterioration of the abrasive performance of the PCD sintered diamond dental burs. Borges et al. [64] also reported that the significant loss of diamond particles occurred during cutting with the commercial sintered diamond bur. In addition, the nickel binder shows major defects generated by pulled-out particles.

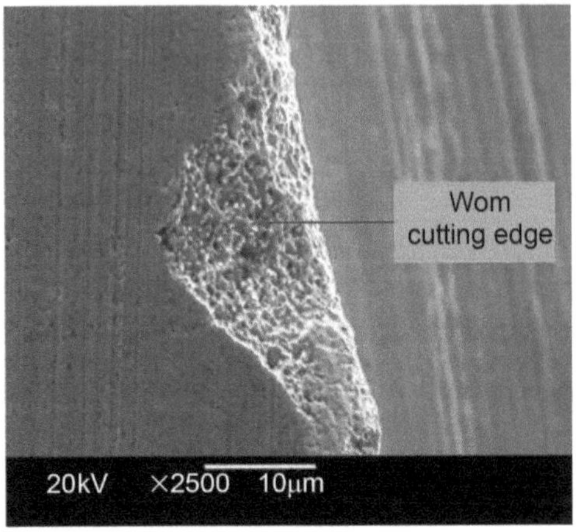

FIGURE 13.26

Worn cutting edge uncoated WC–Co dental bur after drilling borosilicate glass.

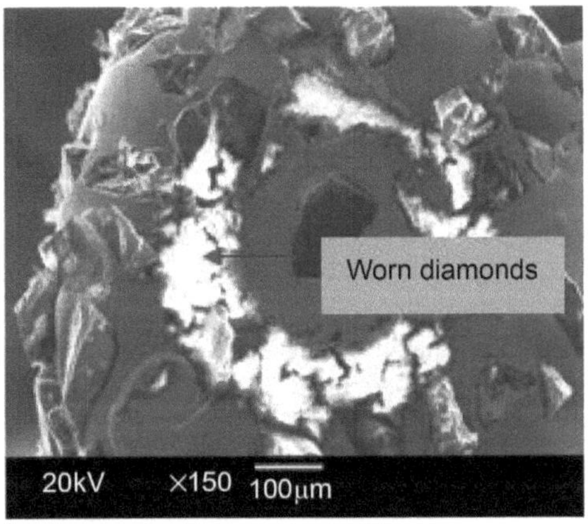

FIGURE 13.27

PCD sintered bur after drilling on borosilicate glass.

Figs. 13.28 and 13.29 show SEM images of a CVD diamond-coated bur after drilling experiments on borosilicate glass and acrylic teeth, respectively, for 5 min at a cutting speed of 30,000 rpm.

After machining, it is evident that the diamond films are still intact on the pretreated WC substrate, and diamond coating displayed good adhesion. Also, there is no indication of diffusion wear after the initial test for 50 holes. However, the machined materials such as glass pieces erode the cutting edge of the diamond dental bur as adhesive wear was observed (Fig. 13.28). After

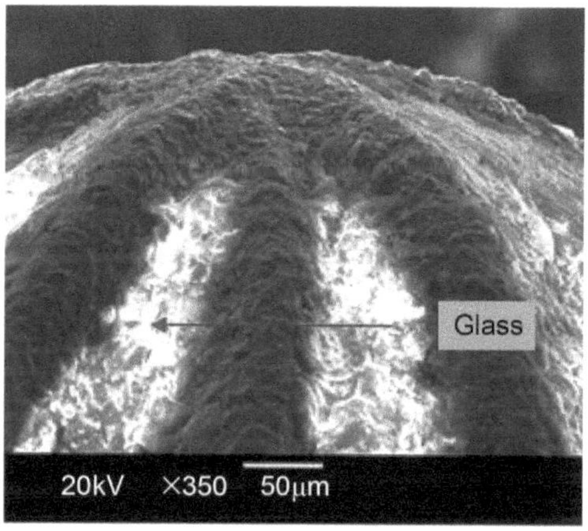

FIGURE 13.28

Cutting edge of dental bur after drilling on borosilicate glass.

FIGURE 13.29

Magnified image of the CVD diamond-coated dental bur before (A) after (B) drilling acrylic tooth material.

conducting experiments on acrylic teeth, it was concluded that the mechanism of wear involves adhesion as well as abrasion. Fig. 13.29 shows that inorganic fillers from acrylic teeth adhered to the cutting tool surface in localized areas when increased rate of abrasion was used.

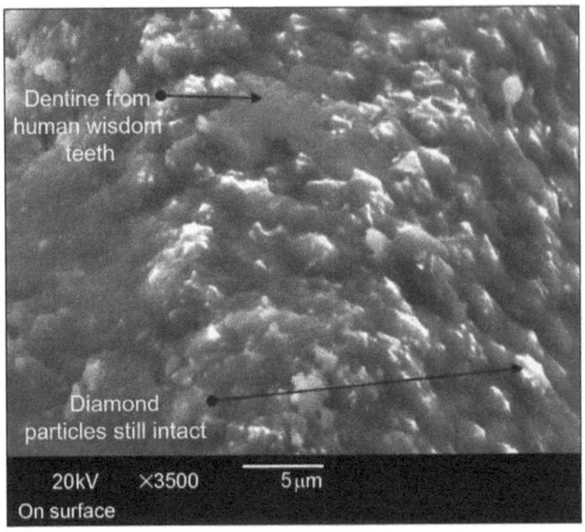

FIGURE 13.30

SEM micrograph of diamond-coated dental bur at a cutting speed of 250,000 rpm.

Natural human teeth were also drilled using a diamond-coated dental bur. Previous studies have indicated that teeth should not be used for reduction tests because of the differences in hardness between enamel and dentine (Knoop hardness data: enamel, 250–500 kg/mm^2; dentine, 50–70 kg/mm^2). Cuts were made in the central groove of the teeth with the burs.

This permitted cutting three grooves in each tooth. Fig. 13.30 shows the SEM of a diamond-coated WC–Co dental bur after testing on extracted teeth. It is evident from the micrograph that the tooth materials such as dentine clog up on the bur reducing its abrasive performance.

An area where this could be important is the surface treatment of small dental tools such as dental burs, which are used in the dental laboratory and clinical surgery for removing unwanted material from teeth. The negatively bias-assisted and diamond-coated WC–Co dental burs were tested with human teeth in order to observe their adhesive strength of diamond particles on the surface. The coated burs have been tested on extracted human wisdom teeth, which have a difference in hardness between enamel and dentine. The teeth were cut in a bench device using an ultrahigh-speed hand piece (air rotor). A frequency meter monitored the speed of the hand piece, which was between 200,000 and 250,000 rpm. Water-cooling device was used to prevent the tool overheating. The SEM micrograph of tested dental bur shows that the diamond film was still intact after the application of highly abrasive cutting at a speed of 250,000 rpm for 2 min (Fig. 13.30). These results are extremely encouraging and clearly demonstrate the extreme toughness and durability of the diamond films using our modified HFCVD system.

A series of machining experiments were also conducted using uncoated, HFCVD dental burs and sintered diamond burs when machining extracted human

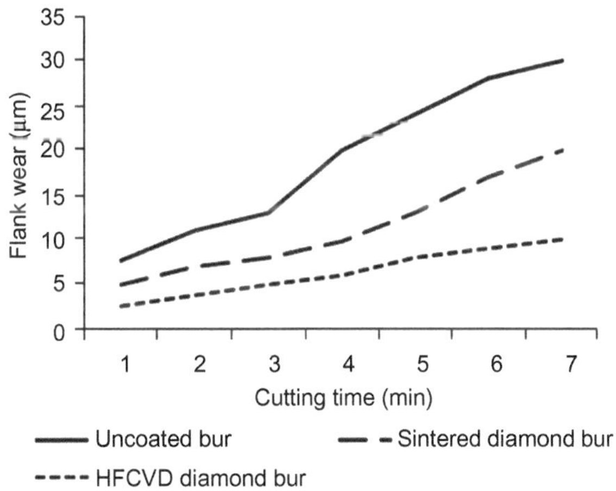

FIGURE 13.31

Flank wear of burs machining borosilicate glass.

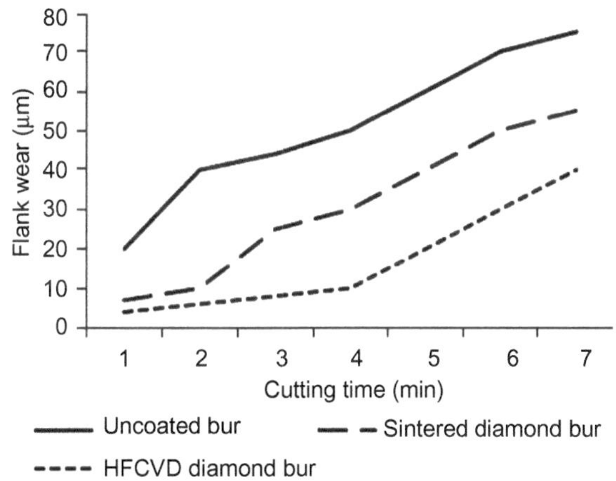

FIGURE 13.32

Flank wear of burs machining acrylic tooth material.

teeth, acrylic teeth, and borosilicate glass. The teeth were machined using the apparatus shown in Figs. 13.19–13.21. The life of the burs in the machining sense was measured by comparing the amount of flank wear exhibited by each type of dental bur. The flank wear was measured at time intervals of 2, 3, 4, 5, 6, and 7 min machining duration. Figs. 13.31–13.33 show the flank wear measurements for each bur machining different dental materials.

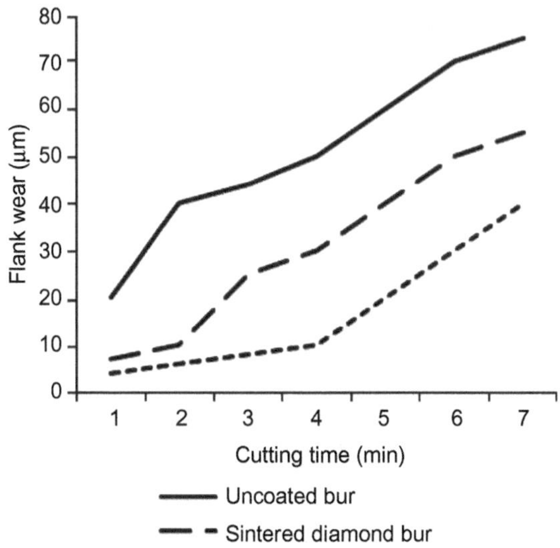

FIGURE 13.33

Flank wear of burs machining human tooth material.

Again, the dental burs were examined using optical and SEM techniques and were found to observe similar trends as burs associated with drilling experiments close.

13.5 CONCLUSIONS

The teeth reside in a harsh environment and get worn and damaged through prolonged use. In order to repair the damage, add fillings, and implants, a wide range of dental burs are employed by dentists and dental technicians and these are described briefly. A new vertical filament HFCVD process for diamond deposition onto the burs has been described. The effects of process parameters on the properties of diamond films and performance of diamond coated and untreated dental burs, such as wear and lifetime, have been investigated. It has been demonstrated that HFCVD coated burs gave superior performance and reduced wear compared to uncoated and conventional burs.

REFERENCES

[1] N.M. Jedynakiewicz, A Practical Guide to Technology in Dentistry, Wolfe Publishing, London, 1992.

[2] R. Simonsen, Dentistry in the 21st Century: A Global Perspective, Quintessence Publishing Co., Hanover Park, IL, 1989.
[3] C. Hollins, Levison's, Textbook for Dental Nurses, Blackwell Muckgaurd, 10th ed., John Wiley & Sons, UK, 2008.
[4] M. Ring, Behind the Dentist's Drill, Invention & Technology, Fall, 1995, p. 25.
[5] A.W. Roger, Textbook of Anatomy, Churchill Livingstone, Edinburgh, 1992.
[6] I. Watanabe, C. Ohkubo, J.P. Ford, M. Atsuta, T. Okabe, Cutting efficiency of air-turbine burs on cast titanium and dental casting alloys, Dent. Mater. 16 (2000) 420.
[7] J.F. McCabe, Applied Dental Materials, seventh ed., Blackwell Scientific Publication, Oxford, 1992.
[8] BS 6828-4: 1987; EN 28323: 1991; ISO 3823-2; 1986. Dental Rotary Instruments. Part 4. Specification for Steel and Carbide Finishing Burs.
[9] BS 6828-9: 1987; EN 28325: 1989; ISO 8325; 1985. Dental Rotary Instruments. Part 9. Methods of Test.
[10] M.S. Pines, A. Schulman, Characterization of wear of tungsten carbide burs, J. Am. Dent. Assoc. 99 (1979) 831.
[11] G.A. Lammie, The effects of restorative materials on plaque composition, Dent, Rec 71 (1972) 286.
[12] H. Sein, W. Ahmed, M.J. Jackson, R. Woodwards, R. Polini, Performance and characterisation of CVD diamond coated, sintered diamond and WC–Co cutting tools for dental and micromachining applications, Thin Solid Films 447–448 (2004) 455.
[13] S.R. Bradbury, D.B. Lewis, P.M. Archer, W. Ahmed, Impact of surface engineering technologies on the performance and life of multipoint cutting tools, Surf. Coat. Technol. 91 (1997) 192.
[14] R. Glenner, A. Davis, S. Burns, The American Dentist, Pictorial Richard Histories Publishing Co., Missoula, MT, 1990.
[15] H. Sein, W. Ahmed, C. Rego, Application of diamond coatings onto small dental tools, Diamond Relat. Mater. 11 (2002) 731.
[16] D.B. Dietz, P.M. Di Fiore, J.K. Bahcall, E.P. Lautenschlager, Effect of rotational speed on the breakage of nickel-titanium rotary files, J. Endod. 26 (2000) 68–71.
[17] M.B. Rowan, J.I. Nicholls, J. Steiner, Torsional properties of stainless steel and nickel-titanium, steel and nickel-titanium endodontic files, J. Endod. 22 (1996) 341–345.
[18] A.A. Caputo, J.P. Standlee, Biomechanics in Clinical Dentistry, Quintessence Book, Chicago, IL, 1987.
[19] M.N. Yoder, in: K.E. Spear, J.P. Dismukes (Eds.), Synthetic Diamond: Emerging CVD Science and Technology, first ed., John Wiley & Sons, New York, 1994.
[20] P. Harris, High-resolution electron microscopy studies of a microporous carbon produced by Arc-evaporation, J. Chem. Soc. Faraday Trans. 90 (18) (1994) 2799–2802.
[21] J.E. Field, The Properties of Natural and Synthetic Diamond, Academic Press, London, 1992.
[22] F.P. Brundy, H.T. Hall, H.M. Strong, R.H. Wentorf, Man-made diamonds, Nature 176 (1955) 51.
[23] H. Liu, D.S. Dandy, D. Haber, Diamond Chemical Vapour Deposition, William Andrew Publishing, USA, 1997.
[24] K. Nassau, J. Nassau, The history and present status of synthetic diamond, J. Cryst. Growth 46 (1979) 157.

[25] E. Bruton, Diamonds, NAG, London, 1978, p. 1.
[26] W.G. Eversole, US Patent 3030188, April 1962.
[27] J.C. Angus, Diamond and diamond-like films, Thin Solid Films 216 (1992) 126.
[28] P.W. May, Diamond Thin Films: A 21st Century Material, Philos. Trans. R. Soc. London A 358 (2000) 473–495.
[29] S. Matsumoto, Y. Sato, M. Kamo, N. Setaka, Vapor deposition of diamond particles from methane, Jpn. J. Appl. Phys. 21 (1982) L183.
[30] S. Matsumoto, Y. Sato, M. Tsutsumi, N. Setaka, Growth of diamond particles from methane-hydrogen gas, J. Mater. Sci. 17 (1982) 3106.
[31] M. Kamo, Y. Sato, M. Matsumoto, N. Setaka, Diamond synthesis from gas phase in microwave plasma, J. Cryst. Growth 62 (1983) 642.
[32] F. Kobashi, K. Nishimura, Y. Kawate, J.T. Horiuchi, Morphology and growth of diamond films, J. Vac. Sci. Technol., A: Vac. Surf. Films 6 (1988) 1816–1818.
[33] B.V. Spitsyn, Growth of Diamond on Diamond Surfaces, Institute of Physical Chemistry, Moscow, 1973 (Dr. Sciences Thesis).
[34] Y. Liou, A. Inspector, R. Weimer, R. Messier, , Studies of diamond on engineering substrates, Appl. Phys. Lett. 55 (1989) 631.
[35] K. Suzuki, A. Sawabe, H. Yasuda, T. Inuzuka, Growth of diamond thin films by dc plasma chemical vapor deposition, Appl. Phys. Lett. 50 (1987) 728.
[36] F. Akatsuka, Y. Hirose, High rated synthesis of diamond film by arc electric discharge, J. Cryst. Crowth. 15 (1) (1988) 60.
[37] K. Suzuki, A. Sawabe, T. Inuzuka, Growth of diamond thin films by dc plasma chemical vapor deposition and characteristics of the plasma, Jpn. J. Appl. Phys. 29 (1990) 153.
[38] W.A. Yarbrough, R. Messier, Current issues and problems in the chemical vapor deposition of diamond, Science 247 (1990) 688.
[39] R. Messier, A.R. Badzian, T. Badzian, K.E. Spear, P.K. Bachmann, R. Roy, From diamond-like carbon to diamond coatings, Thin Solid Films 153 (1987) 1.
[40] J.C. Angus, C.C. Hayman, Low-pressure, metastable growth of diamond and "diamondlike" phases, Science 241 (1988) 913.
[41] K.E. Spear, Diamond–ceramic coating of the future, J. Am. Ceram. Soc. 72 (1989) 171.
[42] P.W. May, C.A. Rego, R.M. Thomas, M.N.R. Ashford, K.N. Rosser, CVD diamond wires and tubes, Diamond Relat. Mater. 3 (1994) 810–813.
[43] J.E. Butler, R.L. Woodin, Thin film diamond growth mechanisms, Philos. Trans. R. Soc. London A 342 (1993) 209.
[44] A. Afzal, Effects of Nitrogen on CVD Diamond Growth, Manchester Metropolitan University, UK, 1999 (Ph.D. Thesis).
[45] I.U. Hassan, Biased Enhanced Nucleation of CVD Diamond Films, Manchester Metropolitn University, UK, 2001 (Ph.D. Thesis).
[46] B. Lux, R. Haubner, Diamond deposition on cutting tools, Ceram. Int. 22 (1996) 347.
[47] R.C. Weast (Ed.), CRC Handbook of Chemistry and Physics, CRC Press, Boca Raton, FL, 1989–90.
[48] R. Haubner, A. Lindlbauer, B. Lux, Diamond deposition on chromium, cobalt and nickel substrates by microwave plasma chemical vapour deposition, Diamond Relat. Mater. 2 (1993) 1505.
[49] I. Endler, K. Bartsch, A. Leonhardt, H.J. Scheibe, H. Ziegele, I. Fuchs, et al., Preparation and wear behaviour of woodworking tools coated with superhard layers, Diamond Relat. Mater. 8 (1999) 834–839.

[50] B.V. Spitsyn, L.L. Bouilov, B.V. Deryagin, Vapor growth of diamond on diamond and other surfaces, J. Cryst. Growth 52 (1981) 219.

[51] R.B. Jackman, J. Beckman, J.S. Foord, Diamond chemical vapour deposition from a capacitively coupled radio frequency plasma, Appl. Phys. Lett. 66 (1995) 1018.

[52] T. Leyendecker, O. Lemmer, A. Jurgens, S. Esser, J. Ebberink, Industrial application of crystalline diamond-coated tools, Surf. Coat. Technol. 48 (1991) 253.

[53] M. Murakawa, S. Takeuchi, Mechanical applications of thin and thick diamond films, Surf. Coat. Technol. 49 (1991) 359.

[54] T. Yashiki, T. Nakamura, N. Fujimori, T. Nakai, Practical properties of chemical vapour deposition diamond tools, Surf. Coat. Technol. 52 (1992) 81.

[55] J. Karner, M. Pedrazzini, I. Reineck, M.E. Sjöstrand, E. Bergmann, CVD diamond coated cemented carbide cutting tools, Mater. Sci. Eng. A 209 (1–2) (1996) 405–413.

[56] D.Y. Wang, Y.H. Song, J.J. Wang, R.Y. Cheng, Implementation of large-scale deposition of diamond films by combustion synthesis, Diamond Relat. Mater. 2 (1993) 304.

[57] A. Inspector, C.E. Bauer, E.J. Oles, Superhard coatings for metal cutting applications, Surf. Coat. Technol. 68/69 (1994) 359.

[58] A. Inspector, E.J. Oles, C.E. Bauer, Theory and practice of diamond coated cutting tools, metal hard materials, Int. J. Refract. 15 (1997) 49.

[59] K. Kanda, S. Takehana, S. Yoshida, R. Watanabe, S. Takano, H. Ando, et al., Application of diamond-coated cuttings tools, Surf. Coat. Technol. 73 (1995) 115.

[60] S. Amirhaghi, H.S. Reehal, R.J.K. Wood, D.W. Wheeler, Diamond coatings on tungsten carbide and their erosive wear properties, Surf. Coat. Technol. 135 (2001) 126.

[61] H. Sein, W. Ahmed, M.J. Jackson, N. Ali, J. Gracio, Stress distribution in diamond films grown on cemented WC–Co dental burs using modified hot-filament CVD, Surf. Coat. Technol. 163–164 (2003) 196.

[62] A. Gicquel, E. Anger, M.F. Ravet, D. Fabre, G. Scatena, Z.Z. Wang, Diamond deposition in a bell-jar reactor: influence of the plasma and substrate parameters on the microstructure and growth rate, Diamond Relat. Mater. 2 (1993) 417–424.

[63] M.C. Polo, W. Wang, G. Sanshez, J. Andujar, J. Esteve, Nucleation and initial growth of bias-assisted HFCVD diamond on boron nitride films, Diamond Relat. Mater. 6 (1997) 579.

[64] C.F.M. Borges, P. Mange, M. Dent, E. Pfender, D. Ring, J. Heberlein, Dental diamond burs made with a new technology, J. Prosthet. Dent. (1999) 73–79.

[65] S. Kamiya, H. Takahashi, R. Polini, E. Traversa, Quantitative determination of the adhesive fracture toughness of CVD diamond to WC–Co cemented carbide, Diamond Relat. Mater. 9 (2000) 191–194.

[66] H. Sein, W. Ahmed, M. Jackson, R. Polini, I.U. Hassan, M. Amar, et al., Enhancing nucleation density and adhesion of polycrystalline diamond films deposited by HFCVD using surface treatments on co-cemented tungsten carbide, Diamond Relat. Mater. 13 (4–8) (2004) 601.

[67] H. Sein, W. Ahmed, M. Jackson, R. Woodwards, R. Polini, Performance and characterisation of CVD diamond coated, sintered diamond and WC–Co cutting tools for dental and micromachining applications, Thin Solid. Films 447–448 (2004) 455.

[68] A.N. Jones, W. Ahmed, I.U. Hassan, H. Sein, C.A. Rego, Pulsed biased growth of nanocrystalline diamond by HFCVD, Surf. Eng. 20 (3) (2004) 1.

[69] N. Ali, G. Cabral, V.F. Neto, H. Sein, W. Ahmed, J. Gracio, Surface engineering of WC–Co used in dental tools technology, J. Mater. Sci. Technol. 19 (2003) 1273–1278.

[70] R. Polini, F. Casadei, P. D'Antonio, E. Traversa, Dry turning of alumina/aluminum composites with CVD diamond coated Co-cemented tungsten carbide tools, Surf. Coat. Technol. 166 (2003) 127–134.
[71] H. Sein, W. Ahmed, N. Ali, J. Gracio, R. Woodwards, Diamond films grown on cemented WC–Co dental burs using an improved CVD method, Diamond Relat. Mater. 12 (8) (2003) 1300.
[72] W. Ahmed, H. Sein, H. Rajab, M. Jackson, Chemical vapor deposition of diamond coatings onto dental burrs, J. Chem. Educ. 80 (6) (2003) 636.
[73] H. Sein, W. Ahmed, C.A. Rego, A.N. Jones, M. Amar, M. Jackson, et al., Chemical vapour deposition diamond coating on tungsten carbide dental cutting tools, J. Phys.: Condes. Matter 15 (2003) S2961.
[74] A.N. Jones, W. Ahmed, I.U. Hassan, C.A. Rego, H. Sein, M. Amar, et al., The impact of inert gases on the structure, properties and growth of nanocrystalline diamond, J. Phys.: Condes. Matter. 15 (2003) S2969.
[75] H. Sein, W. Ahmed, M. Jackson, N. Ali, J. Gracio, Stress distribution in diamond films grown on cemented WC–Co dental burs using modified hot-filament CVD, Surf. Coat. Technol. 163–164 (2003) 196.
[76] M. Amar, W. Ahmed, H. Sein, A.N. Jones, C.A. Rego, Chemical vapour deposition of diamond coatings onto molybdenum dental tools, J. Phys.: Condes. Matter. 15 (2003) S2977.
[77] H. Sein, W. Ahmed, I.U. Hassan, N. Ali, J. Gracio, M.J. Jackson, Chemical vapour deposition of microdrill cutting edges for micro-and nanotechnology applications, J. Mater. Sci 37 (2002) 5057.
[78] H. Sein, W. Ahmed, C.A. Rego, Application of diamond coatings onto small dental tools, Diamond Relat. Mater. 11 (3–6) (2002) 731.
[79] H. Sein, M. Jackson, W. Ahmed, C.A. Rego, Deposition of diamond coatings to micro-machining cutting tools, New Diamond Front. Technol. 12 (2002) 381–390.

FURTHER READING

J. Silvaggio, M.L. Hicks, Effect of heat sterilization on the torsional properties of rotary nickel-titanium files, J. Endod. 23 (1997) 731.

CHAPTER

14

Nanomechanical characterization of mineralized tissues in the oral cavity

Karthikeyan Subramani[1], Yee L. Chan[2], Alfonso H.W. Ngan[2] and Nigel M. King[2]

[1]*Roseman University of Health Sciences, Henderson, NV, United States*

[2]*The University of Hong Kong, Pokfulam, Hong Kong*

CHAPTER OUTLINE

14.1 Introduction 289
14.2 Basic Data Analysis Protocol for Nanoindentation 290
14.3 Nanoindentation of Oral Mineralized Tissues 292
14.3.1 Sample Preparation 292
14.3.2 Hydration 292
14.3.3 Indenter Tips 294
14.3.4 Load Function and Data Analysis 295
14.3.5 Microstructural Influence 297
14.4 Summary 302
References 303
Further Reading 306

14.1 INTRODUCTION

Enamel, dentin, cementum, and bone are the four main types of mineralized tissues in the oral cavity. They are produced and maintained by different mechanisms; hence, they differ from each other in the arrangement and quantity of their constituent materials and also in their mechanical properties. However, despite the long history of dental research, the mechanical properties of these mineralized tissues have not been well characterized. This is because the tissues are small in size and complex in shape, for example, conventional characterization techniques such as compressive or tensile tests cannot be easily conducted. Nanoindentation, with its relative ease in sample preparation, high spatial resolution, and ability to make nano/microscale measurements, is able to overcome these difficulties. The

Emerging Nanotechnologies in Dentistry. DOI: http://dx.doi.org/10.1016/B978-0-12-812291-4.00014-5

advent of nanoindentation in the 1990s has led to a surge in the number of investigations on the mechanical properties of the oral mineralized tissues, which have contributed greatly to our understanding of their structure–property relationship [1]. The nanomechanical properties have provided credible explanations to both the superior performance in healthy tissues [2–4] and the degraded performance associated with pathological conditions [5–7]. Concurrent with our increasing knowledge on the oral tissues, it has become apparent that there are large discrepancies between their mechanical behavior and those of most synthetic materials, in terms of viscoelasticity, anisotropy, and effects due to the hierarchical structures, etc. As such, the conventional method for interpreting nanoindentation data is inadequate for biological tissues. Consequently, different strategies have been developed to overcome some of the problems encountered in nanoindenting oral mineralized tissues and selected examples are discussed in the following sections.

14.2 BASIC DATA ANALYSIS PROTOCOL FOR NANOINDENTATION

Currently, most of the nanomechanical experiments published in the literature concern applications of the nanoindentation technique, in either its basic mode involving depth-sensing indentation on a flat specimen surface, or other modified modes involving using the nanoindenter as a force-application device to deform specimens with more complex shapes such as microcantilevers or micropillars. Indentation testing is a well-established technique whereby a hard tip of a known geometry is pressed against a flat specimen, and "depth-sensing indentation" which refers to an indentation procedure during which the force and tip displacement data are continuously logged. "Nanoindentation" refers to depth-sensing indentation with depth resolution in the nanometric regime. Before discussing issues pertinent to oral mineralized tissues, it is useful to briefly summarize here the basic principles of nanoindentation and the conventional data analysis method developed for synthetic, hard materials. Fig. 14.1A and B illustrates a typical nanoindentation test and the associated load–displacement graph, respectively. In the most basic test mode, the indenter tip is made to slowly approach the surface to be indented, and upon contacting the latter, a triangular or trapezoidal load function is deployed for indenting the specimen. The sample is loaded, usually at a constant rate, until a maximum load is reached. A holding period may be applied at the maximum load, followed by unloading at a preset rate. The load–displacement data obtained in this quasi-static test mode is most frequently analyzed with the Oliver and Pharr method [8], which is contained in the ISO14577 standard "Metallic materials-Instrumented indentation test for hardness and materials parameters." The contact stiffness, S, maximum load, $P_{\max}$, and the maximum penetration depth, $h_{\max}$ are obtained from the load–displacement data at the onset of unloading as seen in Fig. 14.1B and are used for calculating the contact depth h_c via

$$h_c = h_{\max} - \varepsilon \frac{P_{\max}}{S_e}. \tag{14.1}$$

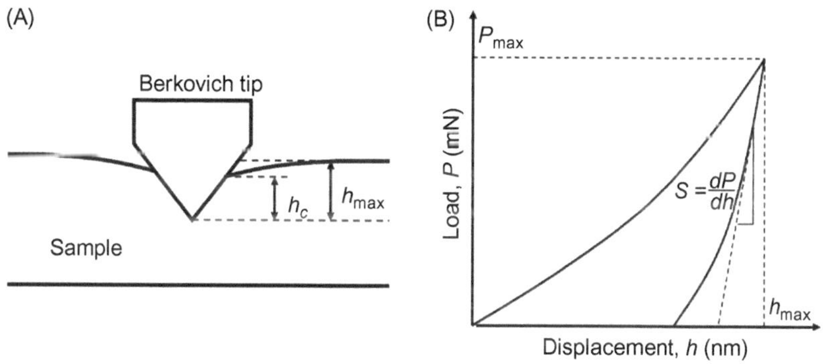

FIGURE 14.1

Schematic representation of: (A) the tip-specimen contact at the maximum load P_{max} (h_{max}, maximum indentation depth; h_c, contact depth). (B) A typical load–displacement curve generated from nanoindentation with a sharp Berkovich tip.

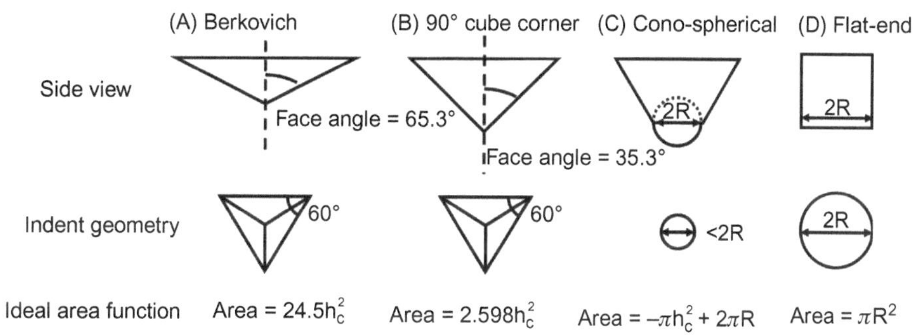

FIGURE 14.2

Schematic representation of common indenter tips with their indent geometries and ideal area function. (A) Berkovich tip, (B) cube corner tip, (C) cono-spherical tip, and (D) flat-end tip.

h_c is then used to find the total tip-sample contact area A_c, through an area function of the tip. Fig. 14.2 shows the ideal area functions of different tip geometries, but for accurate measurements, it is important to calibrate the actual area function of a given tip since deviation from the ideal geometry may occur as a result of bluntness at the very end of the tip. A_c can then be used for calculating hardness according to

$$H = \frac{P_{max}}{A_c}, \tag{14.2}$$

and E_r can be calculated as

$$E_r = \frac{\sqrt{\pi}}{2} \frac{S}{\sqrt{A_c}}. \tag{14.3}$$

The Young modulus E of the sample can be obtained using the relation

$$\frac{1}{E_r} = \left(\frac{1-\nu^2}{E}\right)_{\text{sample}} + \left(\frac{1-\nu^2}{E}\right)_{\text{indenter}}. \tag{14.4}$$

14.3 NANOINDENTATION OF ORAL MINERALIZED TISSUES

14.3.1 SAMPLE PREPARATION

One of the concerns when working with oral mineralized tissues is their small sizes. To overcome this problem, the tissues are sometimes embedded in resins such as poly methyl methacrylate to facilitate handling during processing, as well as to provide mechanical support during indentation. Although infiltration of the embedding material could be prevented by a careful choice of the resin, nanoindentation should only be conducted in areas of the tissues away from the resin–tissue interfaces, since the presence of the resin may alter the mechanical properties of the tissues [9]. In addition, the mechanical properties of the mineralized tissues could be affected, to varying extents when embedded in different resins. This makes the choice of the embedding material crucial if valid comparisons are to be made between different studies [10].

Nanoindentation requires a flat surface with surface roughness <10% of the indentation depth [11]. Since the surfaces exposed by most cutting techniques do not satisfy this requirement, the specimen surfaces are normally further smoothened with mechanical polishing or microtome sectioning. In mechanical polishing, the specimens are first polished with silicon carbide paper to achieve a relatively flat surface before they are fine polished to fulfill the stringent surface requirement. Fine polishing is normally conducted by lapping with a slurry containing diamond or aluminum oxide particles ranging from 0.05 to 6 μm in size. Alternatively, in microtome sectioning, the specimen is first sectioned using a glass knife until a relatively flat surface is achieved. The specimen surface is then further trimmed down with a diamond knife to decrease the surface roughness in preparation for nanoindentation. These ultrasectioned surfaces are reputed to possess smaller surface roughness and less subsurface damages in comparison to mechanical polishing. These characteristics are valuable for investigating localized variations in mechanical properties of the mineralized tissues since the natural properties are better preserved and some of the finer details in the hierarchical structure of the tissues may be more readily revealed [12,13].

14.3.2 HYDRATION

As with most other biological tissues, the mineralized tissues in the oral cavity are hydrated in their natural state. Thus, water is an important constituent of the mineralized tissues and it is well proven that major alterations in their mechanical properties can result from dehydration [13–16]. Therefore, at any stage during the storage, preparation, and experiment, the specimens have to remain hydrated by a suitable fluid. When an inappropriate fluid is used, problems such as demineralization may arise which may induce changes in the mechanical properties of the tissues. Some of the common fluids used for such a purpose include Hank's balanced salt solution (HBSS), calcium chloride ($CaCl_2$) solution, and distilled water [17,18]. Relative to $CaCl_2$ and deionized water, HBSS has been demonstrated to be able to prevent surface demineralization in both dentin and enamel for at least 14 days [18], so it is frequently used in nanoindentation investigations [19–23].

Several strategies have been devised to keep the specimens hydrated during nanoindentation. The simplest method would be to perform indentation within a short time frame after removal of the specimen from the storage medium, such that no significant dehydration could occur during the experiments [24]. Similar techniques include hydrating the specimens from the sides during indentation [25] or applying a fluid to the surface just prior to indentation [7,22,23]. Some of the more sophisticated methods involve hardware modifications such as the use of fluid cells and irrigation systems [26,27].

The presence of a fluid medium that covers the specimen during nanoindentation, on one hand, maintains the hydration of the biological tissues; however, conversely it may induce undesirable surface tension forces during the recording of data [13,28]. The nanoindenter records the total force on the tip comprising the surface tension force at the fluid meniscus, in addition to the force from the contact with the sample surface. Furthermore, depending on the protocol used to define the zero reference for the displacement data, the latter may be mistaken to be the air-fluid interface, instead of the fluid-specimen interface. The problem of incorrectly identifying the contact point, as shown schematically in Fig. 14.3, may be overcome by fitting the initial portion of the load displacement curve with the Hertzian contact relationship whilst subtracting the meniscus pulling force from the fluid [29] i.e.,

$$P = A(h - h_a)^{3/2} - Kh. \tag{14.5}$$

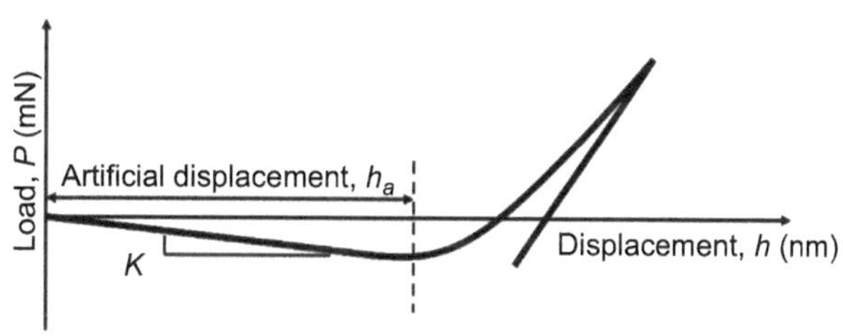

FIGURE 14.3

Illustration of the load displacement curve obtained from a fluid-covered specimen surface. (h_a, artificial drift; K, constant arising from fluid effects.)

The fluid will also apply an additional traction force on the indenter tip at the onset of unloading at which the mechanical properties are measured (cf. Eqs. (14.1) to (14.3)), but the effect of such a force can be removed by the same method for correcting viscoelastic effects which will be described in Section 3.4.

14.3.3 INDENTER TIPS

Most of the tips used in nanoindentation are made of diamond because the high stiffness ensures that the deformation measured by the nanoindenter is mostly contributed by the specimen. The tip geometry is the most important consideration when choosing a suitable tip for indenting biological tissues. Four major kinds of tips are available, namely, Berkovich, cube corner, cono-spherical, and flat-end tips (Fig. 14.3). The Berkovich tip is the most commonly used indenter tip installed in commercial nanoindenters. It has a relatively large face angle of 65.3° which makes it less likely to be damaged under repeated use. An important feature of the Berkovich tip is that the face angle gives the same projected area to depth ratio as the Vickers indenters commonly used in micro-/macroindentation. This allows the conventional Vickers hardness scale to be applicable in analysis of the nanoindentation data.

Similar to the Berkovich tip, the cube corner tip is a kind of three-sided pyramidal sharp tip. The cube corner tip has a face angle of 35.3° and, therefore, is sharper than the Berkovich tip. The main advantage of this shaper tip is that it will create plastic deformation even at shallow indentation depths and ensure that accurate hardness measurements can still be made even when the indent size is small. The cube corner tip is often used for measuring fracture toughness, K_{IC}, of materials since the stress field created by the tip geometry has more cutting action in comparison to the Berkovich tip. Radian and median cracks, therefore, are more readily developed around the indents made by the cube corner tip. The crack length, c, is measured from the center of the indent to the tip of the crack on the specimen surface. The fracture toughness K_{IC} of the material is then calculated using the empirical relation

$$K_{IC} = \alpha \left(\frac{E}{H}\right)^{1/2} \left(\frac{P}{c^{3/2}}\right). \tag{14.6}$$

For a cube corner tip, the value for α is known to range from 0.032 to 0.040 [30].

The sharp tips described above will create plastic deformation in the specimen even at small penetration depths and can only produce a constant nominal strain value during indentation. In contrast, the cono-spherical tip is blunt and produces almost no plastic deformation on the specimen during the initial part of loading. The strain induced by the tip geometry also varies with indentation depth. These characteristics allow the cono-spherical tips to capture the smooth transition from

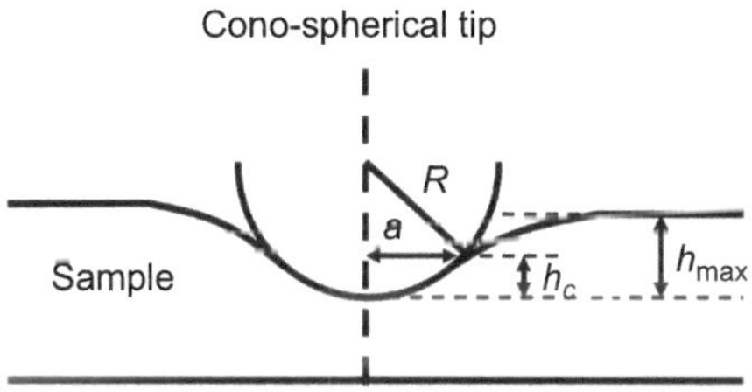

FIGURE 14.4

Schematic representation of the tip-specimen contact at the maximum load P_{max} for a cono-spherical tip. (h_{max}, maximum indentation depth; h_c, contact depth.)

elastic to plastic contact during indentation. The mean contact stress during indentation for $h_{max} \ll R$ is calculated as

$$P_m = \left(\frac{4E_r}{3\pi}\right)\frac{a}{R}, \tag{14.7}$$

and the corresponding indentation strain is calculated as

$$\varepsilon = 0.2\left(\frac{a}{R}\right). \tag{14.8}$$

In the above, R refers to the radius of the cono-spherical tip and a is the radius of the indentation area [31] (Fig. 14.4).

For the flat-end tips, only the flat surface at the end of the tip remains in contact with the specimen surface throughout the experiment such that A_c is always a constant value. However, during an actual indentation, there is often difficulty in aligning the sample perfectly parallel to the tip surface, leading to deviation from the ideal condition. It is worth noting that the elastic deformation of a flat-end tip cannot be described by the term $[(1-\nu^2)/E]_{tip}$ in Eq. (14.4), but if the specimen is much softer than the tip, the deformation of the latter can be ignored anyway so that $E_r = [E/(1-\nu^2)]_{sample}$, i.e., Eq. (14.4) is still applicable in the limit $E_{tip} \to \infty$. Both the cono-spherical and flat-end tips are useful for soft materials such as demineralized dentin, since their stress field is much milder than the sharp tips.

14.3.4 LOAD FUNCTION AND DATA ANALYSIS

It is sometimes desirable to have a constant strain-rate condition when investigating the deformation behavior of viscoelastic materials since their mechanical properties are strain-rate dependent. The relationship between the indentation loading rate, $\dot{P}$, and strain rate, $\dot{\varepsilon}$, can be derived by differentiating the mean contact pressure during indentation, i.e., Eq. (14.2), with respect to time to obtain [32]

$$\frac{\dot{h}}{h} = \frac{1}{2}\left(\frac{\dot{P}}{P} - \frac{\dot{H}}{H}\right) = \dot{\varepsilon}. \tag{14.9}$$

Provided that the value of H is invariant with time, $\dot{\varepsilon}$ produced by the conventional triangular or trapezoidal load function will decrease with an increase in P since $\dot{P}$ is a constant. Therefore, in order to obtain a constant $\dot{\varepsilon}$, $\dot{P}/P$ has to be made a constant, say, k. In cases where there is no preloading, integrating with respect to time yields

$$P = \exp(kt). \tag{14.10}$$

Therefore, P has to be applied in the form of an exponential time function in order to obtain a constant $\dot{\varepsilon}$.

Bone, dentin, cementum, and enamel exhibit different degrees of viscoelasticity depending on their relative amount of organic material. The most obvious effect of viscoelasticity on the nanoindentation data is the presence of a "nose" in the unloading curve due to the superimposition of the downward viscoelastic creep onto the upward elastic recovery upon unloading (Fig. 14.5A). The value of S would appear larger than if the material response is purely elastic and, in extreme cases, S could become negative. One of the practical ways devised to overcome

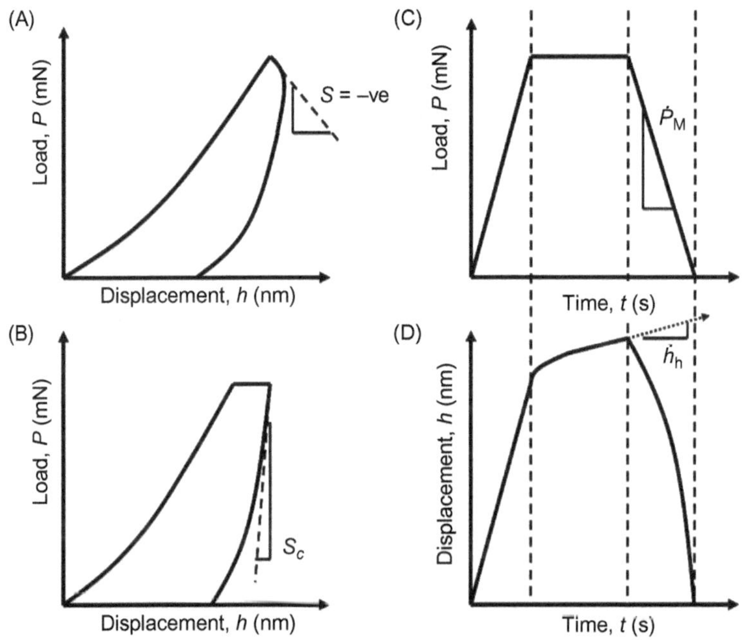

FIGURE 14.5

Schematic representations of the data obtained from nanoindentation of viscoelastic materials. The load–displacement curve for a (A) triangular load function, (B) trapezoidal load function with sufficient hold and fast unloading rate. (C) The load–time graph and (D) the displacement–time graph for the trapezoidal load function. (S_c, apparent contact stiffness, $\dot{P}_u$, unloading rate; $\dot{h}_h$, viscoelastic creep rate at holding.)

this problem is to have a sufficient hold at the maximum load before unloading, coupled with a fast unloading rate (Fig. 14.5B–D). The viscoelastic effect in the load displacement data could be further reduced by modifying the contact stiffness term in the Oliver and Pharr method with the relation [33]

$$\frac{1}{S_e} = \frac{1}{S_c} - \frac{\dot{h}_h}{\dot{P}_u}. \tag{14.11}$$

The three parameters required for eon (14.11) are shown schematically in Fig. 14.5B–D.

For viscoelastic materials, the apparent stiffness is the result of both their elastic and viscous components i.e.,

$$S_c = \frac{\dot{P}_u}{\dot{h}_u} = \frac{\dot{P}_u}{\dot{h}_e + \dot{h}_v}. \tag{14.12}$$

In eon (14.11), $\dot{h}_h$ is the viscoelastic creep rate during the holding period as a result of the viscous component in the biological tissues and approximates the rate of downward viscoelastic creep rate on unloading $\dot{h}_v$. $\dot{h}_h$ is numerically subtracted from the apparent contact stiffness to achieve the corrected contact stiffness,

$$S_e = \frac{\dot{P}_u}{\dot{h}_e} \tag{14.13}$$

which is reflective of the elastic property of the material. With S_e, Eqs. (14.1)–(14.4) can now be used for computing h_c, H, E_r, and E that are free from viscoelastic effects.

14.3.5 MICROSTRUCTURAL INFLUENCE

Biological tissues are a challenging class of materials for mechanical characterization since they have hierarchical structures with important features down to the nanometer or micrometer scale. To fulfill the load-bearing function of mineralized tissues, their mechanical properties often vary gradually across the entire tissue, as in the dentin-enamel junction, such that the resulting stress distribution is less likely to cause failure than in a situation with a sharp interface. Such localized variations in mechanical properties of mineralized tissues were attributed to the tissues' microstructural variation such as mineral density, collagen arrangement, and protein content [34–37]. There are also some small structures distributed throughout the volume of the tissues to modify the function of the material and they may create marked differences in the mechanical properties within the matrix tissue, as in the case of the highly mineralized peritubular dentin located within the intertubular dentin. The measurements made with nanoindentation often have a large scatter in the values as a result of the microstructure.

To elucidate the nanoscale differences across the microstructures, extremely precise control over the location of indentation is required, and this can be done in certain commercial nanoindenters which allow nanoindentation to be performed on the same platform as atomic force microscopy (AFM). Typically, this involves operating the nanoindenter tip as a (very blunt) AFM tip when in the microscopy mode. Such a combination was used to identify the thin prism sheaths sandwiched amidst the prisms, and the measured elastic modulus and hardness values demonstrated a transition across these two neighboring regions in enamel [37].

Despite the fact that most nanoindentation can be conducted at penetration depths of 100 nm, the horizontal dimensions are actually much larger at approximately 1 μm due to the large face angle of the Berkovich tip. In cases where the sizes of the microstructures are smaller than such a resolution limit of the nanoindenter, a commercial AFM can be operated as a nanoindenter to produce much smaller indents. The process for AFM indentation is shown schematically in Fig. 14.6. To calculate the elastic modulus from the AFM data, the tip-sample interaction and the deflection of the cantilever of the AFM tip are treated as two elastic springs in series in the early stage of the indentation, during which the contact between the tip and surface is treated as purely elastic [38]. The ratio K between the specimen height and the diode-electric signal, which measures the AFM cantilever's deflection, can be obtained, and the reduced modulus E_r can be calculated as

$$E_r = \frac{\alpha}{K/A - 1}. \tag{14.14}$$

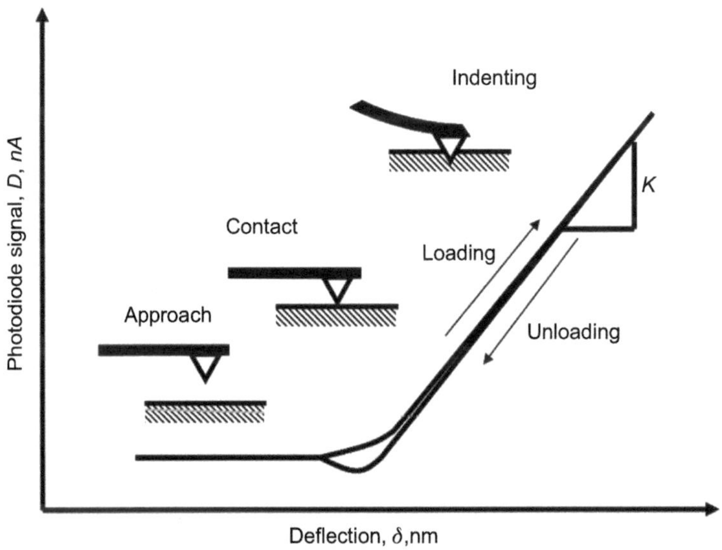

FIGURE 14.6

Illustration of the photodiode signal—cantilever deflection curve obtained from an AFM indentation.

Prior to indentation on the material of interest, the same cantilever tip has to indent on two reference samples each with a known elastic modulus, in order to solve for the cantilever sensitivity, A, and the cantilever tip constant, α [38]. If the sample's deformation is viscoelastic instead of purely elastic, Eq. (14.14) would not be applicable, but it was shown that in this case, E_r can be measured by a load schedule involving a rate jump, such as the connection between loading and unloading in Fig. 14.6 [39]. E_r would still be given by an equation of the same form as Eq. (14.14), albeit that K is now the ratio between the change in the sample's velocity across the jump, and the change in the rate of the photodiode signal across the jump. It is noteworthy that a critical requirement for AFM indentation is that the deformation at the tip-sample contact is significant compared with the deflection of the AFM cantilever. The ratio α/E_r in Eq. (14.14) in fact, measures the tip-sample deformation compared to the cantilever deflection, and this provides a useful guideline for the choice of the cantilever-tip property α [38]. In addition, a further requirement is that $E_{\text{sample}} \ll E_{\text{tip}}$ so that an appreciable amount of the deformation at the tip-sample contact is due to the material of interest, rather than due to the tip. For commercially available silicon tips, their elastic modulus is $\sim$169 GPa [40] which may not satisfy the above requirement if they are to indent enamel which may have E as high as 115 GPa [41]. In this case, other available options include tips made of Si_3N_4 which has an elastic modulus $>$200 GPa [42].

Enamel, dentin, cementum, and bone are anisotropic materials. The loading direction with respect to the microstructures in the tissues, therefore, has an effect on the measured properties. For most cases, the orientation of the microstructure with respect to the indentation surface is already determined during the initial sectioning process. Full understanding of the deformation behavior of the oral tissues requires independent characterization of the elastic properties along the different axes. However, as the stress field under a Berkovich tip is multiaxial, the deformation response elicited would represent some kind of unknown average over different directions. A two-step technique could be used for characterizing the mechanical properties of anisotropic materials using nanoindenters. The elastic constants, c_{ijkl}, of the biological tissue in question are first characterized by ultrasonic means. An effective indentation modulus can then be calculated from the nanoindentation derived data using the c_{ijkl} [43]. A relatively easier, and more direct, method for investigating anisotropic effects in mineralized tissues involves making and testing specimen volumes of small (e.g., micron) dimensions and of known orientations. One method for producing such samples is focused ion beam (FIB) milling. Using this technique, micron-sized cantilevers which can be orientated in any desired direction can be produced as shown schematically in Fig. 14.7 [22]. The free ends of these microcantilevers are loaded by the tip of a nanoindenter until the cantilevers fail. From the load–displacement data, E can be calculated as

$$E = \frac{SL^3}{3I} \tag{14.15}$$

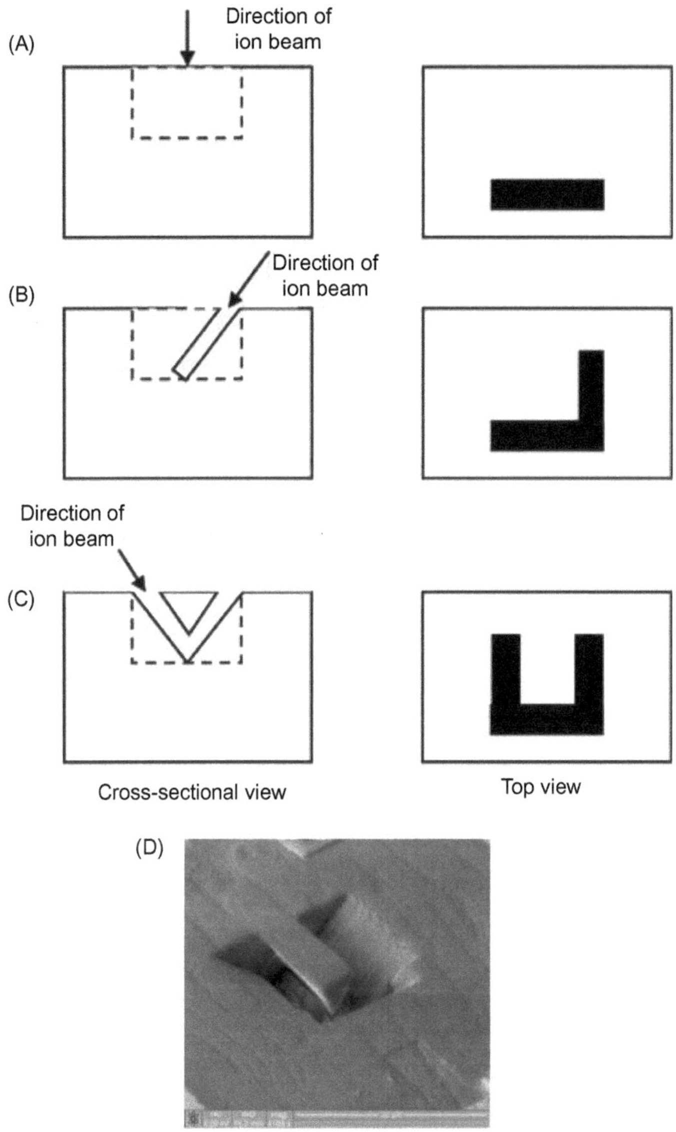

FIGURE 14.7

Milling scheme for the microcantilevers, (A) A $3 \times 7\ \mu m$ trench is first milled, (B) sample is tilted to 45° with respect to the gallium ion beam and a $3 \times 12\ \mu m$ trench is milled, and (C) a 180° rotation with respect to the specimen normal is made and a $3 \times 12\ \mu m$ trench is milled. (D) The final product is a cantilever with a triangular cross-section.

where L is the length between the loading point and the fixed end of the cantilever, $I = wd^3/36$ is the second moment of area of the cantilever's triangular cross-section, w is the width of the cantilever, d is the height of the cantilever's triangular cross-section, and S is the slope of load–deflection curve. This technique also enables the localized flexural strength, σ_{cant}, to be evaluated as

$$\sigma_{cant} = \frac{PLd}{3I}, \tag{14.16}$$

where P is the load at fracture.

Due to their microstructure, the mechanical properties of both enamel and bone exhibit depth dependency when they are indented. Fig. 14.8 shows the typical nanoindentation results obtained from human enamel. Both the hardness and elastic modulus decrease as the indentation depth increases, and their values begin to plateau only after the indent size exceeds that of the prism underneath the indenter tip. This phenomenon is attributed to the continuously changing microstructure as probed by the indenter tip as it penetrates deeper into the enamel. Upon its initial contact with the specimen surface, the indenter tip only interacts with the prism immediately under it. However, as the penetration goes deeper, the indenter begins to interact with the adjacent prism sheaths and prisms, and the measured mechanical properties then reflect the total contribution of both the prisms and prism sheaths [44]. Despite the fact that both bone and dentin are similar in composition, their deformation behaviors differ slightly at the nanoscale.

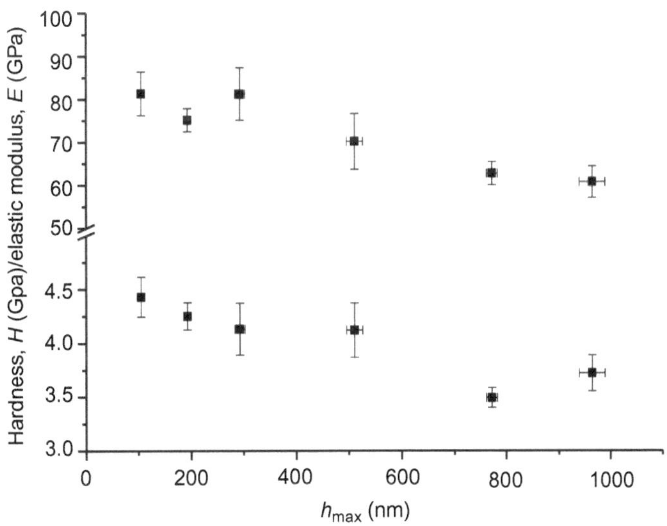

FIGURE 14.8

Human enamel elastic modulus E and hardness H data plotted as a function of the maximum indentation depth h_{max}.

Theoretically, in these multicomponent materials, there should be a certain length scale under which the constituents of the biological tissue no longer function together as a composite, but instead, would behave as individual elements. Under this length scale, the scatter of the recorded data obtained from repeated measurements on random sites would begin to increase drastically due to the large difference in mechanical properties among the constituents. In bone, this value has been estimated to be approximately 600 nm while, in dentin, this value was found to be in the region of 100 nm [45]. The arrangement of the constituent materials is therefore decisive to the mechanical properties of the oral mineralized tissues.

14.4 SUMMARY

Nanoindentation is a proven technique for characterizing the mechanical properties of materials at the nano-/microscale. However, accurate nanoindentation measurements on the oral mineralized tissues cannot be achieved using the conventional technique developed for hard, synthetic materials. Major areas for consideration when conducting nanoindentation on mineralized tissues include:

1. Sample preparation
   a. The specimens are often small in size and may need to be embedded in resins before they are processed for nanoindentation. However, the mechanical properties of the tissues will be affected, to different extents, depending on the resin used for embedding.
   b. The high surface smoothness requirement of nanoindentation can be achieved in two ways: microtome sectioning and mechanical polishing. Microtome sectioning produces a specimen surface of lower surface roughness with less subsurface damage in comparison to mechanical polishing. Both of these features are advantageous when attempting to delineate the differences in the nanoscale structures of the mineralized tissues.
2. Hydration
   a. The specimens have to be hydrated with an appropriate fluid throughout storage, processing, and experimentation to preserve the mechanical properties.
   b. The presence of fluid during an experiment may induce errors in the data recorded by the nanoindenter; therefore, these data have to be corrected for the effects of the fluid with appropriate techniques.
3. Indenter tips
   a. There are four major types of nanoindenter tips, namely, Berkovich, cube corner, cono-spherical, and flat-end tips.
   b. Berkovich tips are most commonly used and they have the same projected area to depth ratio as Vickers indenter tips.

   c. Cube corner tips are sharper than the Berkovich tips and they are often used for small structures and also for characterizing the fracture toughness of materials.
   d. Cono-spherical tips have the unique ability to capture the smooth transition from elastic to plastic contact since the strain field they produce is not self-similar. Depending on the tip radius, these tips can be used for either hard or soft materials.
   e. Flat-end tips have a defined contact area that is suitable for soft materials.
4. Load function
   a. The strain rate associated with conventional triangular or trapezoidal load functions decreases with increased loading. If a constant strain rate is required, the load function has to be in exponential form.
   b. A trapezoidal load function with a long hold period at the maximum load and fast unloading can minimize the viscoelastic effects in the measured contact stiffness.
   c. To fully remove the viscoelastic effect, the data need to be analyzed by a suitable method.
5. Microstructural influence
   a. The oral mineralized tissues are heterogeneous and so large scatter may exist in the measured data.
   b. AFM-based indentation techniques could be used for investigating the mechanical properties of small structures in the mineralized tissues.
   c. Nanoindentation can be coupled with ultrasonic measurements, or FIB milling to investigate the anisotropic properties of the mineralized tissues.

REFERENCES

[1] B. Vanmeerbeek, G. Willems, J.P. Celis, J.R. Roos, M. Braem, P. Lambrechts, et al., Assessment by nano-indentation of the hardness and elasticity of the resin–dentin bonding area, J. Dent. Res. 72 (10) (1993) 1434–1442.

[2] L.H. He, M.V. Swain, Understanding the mechanical behaviour of human enamel from its structural and compositional characteristics, J. Mech. Behav. Biomed. Mater. 1 (1) (2008) 18–29.

[3] V. Imbeni, J.J. Kruzic, G.W. Marshall, S.J. Marshall, R.O. Ritchie, The dentin-enamel junction and the fracture of human teeth, Nat. Mater. 4 (3) (2005) 229–232.

[4] G.W. Marshall Jr., M. Balooch, R.R. Gallagher, S.A. Gansky, S.J. Marshall, Mechanical properties of the dentinoenamel junction: AFM studies of nanohardness, elastic modulus, and fracture, J. Biomed. Mater. Res. 54 (1) (2001) 87–95.

[5] E.K. Mahoney, R. Rohanizadeh, F.S.M. Ismail, N.M. Kilpatrick, M.V. Swain, Mechanical properties and microstructure of hypomineralised enamel of permanent teeth, Biomaterials 25 (20) (2004) 5091–5100.

[6] Z.H. Xie, E.K. Mahoney, N.M. Kilpatrick, M.V. Swain, M. Hoffman, On the structure–property relationship of sound and hypomineralized enamel, Acta Biomater. 3 (6) (2007) 865–872.

[7] L. Angker, M.V. Swain, N. Kilpatrick, Characterising the micro-mechanical behaviour of the carious dentine of primary teeth using nano-indentation, J. Biomech. 38 (7) (2005) 1535–1542.
[8] W.C. Oliver, G.M. Pharr, An improved technique for determining hardness and elastic-modulus using load and displacement sensing indentation experiments, J. Mater. Res. 7 (6) (1992) 1564–1583.
[9] J.-Y. Rho, T.Y. Tsui, G.M. Pharr, Elastic properties of human cortical and trabecular lamellar bone measured by nanoindentation, Biomaterials 18 (20) (1997) 1325–1330.
[10] M. Erik, A. Sailaja, Q. Yi-Xian, The effects of embedding material, loading rate and magnitude, and penetration depth in nanoindentation of trabecular bone, J. Biomed. Mater. Res. A 79A (1) (2006) 86–93.
[11] S. Habelitz, S.J. Marshall, G.W. Marshall, M. Balooch, Mechanical properties of human dental enamel on the nanometre scale, Arch. Oral Biol. 46 (2) (2001) 173–183.
[12] S.P. Ho, H. Goodis, M. Balooch, G. Nonomura, S.J.S.J. Marshall, G. Marshall, The effect of sample preparation technique on determination of structure and nanomechanical properties of human cementum hard tissue, Biomaterials 25 (19) (2004) 4847–4857.
[13] J.-Y. Rho, G.M. Pharr, Effects of drying on the mechanical properties of bovine femur measured by nanoindentation, J. Mater. Sci.: Mater. Med. 10 (8) (1999) 485–488.
[14] M.W. Jameson, J.A.A. Hood, B.G. Tidmarsh, The effects of dehydration and rehydration on some mechanical properties of human dentine, J. Biomech. 26 (9) (1993) 1055–1065.
[15] J.J. Kruzic, R.K. Nalla, J.H. Kinney, R.O. Ritchie, Crack blunting, crack bridging and resistance-curve fracture mechanics in dentin: effect of hydration, Biomaterials 24 (28) (2003) 5209–5221.
[16] G. Guidoni, J. Denkmayr, T. Schöberl, I. Jäger, Nanoindentation in teeth: influence of experimental conditions on local mechanical properties, Philos. Mag. 86 (33–35) (2006) 5705–5714.
[17] M.B. Gustafson, R.B. Martin, V. Gibson, D.H. Storms, S.M. Stover, J. Gibeling, et al., Calcium buffering is required to maintain bone stiffness in saline solution, J. Biomech. 29 (9) (1996) 1191–1194.
[18] S. Habelitz, G.W. Marshall, M. Balooch, S.J. Marshall, Nanoindentation and storage of teeth, J. Biomech. 35 (7) (2002) 995–998.
[19] S.P. Ho, P. Senkyrikova, G.W. Marshall, W. Yun, Y. Wang, K. Karan, et al., Structure, chemical composition and mechanical properties of coronal cementum in human deciduous molars, Dent. Mater. 25 (10) (2009) 1195–1204.
[20] H. Li-Hong, V.S. Michael, Nanoindentation creep behavior of human enamel, J. Biomed. Mater. Res. A 91A (2) (2009) 352–359.
[21] S. Park, J.B. Quinn, E. Romberg, D. Arola, On the brittleness of enamel and selected dental materials, Dent. Mater. 24 (11) (2008) 1477–1485.
[22] Y.L. Chan, A.H.W. Ngan, N.M. King, Use of focused ion beam milling for investigating the mechanical properties of biological tissues: A study of human primary molars, J. Mech. Behav. Biomed. Mater. 2 (4) (2009) 375–383.

[23] Chan, Y.L., A.H.W. Ngan, and N.M. King, Degraded prism sheaths in the transition region of hypomineralized teeth, J. Dent., 38 (3), 2010, 237-244.
[24] Y. Yuan, R. Verma, Measuring microelastic properties of stratum corneum, Colloids Surf. B: Biointerfaces 48 (1) (2006) 6–12.
[25] D.M. Ebenstein, L.A. Pruitt, Nanoindentation of soft hydrated materials for application to vascular tissues, J. Biomed. Mater. Res. A 69A (2) (2004) 222–232.
[26] P.K. Zysset, X. Edward Guo, C. Edward Hoffler, K.E. Moore, S.A. Goldstein, Elastic modulus and hardness of cortical and trabecular bone lamellae measured by nanoindentation in the human femur, J. Biomech. 32 (10) (1999) 1005–1012.
[27] C.E. Hoffler, K.E. Moore, K. Kozloff, P.K. Zysset, M.B. Brown, S.A. Goldstein, Heterogeneity of bone lamellar-level elastic moduli, Bone 26 (6) (2000) 603–609.
[28] A.B. Mann, J.B. Pethica, Nanoindentation Studies in a liquid environment, Langmuir 12 (19) (1996) 4583–4586.
[29] B. Tang, A.H.W. Ngan, Nanoindentation measurement of mechanical properties of soft solid covered by a thin liquid film, Soft Mater. 5 (4) (2007) 169–181.
[30] T. Zhang, Y. Feng, R. Yang, P. Jiang, A method to determine fracture toughness using cube-corner indentation, Scr. Mater. 62 (4) (2010) 199–201.
[31] S.F. Ang, T. Scholz, A. Klocke, G.A. Schneider, Determination of the elastic/plastic transition of human enamel by nanoindentation, Dent. Mater. 25 (11) (2009) 1403–1410.
[32] B.N. Lucas, W.C. Oliver, Indentation power-law creep of high-purity indium, Metall. Mater. Trans. A: Phys. Metall. Mater. Sci. 30 (3) (1999) 601–610.
[33] G. Feng, A.H.W. Ngan, Effects of creep and thermal drift on modulus measurement using depth-sensing indentation, J. Mater. Res. 17 (3) (2002) 660–668.
[34] M. Balooch, I.-C. Wu-Magidi, A. Balazs, A.S. Lundkvist, S.J. Marshall, G.W. Marshall, et al., Viscoelastic properties of demineralized human dentin measured in water with atomic force microscope (AFM)-based indentation, J. Biomed. Mater. Res. 40 (4) (1998) 539–544.
[35] A. Braly, L.A. Darnell, A.B. Mann, M.F. Teaford, T.P. Weihs, The effect of prism orientation on the indentation testing of human molar enamel, Arch. Oral Biol. 52 (9) (2007) 856–860.
[36] H. Fong, M. Sarikaya, S.N. White, M.L. Snead, Nano-mechanical properties profiles across dentin-enamel junction of human incisor teeth, Mater. Sci. Eng.: C 7 (2) (2000) 119–128.
[37] J. Ge, F.Z. Cui, X.M. Wang, H.L. Feng, Property variations in the prism and the organic sheath within enamel by nanoindentation, Biomaterials 26 (16) (2005) 3333–3339.
[38] B. Tang, A.H.W. Ngan, J.B. Pethica, A method to quantitatively measure the elastic modulus of materials in nanometer scale using atomic force microscopy, Nanotechnology 19 (49) (2008).
[39] A.H.W. Ngan, B. Tang, Response of power-law-viscoelastic and time-dependent materials to rate jumps, J. Mater. Res. 24 (3) (2009) 853–862.
[40] G. Simmons, H. Wang, Single Crystal Elastic Constants and Calculated Aggregate Properties: A Handbook, The MIT Press, Cambridge, 1971.
[41] L.H. He, M.V. Swain, Influence of environment on the mechanical behaviour of mature human enamel, Biomaterials 28 (30) (2007) 4512–4520.

[42] K.B. Gavan, H.J.R. Westra, E.W.J.M. van der Drift, W.J. Venstra, H.S.J. van der Zant, Size-dependent effective Young's modulus of silicon nitride cantilevers, Appl. Phys. Lett. 94 (23) (2009) 233108-3.
[43] J.G. Swadener, J.Y. Rho, G.M. Pharr, Effects of anisotropy on elastic moduli measured by nanoindentation in human tibial cortical bone, J. Biomed. Mater. Res. 57 (1) (2001) 108–112.
[44] J. Zhou, L.L. Hsiung, Depth-dependent mechanical properties of enamel by nanoindentation, J. Biomed. Mater. Res. A 81A (1) (2007) 66–74.
[45] M.L. Oyen, C.C. Ko, Indentation variability of natural nanocomposite materials, J. Mater. Res. 23 (3) (2008) 760–767.

FURTHER READING

J. Xu, J.Y. Rho, S.R. Mishra, Z. Fan, Atomic force microscopy and nanoindentation characterization of human lamellar bone prepared by microtome sectioning and mechanical polishing technique, J. Biomed. Mater. Res. A 67A (3) (2003) 719–726.

CHAPTER

15

Nanoindentation techniques for the determination of mechanical properties of materials in dentistry

Jaraslov Menčík
University of Pardubice, Pardubice, Czech Republic

CHAPTER OUTLINE

15.1 Introduction 307
15.2 Basic Information From the Load–Displacement Curves 309
15.2.1 Hardness and Elastic Modulus 309
15.2.2 Harmonic Contact Stiffness 310
15.2.3 Work of Indentation and Other Information From *P–h* Curves 311
15.2.4 Indenter Calibration 312
15.3 Characterization of Inelastic Properties 313
15.3.1 Stress–Strain Diagram 313
15.3.2 Yield Stress 315
15.4 Determination of Properties in Nonhomogeneous Bodies 315
15.4.1 Surface Layers and Coatings 315
15.4.2 Multiphase Microstructure 317
15.5 Characterization of Time-Dependent Load Response 317
15.6 Resistance Against Crack Propagation 320
15.7 Scratch Tests for the Evaluation of Friction and Wear Resistance 322
15.8 Devices for Nanoindentation 322
Acknowledgments 323
References 324

15.1 INTRODUCTION

A good knowledge of mechanical properties of teeth is necessary for their effective repairs and for the development of advanced restorative materials. The choice of suitable methods is limited because of a very small size of a tooth. A modern

Emerging Nanotechnologies in Dentistry. DOI: http://dx.doi.org/10.1016/B978-0-12-812291-4.00015-7

test method which enables measurement of important mechanical properties is nanoindentation [1,2]. With this method, a hard tip is pressed into the specimen, and its penetration is measured as a function of load and time. Special devices, nanoindenters, work with loads ranging from several tens of μN to N and penetrations from tens of nm (1 nm = 10^{-9} m) to several tenths of a millimeter. The loaded area can be as small as a tiny fraction of an mm^2. This enables the study of macroscopic properties of a tooth, as well as the properties of its parts and their microstructural components [2,3].

This chapter is closely related to Chapter 13, Surface Engineering of Dental Tools with Diamond for Enhanced Life and Performance which has explained the basic nanoindentation procedures. In the following text, some of the topics will be treated in more detail, and also other possibilities will be shown, such as the use of elastic and plastic work of indentation, determination of the onset of permanent deformations and construction of stress−strain curves, or the study of properties in nonhomogeneous materials. Attention will also be devoted to the characterization of materials with time-dependent load response and evaluation of the resistance against crack propagation. But first, basic terms will be explained for the readers with only moderate knowledge from mechanics of materials.

The internal response of material to load is characterized by means of *stress* σ (force per unit area, N/mm^2 = MPa) and *strain* ε, which expresses the relative elongation of material ($\Delta L/L$, nondimensional or expressed in %). For relatively low stresses, the deformations are elastic and reversible, and the strain is directly proportional to stress, $\varepsilon = \sigma/E$. The proportionality constant E, *elastic modulus*, characterizes the stiffness of material and corresponds to the stress, which would lengthen the specimen by 100%. If the stress exceeds some limit value, the body either breaks (brittle materials, after exceeding the *strength* σ_U) or starts to deform faster (ductile materials, after exceeding the *yield strength* σ_Y). In this case, permanent deformations remain in the body after unloading. Resistance against plastic deformations can be characterized by hardness H, measured by pressing an indenter into the specimen and calculating the ratio of the load and area of the imprint. Stresses also appear in the body due to imposed deformations or temperature changes, if dilatations are prevented. In *elastic-plastic materials*, like metals and ceramics, deformations occur nearly simultaneously with the load. In polymers or biological tissues, deformations depend on the load and also on its duration and time course. Such materials are called *viscoelastic*. Under constant load, their deformations grow gradually, slower under lower stress. They can be reversible (delayed elastic deforming) or irreversible (creep). A complementary property of viscoelastic materials is *relaxation* of stresses and forces in cases of imposed constant deformation.

Two other important phenomena are fracture and fatigue. *Fracture* in a relatively brittle material, such as enamel, proceeds by propagation of a crack across the body. The crack can be present as an internal defect, or can be nucleated in a localized contact with a hard body or in fatigue processes. Under load, stresses are highly concentrated at the crack tip. The seriousness of a crack is characterized by the *stress intensity factor* K_I, which simultaneously considers the effect of stress and the crack size; $K_I = \sigma Y a^{1/2}$; σ is the nominal stress, a is the characteristic crack dimension, and Y is a nondimensional geometric parameter. The unit of K_I is MPa m$^{1/2}$. A crack propagates fast if the crack intensity factor exceeds critical value K_{IC}, called *fracture toughness*. K_{IC} characterizes the resistance to fast crack growth, and is determined experimentally. In some cases, cracks grow (very slowly) even if the stress intensity factor is lower than K_{IC}. This occurs in *fatigue processes*, which can lead to delayed fracture (from minutes to years). Here, the crack grows due to the creation and cumulation of submicroscopic defects in cyclic loading (especially in metallic materials) or due to chemical action of the environment at the crack tip in the stressed material (stress-enhanced corrosion, typical of silicate ceramics, and some other materials). The velocity of subcritical crack growth can also be expressed as a function of stress intensity factor.

15.2 BASIC INFORMATION FROM THE LOAD–DISPLACEMENT CURVES

15.2.1 HARDNESS AND ELASTIC MODULUS

Nanoindentation, called also instrumented indentation, provides information from the indenter load and displacement. This can yield much more information than the common hardness tests based on the observation of imprints, though observation always brings useful information. Most nanoindentation measurements use a simple load–unload cycle (Fig. 15.1). The basic material characteristics are hardness H and elastic modulus E, introduced in Section 13.2. Hardness in nanoindentation tests is defined as the mean contact pressure p_m,

$$H = p_m = P/A; \tag{15.1}$$

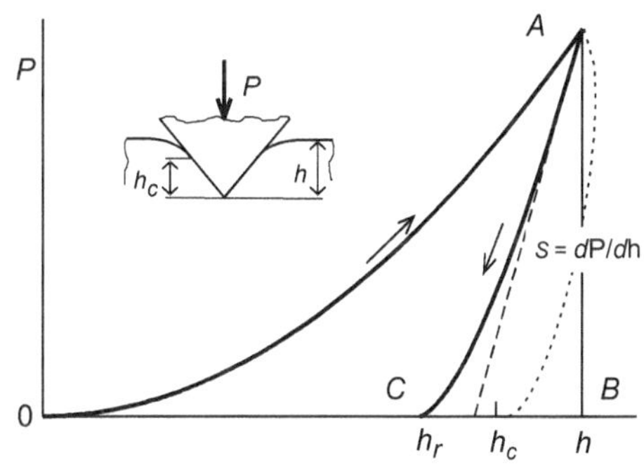

FIGURE 15.1

Load–depth curves of an indentation test (a schematic). P, load; h, total depth of penetration; h_c, contact depth; h_r, residual depth after unloading; S, contact stiffness. Dotted curve at the right—unloading curve typical of viscoelastic materials.

P is the nominal load and A is the projected contact area, determined from the contact depth h_c, which can be calculated from the total penetration h, the indenter load and contact stiffness S at the beginning of unloading as [4]

$$h_c = h - \varepsilon P/S, \tag{15.2}$$

where ε is a constant ($\varepsilon = 0.75$ for a spherical or pointed indenter). The stiffness S ($=\mathrm{d}P/\mathrm{d}h$) is obtained from the upper part of the unloading curve, usually fitted by regression function [4]:

$$P = k(h - h_p)^m; \tag{15.3}$$

k, h_p, and m are constants. Contact stiffness and area (see further) also serve for the determination of the reduced (=sample + indenter) modulus E_r:

$$E_r = \pi^{1/2} S/(2\beta A^{1/2}); \tag{15.4}$$

β is the correction factor for indenter shape ($\beta = 1$ for circular contact and 1.05 for Berkovich indenter). E_r is related to the elastic (Young) modulus E and Poisson's ratio ν of the sample (no subscript) and indenter (subscript i) as

$$1/E_r = (1 - \nu^2)/E + (1 - \nu_i^2)/E_i. \tag{15.5}$$

Eqs. (15.1)–(15.5) are valid for elastic as well as elastic-plastic indentation and any indenter shape. For elastic contact with a spherical indenter, also Hertz' formulae [5] may be used:

$$h = \sqrt[3]{\frac{9P^2}{16RE_r^2}}, \quad h_c = h/2, \tag{15.6}$$

$$p_m = \frac{1}{\pi}\sqrt[3]{\frac{16PE_r^2}{9R^2}}; \tag{15.7}$$

R is the indenter radius and p_m is the mean contact pressure. Sometimes, p_m is denoted as hardness, but the term mean contact pressure is more general, suitable for elastic as well as plastic deformations, while hardness is more often used to characterize the material resistance to permanent deformations.

The values of enamel hardness and elastic modulus, measured by nanoindentation [2,3,6–8], vary between 3–4 GPa for H and 70–120 GPa for E; a role is played by the age, position on the tooth and other factors. Hardness obtained under high loads is often significantly lower, also the modulus and hardness of dentin or other parts of a tooth.

15.2.2 HARMONIC CONTACT STIFFNESS

Nanoindentation tests mostly use monotonic loading and unloading (Fig. 15.1), and the contact stiffness S($=\mathrm{d}P/\mathrm{d}h$) is determined from the regression function fitted to the upper part of the unloading curve [4]. Some devices also enable continuous stiffness measurement (CSM or Dynamic Mechanical Analysis (DMA)

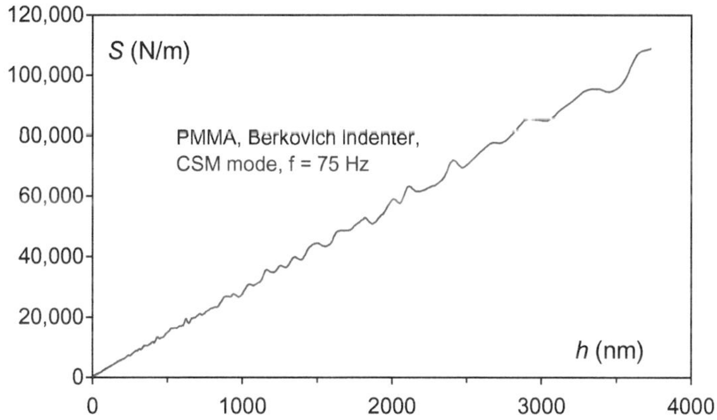

FIGURE 15.2

Harmonic contact stiffness S as a function of indenter penetration h into polymethyl-methacrylate.

mode). In this mode, a small harmonic signal (amplitude several nm or a fraction of a millinewton) is added [4] to the monotonously increasing basic load, and the harmonic contact stiffness S_f is defined [9] as the ratio of the load and displacement amplitudes of these oscillations,

$$S_f = \Delta F / \Delta h; \tag{15.8}$$

the subscript f denotes the exciting frequency. S_f is measured continuously from zero to the nominal load (Fig. 15.2). This significantly simplifies the measurement in cases, where the properties and load response change with indenter depth, for example in enamel. In these measurements, the "monotonic" contact stiffness dP/dh must be obtained from unloading curve in addition to S_f, as it is necessary for the determination of contact depth [9]. The CSM mode is also suitable in the study of materials with time-dependent response, e.g., polymers [10]. For this purpose, *dynamic elastic modulus* $E_{r,f}$ can be determined from Eq. (15.4) with S replaced by S_f. Another useful quantity is the shift between the load and displacement amplitudes (*phase angle* ψ), which informs about the internal friction in the material.

15.2.3 WORK OF INDENTATION AND OTHER INFORMATION FROM *P–H* CURVES

The load–unload P–h curves in depth-sensing tests also bear information about energies involved in the process, such as the total work of indentation W_{tot} (area under the curve 0AB0 in Fig. 15.1), the elastic work W_{el}, released during unloading (area CABC), and the plastic work W_{pl} (area 0AC0), giving the energy dissipated during indentation. The energy components can be used for the characterization of plastic properties of the material. As the work depends on the indenter load, nondimensional quantities are useful, for example, *plasticity index* [11,12],

$$\omega_{pl} = W_{pl}/W_{tot}. \tag{15.9}$$

This index changes from 0 for purely elastic response to 1 if all energy is dissipated. For a homogeneous material and pointed indenter, the load–unload curves for different nominal loads are self-similar, so that the plasticity index does not depend on the load, and can be used as material characteristic. He and Swain [6,13] used the plasticity index in the study of irreversible processes in enamel.

The P–h curves are not always as smooth as in Fig. 15.1. Discontinuities, small jumps or change of slope indicate irreversible changes in the specimen, e.g., nucleation of cracks or plastic flow. Analysis of the pertinent parts of a P–h diagram enables one to measure the work spent in the individual processes, such as formation and propagation of cracks or delamination, and to use it for the determination of specific energies. If the energy is divided by the imprint volume, energy density is obtained, which can also serve as a material characteristic.

15.2.4 INDENTER CALIBRATION

The basic indenter shapes, described in Section 13.3.3, are simple (a pyramid or a sphere). Pointed indenters (Fig. 15.3A), such as Berkovich or Vickers, create self-similar stress field, where the measured value of some property is the same for any depth in a homogeneous specimen. This is useful for larger depths of penetration and for the study of properties distribution in nonhomogeneous specimens. The stresses and strains under a spherical indenter (Fig. 15.3B) increase gradually. This enables the study of elastic as well as elastic-plastic response, determination of yield strength, and construction of stress–strain curves.

In ideal case, the contact area is related to the contact depth as $A = 24.5\ h_c^2$ for Berkovich or Vickers pyramid, and $A \approx 2\pi R h_c$ for a spherical tip and small depth of penetration. However, the tip shape of small indenters is often not ideal, and

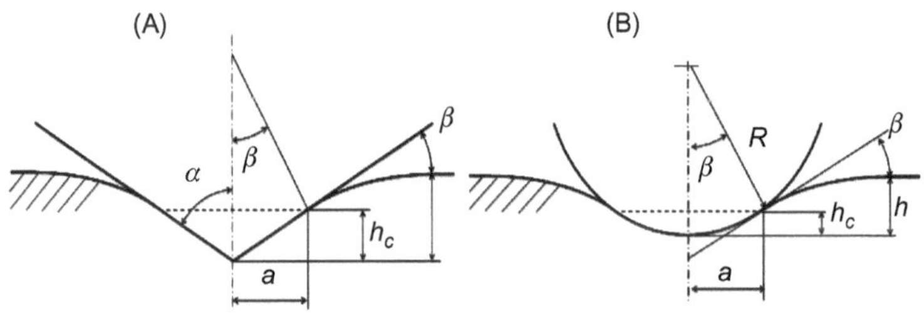

FIGURE 15.3

Contact geometry: (A) pointed indenter, (B) spherical indenter.

this can cause significant errors in the measurement. Calibration is therefore necessary, giving the mean contact radius as a function of the distance from the tip, $a = a(h_c)$. In such calibration, the indenter is pressed into a suitable reference material (usually fused silica), and the load and displacement are measured for the chosen depth range, $P = P(h)$. Also the contact stiffness is determined as a function of depth, $S(h)$, either using continuous stiffness measurement mode or by fitting the unloading curves for a series of nominal loads and making the derivatives ($S = \mathrm{d}P/\mathrm{d}h$). Then, contact depths are calculated via Eq. (15.2) as a function of indenter penetration, $h_c(h)$. Finally, the contact radius for these depths is obtained (using the known elastic modulus of the reference material) as

$$a(h_c) = S(h_c)/(2E_{r,\mathrm{ref}}), \tag{15.10}$$

as it follows from Eq. (15.4). The calibration curve is usually approximated by a simple analytical expression, such as [14] $a = kh_c^m$ or a polynomial. Also the calibration curve for contact area, $A(h_c)$, can be created in a similar way [4].

15.3 CHARACTERIZATION OF INELASTIC PROPERTIES

15.3.1 STRESS–STRAIN DIAGRAM

In common solids, including teeth, low stresses cause elastic reversible deformations, with strains proportional to the stress. If the stress exceeds the yield strength σ_Y, the material deforms faster, as irreversible deformations appear in addition to the elastic ones. Yield strength, which is related to hardness, is thus an important material parameter. Further information is provided by the stress–strain diagram $\sigma(\varepsilon)$, showing how strain develops with stress in the elastic as well as plastic range. Such diagram can be obtained from indentation tests. As the stress field under indenter is nonhomogeneous, it must be characterized by some representative stress and strain, σ_{rep} and $\varepsilon_{\mathrm{rep}}$. The mean contact pressure p_m is very suitable for σ_{rep}, while the expression for representative strain will depend also on the indenter shape.

A pointed indenter, Fig. 15.3A, is not sufficient for the construction of $\sigma(\varepsilon)$ diagrams, as the distribution of stresses beneath it is similar for any depth, and only one value of p_m is obtained for a homogeneous material. The representative strain was defined by Tabor [15] as $\varepsilon_{\mathrm{rep}} = k \tan \beta$, where k is a constant and β is the angle between the indenter and specimen surface (Fig. 15.3A). Tabor has recommended $k = 0.2$, based on the comparison of indentation and tensile tests on metals. For Berkovich or Vickers indenters, $\beta = 19.7°$, so that the measurements with such indenter correspond to a rather high characteristic strain, $\varepsilon_{\mathrm{rep}} \approx 7\%$.

Under a spherical indenter (Fig. 15.3A), the mean contact pressure and strains increase with the depth of penetration and can be used for the construction of stress–strain (p_m–$\varepsilon_{\mathrm{rep}}$) curves [15–19]. The representative strain is usually expressed as the ratio of contact radius a and indenter radius R; $\varepsilon_{rep} = k(a/R)$; often, $k = 0.2$ is assumed. For small depths,

$$p_m = \frac{F}{\pi(2Rh_c - h_c^2)} \approx \frac{F}{2\pi R h_c}, \tag{15.11}$$

$$a = \sqrt{2Rh_c - h_c^2} \approx \sqrt{2Rh_c}. \tag{15.12}$$

For elastic deformations, the contact depth is obtained from indenter displacement as $h_c = h/2$.

If the penetration is larger, or if the deformations are elastic-plastic, a more complex procedure must be used. First, the dependence of contact depth on the indenter displacement, $h_c(h)$, is obtained from Eq. (15.2). The contact stiffness $S(h)$ for these calculations can be measured either directly using the CSM, or calculated from the unloading curves for several loads and depths. Then, the contact radius a is assigned to the individual depths using the indenter calibration function $a(h_c)$. The mean contact pressure and representative strain are then calculated as

$$p_m = P/A = P/(\pi a^2), \quad \varepsilon_{rep} = 0.2\ a/R, \tag{15.13}$$

with all quantities expressed as functions of contact depth h_c or penetration h. Finally, the stress–strain diagram is obtained by plotting the mean contact pressure p_m as a function of representative strain ε_{rep} (Fig. 15.4). However, if the indenter is not perfectly spherical and its radius R varies, it is better to use the same definition as for conical indenters, $\varepsilon_{rep} = 0.2 \tan\beta$ (the difference between $\tan\beta$ and $a/R = \sin\beta$ is negligible for small β). For the known calibration function $a = f(h_c)$, $\tan\beta$ can be obtained [14] as the reciprocal of its derivative with respect to h_c, $\tan\beta = dh_c/da$. Examples of stress–strain curves for enamel and other materials used in dentistry are shown in Refs. [6,16,19].

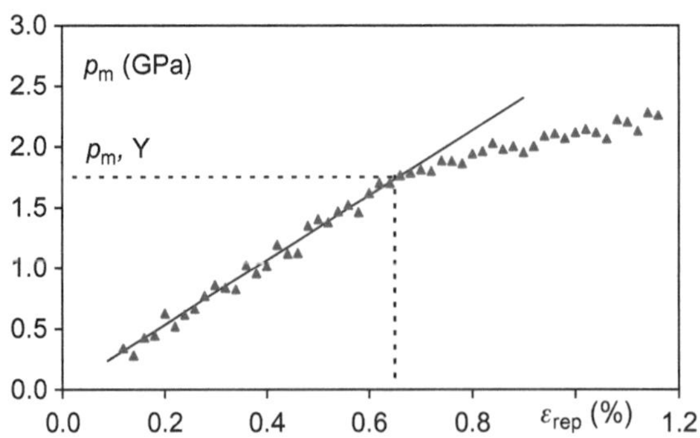

FIGURE 15.4

Mean contact pressure p_m as a function of representative strain ε_{rep} in enamel—a schematic; $p_{m,Y}$ denotes the onset of irreversible deformations.

15.3.2 YIELD STRESS

For relatively low stresses, p_m is directly proportional to ε_{rep}. As soon as irreversible deformations appear, the representative strain grows faster. The onset of plastic flow can thus be identified as the point where the p_m–ε_{rep} curve starts deviating from linearity [14–18] (Fig. 15.4). The corresponding mean contact pressure, $p_{m,Y}$, is sometimes denoted as yield stress. However, the actual yield stress σ_Y is different. Plastic deforming starts [5,18] at $p_m = 1.1\sigma_Y$, in a very small volume beneath the surface. With increasing load, this volume grows, together with p_m, but the load–displacement curve is influenced much later [18], for $p_m \geq 1.6\sigma_Y$. Nevertheless, the yield contact pressure, p_{mY}, may also be used as a material characteristic.

The onset of irreversible processes can also be revealed [16] from the diagram showing the ratio h_c/h as a function of mean contact pressure p_m. For elastic deformations and spherical indenter, $h_c/h \approx 0.5$, while for elastic-plastic deformations $h_c/h > 0.5$. The yield contact pressure p_{mY} corresponds to the point from which the ratio h_c/h starts growing [16].

15.4 DETERMINATION OF PROPERTIES IN NONHOMOGENEOUS BODIES

Foods and drinks act chemically on enamel and can change its mechanical properties in a thin surface layer. Changes are also caused by various treatment, such as teeth whitening. Surface modifications or special coatings are used to improve the properties of components for restorative dentistry. Another problem is the study of materials consisting of several phases, such as enamel (tiny hard rods surrounded by softer organic matter) or polymeric resins with a filler in form of small particles. Nanoindentation can provide quantitative information about mechanical properties in all these cases. However, the preparation and evaluation of the measurements must consider that the obtained values (e.g., hardness) are influenced not only by the properties of the small indented volume, but also by the material around it.

15.4.1 SURFACE LAYERS AND COATINGS

The situation is simple if the change of properties is gradual [20], caused, e.g., by chemical action of a drink or some surface treatment. The first information is obtained by making several tests with various loads, and plotting the measured quantity as a function of depth, e.g., $H(h)$ or $E(h)$, generally $X(h)$. The series of $X(h)$ values is then fitted by a smooth curve, and the surface value is obtained by its extrapolation to $h = 0$, Fig. 15.5. Pointed indenters are better, as the impression shape does not depend on the depth of penetration. At least 4–5 values are necessary in order to reveal trends in the property distribution.

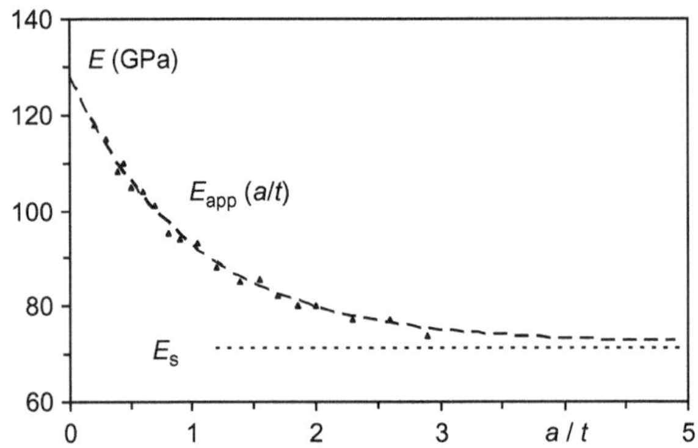

FIGURE 15.5

Apparent elastic modulus E_{app} as a function of relative depth of indenter penetration into a coating (a schematic). E_s, substrate modulus; a, contact radius; t, coating thickness. The coating modulus E_c is obtained by extrapolation of $E_{app}(a/t)$ for $a/t \rightarrow 0$.

The devices working with small harmonic oscillations (CSM or DMA mode) are better, as they can yield the necessary series of $X(h)$ data in one test.

More complex is the situation with coatings, where the change of properties at the interface is sudden. A rule of thumb says that if the indentation depth is less than 1/10 of the coating thickness, only its property is measured. However, this holds only for hardness, especially if the layer is softer than the substrate. The apparent (as measured) elastic modulus is influenced by the substrate properties beginning from the minimum loads. Again, it is necessary to make a series of tests for various depths, and extrapolate the $X(h)$ values to $h=0$. It is reasonable to work with nondimensional penetration ξ, defined as h/t or a/t, where h, a, and t are depth, contact radius, and coating thickness, respectively. The fitting function has general form [21,22]

$$X(\xi) = X_c\ \Phi(\xi) + X_s[1 - \Phi(\xi)], \tag{15.14}$$

where X_c and X_s is the property of the coating and the substrate, $\Phi(\xi)$ is a nondimensional weight function, which decreases from 1 for $\xi = 0$ (or $h = 0$) to zero for $h \gg t$. Knowing the substrate property X_s from independent measurements, one can obtain the coating property X_c as a regression constant by fitting the $X(\xi)$ values by Eq. (15.9) with a suitable function Φ.

Several weight functions have been proposed. For elastic modulus, Doerner and Nix [23] used a simple exponential function,

$$\Phi(\xi) = \exp[-(c\xi)]; \tag{15.15}$$

c is a constant obtained by fitting the measured $X(\xi)$ values. Very good results were received [21] with a more complex weight function $\Phi(\xi)$ proposed by Gao et al. [24]. An example of the determination of elastic modulus of a metallic coating on a glass substrate is shown in Fig. 15.5.

When determining hardness, two cases can be distinguished: soft coating on a hard substrate and a hard coating on a more compliant substrate. While the plastic deformations are limited to the coating in the former case, a hard coating on a soft substrate deforms elastically and plunges into plastically deformed substrate, and, eventually, breaks. The simplest empirical approximations are Eq. (15.15) for hard films on softer substrates, and

$$\Phi(\xi) = \exp[-(c'\xi)^2] \quad (15.16)$$

for soft films on hard substrates; c' is a constant, $\xi = h/t$; X_c, X_s correspond to H_c and H_s. More approximations can be found in Refs. [21,22,25–28].

15.4.2 MULTIPHASE MICROSTRUCTURE

Several kinds of values can be obtained by nanoindentation, depending on the imprint size. If this is relatively large compared to the size of individual phases, only average properties are measured. If the device enables very low loads and good positioning of the indenter, it is possible to study single components. (One must be aware that even the pointed indenters have a rounded tip, with radius several tens of nm.) It is also possible to make a series of indents for various forces and see, how the apparent $X(h)$ values gradually change due to increasing influence of the surrounding material. In this way, properties of tooth enamel were studied [3,6]. At very low loads, the indenter touched only one (relatively stiff) enamel rod. Under increasing load, more rods appeared beneath the indenter, but also more of softer interrod material, and the apparent hardness decreased. Such effects can contribute to the dependence of measured values on the imprint size (so-called indentation size effect).

Information can also be obtained by statistical analysis of many indentations, arranged in a grid over some area. Due to random influences, the measured values vary from a test to test. In a homogeneous material, their histogram is a bell-like curve (e.g., gauss, log-normal or Weibull distribution). The presence of two "hills" indicates a two-phase structure, for example enamel rods and interrod material. Using statistical tools, one can decompose the original group of values into two distributions [29] and determine the typical values of each. The situation is easier, if the volume fractions of individual components are known. A role is also played by their mutual arrangement. With more phases, the decomposition becomes more difficult or impossible.

15.5 CHARACTERIZATION OF TIME-DEPENDENT LOAD RESPONSE

Dentin and other biological tissues, as well as enamel or resins used for dental restorations, belong to viscoelastic materials. At the instant of loading, they deform elastically, but under load the deformations continue growing to a small extent, Fig. 15.6. After unloading they gradually diminish. At a first approximation, these materials can be considered as elastic and characterized by Young's modulus E. In a more detailed analysis or in the nanoindentation determination of properties, their viscoelastic behavior must be taken into account. In common load–unload tests, the unloading part of the P–h curve is more distorted than with elastic materials (dotted curve in Section 15.1). As a consequence, the apparent contact stiffness S is higher than the actual value. This could lead to an error in the determination of contact depth and area, as well as of the elastic modulus and hardness.

The influence of viscoelastic effects on the unload curve can be reduced in various ways. Often, a dwell is inserted between the loading and unloading, so long that the indenter velocity becomes negligible. Also a quick unloading is recommended [30] and the use of effective contact stiffness [31], explained in detail in Section 13.3.4. A disadvantage is that the indenter depth h at the beginning of unloading is larger than at the end of load increase. This results in larger contact area and lower apparent hardness. Moreover, it is generally insufficient to characterize materials which gradually flow under load, only by a single value of hardness or elastic modulus. The time-dependent properties are better described by universal rheologic (spring-and-dashpot) models, which can also be used in computer codes for the finite element analysis and modeling.

The parameters in these models can be obtained by fitting the time course of indenter penetration by a suitable creep function. Generally, the relationship between the load P and deformation h can be expressed as [5,32–35],

$$h^m(t) = K\psi(P, J, t); \tag{15.17}$$

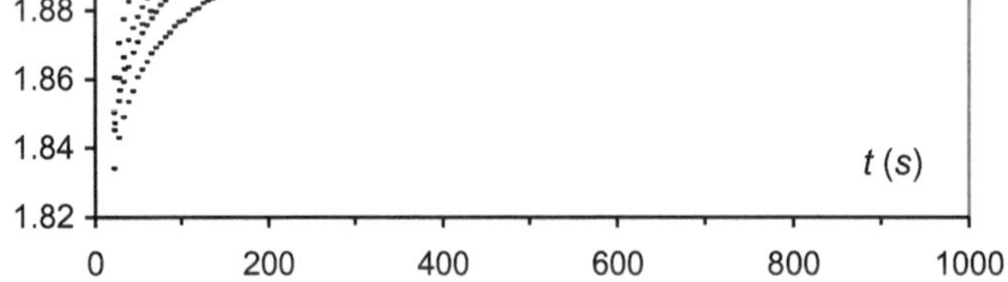

FIGURE 15.6

Penetration of Berkovich indenter into human tooth enamel under constant load 250 mN (measurements done by Dr. Li Hong He). H, depth; t, time. The "instantaneous" penetration during the initial load increase period 20 s was about 1.80 μm.

where ψ (P, J, t) is a response function depending on the load, material and time. J is the creep compliance function, which expresses the material response to the step-load of unit magnitude. For the simplest case of instant elastic deformation accompanied by reversible delayed deforming,

$$J(t) = C_0 + C_1[1 - \exp(-t/\tau_1)], \tag{15.18}$$

with the constants C_0, C_1, and τ_1. C_0 is the elastic compliance (reciprocal of the elastic modulus), C_1 expresses the extent of delayed deforming, and the retardation time τ_1 characterizes its duration. In Eq. (15.18), the time-dependent processes end at about $5\tau_1$. For long processes, creep compliance function usually contains more exponential terms, $C_j[1-\exp(-t/\tau_j)]$. K and m in Eq. (15.17) are constants, depending on the geometry of loading. For pointed indenters, $m = 2$ and $K = \pi/(2 \tan \alpha)$; α is the semi angle of the equivalent cone (70.3° for Berkovich and Vickers). For a spherical indenter, $m = 3/2$ and $K = 3/(4\sqrt{R})$, where R is the tip radius. Pointed indenters nearly always cause permanent deformations. As the deformations of teeth in biting are mostly reversible, the corresponding viscoelastic properties are better studied using spherical indenter and low contact pressure (sufficiently lower than the hardness obtained by a pointed indenter).

Indentation tests for obtaining viscoelastic parameters usually proceed as follows. The indenter is loaded quickly to the nominal load P, which is then held constant. After the time t, the indenter is unloaded—to zero in the simplest case, or to a very low-load P_{low}, which is held constant some time and unloaded to zero. The material constants are determined from the creep during the dwell under nominal load, while the back-creep during the low-load dwell can serve for a check of the number of elements in the model. A universal function for fitting the indenter penetration under constant load is [34,35]

$$h^m(t) = P\,K\left[B_0 + c_v t - \sum D_j \exp(-t/\tau_j)\right]. \tag{15.19}$$

This function characterizes the instantaneous elastic as well as plastic deformations (the term B_0), irreversible viscous flow ($c_v t$) and delayed reversible deforming, $D_j \exp(-t/\tau_j)$; the number of D_j terms depends on the duration of creep processes. If no irreversible viscous flow occurs, the term $c_v t$ is omitted. The determinations of parameters proceed in three steps [34,35]:

Step 1. Calculation of regression constants B_0, c_v, D_j and retardation times τ_j in the proposed model by fitting the $h^m(t)$ data from the creep period (for $t \geq t_R$) by Eq. (15.19). A curve fitter or a solver combined with the least-squares method is necessary.

Step 2. Calculation of the constants C_j from D_j as $C_j = D_j/\rho_j$; ρ_j are the ramp correction factors [33], considering the fact that the load increase to P was not instantaneous, but lasted t_R:

$$\rho_j = (\tau_j/t_R)[\exp(t_R/\tau_j) - 1] \tag{15.20}$$

Step 3. Determination of the constant C_0 as

$$C_0 = B_0 + c_v t_R/2 - \sum C_j. \qquad (15.21)$$

If no permanent deformations occurred, C_0 is related to the reduced modulus as $C_0 = 1/E_r$.

Fig. 15.6 shows penetration of a Berkovich indenter into enamel. Eq. (15.19) fitted the experimental data very well [34]. After unloading, the deformations decreased, but small permanent imprint remained [34,35].

Besides these measurements, also common load–unload tests can be done in order to obtain the "elastic" modulus and hardness, with the corrections from Section 13.3.4, if necessary. The time-dependent component of deformation is important in some cases, while sometimes it may be neglected. Simple, but very illustrative information about the significance of viscoelastic processes is the ratio of the time-dependent component of penetration to the total penetration. Useful also is the ratio of the residual deformation (after unloading) to the total deformation under load. For more, see Ref. [36].

15.6 RESISTANCE AGAINST CRACK PROPAGATION

Fracture of brittle materials, such as enamel, proceeds by crack propagation. The material resistance against propagation of an existing crack can be characterized by fracture toughness K_{IC}, defined as the value of stress intensity factor at the onset of fast crack growth. K_{IC} is usually determined by breaking a standard specimen with a preformed crack, but it can also be calculated from the length of

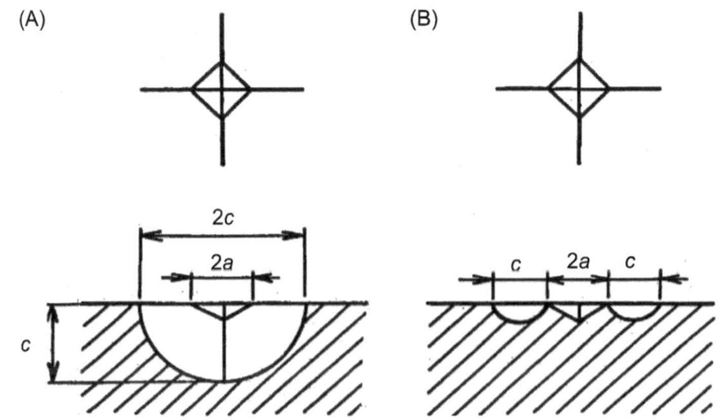

FIGURE 15.7

Radial cracks in imprints made by Vickers indenter: (A) central cracks, (B) Palmqvist's cracks.

cracks created by an indenter (Fig. 15.7). This is advantageous for very small specimens, such as teeth. A semiempirical formula [37] is (see also Section 13.3.3):

$$K_{IC} = k(E/H)^{1/2}P/c^{3/2}; \tag{15.22}$$

P is the indenter load, leading to the formation of cracks, c is the crack half-length, measured from the imprint center, E is elastic modulus, H is Vickers hardness, and k is an empirical constant. Eq. (15.22) was proposed for central cracks of semielliptical shape (Fig. 15.7A), with [7] $k = 0.016$ for Vickers and Berkovich indenter. Also cube-corner indenters are used, as they are sharper than Vickers or Berkovich and nucleate cracks at lower forces. For these indenters [38], $k = 0.032 - 0.040$, or even [7] 0.054. Sometimes, shallow Palmqvist's cracks appear (Fig. 15.7B), for which [39,40]

$$K_{IC} = k'(E/H)^{2/5}P/(ac^{1/2}); \tag{15.23}$$

a is the half-length of the imprint diagonal and c is the crack length measured from the imprint corner. The constant k' was recommended [8] as $k' = 0.0084$.

The length of the cracks is measured on the specimen surface after the test. With an unknown material, it can be difficult to decide from this observation whether the cracks are central or Palmqvist. In such case, it is possible to color the crack surfaces using a dye penetrant, and look at them after breaking the specimen.

Fracture toughness K_{IC} of human enamel, measured by indentation, was in the range [41] 0.4–1.6 MPa m$^{1/2}$, depending on the tested place and indenter orientation. Also other factors play a role. For example, old enamel is more brittle than young. Dentin is much tougher than enamel, and the attempts to nucleate cracks in it by nanoindentation were unsuccessful so far.

A disadvantage of this method is that elastic modulus and hardness must be known in addition to the crack nucleation and measurement. This needs two groups of tests. Recently, Zhang et al. [38] recommended another approach. They have shown that a direct proportionality exists between the ratio of hardness and elastic modulus (H/E) and the ratio of the elastic and total work (W_{el}/W_{tot}) in a load–unload indentation cycle. These works correspond to the areas under the force–displacement curve (CABC in Fig. 15.1 for W_{el} and 0AB0 for W_{tot}), and can be obtained easily from the P–h data. This enables the determination of fracture toughness in one load–unload test, using the following modification [38] of Eq. (15.22):

$$K_{IC} = \lambda(W_{tot}/W_{el})^{1/2}\,P/c^{3/2}, \tag{15.24}$$

with the constant $\lambda = 0.0695$ for cube-corner indenter. However, this value was obtained by the calibration on three silicate glasses and Si only, and its validity for enamel and other hard tissues has not been verified yet. Moreover, a recent paper [42] has brought a review of various indentation methods for the determination of fracture toughness, and its conclusions were rather skeptic regarding their accuracy. For similar reason, some authors recommend the term *indentation toughness* instead of *fracture toughness* [43]. Nevertheless, indentation methods do characterize the material resistance against crack propagation, and can be, at least, useful for comparison or ranking of various materials or treatments.

15.7 SCRATCH TESTS FOR THE EVALUATION OF FRICTION AND WEAR RESISTANCE

One cause of gradual teeth deterioration is the wear due to the friction between the tooth and another object, such as the opposite tooth, a hard teeth brush, piece of food, or a foreign particle. The main mechanisms of teeth wear are attrition, abrasion, and erosion. The knowledge of factors that influence the wear rate is important for the development of teeth-care tools and treatments, for the development of restorative materials, including their surface modifications or coatings, but also in the development of foods and beverages.

Wear can be studied in two ways. In standard wear tests, a pin of a suitable shape slides along the test sample in rotating or reciprocating way. The normal and tangential (friction) force, P_n and P_t, are measured and used for the determination of the coefficient of friction, $\mu = P_t/P_n$. Also the rate of material removal is measured, and the appearance of the surface is observed under a microscope. This "macro" approach is suitable for the study of the influence of various factors on the friction, material loss, and the time to reaching a certain damage.

The details of wear processes and their mechanisms on micro- or nanolevel are better studied via scratch tests. These can be performed by nanoindenters (scratch testers), which enable not only the movement of the indenter into the specimen, but also their mutual sliding. In the test, a suitable stylus (or indenter) is moved along the surface, usually under increasing load, and the normal and tangential force is measured. A videomicroscope enables observation of the scratch trace on the surface. Continuous recording of both force components as functions of indenter penetration and path enables one to determine not only the coefficient of friction but also the onset of irreversible processes in the material, demonstrated by discontinuities or other changes in the force–depth curves. Moreover, the characteristic points on these curves can be assigned to the corresponding places on the scratch trace on the sample, which can then be studied in detail using an electron microscope or atomic force microscope (AFM). Scratch tests are also very important in the development of special coatings, as they can reveal the instant of coating fracture or delamination [22].

Scratch tests have been used in industry long time, but gradually they become common also in dental research; see, e.g., Refs. [44–46].

15.8 DEVICES FOR NANOINDENTATION

Various commercial devices are available for nanoindentation measurements in dentistry. They usually consist of a frame, which carries the measuring head and enables its positioning over a small table with the sample. The measuring head carries the indenter and electromagnetic, electrostatic or piezoelectric actuators, and sensors for its movement and loading. A videomicroscope enables the observation of specimen surface and imprints and facilitates the indenter positioning.

The device is usually placed in a cabinet, protecting against temperature variations. This is very important with regard to the extreme sensitivity of displacement sensors. Some devices enable coupling with AFM. Still there are no small light-weight devices for nanoindentation measurements on teeth in situ, i.e., in mouth, but the development is in progress also in this direction.

Nanoindenters can work over a very wide range of loads (from μN to hundreds of mN) and displacements (from nm to hundreds of μm), depending on the range used. A part of the equipment is software, serving for the control of indenter load and displacement and for gathering the load–depth–time data and their processing. A test sequence (number of indents, indenter positions, loads, load rates, test duration, etc.) can be programed in advance. Usually the software yields the values of common quantities, such as hardness and elastic modulus, etc., and also all basic data (load–depth–time) in a txt file, so that the user can process them according to his/her special needs.

The devices differ in parameters and price. As they have been continuously improved, no particular data will be given here; the reader is referred to the webpages of individual manufacturers, given below. The short list contains only those mentioned in the literature most often, and does not mean advertising—the arrangement is alphabetical. Some other devices could be found via web search. There are also special manufacturers of indenters and other accessories; the nanoindenter providers can help with their choice.

Manufacturers and names of some nanoindentation devices

Agilent Technologies, Inc. http://www.agilent.com/find/nano

Nano Indenter (various types; some enable also scratch testing)

CSM Instruments SA http://www.csm-instruments.com

Nanoindentation testers (NHT, UNHT, TTX), scratch testers, tribometers

Hysitron, Inc. http://www.hysitron.com

Triboindenter, Picoindenter, Triboscope, etc.

Micro Materials Ltd http://www.micromaterials.co.uk

NanoTest (for nanoindentation and also wear study via a nanoscratch module)

Shimadzu Corporation http://www.shimadzu.com/products/test/hardness/index.html

Dynamic Ultra Micro Hardness Testers (DUH)

Umis (IBIS) http://www.ibisonline.com.au

Ultra Micro Indentation System

ACKNOWLEDGMENTS

The author expresses his thanks to the following institutions for enabling him the study and experiments mentioned in this chapter: Research Centre of Karlsruhe (KIT), Germany, Centre de Recherche Public Henri Tudor, Luxembourg, University of Otago, Dunedin, New Zealand, and University of Pardubice, Czech Republic. Indentation measurements on enamel, done by Dr. Li Hong He of University of Otago, are gratefully appreciated.

REFERENCES

[1] A.C. Fischer-Cripps, Nanoindentation, Springer-Verlag, New York, 2004.

[2] L. Angker, M.V. Swain, Nanoindentation: application to dental hard tissue investigations, J. Mater. Res. 21 (8) (2006) 1893–1905.

[3] S.F. Ang, E.L. Bortel, M.V. Swain, A. Klocke, G.A. Schneider, Size-dependent elastic/inelastic behavior of enamel over millimeter and nanometer length scales, Biomaterials 31 (2010) 1955–1963.

[4] W.C. Oliver, G.M. Pharr, An improved technique for determining hardness and elastic modulus using load and displacement sensing indentation experiments, J. Mater. Res. 7 (6) (1992) 1564–1583.

[5] K.L. Johnson, Contact Mechanics, Cambridge University Press, Cambridge, 1985.

[6] L.H. He, M.V. Swain, Understanding the mechanical behaviour of human enamel from its structural and compositional characteristics, J. Mech. Behav. Biomed. 1 (2008) 18–29.

[7] G.W. Marshall Jr, M. Balooch, R.R. Gallagher, S.A. Gansky, S.J. Marshall, Mechanical properties of the dentinoenamel junction: AFM studies of nanohardness, elastic modulus, and fracture, J. Biomed. Mater. Res. 54 (1) (2001) 87–95.

[8] S. Park, J.B. Quinn, E. Romberg, D. Arola, On the brittleness of enamel and selected dental materials, Dent. Mater. 24 (2008) 1477–1485.

[9] J. Menčík, G. Rauchs, J. Bardon, A. Riche, Determination of elastic modulus and hardness of viscoelastic-plastic materials by instrumented indentation under harmonic load, J. Mater. Res. 20 (10) (2005) 2660–2669.

[10] E.G. Herbert, W.C. Oliver, G.M. Pharr, Nanoindentation and the dynamic characterization of viscoelastic solids, J. Phys. D: Appl. Phys. 41 (2008) 1–9.

[11] J. Menčík, M.V. Swain, Characterisation of materials using micro-indentation tests with pointed indenters, Mater. Forum 18 (1994) 277–288.

[12] J. Menčík, Determination of mechanical properties by instrumented indentation, Meccanica 42 (2007) 19–29.

[13] L.H. He, M.V. Swain, Energy absorption characterization of human enamel using nanoindentation, J. Biomed. Mater. Res. 81A (2006) 484–492.

[14] L.H. He, N. Fujisawa, M.V. Swain, Elastic modulus and stress–strain response of human enamel by nanoindentation, Biomaterials 27 (2006) 4388–4398.

[15] H. Tabor, Hardness of Metals, Clarendon Press, Oxford, 1951.

[16] L.H. He, M.V. Swain, Nanoindentation derived stress–strain properties of dental materials, Dent. Mater. 23 (2007) 814–821.

[17] E.G. Herbert, W.C. Oliver, G.M. Pharr, On the measurement of yield strength by spherical indentation, Philos. Mag. 86 (33–35) (2006) 5521–5539.

[18] S.D. Mesarovic, N.A. Fleck, Spherical indentation into elastic-plastic solids, Proc. R. Soc. London A 455 (1999) 2707–2728.

[19] S.F. Ang, T. Scholz, A. Klocke, G.A. Schneider, Determination of the elastic/plastic transition of human enamel by nanoindentation, Dent. Mater. 25 (2009) 1403–1410.

[20] J. Zhou, L.L. Hsiung, Depth-dependent mechanical properties of enamel by nanoindentation, J. Biomed. Mater. Res. 81A (1) (2007) 66–74.

[21] J. Menčík, D. Munz, E. Quandt, E.R. Weppelmann, M.V. Swain, Determination of elastic modulus of thin layers using nanoindentation, J. Mater. Res. 12 (9) (1997) 2475–2484.

[22] J. Menčík, Mechanics of Components with Treated or Coated Surfaces, Kluwer Academic Publishers, Dordrecht, 1996.
[23] M.F. Doerner, W.D. Nix, A method for interpreting the data from depth-sensing indentation instruments, J. Mater. Res. 1 (4) (1986) 601–609.
[24] H. Gao, C.H. Chiu, J. Lee, Elastic contact versus indentation modelling of multilayered materials, Int. J. Solids Struct. 29 (1992) 2471–2492.
[25] J. Menčík, Simple models for characterization of mechanical properties by nanoindentation, Advances in Nanotechnology, Vol. 5, Nova Science Publishers, Hauppauge, NY, 2011 [Chapter 4].
[26] Z. Bartul, J. Trenor (Eds.), Advances in Nanotechnology, Vol. 5, Nova Science Publishers, Hauppauge, NY, 2011.
[27] B. Jönsson, S. Hogmark, Hardness measurement of thin films, Thin Solid Films 114 (1984) 257–269.
[28] A.M. Korsunsky, M.R. McGurk, S.J. Bull, T.F. Page, On the hardness of coated systems, Surf. Coat. Technol. 99 (1–2) (1998) 171–183.
[29] J. Němeček, J. Lukeš, On the evaluation of elastic properties from nanoindentation of heterogeneous systems, Chem. Listy 104 (2010) s279–s282.
[30] Y.T. Cheng, C.M. Cheng, Relationships between initial unloading slope, contact depth, and mechanical properties for conical indentation in linear viscoelastic solids, J. Mater. Res. 20 (4) (2005) 1046–1053.
[31] G. Feng, A.H.W. Ngan, Effects of creep and thermal drift on modulus measurement using depth-sensing indentation, J. Mater. Res. 17 (3) (2002) 660–668.
[32] E.H. Lee, J.R.M. Radok, The contact problem for viscoelastic bodies, Trans. ASME, Ser. E, J. Appl. Mech. 27 (1960) 438–444.
[33] M.L. Oyen, Analytical Techniques for Indentation of Viscoelastic Materials, Philos. Mag. 86 (33–35) (2006) 5625–5641.
[34] J. Menčík, L.H. He, M.V. Swain, Determination of viscoelastic-plastic material parameters of biomaterials by instrumented indentation, J. Mech. Behav. Biomed. 2 (2009) 318–325.
[35] J. Menčík, L.H. He, J. Němeček, Characterization of viscoelastic-plastic properties of solid polymers by instrumented indentation, Polym. Test. 30 (2010) 101–109.
[36] J. Menčík, Determination of parameters of viscoelastic materials by instrumented indentation. Part 3: rheological model and other characteristics, Chem. Listy 104 (2010) s275–s278.
[37] B.R. Lawn, A.G. Evans, D.B. Marshall, Elastic/plastic indentation damage in ceramics: The median/radial crack system, J. Am. Ceram. Soc. 63 (9–10) (1980) 574–581.
[38] T. Zhang, Y. Feng, R. Yang, P. Jiang, A method to determine fracture toughness using cube-corner indentation, Scr. Mater. 62 (2010) 199–201.
[39] K. Niihara, R. Morena, D.P.H. Hasselman, Evaluation of KIC of brittle solids by the indentation methods with low crack to indent ratios, J. Mater. Sci. Lett. 1 (1) (1982) 13–16.
[40] K. Niihara, A fracture mechanics analysis of indentation-induced Palmqvist crack in ceramics, J. Mater. Sci. Lett. 2 (5) (1983) 221–223.
[41] H.H.K. Xu, D.T. Smith, S. Jahanmir, E. Romberg, J.R. Kelly, V.P. Thompson, et al., Indentation damage and mechanical properties of human enamel and dentin, J. Dent. Res. 77 (3) (1997) 472–480.

[42] J.J. Kruzic, D.K. Kim, K.J. Koester, R.O. Ritchie, Indentation techniques for evaluating the fracture toughness of biomaterials and hard tissues, J. Mech. Behav. Biomed. 2 (2009) 384–395.
[43] V. Imbeni, J.J. Kruzic, G.W. Marshall, S.J. Marshall, R.O. Ritchie, The dentin-enamel junction and the fracture of human teeth, Nat. Mater. 4 (3) (2005) 229–232.
[44] S. Gao, S. Huang, L. Qian, H. Yu, Nanoscratch resistance of human tooth enamel treated by Nd:YAG laser irradiation, J. Eng. Tribol. 224 (2010) 529–537.
[45] G. Guidoni, M.V. Swain, I. Jäger, Nano-scale sliding contact deformation behaviour of enamel under wet and dry conditions, J. Mater. Sci.: Mater. Med. 21 (2010) 1195–1203.
[46] Y.R. Jeng, T.T. Tin, T.Y. Wong, H.J. Chang, D.B. Shieh, Nano-mechanical properties of fluoride-treated enamel surfaces, J. Dent. Res. 87 (7) (2008) 381–385.

CHAPTER

Nanocharacterization techniques for dental implant development

16

Dermot Brabazon
Dublin City University, Dublin, Ireland

CHAPTER OUTLINE

16.1 Measurement of the Topology of Nanostructures ... 327
16.1.1 Field Emission Scanning Electron Microscope ... 327
16.1.2 Scanning Probe Microscopy ... 330
16.1.3 Confocal Microscopy and Interferometry ... 333
16.2 Measurement of Nanostructure Internal Geometries ... 336
16.2.1 Transmission Electron Microscope ... 336
16.2.2 Focused Ion Beam ... 338
16.2.3 X-Ray Diffraction ... 338
16.2.4 Mercury Porosimetry ... 340
16.3 Measurement of Composition of Nanostructures ... 342
16.3.1 Energy Dispersive X-Ray Spectroscopy ... 342
16.3.2 X-Ray Photoelectron Spectroscopy ... 343
16.3.3 Secondary Ion Mass Spectroscopy ... 345
16.3.4 Auger Electron Spectroscopy ... 346
16.4 Measurement of the Mechanical Properties of Nanostructures ... 347
16.4.1 Nanoscratch Testing ... 347
16.4.2 Nanohardness Test ... 348
16.5 Conclusions ... 349
References ... 350

16.1 MEASUREMENT OF THE TOPOLOGY OF NANOSTRUCTURES

16.1.1 FIELD EMISSION SCANNING ELECTRON MICROSCOPE

Scanning electron microscopes (SEMs) have been used by researchers since 1935 to examine micrometer-scale structures and more often recently for the examination of nanoscale structures [1–4]. This is a versatile technique with which

Emerging Nanotechnologies in Dentistry. DOI: http://dx.doi.org/10.1016/B978-0-12-812291-4.00016-9

relatively large samples can be visualized, dimensional measurements can be taken, and compositional analysis can be performed. The SEM works by initially firing primary electrons at the sample to be imaged. Electrons are dislodged from the atoms at the surface of the sample and are attracted to a positively charged detector grid. These electrons are known as secondary electrons. When a set pattern of primary electron beam scanning is used over the surface, recording of the secondary electrons allows the surface topology to be interpreted and displayed. Spatial resolution within a given SEM depends on the primary electron beam spot size and the volume of material with which the electrons interact. Under good conditions, such as high accelerating voltage (e.g., 30 kV), well-aligned apertures, well-corrected astigmatism, small spot size (small probe current), and no sample charging, resolutions of 3 nm can typically be achieved. Conventionally, tungsten and carbon elements were used in SEMs. More recently to achieve longer cathode gun lives LaB_6 elements have been adopted. Primary electrons that are bounced back off the surface are known as back-scattered electrons (BSE). The energy of these electrons is directly related to the density of the atoms from which they are repelled and therefore their recording allows the variation of surface composition to be visualized.

A field emission cathode in the electron gun of a SEM provides narrower probing beams resulting in both improved spatial resolution and less sample charging. Such systems are designated as field emission SEMs (FESEMs). In order to achieve this increased electron focusing, a different gun design is required. In this design, electrons are expelled by applying a high electric field very close to the filament tip. The size and proximity of the electric field to the electron reservoir in the filament controls the degree to which electrons tunnel out of the reservoir. One type of field emission gun commonly used is known as the Schottky in-lens thermal FESEM electron gun. Cold gun alternatives are available for even finer FESEM resolution; however, these suffer rapid degradation and can therefore lead to expensive operation due to relatively frequent placement. The field emission guns have higher stability, can allow higher current, and hence provide a smaller spot size. Under good operating conditions a typical FESEM resolution of 1 nm is achievable. Elements that add to improved operation and FESEM resolution include designs with a beam booster to maintain high beam energy, an electromagnetic multihole beam aperture changer, a magnetic field lens, and a beam path have been designed to prevent electron beam crossover.

16.1.1.1 FESEM case studies

Researchers have investigated the potential beneficial effects of well-adhered TiO_2 nanolayers on dental implants for increased biocompatibility and have utilized FESEM to aid characterization of the nanostructures produced [5,6]. It was shown using FESEM that a nanotubular amorphous fluoride containing TiO_2 layer could be produced on a microroughened titanium surface [6]. Potential benefits of such a surface include increased surface area that could be used for drug delivery and growth factors as well as better matched interfacial elastic modulus that should enhance osseointegration as compared with conventional dental and orthopedic implants with smooth or acid-etched surfaces [6]. This latter work presents a good example of where multiple characterization techniques can be useful to elucidate information and understand new material developments. Given the potential benefits to patients of more readily accepted and longer lasting implants, the importance of full characterization of possible new dental materials or processes can be understood. In the work of Kim et al. [6], an FESEM (S-4700, Hitachi, Japan) with an energy dispersive X-ray spectrometer (EDS), a transmission electron microscope (TEM), an X-ray photoemission spectrometer (XPS), an X-ray diffractometer (XRD), a profilometer, a contact angle measurement device, a nanoindentation device, and a scratch resistance measurement device were used to characterize the nanoporous layer. A machine smoothened and a restorable, blast-media-roughened surface was anodized in order to produce a nanotubular surface. Such a TiO_2 layer has been reported to efficiently improve the cellular activities on titanium in vitro and bone–implant bonding properties in vivo. Roughened implants affect the rate of osseointegration and biomechanical fixation. Roughened titanium dental implants exhibit higher bone–implant contact area and greater pullout strengths. From the FESEM images, it was found that the nanotubes, of about 100 nm in diameter and 500 nm in length, were closed at the substrate end, open at the end exposed to electrolyte, and were attached to each other via their side walls (Fig. 16.1).

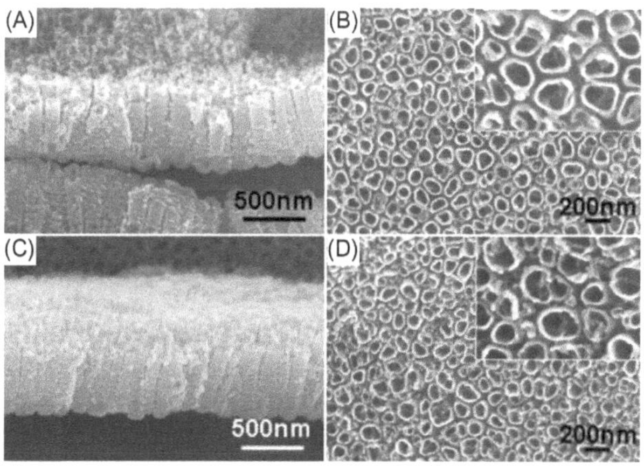

FIGURE 16.1

FESEM of TiO_2 nanotubes (A) machined transverse view, (B) machined plan view, (C) restorable blast media transverse view, and (D) restorable blast media side view [6].

16.1.2 SCANNING PROBE MICROSCOPY

Scanning probe microscopy (SPM) is a general term that covers a wide range of techniques within which a physical probe is passed over a surface via piezoelectric actuators in order to reproduce the surface features. The first of these, scanning tunneling microscopy (STM), techniques was invented in 1981.

16.1.2.1 Scanning tunneling microscope

In 1986, Binnig and Rohrer won the Nobel Prize in Physics for their work to develop the STM technique for imaging surfaces with atomic precision [7]. An STM consists of a tip [typically tungsten, platinum–iridium, gold, or carbon nanotube (CNT)] sharpened to one atom width which is scanned over the surface to be measured. Piezoelectric actuators are used to control the scanning of the tip in the x, y (lateral), and z (normal) directions with respect to the surface. This technique is used to image conductive or semiconductive materials' surfaces and provides the highest resolution of many of the surface imaging techniques. Well-operated and constructed STMs can be used to provide for 0.1-nm lateral resolution and 0.01-nm-depth resolution. In order to achieve these resolutions, vibration control of the sample and within the measurement system is crucial. Spring-based and other vibration isolation systems are often used for this purpose. The sample can be measured in air, liquid, or alternate gas environments. However, to avoid sample contamination, to achieve more stable operating conditions, and to achieve higher resolutions, the sample is often measured under vacuum.

During operation of an STM, the tip is brought close to the surface under coarse control and then to an equilibrium position between tip attraction and repulsion that is typically within a range of 4–7 Å for the surfaces. When the conducting tip is brought this close to the surface, a bias between the two can allow electrons to tunnel between them. At low voltages, this current provides a record of the surface height. There are two modes of operation for the STM, the constant height and the constant current mode. In constant current mode, as the tip is scanned laterally over the surface, the tip height is moved in order to keep the current constant. The movement of the tip height is used to profile the surface. In constant height mode, the height and tip voltage are held constant. The change in current required to keep the voltage constant is related to local charge density to provide a record of the surface profile. Constant height mode is faster than constant current as piezoelectric movements in constant current require more time than the electronic control to keep voltage constant in constant height mode. This can provide a significant advantage as typical STM measurements for area profile recording can be very time consuming. Appropriate sample handling and preparation procedures need to be developed for each material type to be examined [8–11]. Organic molecules can be imaged with STM when placed on a conducting substrate. It is important to minimize resolution loss due to unwanted mixing between molecular and substrate electronic levels. Better resolution can typically be achieved at ultrahigh vacuum (UHV); however, samples can provide difficulties with ultralow vapor pressures or decomposition or which denature at elevated preparation temperatures. The benefits of using passivated substrates have been demonstrated for the capturing of organic molecules to allow examination with UHV STM.

16.1.2.2 Atomic force microscope

The atomic force microscope (AFM), also called the scanning force microscopy (SFM), was developed in 1986, subsequent to the STM [12]. Similar in operation to the STM, the AFM involves scanning a sharp tip across a sample surface while monitoring the tip–sample interaction to allow the reconstruction of the three-dimensional surface topography. A typical AFM can provide resolutions of the order of 1 nm laterally and 0.07 nm (subangstrom) vertically and can image insulators as well as conductors. UHV AFM resolution is comparable to that available from STM and TEM. An AFM consists of a sharp tip on the end of a flexible cantilever that is moved across a sample surface by piezoelectric actuators. The cantilever is typically made from Si or Si_3N_4 with a tip curvature radius of a few nanometers and typically has a spring stiffness, k, in the range of 10–100 N/m. Displacement of the tip is recorded by a noncontact laser displacement measurement. A laser light directed onto the cantilever above the laser tip is recorded on a photodetector area which allows calculation of displacement via signal strength measurement or triangulation. A feedback loop maintains a constant tip–surface interaction force by vertically moving the scanner to maintain a constant photodetector difference signal. The distance the scanner moves vertically is recorded at each x, y position that allows the surface information to be presented and analyzed. A complicated set of forces can be present at the tip–sample interaction. For a surface under ambient conditions, when the tip touches the surface a repulsive force is present, with the tip at a small distance from the surface attractive forces can be present as well as van der Waals force and capillary force arising from condensation of water vapor in the contact area. Operating modes can be roughly classified as contact, noncontact, or dynamic.

In contact mode, the scanning tip is dragged across the sample surface and the tip deflection monitored. Using Hook's law, the force between the tip and the surface is automatically kept constant during scanning (typically between 0.1 and 100 nN). Lower stiffness cantilevers (spring constant, $k < 0.1$ N/m) are used in this mode to amplify the deflection signal. Contact mode may not be suitable for soft materials which can be easily deformed or damaged, such as for polymer or molecular imaging. When scanning is performed in the region where the tip is attracted to the surface, the scanning is termed noncontact mode. In this region, the cantilever bends toward the sample. If an oscillatory tip displacement is sufficiently large to pass through both regions, the probe experiences both attractive and repulsive forces. This mode is known as dynamic, intermittent, or tapping mode. Tapping mode was developed for investigation of soft materials [13]. In this mode, the cantilever oscillates near its resonant frequency and slightly taps the surface during scanning. The tip rapidly moves in and out of the sample surface with an amplitude that is sufficiently high to overcome adhesion forces so that it stays in contact only for a short fraction of the oscillation period. Depending on the cantilever type, the frequency typically varies from 50 to 500 kHz, and amplitudes up to 100 nm are used. The laser spot deflection is used to measure the amplitude of cantilever oscillation and a feedback loop maintains a constant oscillation amplitude by adjustments to the servo that adjusts the cantilever height. In addition to the favorable imaging conditions and high resolution, the AFM offers a variety of new contrast mechanisms which can be used to provide information on differences in sample friction, adhesion, elasticity, hardness, electric fields, magnetic fields, carrier concentration, temperature distribution, spreading resistance, and conductivity.

16.1.2.2.1 AFM case studies

A discussion of surface roughness measurement applied to bioengineering applications and a comparison of AFM, stylus profilometry, and optical profilometry has previously been presented [14]. Mechanical, chemical, or electrochemical surface modification techniques are commonly used to provide surface textures that are deemed to be more biocompatible or osteoconductive. Surface chemistry modification, most commonly known as organic treatments, can also be used such as immobilizing peptides or proteins onto an implant surface. These surfaces can be tailored as biomimetic surfaces that control or aim to trigger cascading events that would occur when cells reach the surface. In the work of Wang et al. [15], a modified AFM setup (cyto-detacher) was used to measure cell adhesion forces. Another AFM was used in this work to image the surface topology before and after cell incubation in a bioactivity assay. Surgical grade Ti−6Al−4V alloy was used without any surface treatment and with modification by oxidizing at 400°C in an atmosphere furnace for 45 min. Roughness (Ra) values of 1.9 ± 0.1 and 2.8 ± 0.1 nm were measured, respectively, for these surfaces. Lateral and vertical resolutions of 2 and 0.01 nm were reported for this AFM work. Fig. 16.2 shows

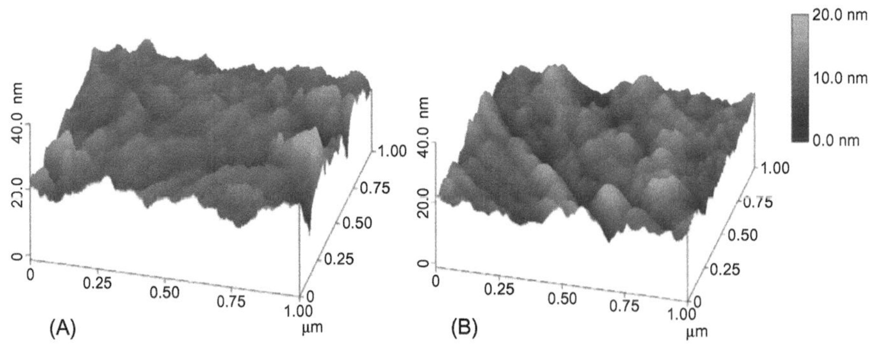

FIGURE 16.2

Tapping AFM scans of 1 × 1 μm area at 1 Hz of the (A) no heat-treated surface with an Ra of 1.9 nm and (B) the heat-treated surface with an Ra of 2.8 nm [15].

the AFM-recorded surface profiles for the nonheat-treated surface and the surface after heat treatment. In this work, an SPM IIIa Dimension 3000 (Digital Instruments Inc., New York, USA) instrument was used with scanning conducted using etched silicone tips (NSC15, MikroMasch Co., Spain) with the spring constant range capability of 20–100 N/m.

16.1.3 CONFOCAL MICROSCOPY AND INTERFEROMETRY

16.1.3.1 Confocal microscopy

In confocal microscopy, two focusing arrangements are used to focus on the point in a sample to be imaged [16,17]. One focuses laser light through an objective lens to the point of interest, and the other focuses the reflected light to the imaging sensor. Light from the point to be imaged is passed through a pinhole such that all extraneous out-of-focus light is removed. This allows lateral resolutions approximately 1.4 times greater than in conventional microscopes to be achieved with confocal microscopy. The depth of the focal plane depends on the specimen's optical properties and importantly on the squared value of the objective lens' numerical aperture. Three-dimensional reconstructions of cells and surfaces can be achieved with this technique. To achieve this, the sample is scanned such that one 2D slice is recorded, the focal plane of the sample is then moved a prescribed amount where the next 2D slice is recorded, and this sequence is repeated until the required volume is scanned. Image processing software is then used to process the collected data to reconstruct the three-dimensional object. Confocal microscopes are most often used to image biological systems and semiconductor surfaces [18–20].

Three variations of scanning are available in confocal microscope systems. In the conventional confocal laser scanning microscope, the sample is raster scanned which results in a scanning rate of about three frames per second. Such systems provide the highest spatial resolution, however for higher temporal resolution, the spinning disk (Nipkow) and the programmable array microscope (PAM) systems can provide rates of 30 frames per second [21]. In a Nipkow disk system, a thin disk with hundreds of spirally patterned pinholes is spun in the light path to the objective lens. The pinholes only allow perpendicularly oriented rays of light to penetrate, which allows high scanning speeds independent of the laser scanning speed. PAM is a variation on this, whereby an acousto- or electrooptical filter can be patterned to automatically produce the pinhole pattern required. Such a system can allow, e.g., up to 1000 beams to simultaneously scan the entire field at millisecond scan speeds. High frequency scanning has the added advantage of reducing photon exposure to sensitive samples which may be damaged due to photobleaching or phototoxicity.

Fluorescent dyes are often added to a surface or flurophores to cellular systems to enable enhanced imaging with confocal fluorescent microscopy. Various excitation laser wavelengths are available for these systems ranging from 442 to

647 nm depending on the flurophore excitation and emission wavelengths used. Reflected and fluoresced light waves are emitted from the sample. A beam splitter can be selected to reflect only the fluoresced light to the detector, which provides enhanced signal-to-noise ratio. As described earlier, the pinhole is also used in this setup to eliminate the out-of-focus signal and record only the light from the region of interest.

16.1.3.1.1 Confocal microscopy case studies

In a recent study, Manara et al. [22] examined an electrochemically assisted process for the deposition of hydroxyapatite (HA)/collagen coatings on titanium plate. The produced coatings are biomimetic bone-like composites made of self-assembled collagen fibrils and carbonate HA nanocrystals. As well as studying these structures chemically, structurally, and morphologically, these workers also evaluated their ability to bind fibronectin (FN) in order to test their bioactivity. FN (a high-molecular weight glycoprotein) aids the development of the cytoskeleton by providing binding sites for cell surface receptors and other extracellular matrix components. FN has been identified as a potential communication network directing bone growth and is also involved in wound healing. Manara et al. [22] used laser scanning confocal microscopy to characterize the degree of FN adsorption on the various composite coatings that they produced. Confocal images showed collagen–FN fiber bundles more than 50 μm in size and 20 μm deep after 5 min of collagen deposition (Fig. 16.3). It was seen from this work that when the apatite phase was intergrown with collagen fibrils, higher FN adsorption was recorded. The authors pointed out that this can be accounted for as FN has at least three receptors for collagen located close to the amino terminus of the molecule. It was further noted that FN adsorption was irregular following the specific topography of the sample.

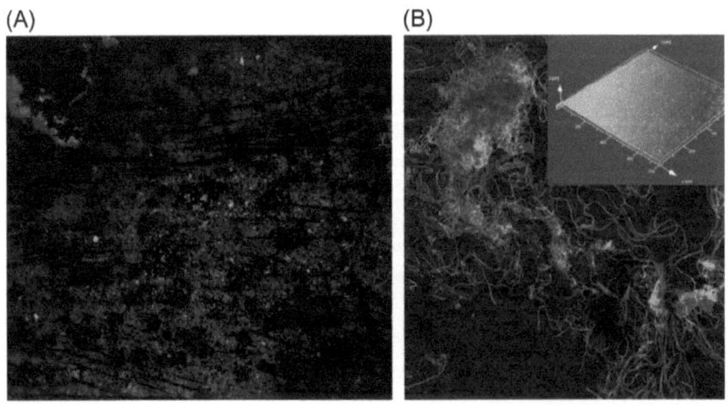

FIGURE 16.3

(A) HA coating preadsorbed with fluorescein isothiocyanate tagged fibronectin (FITC–FN) and (B) collagen–FN network bundles after 5 min of collagen deposition with insert showing topographical view of bound FITC–FN [22].

Wang et al. [23] presented work evaluating the mechanical properties and corrosion behaviors of oriented $Ca(PO_3)_2$ glass. In order to produce these structures, polycrystalline Al_2O_3 plate was inserted into the $Ca(PO_3)_2$ glass melt. The crystallographic *c*-axis of most crystals were oriented in the direction perpendicular to the surface of Al_2O_3 plate. The plate was then cut to extract crystal samples oriented perpendicular and parallel to the plate. The influence of crystal alignment on corrosion behaviors in simulated body fluid was investigated by measuring and comparing the changes of surface roughness from three-dimensional surface topography which was measured using confocal laser scanning microscopy. By measuring surface area roughness from the topology measurements, it was found that crystals aligned parallel to the plate were attacked more strongly than crystals aligned perpendicular to the plate. This new information indicates that directionality of these coatings must be taken into account if these are to be used on implant surfaces.

16.1.3.2 Interferometry

Interferometers are used to measure the interference pattern of a specific light source from the surface to be profiled. A common interferometer is the white-light scanning interferometer or the coherence scanning interferometer [24–29]. This system is used to capture the intensity of interference patterns from the surface during scanning in the vertical direction. The surface can then be reconstructed digitally by using the shape of the white-light interferogram, the localized phase of the interferogram, or a combination of both shape and phase. From recorded peak fringe contrast as a function of wavelength, an interference function which is dependent on optical path difference can be employed to map the surface. Captured data are processed in a variety of ways depending on the interferometer system manufacturer, including being Fourier-transformed into frequency space, subject to crosscorrelation methods, or analysis in the spatial domain.

16.1.3.2.1 Interferometry case studies

Interferometers have been used to quantify quality of implant laser micromachining [30] and implant material degradation [31] and to evaluate the surface properties of commercial implant screws [32]. In this later work of Svanborg et al. [32], different commercial screw-shaped implants with different surface modifications were investigated. Two of these commercial implants tested were OsseoSpeed and Lifecore. OsseoSpeed implants had a blasted surface and the Lifecore had a turned surface. The Lifecore surface was found to be less rough on the micrometer and nanometer levels. The surface roughness was examined using a white-light interferometer (MicroXAM, Phaseshift, AZ, USA). A 50 × objective and a zoom factor of 0.62 were used with a measurement area of 264 × 200 μm and the vertical measuring range of 100 μm. The maximal resolution of the technique was reported as 0.3 μm horizontally and 0.05 nm vertically.

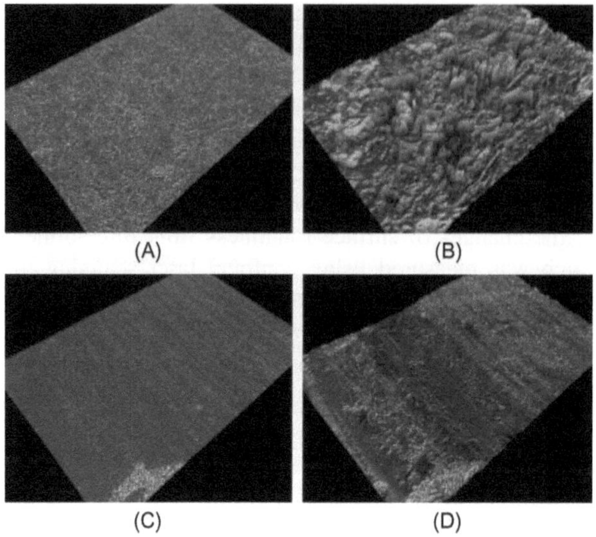

FIGURE 16.4

Interferometer recorded surface topography of blaster surface with (A) a 1 × 1 μm Gaussian filter and (B) with a 1 × 1 μm Gaussian filter low pass; and a turned surface with (C) a 1 × 1 μm Gaussian filter and (D) a 1 × 1 μm Gaussian filter low pass [32].

The work of Svanborg et al. showed the use of the three-dimensional roughness parameters (e.g., Sa rather than Ra) that allow for better averaging of the roughness of the surface and also the importance of surface profile data filters. To be able to describe the surface topography, the roughness, the waviness, and shape must be taken into consideration. The standard filter used to separate roughness from waviness and shape for digital three-dimensional measurements on a micrometer level is a high-pass Gaussian filter. A filter size of 50 × 50 μm was used for evaluation of the micrometer roughness. However, to evaluate the roughness at the nanometer level, the micrometer roughness must be regarded as waviness and a suitable filter size to eliminate it must be identified. A 1 × 1 μm high-pass Gaussian filter was found to be useful for identifying nanometer roughness with respect to height deviation. A low-pass filter was also applied to show how much of the shape, waviness, and roughness were removed by the high-pass filter. Fig. 16.4 shows the effects of filter choice for the OsseoSpeed and less rough Lifecore surfaces.

16.2 MEASUREMENT OF NANOSTRUCTURE INTERNAL GEOMETRIES

16.2.1 TRANSMISSION ELECTRON MICROSCOPE

TEM is an established characterization technique, which is capable of providing both image mode and diffraction mode information from a single sample [33]. It is regarded as one of the main techniques for nanomaterial characterization, largely due to its high lateral, spatial resolution which is in the region of 0.08 nm [34]. A feature of nanomaterials is that specific properties, e.g., color, can be related to a particle's size. Agglomeration of nanoparticles or failure to isolate individual nanostructures is likely to result in anomalous property characterization. Characterizing the elastic or mechanical properties of individual nanoparticle/nanotube/nanofibers is a challenge to many existing testing and measurement techniques. It is difficult to pick up samples and difficult to clamp samples, in order to test for tensile strength or creep, for example [35].

16.2.1.1 TEM case studies

TEM has shown itself to be capable of meeting such challenges. It is commonly used specifically for its ability to isolate and examine individual nanoparticles. This approach reduces the potential for agglomeration which can be a problem with wet-based laser scattering techniques. Nanoparticles attached to CNTs have been imaged, with remarkable clarity [36]. The select area electron diffraction patterns revealed very clear diffraction rings owing to the polycrystalline nature of the iron oxide nanoparticles [36]. TEM's electron imaging and diffraction options allow property–structure relationships of nanostructures to be understood. It has the resolution to differentiate between nanotubes with subtle nanoscale structural patterns. Interlayer distances of about 0.34 nm have been measured and imaged clearly, consistent with the (0 0 2) plane lattice parameter of graphited carbon [37].

TEM is well renowned for its high-resolution-imaging capabilities. However, in recent years TEM has also gained acceptance as a viable means of measuring the elastic and mechanical properties of nanostructures [38]. Wang et al. [35] demonstrated a technique to measure the mechanical strength of single CNTs using in situ TEM. This was done by using an externally applied voltage to induce a charge in the nanotube [35]. The electrostatic force resulted in a deflection of the nanotubes (Fig. 16.5).

Cyclical loading of the nanotube was achieved by subjecting the tube to positive and negative voltages. This approach was used to apply a load to an individual nanotube, thus allowing a direct measurement of its elastic limit [35]. Wang

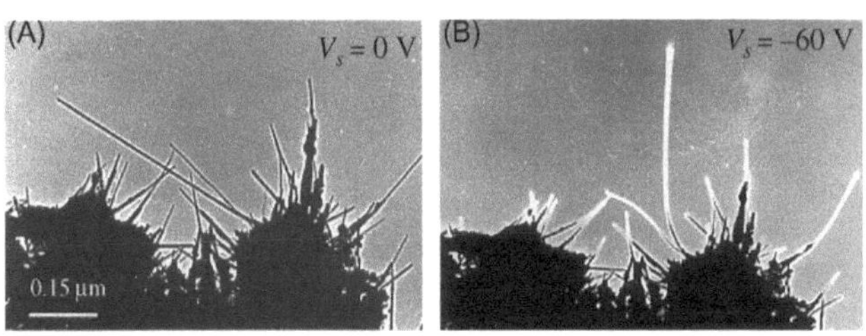

FIGURE 16.5

Electrostatic deflection of a CNT induced by a constant field of (A) 0 V and (B) −60 V across the electrodes [35].

et al. [35] also employed their in situ TEM setup to measure the mass of particles as small as 22 ± 6 fg (1 fg $= 10^{-15}$ g). This nanobalance worked by attaching the mass to be measured to the tip of a nanotube. The change in the resonance frequency of the altered nanotube enabled the mass of the particle to be calculated [35]. The conductance of CNTs is of great interest, partly due to the possibility of using CNTs as interconnects for molecular devices without heat dissipation. AFM studies have shown that conductance effects are related to defects in the CNT which is in turn related to the production method used to make the CNTs. Wang et al. [35] utilized their in situ TEM setup to repeat this experiment by measuring the conductance of a CNT in contact with liquid mercury.

Nanotube alignment is important, because in addition to mechanical properties, functional properties, such as electrical, magnetic, and optical properties, of polymer/CNT nanocomposites are linked directly to the alignment of CNTs in the matrix. Jose et al. [39] successfully synthesized surface-modified multiwalled carbon nanotubes (MWNTs) via electrospinning, using a rotating mandrel. They used TEM to successfully verify high nanotube alignment.

16.2.2 FOCUSED ION BEAM

Focused ion beams (FIBs) have become a popular tool for surface modification of materials and functional structure prototyping at the micro- and nanoscale. Modern FIBs have spot sizes of less than 5 nm and are produced by using electrostatic lenses to focus the image of a point source, often gallium liquid metal ion source, onto the substrate, and to deflect it in a precise fashion. For a comprehensive review of developments in FIB implantation and sputtering, FIB gas-assisted etching, and FIB-induced deposition, the reader is referred to Stanishevsky [40–43].

16.2.2.1 FIB case studies

Moghadam et al. [44] have recently reported on the use of FIB for analysis of pharmaceutical microspheres, which had been prepared by a double-emulsion, solvent-evaporation technique (Fig. 16.6).

Normally, cross-sectioning is required to examine such microspheres. However, in this case it was possible to remove the outer surfaces of the microspheres, layer by layer. Despite having nonporous surfaces, the microspheres were found to become increasingly porous toward the center of the particle. FIB was also used to extract a sample from the center of a microsphere, which was then used for TEM analysis. Fig. 16.7 shows the initial stages of the FIB sample extraction.

16.2.3 X-RAY DIFFRACTION

X-ray diffraction (XRD) is a powerful technique used to uniquely identify the crystalline phases present in materials and to measure the structural properties

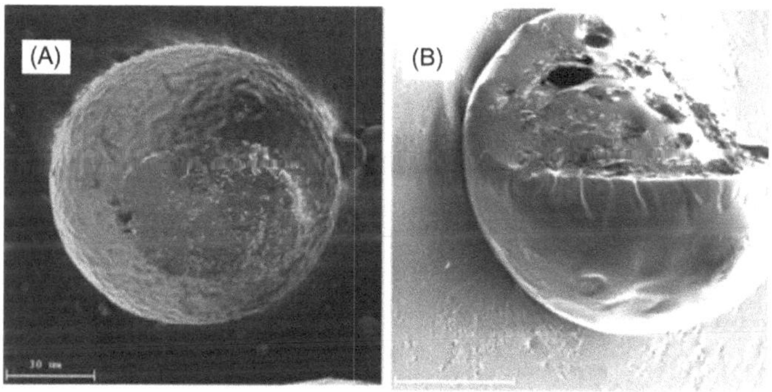

FIGURE 16.6

Internal structure of a PLA (poly D,L-lactide) microsphere: (A) intact microsphere, scale bar is 30 μm and (B) after milling top half using FIB milling, scale bar is 10 μm [44].

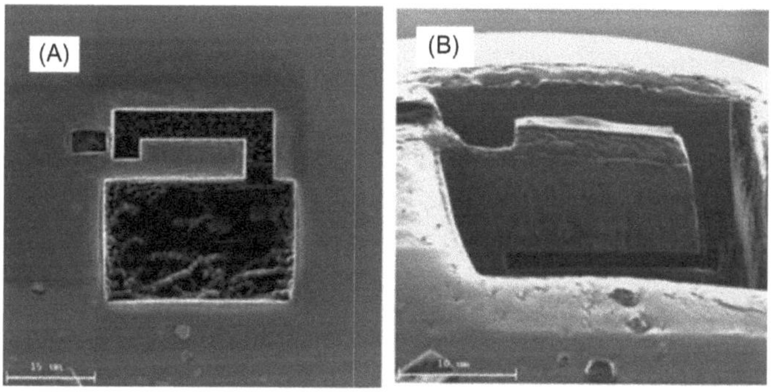

FIGURE 16.7

Secondary ion images of a PLA microsphere at different stages of FIB sample preparation: (A) top view of ion-milled trenches, scale bar is 15 μm and (B) fully undercut sample, scale bar is 10 μm [44].

(strain state, grain size, epitaxy, phase composition, preferred orientation, and defect structure) of these phases. XRD is noncontact and nondestructive. XRD is most sensitive to high-Z elements; as a consequence, the sensitivity of XRD depends on the material of interest [45]. The regular array of atoms in a crystalline material forms a three-dimensional diffraction grating for waves with a wavelength around that of the distance between the atoms. When waves enter a crystal, they are scattered in all directions by the atoms. In certain directions, these waves can interfere destructively. In other directions, constructive interference will occur resulting in peaks in X-ray intensity. The diffraction pattern that results is a map of the reciprocal lattice of the crystal and can be used to determine the structure of the crystal [46]. Bragg's law is the basis for crystal diffraction:

$$n\lambda = 2d \sin \theta \tag{16.1}$$

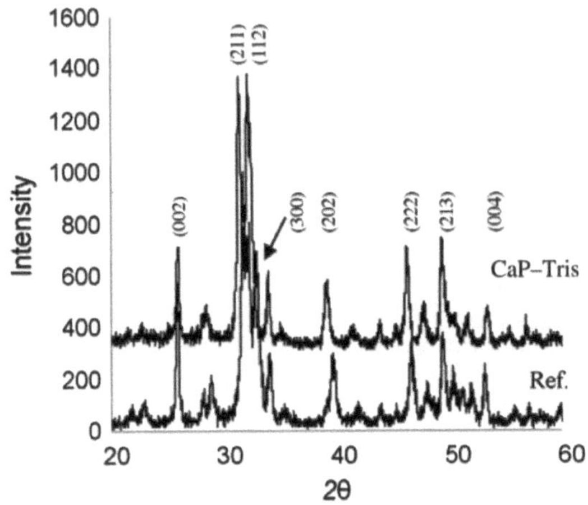

FIGURE 16.8

XRD patterns of HA reference and synthesized using CaP—Tris solution [48].

where n is an integer known as the order of diffraction, λ the X-ray wavelength, d the spacing between two consecutive scattering planes, and θ the angle between the atomic planes and the incident (and diffracted) X-ray beam [47].

16.2.3.1 XRD case studies

Cengiz et al. [48] used XRD to characterize nanoparticles of a novel calcium phosphate growth medium referred to as CaP—Tris. This was prepared by mixing Tris ($(CH_2OH)_3CNH_2$), HCl, K_2HPO_4, and $CaCl_2$ in deionized water. This new formulation was compared with a reference HA powder. Fig. 16.8 shows the XRD patterns of reference HA and CaP—Tris. As can be seen, the XRD patterns are almost identical in terms of peak position and relative intensity. Thus, HA nanoparticles have been synthesized by a new precipitation method and CaP—Tris calcium phosphate growth medium.

16.2.4 MERCURY POROSIMETRY

Mercury porosimeter is a device which is capable of generating suitably high pressures and measuring simultaneously both the pressure and volume of mercury taken up by a porous material [49]. Mercury does not wet most substances and will only penetrate pores when forced to do so under high pressure. Entry of mercury into pores requires applying pressure in inverse proportion to pore size. In other words, large pores will fill first, with smaller pores filling at increasingly higher pressures. Eq. (16.2), known as the Washburn equation, is the basis of the mercury porosimeter method for measuring pore size distribution:

$$D = \frac{-4\gamma \cos \theta}{P} \tag{16.2}$$

where D is the pore diameter, γ the surface tension, θ the contact angle, and P the applied pressure. Mercury exhibits a high contact angle against most solids. Reported contact angles vary, with 130° being the most widely used value. Liquid mercury has a high surface tension; usually its value is taken to be 485 mN/m (485 dyn/cm) [50]. High-pressure mercury porosimeters can normally attain maximum pressures of 207 MPa (30,000 psia) or 414 MPa (60,000 psia). The measurable pore size ranges from a maximum of 360 μm to a minimum of 6 nm (for the 30,000-psia system) or 3 nm (for the 60,000-psia system) [50]. In principle it would be possible to explore in the pore range below 3 nm if a porosimeter with sufficiently high pressures could be made [49]. In practice BET surface area analysis is normally employed for area analysis where pore size is below 2 nm. In mercury porosimetry, an apparatus is used to evacuate the sample and then to surround the sample with mercury. Evacuation is achieved by exposing the sample to a vacuum. The sample to be analyzed is contained inside a penetrometer, which is a long glass capillary tube, the sample end being bulb shaped. Using the vacuum control on the filling apparatus, gases and vapors are removed from the sample. The vacuum valve is closed and the penetrometer titled so that the stem end is immersed in mercury. The vent control valve is then slowly opened such that air fills the mercury chamber. Mercury is forced up the capillary stem and into the bulb. The filled penetrometer is then removed and inserted in the high-pressure porosimeter for pore analysis.

The construction of the glass penetrometer is the key to pore measurement. A metal sheath fits over the capillary section. A metal seal attaches to the sample end of the penetrometer as a base electrode. The construction is thus mercury–glass–metal, or conductor–insulator–conductor. In this way a coaxial capacitor is created [50]. The capacitance changes as a result of the change in mercury level within the penetrometer. The mercury level will change as porous samples are filled with mercury under increasingly high pressure. In a similar way, a normal mercury thermometer will change mercury level, indicating a temperature change. Some porosimetry systems operate on the basis of a wire dipped remotely into the mercury; the change in electrical resistance of the wire being used as a means of measuring the volume of mercury taken up by the pores [49].

16.2.4.1 Mercury porosimetry case studies

Mercury porosimetry has been used in many dental materials' research applications including to elucidate the efficacy of copolymer of polylactic/polyglycolic acid at promoting bone healing and new bone formation in postextraction sockets [51], measure the porosity of dental luting cements [52], and to determine the porosity and pore size distribution of demineralized dentin [53].

In the latter work, a mercury porosimeter (Micromeritics Autopore IV, Micromeritics, Georgia, USA) was used to measure porosity in demineralized dentin. In restorative dentistry, to obtain an efficient infiltration of superficially demineralized dentin, designed porosity is required. In order to have the highest porosity, pretreatment of dentin can be performed and mercury porosimetry can

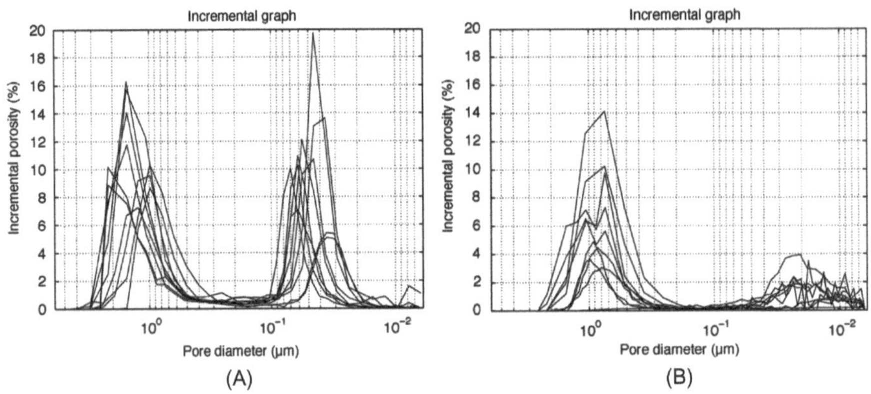

FIGURE 16.9

Incremental porosity versus pore size for demineralized (A) lyophilized-treated and (B) HMDS-dried dentin specimens [53].

be used to evaluate the effect of different pretreatments on demineralized dentin. In the work of Vennat et al. [53], after weighing the assembly including penetrometer, mercury, and sample (to obtain the sample volume using the porosimeter as a pycnometer), pressures of between 0.20 and 200 MPa were applied.

Measurements of intruded volume of mercury versus applied pressure were obtained, and the pressures were converted into pore sizes using the Washburn equation. The higher intrusion pressure of 200 MPa corresponded to a pore diameter around 0.01 μm. Fig. 16.9 shows the incremental porosity versus pore size obtained for demineralized lyophilized-treated and hexamethyldisilazane (HMDS)-dried dentin specimens. The findings of this work indicated that freeze–drying with lyophilization drying technique was more reliable than the HMDS drying with less shrinkage and less overall internal binding of collagen fibers. The results showed two types of pores corresponding either to tubules and microbranches or to interfibrillar spaces created by demineralization. Global porosity varied from 59% for the HMDS-dried samples to 70% for the freeze–dried samples.

16.3 MEASUREMENT OF COMPOSITION OF NANOSTRUCTURES

16.3.1 ENERGY DISPERSIVE X-RAY SPECTROSCOPY

EDS is one of the most common spectroscopy techniques used as SEMs from the 1960s, which have been commonly equipped with this chemical analytical device [54,55]. In this technique, electromagnetic radiation is bombarded into a material surface which causes electrons from inner atomic shells to be ejected and subsequently filled with electrons from higher energy levels. Electromagnetic radiation used to excite the sample is usually a focused high-energy stream of electrons, protons, or X-rays. In a typical SEM, a stream of electrons is used. Electron transitions from the higher energy shells to lower energy shells causes X-rays to be emitted. A detector of SiLi or more commonly now a silicon drift detector is used to collect and count the number of X-rays emitted at each energy level. The energy level characterizes the element from which the X-ray was emitted while the count of the number of X-rays with this energy level is used to characterize the amount of the element that is present. A typical spectrum presents the count of the X-rays versus the energy level of the X-rays. New EDS systems come precalibrated to allow automatic detection and quantification of the elements present within the sample.

Care must be taken when interpreting EDS results. For example, wrong elements can often be detected where energy levels emitted from different elements overlap. X-rays can be generated from K, L, or M energy-level shells in a typical element. Therefore, overlapping of energy levels detected can occur, e.g., when Ti and Ba (Ti–Kα and Ba–L), or Mn and Fe (Mn–Kβ and Fe–Kα), or Mn and Cr (Mn–Kα and Cr–Kβ) are present. Some knowledge of the sample elemental chemistry or knowledge gained from other analytical techniques is often useful for element identification. Wavelength dispersive X-ray spectroscopy (WDS) is similar to EDS but analyzes the diffraction patterns from the material–radiation interaction in order to identify one element at a time. WDS provides greater spectral resolution. Often the use of EDS followed by WDS can provide further definition of sample elemental content. The interaction volume from which X-rays are emitted due to the primary electron bombardment is in the shape of a tear drop beneath the surface. The accelerating voltage used and the density of the material define the volume size. The depth from which X-rays are emitted to the detector is usually from 1 to 5 μm and can be calculated from the empirical expression $(0.1 \times E^{1.5})/\rho$, where E is the beam energy and ρ the material density [56]. The width of the volume can be approximated from $(0.077 \times E^{1.5})/\rho$.

16.3.1.1 EDS case study

A SEM equipped with EDS and ultrathin window SiLi detector was used for X-ray measurement of electrodeposited HA coatings on Ti–6Al–4V substrates [57]. X-ray spectra were acquired at primary beam energy of 30 kV, current of 5 nA, and acquisition time of 180 s. The stoichiometric molar ratio for pure HA is 1.67. X-ray spectra captured in this work showed CaP ratios of 1.51 corresponding to a Ca-deficient HA.

16.3.2 X-RAY PHOTOELECTRON SPECTROSCOPY

When excess electromagnetic energy is transferred to an electron that is in a further out shell it is called an Auger electron. An analysis of these for chemical identification is known as Auger electron spectroscopy (AES). X-ray photoelectron spectroscopy (XPS) analyzes electron emission of similarly high energy but can be used to measure the chemical or electronic state of surface elements, detect chemical contamination, or map chemical uniformity of biomedical implant surfaces [58–60].

For XPS, the material to be examined is irradiated with aluminum or magnesium X-rays. Monochromatic aluminum Kα X-rays are normally produced by diffracting and focusing a beam of nonmonochromatic X-rays off of a thin disk of crystalline quartz. Such X-rays have a wavelength of 8.3386 Å, corresponding photon energy of 1486.7 eV, and provide a typical energy resolution of 0.25 eV. Nonmonochromatic magnesium X-rays have a wavelength of 9.89 Å, corresponding photon energy of 1253 eV, and a typical energy resolution of 0.90 eV. The kinetic energy of the emitted electrons is recorded. This kinetic energy of the ejected electrons is directly related to the element-specific atomic binding energy of the liberated. A plot of these energies against the corresponding number of electron counts provides the spectrum which indicates the qualitative and quantitative elemental composition. At these higher energies, XPS only analyzes to a depth of 10 nm into the surface. Electrons emitted at greater depths are recaptured or trapped in various excited states within the material. Spectral profiles as up to 1 μm deep can however be obtained by continuous spectral recording during ion etching or from consecutive ion etching and XPS measurement steps.

XPS is usually performed in UHV and typically provides resolutions down to 1000 ppm. With optimum settings and long recording times, resolutions down to 100 ppm can be achieved. Nonmonochromatic X-ray sources can produce a significant amount of heat (up to 200°C) on the surface of the sample as the anode producing the X-rays is typically only a few from the sample. This level of heat when combined with high-energy Bremsstrahlung X-rays can degrade the surface. Organic chemicals are therefore not routinely analyzed by nonmonochromatic X-ray sources.

16.3.2.1 XPS case study

The effect of aluminum addition to titanium nitride (TiN) matrix on the structural, mechanical, and corrosion resistance properties of titanium–aluminum–nitride was studied [61]. Si wafer, AISI 316L stainless steel, and low carbon steel substrates were coated with $Ti_{1-x}Al_xN$ (where x was 0, 0.5, and 1) films by a direct current magnetron sputtering process. Layers were sputtered in pure Argon with a substrate temperature at 400°C, power of 250 W, and a sputtering time of 120 min. The XPS survey spectra of the normally unetched surface of the $Ti_{0.5}Al_{0.5}N$ film on steel exhibited the characteristic Ti2p, Al2p, N1s, and O1s peaks at the corresponding binding energies of 454.5, 74.3, 399.6, and 532.1 eV, respectively. From high resolution XPS measurements of the normal surface of the films, the spin orbit doublet Ti2p1/2 and Ti2p3/2 peaks at binding energies 463.1 and 460.3 eV, respectively, were found in the Ti spectra. The Ti2p3/2 peaks included three components whose peaks centered at 460.3 (I), 456.9 (II), and 454.5 eV (III). These components were associated with TiO_2, TiO_xN_y, and TiN phases, respectively. The chemical bonding in the coating could be understood further by analyzing N1s photoelectron spectra. The N1s spectrum of the unsputtered specimen exhibited a bump at the higher binding energy side. The spectrum were deconvoluted into the components using curve fitting; one part was associated with the Ne–O bonding state at 396.65 eV and the other with the Ne–N bond at 399.6 eV, which indicated some adsorbed nitrogen in the surface of the film. Based on these analytical results, it could be concluded that the Ti–Al–N films were composed of the Al–N and Ti–N phases mainly, on which a thin Ti–Al–O–N surface layer was formed upon exposure of the films to the air after deposition.

16.3.3 SECONDARY ION MASS SPECTROSCOPY

Secondary ion mass spectroscopy (SIMS) is a destructive analytical technique in which material is removed from a surface by ion beam sputtering, and the resultant positive and negative ions are mass analyzed in a mass spectrometer [62]. The technique is element specific and is capable of detecting all elements as well as isotopes and molecular species. Of all the beam techniques, it is the most sensitive with detection limits for some elements in the 10^{14}–10^{15} cm^{-3} range if there is very little background interference signal. Lateral resolution is typically 100 μm but can be as small as 0.5 μm with depth resolution of 5–10 nm [34].

SIMS works by removing material from a sample by sputtering using an ion beam. The mass/charge ratio of the removed ions is analyzed, detected as a mass spectrum, as a count, or displayed on a fluorescent screen [34]. *Static* SIMS employs very low primary ion density of around 1 nA/cm^2 and low primary ion energy (0.5–2 keV) so that a nearly undisturbed monolayer of the surface can be analyzed. *Dynamic* SIMS involves primary ion currents greater than 1 $\mu A/cm^2$ and usually more than one monolayer is removed during the analysis [46]. Dynamic SIMS can produce depth profiles, and quantitative depth profiling is unquestionably the major strength of SIMS. SIMS lends itself to investigations of grain boundary diffusion or diffusion across interfaces. It is a powerful tool for studying the transport processes in ceramics in the temperature range where diffusion distances are too small to be analyzed by serial mechanical sectioning.

16.3.3.1 SIMS case studies

In the work of Gao et al. [63], the effect of CO_2 laser and fluoride treatments to inhibit root caries was investigated. Time of Flight-SIMS (ToFSIMS IV, ION-TOF GmbH, Münster, Germany) was carried out for elemental analysis on the cut surfaces of the treated sections. A significant laser-enhanced fluoride uptake, characterized by the ToF-SIMS analysis was revealed. Regions of 50 × 50 μm were analyzed beneath the root surface and revealed 37% and 400% increments in loosely and firmly bound fluorides. In other work by Balter and Reynard [64], the use of SIMS to aid the study the distribution of Sr at low concentration in bones was shown. This work showed that Sr administration at low dose reduces bone resorption and increases bone formation, leading to an increase of the bone mass.

16.3.4 AUGER ELECTRON SPECTROSCOPY

When an electron or ion is incident on a semiconductor, it may transfer enough energy to an inner-shell electron to eject it from its parent atom. The atom is in an excited state, and, to lower its energy, an electron from a less tightly bound shell may fill the hole while simultaneously emitting a third electron from the atom. This ejected atom is known as an Auger electron [46]. Its energy is related specifically to the electron energy levels involved in the process and, therefore, is characteristic of the atom concerned. Since the Auger process is a three-electron process, hydrogen nor helium cannot be detected since both have less than three electrons. AES has two distinct advantages over energy dispersive X-ray spectroscopy (EDX) analysis.

It is a far more surface-sensitive technique. Escape depths range from less than a nanometer to a few nanometers. In EDX, it can be difficult to analyze small particles on a substrate, because the electron beam passes through the particles and spreads out in the substrate below it [62]. There is the potential for chemical-state information in Auger spectroscopy, e.g., the oxidation state of silicon at a $Si-SiO_2$ interface may be ascertained [46]. EDX lacks the same chemical-state information capability.

AES has found applications in measuring semiconductor composition, oxide film composition, silicides, metallization, particle analysis, and the effects of surface cleaning. AES measurements are made in a high vacuum environment ($10^{-12}-10^{-10}$ Torr) to retard the formation of hydrocarbon contamination layers on the sample surface [34]. Scanning Auger microscopy (SAM) allows surfaces to be mapped for one selected element at a time. In this mode, the electron beam is scanned over a selected area. The Auger intensity is measured at each point of the area [65–67]. SAM requires higher beam currents and is much slower than SEM/EDX [34].

16.3.4.1 AES case studies

Pisarek et al. [68] used AES to investigate chemical and morphological changes to the surface of titanium following exposure to acidic or alkaline solutions. Control of these exposures may increase the effective surface area of the Ti, and thus the biocompatibility. Auger electron microanalysis enabled subtle changes in the modified Ti surface layer to be examined, in terms of local chemical composition. SEM images and AES spectra are shown in Fig. 16.10 for (A) as-received Ti and (B) after pretreatment in NaOH.

Immersion in 5M NaOH and subsequent heating up to 600°C results in a porous, "honeycomb" morphology. In terms of the Auger spectra, the signals from Ti and O are well distinguishable, implying that the Ti is oxidized. The relative peak ratio of O KLL signal to Ti LMM signal is much higher for the NaOH

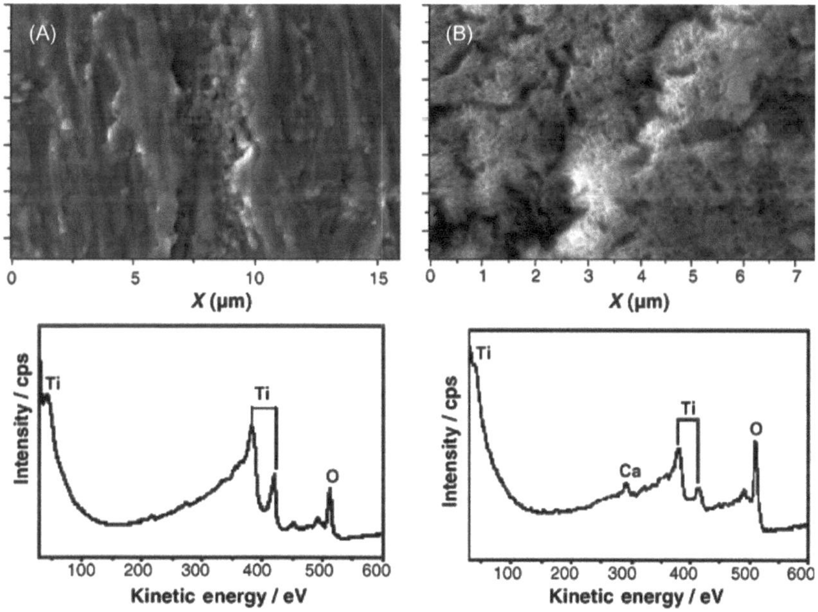

FIGURE 16.10

AES spectra are shown for (A) as-received Ti and (B) after pretreatment in NaOH [68].

pretreated sample than for the as-received sample, indicating a much higher degree of oxidation. The spectrum from AES includes a sequence of peaks with specific energy levels. Reference energy peak bands from oxygen, belled O KLL, and titanium, labeled Ti LMM, can be used to compare against peaks from other energy states. The O KLL peaks, for example, represent the energy of the electrons ejected from the atoms due to the filling of the O 1s state (K shell) by an electron from the L shell coupled with the ejection of an electron from an L shell.

16.4 MEASUREMENT OF THE MECHANICAL PROPERTIES OF NANOSTRUCTURES

16.4.1 NANOSCRATCH TESTING

Microscratch testing is commonly performed to evaluate the adherence of surface coatings. In one study, the adherence of functionally graded HA film on Si substrate and sputter-deposited film with 62% crystallinity on glass substrate were determined using a CSM Microscratch Instrument (MicroPhotonics, Irvine, CA, USA) [69]. Scratch tests were performed under a linearly increasing load, from 0.01 to 3 N. The scratch length was set to 3 mm, and the scratch speed was set to 1 mm/min. A diamond tip (20 μm tip diameter, Rockwell C geometry) was used for these tests. The scratch-tip diameter has a strong influence on the maximum and mean Hertzian pressures. A load of 3 N applied to a circular diamond tip of 20 μm diameter in contact with a flat HA coating surface results in a maximum Hertzian pressure of 70 GPa. The commonly used 200 μm diamond tip would require a load of 300 N to reach the same maximum Hertzian pressure.

Nanoscratch test work is performed similarly, in order to provide a representative measure of coating adhesion. The nanoscratch test may be useful to evaluate mineralized tissue adhesion properties in titanium cultures. The adhesion of mineralized tissue to titanium and polystyrene was assessed using the nanoscratch test by Butz et al. [70]. Osteoblastic cells derived from rat bone marrow were cultured on polystyrene, titanium-coated polystyrene, and titanium disks with either a machined or dual-acid-etched surface. Nanoscratch testing was performed (NST, CSEM Instruments, Neuchatel, Switzerland) on mineralized tissue specimens after 28 days of culturing. The scratch path was monitored by light microscopy until complete delamination of mineralized tissue from the substrate occurred, and the required force was recorded as the critical load. Mean critical load values were 31 mN for polystyrene, 67 mN for titanium-coated polystyrene, 76 mN for machined titanium, and 107 mN for dual-acid-etched titanium surface. Culturing mineralized tissue on titanium, especially on roughened surfaces, was seen to increase the tissue critical load.

16.4.2 NANOHARDNESS TEST

Nanohardness is a widely used technique for measuring the local stress, strain, modulus, and toughness properties of dental materials [71–73]. This is a particularly relevant test for dental materials as an understanding of the brittleness or stiffness an implant or bonding material is important for its successful application. In the field of nanoindentation, the Berkovich tip has been widely used to measure the mechanical properties of bulk materials and thin films. This indenter has a triangular-shaped tip with the total included angle of 142.3° and half angle of 65.35°; this tip can only produce a constant equivalent strain value during a test procedure. To measure the stress–strain response of the material, changing strain values are necessary during the course of the test. A spherical tipped indenter can be used to meet this need [72]. Equivalence exists between conventional stress–strain curves and those generated by plotting hardness versus a/R, contact radius, a, divided by indenter radius, R, when using a spherical tipped indenter. The depth of indenter in contact with the material, h_p, can be determined using the Oliver–Pharr method as $h_p = h_t - 0.75(P_t/S)$, where h_t and P_t are the peak displacement and load at the onset of unloading.

In order to provide materials for dental crown/bridges that are structurally reliable, strong and tough with primarily an elastic response, materials such as toughened ceramics are required [73]. On the other hand, materials with too high a hardness or yield response may damage opposing teeth during occlusal contact. A nanobased indentation system (Ultra Micro-Indentation System, UMIS-2000,

CSIRO, Australia) was used to determine the indentation stress–strain response of two kinds of dental ceramics (Cerec2 Mark II, Sirona Dental Systems, New York, USA; and Vita VM9, VITA Zahnfabrik H. Rauter Gmbh & Co. KG, Germany), one kind of dental alloy (Wiron 99, Bego Bremer Goldschlagerei Wilh. Herbst Gmbh & Co, Bremen, Germany) and healthy enamel [73]. A spherical indenter was used to test the materials with nanometer and micronewton displacement and force resolution. Assuming that the elastic modulus remained constant, a plot of contact pressure versus contact strain, $H-a/R$, of each material was obtained. It was concluded in this work that only the metallic alloy (stiffness of approximately 2 GPa) had similar stress–strain response to enamel. Dental ceramics showed much higher yield stress (Mark II and VM9 approximately 10 and 7 GPa, respectively) response than enamel. From this work, $H-a/R$ curves can be seen to provide a method to compare the mechanical properties of these different dental materials.

16.5 CONCLUSIONS

Nanotechnology has, and will in the future have, a large impact on technology development for developing dental materials and systems. Some of these include drug delivery systems, structural materials, biochemical and biomedical devices, adhesives, and pigments. In recent years, with the advent of many new techniques to fabricate nanostructures and their potential applications, the tools to be able to characterize nanoscale systems are becoming more commonly used and increasingly required as standard research tools. In this chapter, methods to characterize to the micron, nano-, and angstrom scale of topology and internal structure were discussed. Images captured for internal structure or surface topography mapping are often processed with software techniques to extract better resolution and further information from the captured data. Image processing and quantitative morphological measurements are therefore important areas for research to allow the most to be made of these often expensive and time-consuming characterization techniques. Applications and reviews of this software research by which nanoparticles can be characterized have been published [74–76].

Methods of chemical characterization of these structures down to 100 ppm were presented in this chapter. Two other spectroscopy techniques worth mentioning are Raman spectroscopy and Fourier-transform infrared (FTIR) spectroscopy. Raman spectroscopy is commonly used to detect specific molecular bonds. For this technique, the sample to be analyzed is generally illuminated laser light with frequency within the visible, infrared, or ultraviolet ranges. Reflected light is focused through a monochromator filtering out wavelengths close to the laser wavelength to allow the detection of other scattered wavelengths. A very small percentage of scattered photons are scattered by excitation and having a frequency different (usually lower) from the frequency of the incident photons. As this frequency is specific to the chemical bond under investigation it allows identification of the species.

FTIR is similar to Raman spectroscopy in that it also uses molecular bond vibration for chemical species identification. In particular though, in FTIR the resonant frequency of bond vibration after exposure to infrared radiation is detected as an identifier for the species under examination. Fourier-transform techniques are based on measurement of the radiation temporal coherence. In the FTIR technique, this relates to the measurement of the frequencies of vibration of the molecular bonds. The resonant frequencies are determined by the shape of the molecular potential energy surfaces, the masses of the atoms, and associated vibronic coupling. Procedurally, a split beam of infrared light is typically sent though the sample dissolved in a liquid solvent medium and at the same time through a sample of the solvent. Both beams are alternatively recorded in the detector. Differences between the signals are used to measure the frequencies of bond vibrations in the sample to be analyzed. Typical applications of FTIR include forensic chemical analysis, measurement of polymer type, degree of polymerization, polymer degradation, and semiconductor type. Most commercial instruments automatically suggest the type and quantity of species present by comparing the spectra to large databases of reference spectra.

It can be seen from a number of the case studies reported that additional information on the mechanism of fabrication or functionality of nanoscale features can be found by using a number of these techniques simultaneously or consecutively. Examples of these can be seen in the work presented above from Pap et al., Moghadam et al., and Pisarek et al. [6,44,68]. Often one of these techniques is used to back up another. With so many advanced techniques for the characterization of nanostructures available, it is important to understand these and their capabilities so that they can be selected and used correctly. Due to the important technological advancements that can be envisaged from the control and understanding of nanostructures, the development of improved and new nanocharacterization techniques will continue for the foreseeable future.

REFERENCES

[1] M. Knoll, Aufladepotentiel und Sekundäremission elektronenbestrahlter Körper, Z. Tech. Phys. 16 (1935) 467–475.

[2] M. Rieth, Nano-Engineering in Science and Technology: An Introduction to the World of Nano-Design, World Scientific, Singapore. ISBN 9812380736.

[3] Y. Shu, G. Ou, L. Wang, J. Zou, Q. Li, Surface modification of titanium with Heparin-Chitosan multilayers via layer-by-layer self-assembly technique, J. Nanomater. (2011). Available from: http://dx.doi.org/10.1155/2011/423686.

[4] R. Karpagavalli, A. Zhou, P. Chellamuthu, K. Nguyen, Corrosion behavior and biocompatibility of nanostructured TiO_2 film on Ti6Al4V, J. Biomed. Mater. Res. 83A (2007) 1087–1095.

[5] W.-E. Yang, H.-H. Huang, Improving the biocompatibility of titanium surface through formation of a TiO_2 nano-mesh layer, Thin Solid Films 518 (2010) 7545–7550.

[6] H.-S. Kim, Y. Yang, J.-T. Koh, K.-K. Lee, D.-J. Lee, K.-M. Lee, et al., Fabrication and characterization of functionally graded nano-micro porous titanium surface by anodizing, J. Biomed. Mater. Res. B 88B (2009) 427–435.

[7] G. Binnig and H. Rohrer, Scanning tunneling microscopy, IBM J. Res. Dev. 44 (2000) 279–293.

[8] T. Maffeis, D. Yung, L. LePennec, M. Penny, R.J. Cobley, E. Comini, et al., STM and XPS characterisation of vacuum annealed nanocrystalline WO3 films, Surf. Sci. 601 (2007) 4953–4957.

[9] J. Boland, The best of both worlds—a novel approach to macromolecular STM studies, Surf. Sci. 602 (2008) 1–2.

[10] M. Bowker, E. Fourre, Direct interactions between metal nanoparticles and support: STM studies of Pd on TiO_2(1 1 0), Appl. Surf. Sci. 254 (2008) 4225–4229.

[11] A. Visikovskiy, M. Yoshimura, K. Ueda, Pt-induced structures on Si (1 1 0) studied by STM, Appl. Surf. Sci. 254 (23) (2008) 7626–7629.

[12] G. Binning, C. Quate, Ch Gerber, Atomic force microscope, Phys. Rev. Lett. 12 (1986) 930.

[13] Q. Zhong, D. Inniss, K. Kjoller, V. Elings, Fractured polymer/silica fiber surface studied by tapping mode atomic force microscopy, Surf. Sci. 290 (1993) 688–692.

[14] A. Ponche, M. Bigerelle, K. Anselme, Relative influence of surface topography and surface chemistry on cell response to bone implant materials. Part 1: Physico-chemical effects, Proc. Inst. Mech. Eng. H 224 (Part H) (2010) 1471–1486.

[15] C.C. Wang, Y.C. Hsu, M.C. Hsieh, S.P. Yang, F.C. Su, T.M. Lee, Effects of nano-surface properties on initial osteoblast adhesion and Ca/P adsorption ability for titanium alloys, Nanotechnology 19 (2008) 1–10.

[16] P. Michael Conn (Ed.), Techniques in Confocal Microscopy, Academic Press, 2010. ISBN: 0123846587.

[17] M. Minsky, Patent US3013467 A, December 19th 1961.

[18] D.V. Patel, C.N. McGhee, Contemporary in vivo confocal microscopy of the living human cornea using white light and laser scanning techniques: a major review, Clin. Experiment. Ophthalmol. 35 (1) (2007) 71–88.

[19] A. Hoffman, M. Goetz, M. Vieth, P.R. Galle, M.F. Neurath, R. Kiesslich, Confocal laser endomicroscopy: technical status and current indications, Endoscopy 38 (12) (2006) 1275–1283.

[20] S. Astner, K. Swindells, S. Gonzalez, E. Stockfleth, J. Lademann, Confocal microscopy: innovative diagnostic tools for monitoring of noninvasive therapy in cutaneous malignancies, Drug Discov. Today 5 (2008) e81–e91. Available from: http://dx.doi.org/10.1016/j.ddmec.2008.04.002.

[21] M. Fulwyler, Q. Hanley, C. Schnetter, I. Young, E. Jares-Erijman, D. Arndt-Jovin, et al., Selective photoreactions in a programmable array microscope (PAM): photoinitiated polymerization, photodecaging, and photochromic conversion, Cytometry A 67A (2005) 68–75.

[22] S. Manara, F. Paolucci, B. Palazzo, M. Marcaccio, E. Foresti, G. Tosi, et al., Electrochemically-assisted deposition of biomimetic hydroxyapatite–collagen coatings on titanium plate, Inorg. Chim. Acta. 361 (2008) 1634–1645.

[23] K. Wang, C. Russel, C. Liu, Preparation, mechanical properties and corrosion behaviors of oriented $Ca(PO_3)_2$ glass–ceramics, Mater. Chem. Phys. 111 (2008) 106–113.

[24] C. O'Mahony, M. Hill, M. Brunet, R. Duane, A. Mathewson, Characterization of micromechanical structures using white-light interferometry, Meas. Sci. Technol. 14 (10) (2003) 1807–1814.
[25] I. Kassamakov, K. Hanhijärvi, I. Abbadi, J. Aaltonen, H. Ludvigsen, E. Hæggström, Scanning white-light interferometry with a supercontinuum source, Optic. Lett. 34 (10) (2009) 1582–1584.
[26] R. Dai, T.B. Xie, S.P. Chang, A scanning white-light interferometric profilometer for smooth and rough surface, J. Key Eng. Mater. 364–366 (2008) 364–370.
[27] J.M. Coupland, J. Lobera, Measurement of steep surfaces using white light interferometry, Int. J. Strain Mech. 46 (1) (2010) 69–78.
[28] R.S. Sirohi, Optical Methods of Measurement: Wholefield Techniques, second ed., CRC Press, 2017. ISBN: 1574446975.
[29] D. Wang, J. Mao, B. Zhou, X.-F. Liao, S.-Q. Gong, Y. Liu, et al., A chimeric peptide that binds to titanium and mediates MC3T3-E1 cell adhesion, Biotechnol. Lett. 33 (2011) 191–197.
[30] R.A. Delgado-Ruız, J.L. Calvo-Guirado, P. Moreno, J. Guardia, G. Gomez-Moreno, J.E. Mate-Sanchez, et al., Femtosecond laser microstructuring of zirconia dental implants, J. Biomed. Mater. Res. B 96B (2011) 91–100.
[31] L. Hjalmarsson, J.I. Smedberg, A. Wennerberg, Material degradation in implant-retained cobalt-chrome and titanium frameworks, J. Oral. Rehabil. 38 (2011) 61–71.
[32] L.M. Svanborg, M. Andersson, A. Wennerberg, Surface characterization of commercial oral implants on the nanometer level, J. Biomed. Mater. Res. B 92B (2010) 462–469.
[33] C.R. Brundle, C.A. Evans Jr., S. Wilson (Eds.), Encyclopaedia of Materials Characterization, Surfaces, Interfaces, Thin Films, Butterworth-Heinemann, 1992 (Transmission electron microscopy).
[34] D.K. Schroder, Semiconductor Material and Device Characterization, third ed., John Wiley & Sons, New Jersey, 2006.
[35] Z.L. Wang, P. Poncharal, W.A. de Heer, Measuring physical and mechanical properties of individual carbon nanotubes by in-situ TEM, J. Phys. Chem. Solids 61 (2000) 1025–1030.
[36] C. Huigan, Z. Meifang, L. Yaogang, Decoration of carbon nanotubes with iron oxide, J. Solid State Chem. 179 (2006) 1208–1213.
[37] C. He, N. Zhao, C. Shi, J. Li, H. Li, Magnetic properties and transmission electron microscopy studies of Ni nanoparticles encapsulated in carbon nanocages and carbon nanotubes, Mater. Res. Bull. 43 (2008) 2260–2265.
[38] J.P. Salvetat, J.M. Bonard, N.H. Thomson, A.J. Kulik, L. Forro, W. Benoit, et al., Mechanical properties of carbon nanotubes, Appl. Phys. A 69 (1999) 255–260.
[39] M.V. Jose, B.W. Steinert, V. Thomas, D.R. Dean, M.A. Abdalla, G. Price, et al., Morphology and mechanical properties of Nylon 6/MWNT nanofibers, Polymer 48 (2007) 1096–1104.
[40] A.V. Stanishevsky, Focused ion beam nanofabrication, in: H.S. Nalwa (Ed.), Encyclopaedia of Nanoscience and Nanotechnology, vol. 3, American Scientific Publishers, California, USA, 2004.
[41] C.R. Musil, J.L. Bartelt, J. Melngailis, Focused ion beam microsurgery for electronics, IEEE Electron Device Lett. vol. EDL-7 (5) (1986) 285–287.
[42] C. Boit, New physical techniques for IC functional analysis of on-chip devices and interconnects, Appl. Surf. Sci. 252 (2005) 18–23.

[43] R. Schlangen, U. Jkerst, C. Boit, T. Malik, R. Jain, T. Lundquist, Non destructive 3D chip inspection with nano-scale potential by use of backside FIB and backscattered electron microscopy, Microelectron. Reliab. 47 (2007) 1523−1528.

[44] S.H. Moghadam, R. Dinarvand, L.H. Cartilier, The focused ion beam technique: a useful tool for pharmaceutical characterization, Int. J. Pharm. 321 (2006) 50−55.

[45] C.R. Brundle, C.A. Evans Jr., S. Wilson (Eds.), Encyclopaedia of Materials Characterization, Surfaces, Interfaces, Thin Films, Butterworth-Heinemann, An Imprint of Elsevier, Massachusetts, USA, 1992 (X-ray diffraction).

[46] T.E. Jennings, Semiconductor Science: Growth and Characterisation Techniques, Prentice Hall, New Jersey, USA, 1998.

[47] W.R. Runyan, Semiconductor Measurements and Instrumentation, McGraw-Hill, New York, USA, 1975.

[48] B. Cengiz, Y. Gokce, N. Yildiz, Z. Aktas, A. Calimli, Synthesis and characterisation of hydroxyapatite nanoparticles, Colloid Surf. A: Physicochem. Eng. Asp. 322 (2008) 29−33.

[49] S.J. Gregg, K.S.W. Sing, Adsorption, Surface Area and Porosity, Academic Press, New York, 1967.

[50] P.A. Webb, C. Orr, Analytical Methods in Fine Particle Technology, Micromeritics Instrument Corp., 1997.

[51] M. Tschon, M. Fini, G. Giavaresi, L. Rimondini, L. Ambrosio, R. Giardino, In vivo preclinical efficacy of a PDLLA/PGA porous copolymer for dental application, J. Biomed. Mater. Res. B 88B (2009) 349−357.

[52] A.D. Milutinovic-Nikolic, V.B. Medic, Z.M. Vukovic, Porosity of different dental luting cements, Dent. Mater. 23 (2007) 674−678.

[53] E. Vennat, C. Bogicevic, J.-M. Fleureau, M. Degrange, Demineralized dentin 3D porosity and pore size distribution using mercury porosimetry, Dent. Mater. 25 (2009) 729−735.

[54] J. Goldstein, D. Newbury, D. Joy, C. Lyman, P. Echlin, E. Lifshin, et al., Scanning Electron Microscopy and X-Ray Microanalysis, Kluwer Academic/Plenum, 2003.

[55] E. Lifshin, X-ray Characterization of Materials, John Wiley & Sons, New York, USA, 1999. ISBN-10: 3527296573.

[56] P.J. Potts, A Handbook of Silicate Rock Analysis, Blackie, 1987, p. 622.

[57] N. Dumelie, H. Benhayoune, D. Richard, D. Laurent-Maquin, G. Balossier, In vitro precipitation of electrodeposited calcium-deficient hydroxyapatite coatings on Ti6Al4V substrate, Mater. Charact. 59 (2008) 129−133.

[58] S.R. Hejazi, H.R. Hosseini, M. Sasani Ghamsari, The role of reactants and droplet interfaces on nucleation and growth of ZnO nanorods synthesized by vapor−liquid−solid (VLS) mechanism, J. Alloy Comp. 455 (2008) 353−357.

[59] J.T. Grant, D. Briggs (Eds.), Surface Analysis by Auger and X-ray Photoelectron Spectroscopy, IM Publications, England, 2003.

[60] J.W. Robinson, Practical Handbook of Spectroscopy, CRC Press, 1991.

[61] B. Subramanian, R. Ananthakumar, M. Jayachandran, Microstructural, mechanical and electrochemical corrosion properties of sputtered titanium-aluminum-nitride films for bio-implants, Vacuum 85 (2010) 601−609.

[62] A.C. Diebold (Ed.), Handbook of Silicon Semiconductor Metrology, Marcel Dekker (Basel) Inc., 2001.

[63] X.-L. Gao, J.-S. Pan, C.-Y. Hsu, Laser-fluoride effect on root demineralization, J. Dent. Res. 85 (2006) 919−923.

[64] V. Balter, B. Reynard, Secondary ionization mass spectrometry imaging of dilute stable strontium labeling in dentin and enamel, Bone 42 (2008) 229–234.
[65] J.F. Watts, J. Wolstenholme, An Introduction to Surface Analysis by XPS and AES, second ed., John Wiley & Sons, 2003. ISBN: 0470847131.
[66] L. Reimer, P.W. Hawkes, Scanning Electron Microscopy: Physics of Image Formation and Microanalysis, Springer, 2010. ISBN: 3642083722.
[67] J.C. Vickerman (Ed.), Surface Analysis. The Principal Techniques, John Wiley & Sons, 1997.
[68] M. Pisarek, M. Lewandowska, A. Roguska, K.J. Kurzydlowski, M. Janik-Czachor, SEM, scanning Auger and XPS characterisation of chemically pre-treated Ti surfaces intended for biomedical applications, Mater. Chem. Phys. 104 (2007) 93–97.
[69] A. Rabiei, B. Thomas, C. Jin, R. Narayan, J. Cuomo, Y. Yang, et al., A study on functionally graded HA coatings processed using ion beam assisted deposition with in situ heat treatment, Surf. Coat. Technol. 200 (2006) 6111–6116.
[70] F. Butz, H. Aita, K. Takeuchi, T. Ogawa, Enhanced mineralized tissue adhesion to titanium over polystyrene assessed by the nano-scratch test, J. Biomed. Mater. Res. 74A (2005) 164–170.
[71] E. Ho, M. Marcolongo, Effect of coupling agents on the local mechanical properties of bioactive dental composites by the nano-indentation technique, Dent. Mater. 21 (2005) 656–664.
[72] L.H. He, N. Fujisawa, M.V. Swain, Elastic modulus and stress–strain response of human enamel by nano-indentation, Biomaterials 27 (2006) 4388–4398.
[73] L.H. He, M.V. Swain, Nanoindentation derived stress–strain properties of dental materials, Dent. Mater. 23 (2007) 814–821.
[74] M.F. Costa, Application of image processing to the characterisation of nanostructures, Rev. Adv. Mater. Sci. 6 (2004) 12–20.
[75] P. Soille, Morphological Image Analysis, Springer-Verlag, Aachen, Germany, 1999.
[76] R.C. Gonzalez, R.E. Woods, S.L. Eddins, Digital Image Processing Using Matlab, Pearson Prentice Hall, 2004.

CHAPTER

Nanoparticulate drug-delivery systems for oral cancer treatment

17

Karthikeyan Subramani[1] and Waqar Ahmed[2]

[1]*Roseman University of Health Sciences, Henderson, NV, United States*

[2]*University of Lincoln, Lincoln, United Kingdom*

CHAPTER OUTLINE

17.1 Introduction 355
17.2 Cancer-Treatment Techniques 356
17.3 Mechanism of Action of Chemotherapeutic Agents 357
17.3.1 Prevention of Synthesis of Pre-DNA Molecule Building Blocks 357
17.3.2 Chemical Damage of DNA in the Cell Nuclei 357
17.4 Oral Cancer 358
17.5 TNM Classification of Tumors 359
17.6 Management of Oral Cancer 359
17.7 Nanoparticulate-Based Drug Delivery in Cancer Treatment 360
17.7.1 Gold Nanoparticles for Anticarcinogenic Drug Delivery 361
17.7.2 Liposomes in Oral Cancer Treatment 363
17.7.3 Magnetic Nanoparticles in Oral Cancer Treatment 365
17.7.4 Polymeric micelles as Drug-Delivery Systems 366
17.8 Limitations of Nanoparticles and Conclusion 366
References 367

17.1 INTRODUCTION

Nanotechnology has been evolving rapidly with much more potential impacts in the treatment of chronic diseases such as cancer and diabetes, respiratory diseases such as asthma, and ocular diseases, and in gene therapy. Scientific community worldwide has been working toward discovering "nanoscale" solutions to treat these diseases by using nanoparticle-based drug-delivery systems. There has been a remarkable progress over the last decade in the development of nanoparticles as effective drug-carrier systems. The recent advances and applications of such systems toward the treatment of oral cancer are discussed in the following sections.

Emerging Nanotechnologies in Dentistry. DOI: http://dx.doi.org/10.1016/B978-0-12-812291-4.00017-0

17.2 CANCER-TREATMENT TECHNIQUES

Currently, there are numerous techniques that are used for cancer treatment. But each technique has its own limitations and adverse effects [1,2]. Surgical treatment (excision of the tumor) is usually the first choice of treatment preferred by physicians. However, surgical excision is not effective when the cancer cells have infiltrated the nearby vital organs or have spread to distant parts of the body (metastasis). Surgical excision is preferred for the removal of larger tumors. Cryosurgery is another surgical technique that is used for freezing and killing the tumor cells. It is an alternative to surgical excision and is used to treat tumors that have not spread to distant organs and for the treatment of precancerous or noncancerous lesions. Chemotherapy is the treatment of cancer with anticancer drugs. These drugs are used as pills, intravenous injections, or topical applications. Chemotherapeutic drugs may destroy healthy tissue along with cancer cells and carcinomatous tissue (cytotoxicity). The cytotoxic effect of chemotherapeutic drugs is highest in bone marrow, gonads, hair follicles, and digestive tract, all of which contain rapidly proliferating cells. The adverse effects of chemotherapy include fatigue, nausea, vomiting, alopecia (loss of hair), gastrointestinal disturbance, impaired fertility, impaired ovarian function, and bone marrow suppression, resulting in anemia, leucopenia, and thrombocytopenia [3,4]. Another technique of cancer treatment is radiation therapy, which uses radiation energy to destroy cancer cells and reduce the size of tumors. Bone marrow transplantation and peripheral blood stem cell transplantation are done to restore stem cells that are destroyed by high doses of radiation or chemotherapy. Immunotherapy, sometimes called biological therapy, biotherapy, or biological response modifier (BRM) therapy, is a treatment technique that utilizes human body's immune system to destroy cancer cells [5]. The immune system is stimulated by an outside source, such as an antibody, synthetic immune system proteins, or BRMs. BRMs include interferons, interleukins, colony-stimulating factors, monoclonal antibodies, vaccines, gene therapy, and nonspecific immunomodulating agents. Recent research work has been focused on studying gene therapy for cancer treatment. Gene therapy is an experimental treatment that involves introducing genetic material into the cancer cells to destroy the cells [6]. Angiogenesis inhibitors are also currently being evaluated in clinical trials. These are chemicals that inhibit the formation of blood vessels (angiogenesis). Angiogenesis plays an important role in the growth and spreading of cancer cells [7]. New blood vessels act as a source of oxygen and nutrients to the cancer cells, allowing these cells to grow, invade nearby tissue, spread to other parts of human body, and form new colonies of cancer cells. Angiogenesis inhibitors are used to prevent the formation of blood vessels, thereby depleting the cancer cells of oxygen and nutrients. Hyperthermia (also called thermal therapy or thermotherapy) is a type of cancer-treatment technique, in which the cancer cells are exposed to high temperatures (up to 113°F). Research has shown that high temperatures can damage and kill cancer cells with minimal injury to normal tissues [8]. By damaging proteins and functional structures within cells, hyperthermia destroys cancer cells [9]. Hyperthermia may make some cancer cells more sensitive to radiation or harm other cancer cells that radiation cannot damage. It can also enhance the anticarcinogenic effect of certain anticancer drugs. Thus, it is almost used with other forms of cancer therapy, such as radiation and chemotherapy [10]. Currently, hyperthermia is at clinical trial stage. Laser therapy uses high-intensity laser to treat cancer [11]. Laser can be used to shrink or destroy tumors. Laser therapy is most commonly used to treat superficial tumors on the surface of the body or the lining of internal organs. Photodynamic therapy is a type of cancer treatment that uses a drug called a photosensitizer or photosensitizing agent [12]. Photosensitizer is activated by light of a specific wavelength. When photosensitizers are exposed to this specific wavelength, they produce singlet oxygen, which destroys cancer cells. Targeted cancer therapy uses target-specific drugs that invade cancer cells and block the growth and metastasis of cancer cells by interfering with specific molecules involved in carcinogenesis and tumor growth [13]. To overcome the disadvantages of current cancer-treatment techniques, the scientific community has turned toward nanotechnology to develop newer and more effective drug-carrier systems to safely shepherd the anticancer drugs to the cancer cells.

17.3 MECHANISM OF ACTION OF CHEMOTHERAPEUTIC AGENTS

In general, anticarcinogenic chemotherapeutic agents can be divided into three main categories based on their mechanism of action [14].

17.3.1 PREVENTION OF SYNTHESIS OF PRE-DNA MOLECULE BUILDING BLOCKS

DNA building blocks are folic acid, heterocyclic bases, and nucleotides, which are formed naturally within cells. All of these agents work to block some steps in the formation of nucleotides or deoxyribonucleotides (necessary for making DNA). When these steps are blocked, the nucleotides, which are the building blocks of DNA and RNA, cannot be synthesized. Thus, the cells cannot replicate due to impaired DNA synthesis. Examples of drugs in this class include methotrexate, fluorouracil, hydroxyurea, and mercaptopurine.

17.3.2 CHEMICAL DAMAGE OF DNA IN THE CELL NUCLEI

Some chemotherapeutic agents destroy DNA and RNA of cancer cells. They disrupt the replication of DNA and totally halt the replication of DNA or RNA, which may stimulate cancer cell formation. A few examples of drugs in this class include cisplatin and antibiotics such as daunorubicin, doxorubicin, and etoposide.

Disruption of Synthesis or Breakdown of Mitotic Spindles Mitotic spindles serve as molecular railroads with "north and south poles" in the cell when it starts to divide. These spindles are very important, because they help to split the newly copied DNA such that a copy goes to each of the two new cells during cell division. These drugs disrupt the formation of these spindles and, therefore, interrupt cell division. Classic examples of drugs in this class of mitotic disrupters include vinblastine, vincristine, and paclitaxel.

17.4 ORAL CANCER

Oral cancer or cancer of the mouth, a subtype of head and neck cancer, encompasses any cancerous tissue (or) growth located within the oral cavity. It may arise as a primary lesion originating in any of the oral tissues, by metastasis from a distant site of origin, or by extension from a neighboring anatomic structure, such as the nasal cavity or the maxillary sinus. Oral cancer may originate in any of the tissues of the mouth and may be of varied histological types: teratoma (true neoplasms composed of multiple tissues foreign to the site from which they originate), adenocarcinoma derived from a major or minor salivary gland, lymphoma from tonsillar or other lymphoid tissue, or melanoma from the pigment producing cells of the oral mucosa. Oral cancer accounts for approximately 2% of all malignant tumors in the USA and Europe; approximately 30%–40% in the Indian subcontinent and more than 90% are squamous cell carcinomas, originating in the tissues that line the mouth and lips [15]. Oral cancer most commonly involves the tongue. It may also occur on the floor of the mouth, cheek lining, gingiva (gums), lips, or palate (roof of the mouth). Squamous cell carcinoma is a malignant tumor of squamous epithelium (epithelium that shows squamous cell differentiation) lining the oral mucosa. These are malignant and tend to spread rapidly. The main causes of oral cancer are excessive intake of alcohol and tobacco use. Exposure to sunlight is a causative factor for cancer of the lips, similar to skin cancer. Human Papilloma Virus, called HPV, is also one of the risk factors for cause of oral cancer. Immunosuppression, e.g., HIV patients and renal transplant patients have the highest risk factor for developing oral cancer.

Oral cancerous lesions are most often seen as a painless ulcer, although may present as a swelling, an area of leukoplakia (white patch or plaque), erythroplakia (well-demarcated red velvety patch of oral mucosa), erythroleukoplakia (speckled leukoplakia), as malignant change of long-standing benign tumors, or rarely in cyst linings or nonhealing extraction socket. Malignancy should be suspected if any of the above lesions persists for more than 3 weeks. Pain is usually a late feature when the lesion becomes ulcerated or superinfected. Ulcerated lesions are firm with raised edges, with an indurated, inflamed, granular base and are fixed to the surrounding tissues. In the later stages, it involves tongue problems, difficulty in swallowing, pain, and paresthesia (numbness of the affected area).

Table 17.1 TNM Classification of Cancer Staging

T—Size of the Primary Tumor	
T1	<2 cm diameter
T2	2–4 cm diameter
T3	>4 cm diameter
T4	Massive, invading other tissues
N—Cervical Nodes	
N0	Tumor cells absent from regional lymph nodes
N1	Single node with size <3 cm
N2	Single node with size 3–6 cm (N2a), multiple nodes (N2b), or contralateral nodes (N2c)
N3	Multiple nodes with size >6 cm
M—Distant Metastasis	
M0	Present
M1	Absent

T describes the size of the tumor and whether it has invaded nearby tissue, ***N*** *describes regional cervical lymph nodes that are involved, and* ***M*** *describes distant metastasis (spread of cancer from one body part to another).*

17.5 TNM CLASSIFICATION OF TUMORS

The TNM staging system for all solid tumors was devised by Pierre Denoix between 1943 and 1952, using the size and extension of the primary tumor, its lymphatic involvement, and the presence of metastases to classify the progression of cancer (Table 17.1).

17.6 MANAGEMENT OF ORAL CANCER

Suggested management plan (which may vary between operating surgeons) involves the following:

- T1—N0: Surgery or radiotherapy
- Tumor close to bone having radiotherapy: Safest to remove associated bone to prevent osteoradionecrosis (a condition of nonvital bone in the site of radiation)
- T2—N0: Selective neck dissection
- T2, T3: Prophylactic radiotherapy or prophylactic selective neck dissection

Surgical excision (removal) of the tumor is usually recommended if the tumor is small enough, and if surgery is likely to result in a functionally satisfactory result. Radiation therapy with or without chemotherapy is often used in conjunction with surgery, or as the definitive radical treatment, especially if the tumor is inoperable. Surgeries for oral cancers include the following:

- Maxillectomy (removal of the part of affected bone tissue in maxilla or upper jaw)
- Mandibulectomy (removal of the part of affected bone tissue in mandible or lower jaw)

- Glossectomy (tongue removal, can be total, hemi, or partial)
- Radical neck dissection, in which there is removal of lymph nodes and other structures in the head and neck that are likely or proven to be malignant.
- Laryngectomy, where the part or whole of the larynx is removed if it is affected with cancer.
- Combinational, e.g., glossectomy and laryngectomy done together.
- Photodynamic therapy.

Owing to the vital nature of the structures in the head and neck area, surgery for larger cancers is technically demanding. Reconstructive surgery may be required to give an acceptable cosmetic and functional result. Bone grafts and surgical flaps are used to help rebuild the structures removed during excision of the cancer. An oral prosthesis may also be required. Most oral cancer patients depend on a feeding tube for their hydration and nutrition. Some will also get a port for the chemotherapy to be delivered. Many oral cancer patients are disfigured and suffer from many long-term side effects. The after effects often include fatigue, speech problems, trouble maintaining weight, thyroid problems, difficulty in swallowing, memory loss, weakness, dizziness, high-frequency hearing loss, and sinus damage.

Survival rates for oral cancer depend on the precise site and the stage of the cancer at diagnosis. Overall, survival is around 50% at 5 years when all stages of initial diagnosis are considered. Survival rates for T1–N0 cancers are 90%, hence the emphasis on early detection to increase survival outcome for patients. For T2/3–N1, it is 30% and worse for T4. Following treatment, rehabilitation may be necessary to improve movement, chewing, swallowing, and speech. Speech and language pathologists may be involved at this stage.

Chemotherapy is useful in oral cancers when used in combination with other treatment modalities, such as radiation therapy. It is not used alone as a monotherapy. When cure is unlikely, it can also be used to extend life and can be considered palliative but not curative care. Biological agents, such as Cetuximab, have recently been shown to be effective in the treatment of squamous cell head and neck cancers and are likely to have an increasing role in the future management of this condition when used in conjunction with other treatments. Treatment of oral cancer will usually be by a multidisciplinary team, with treatment professionals from the realms of radiation, surgery, chemotherapy, nutrition, dental professionals, and even psychology all possibly involved with diagnosis, treatment, rehabilitation, and patient care.

17.7 NANOPARTICULATE-BASED DRUG DELIVERY IN CANCER TREATMENT

The critical step in cancer treatment is the detection of cancer at its initial stage of carcinogenesis. Results of numerous scientific research studies done in nanotechnology and nanomedicine are inspiring the scientific community to discover new, innovative, noninvasive tools at the nanoscale level for such purposes.

Nanoscale cantilevers [16] and quantum dots [17] are being studied as cancer detection tools at the cellular level. If the tumor has not been detected in its early stage, treatment methods should be devised to eradicate the fully developed cancer cells without harming the normal, healthy cells of human body. The various types of nanoparticles that are currently studied for their use as drug-delivery systems are polymeric micelles, magnetic nanoparticles, colloidal gold nanoparticles, and ceramic nanoparticles [18–20]. These nanoparticulate-based drug-delivery systems can be characterized for their localization in tumor cells by coating them with tumor-specific antibodies, peptides, sugars, hormones, and anticarcinogenic drugs. These nanoparticles have been effectively coupled with the abovementioned anticarcinogenic chemotherapeutic agents and have been tested for their target specificity. These nanoparticles are superior over conventionally available drug-delivery systems, as the chemotherapeutic agents can be targeted to a specified area of the human body by adding nanoscale surface receptors. These receptors specifically recognize the target tissue, and bind to it and release the drug molecules [21]. Thus, healthy cells are spared from cytotoxic effects of the drug. Drugs can also be protected from degradation by encapsulating them with nanoparticle coatings [22]. As nanoparticles are extremely small, they can penetrate through smaller capillaries and are easily taken up by cancer cells. This causes efficient drug accumulation at the target site. The use of biodegradable nanoparticles allows sustained drug release over a period of time [23]. Thus, nanoparticles as drug-delivery systems, with enhanced target specificity, can overcome the limitations of conventional cancer-treatment techniques. There are numerous other nanobiotechnology-based approaches being developed to formulate nanoparticles as carriers of anticarcinogenic agents. These include dendrimers, chitosan nanoparticles, low-density lipoproteins, nanoemulsions, nanolipospheres, nanoparticle composites, polymeric nanocapsules, nanospheres, and nanovesicles (Fig. 17.2). Their applications in nanoencapsulation and targeted drug delivery of anticancer drugs, in combination with radiotherapy, laser therapy, thermotherapy, photodynamic therapy, ultrasound therapy, and nanoparticle-mediated gene therapy, have been extensively reviewed in the literature [24]. The following sections discuss the most promising groups of nanoparticles and their applications in drug delivery for treatment of oral cancer.

17.7.1 GOLD NANOPARTICLES FOR ANTICARCINOGENIC DRUG DELIVERY

Colloidal gold nanoparticles are the most commonly used nanoparticles for anticarcinogenic drug delivery. Colloidal gold nanoparticles are more biocompatible than other nanoparticles. The physical and chemical properties of colloidal gold nanoparticles allow more than one protein molecule to bind to a single particle of colloidal gold. The use of colloidal gold nanoparticles as drug-delivery vectors of tumor necrosis factor (TNF) has been tested in a growing tumor in mice [25]. Although TNF has been evaluated in cancer treatment, it causes adverse effects such as hypotension and, in some cases, organ failure, resulting in death. But recent researches have shown that when coupled with colloid gold particles, therapeutic amounts of TNF can be successfully delivered to destroy the tumor cells in animals [26]. The use of laser to destroy the tumor cells in cancer tissue has been described by a technique using selective nanothermolysis of self-assembling gold nanoparticles [27] (Fig. 17.1). These gold nanoparticles were coated with secondary Ab goat antimouse IgG. This structural configuration showed specific localization in the adenocarcinomatous cells targeted with primary Ab. The targeted photothermal therapy has been tested in vitro toward killing the oral squamous carcinoma cells [28]. Epidermal growth factor or epithelial growth factor (EGF) is a growth factor that plays an important role in the regulation of cell growth, proliferation, and differentiation by binding to its receptor EGF receptor (EGFR). Because of the increased risk of cancer by EGF, inhibiting it decreases cancer risk. Such approaches are aimed at inhibiting the EGFR. In this study [28], two oral squamous carcinoma cell lines and one benign epithelial cell line were incubated with anti-EGFR antibody conjugated gold nanoparticles and then exposed to continuous visible argon ion laser at 514 nm. It was found that the malignant cells required less than half the laser energy to be killed than the benign cells after incubation with anti-EGFR antibody conjugated gold nanoparticles. No photothermal destruction was observed for all types of cells in the absence of nanoparticles at four times energy required to kill the malignant cells with anti-

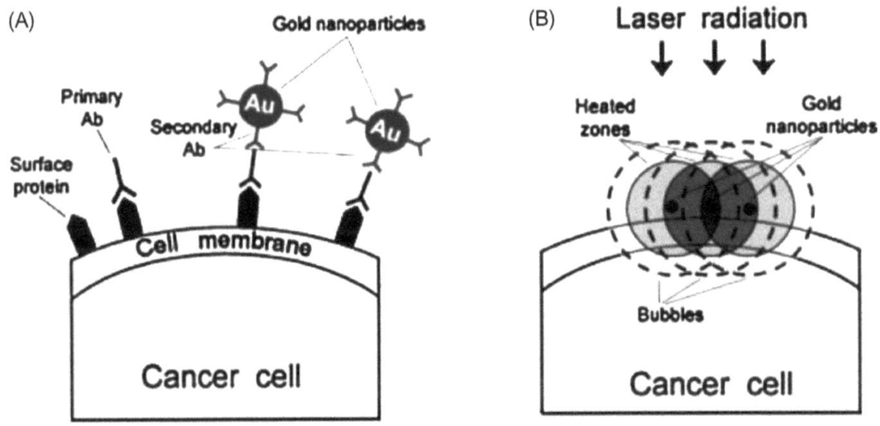

FIGURE 17.1

Principle of selective nanophotothermolysis of cancer cells targeted with gold nanoclusters. (A) The human adenocarcinoma cell targeted with primary antibodies (Abs), which are selectively attached to surface proteins; secondary Ab-goat antimouse immunoglobulin G, conjugated with 40-nm gold nanoparticles, is selectively attached to the primary Abs. (B) The schematics of laser-induced overlapping of heated zones and bubbles from the nanoparticles attached to cell membrane.

From Zharov VP, Galitovskaya EN, Johnson C, et al. Synergistic enhancement of selective nanophotothermolysis with gold nanoclusters: potential cancer therapy. Lasers Surg. Med. 2005; 39:219–226.

EGFR/Au conjugates. Gold nanoparticles can, thus, offer a novel class of selective photothermal agents using laser at low power.

Novel optically active reagents have been used for simultaneous molecular imaging and photothermal cancer therapy. It is desirable to use these agents that are active in the near-infrared region (NIR) of the radiation spectrum to minimize the light extinction by intrinsic chromophores in native tissue. Gold nanorods with suitable aspect ratios (length/width) can absorb and scatter strongly in the NIR region (650–900 nm). In an in vitro study [29], gold nanorods were evaluated as novel contrast agents for both molecular imaging and photothermal cancer therapy. Nanorods were synthesized and conjugated to anti-EGFR monoclonal antibodies and incubated in cell cultures with a nonmalignant epithelial cell line and two malignant oral squamous carcinoma cell lines. The anti-EGFR antibody-conjugated nanorods bound specifically to the surface of the malignant-type cells with much higher affinity due to the over expressed EGFR on the cytoplasmic membrane of the malignant cells. As a result of the strongly scattered red light from gold nanorods in dark field, the malignant cells were clearly visualized and diagnosed from the nonmalignant cells. It was found that, after exposure to continuous red laser at 800 nm, malignant cells required about half the laser energy to be photothermally destroyed than the nonmalignant cells. Thus, both efficient cancer cell diagnostics and selective photothermal therapy of cancer cells were realized at the same time. The potential use of gold nanospheres in fluorescence-based detection and photothermal therapy of oral cancer cells were reported later [30,31]. In a recent study [32], it was shown that nuclear targeting of gold nanoparticles in cancer cells induced DNA damage, causing cytokinesis arrest and apoptosis (cell death). By properly conjugating gold nanoparticles with specific targeting peptides, these nanoparticles were selectively transported to the nuclei of cancer cells, inducing DNA damage and cell death. In another study [33], it was shown that gold nanoparticles enhanced the radiation therapy of a radioresistant and highly aggressive mouse head and neck squamous cell carcinoma model.

17.7.2 LIPOSOMES IN ORAL CANCER TREATMENT

Liposomes are small artificial spherical vesicles made from naturally occurring, nontoxic phospholipids and cholesterol. There are four major types of liposomes [34]. Conventional liposomes are either neutral or negatively charged. Sterically stabilized "stealth" liposomes carry polymer coatings to obtain prolonged circulatory duration. Immunoliposomes have specific antibodies or antibody fragments on their surface to enhance target-specific binding. Cationic liposomes interact with negatively charged molecules and condense them to finer structure, thereby carrying them externally rather than encapsulating the molecules within. Owing to their size, biocompatibility, hydrophobicity, and ease of preparation, liposomes serve as promising systems for drug delivery. Their surfaces can be modified by attaching poly(ethylene glycol) (PEG) units to enhance the circulation time in bloodstream. Liposomes can also be conjugated with ligands or antibodies to improve their target specificity (Fig. 17.2). Liposomes have been shown to be

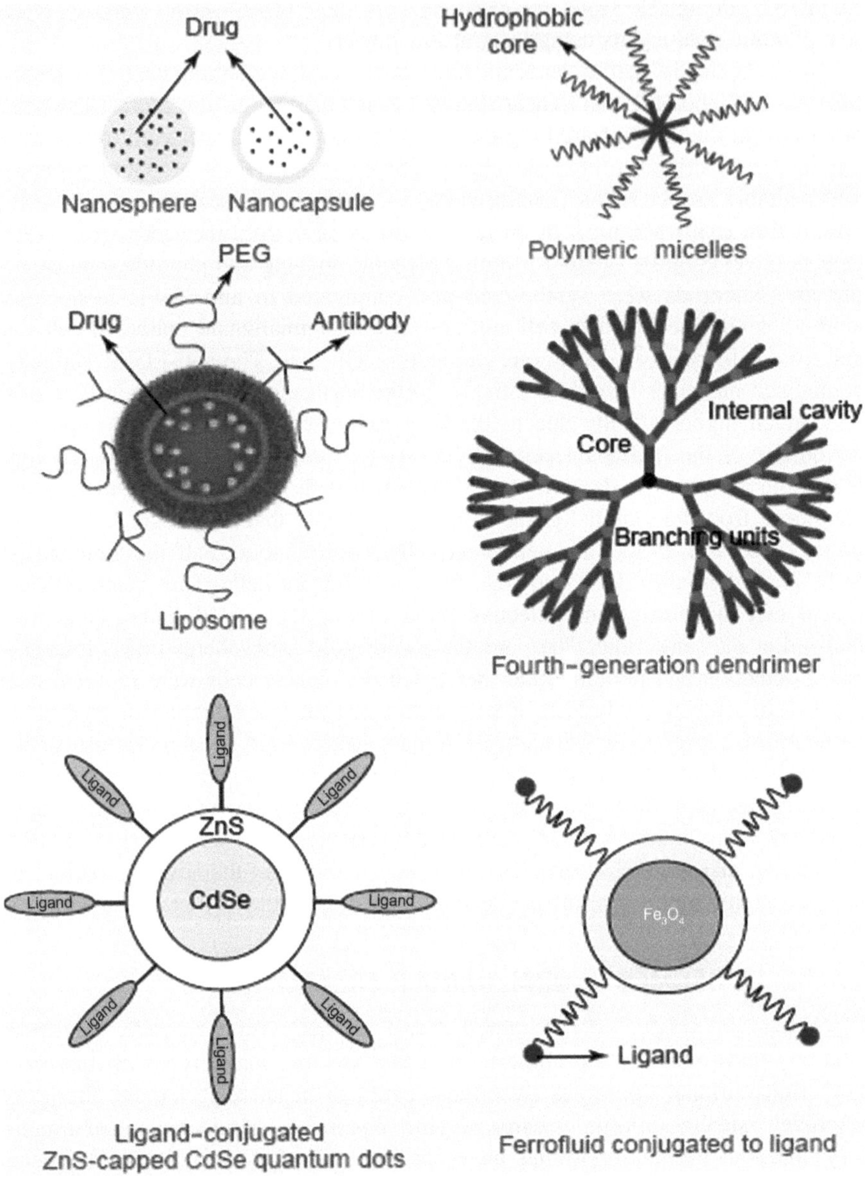

FIGURE 17.2

Schematics of different nanotechnology-based drug-delivery systems. Nanoparticles are small polymeric colloidal particles with a therapeutic agent either dispersed in polymer matrix or encapsulated in polymer. Polymeric micelles are self-assembled block copolymers, which in aqueous solution arrange to form an outer hydrophilic layer and an

(*Continued*)

excellent carriers for target-specific delivery of tumor-specific antibodies, peptides, and anticarcinogenic drugs for treating head and neck carcinoma cells. Drugs like doxorubicin [36], all-trans retinoic acid (atRA) [37], Gefitinib [38], cetuximab [39], tirapazamine [39], tumor-specific antibodies like Cetuximab [38] have been evaluated for their efficacy using liposomes as carriers. Gene therapy is a new therapeutic approach, in which defective genes are replaced with functional ones, or the delivered genes can specifically kill cancer cells. Efficient gene delivery is an important component of gene therapy approaches. Potential safety problems with viral vectors as carriers necessitated the development of efficient nonviral vectors. DNA complexes with synthetic cationic liposomes have become a simple means of transferring DNA into target cells. Over the past few years, gene therapy have been shown to be effective in treating head and neck cancer cells, tongue cancer cells, and squamous cell carcinoma [39–42].

17.7.3 MAGNETIC NANOPARTICLES IN ORAL CANCER TREATMENT

Magnetic nanoparticles are iron oxide particles sheathed with sugar molecules (Fig. 17.2). Therefore, these are not recognized by the immune system [43]. When these particles are brought under the influence of an external magnetic field, they heat up the tumor cells and destroy them without affecting the surrounding healthy tissues. A group of researchers have synthesized biodegradable magnetic nanoparticles by using organic polymers and nanosized magnetites [44]. After the characterization studies, an external magnetic field was used as a guidance system to direct the magnetic nanoparticles to the specified part of the experimental setup. The results of such studies substantiate the theory of targeting magnetic nanoparticles to specific areas of the human body by using an external magnetic field. Iron oxide nanoparticles can also be coated with amino groups to achieve cell-specific delivery of therapeutic agents, for example, to carcinomatous brain cells without unselectively invading the whole brain. Magnetic drug targeting in treatment of squamous cell carcinoma has shown to be effective with the advantage of no adverse clinical side effects. Antineoplastic/cancer drug like

◀ inner hydrophobic core. The miceller core can be loaded with a water insoluble therapeutic agent. Liposomes are lipid structures that can be made "stealth" by PEGylation and further conjugated to antibodies for targeting. Dendrimers are monodispersed symmetric macromolecules built around a small molecule with an internal cavity surrounded by a large number of reactive end groups. Quantum dots are fluorescent nanocrystals that can be conjugated to a ligand and thus, can be used for imaging purposes. Ferrofluids are colloidal solutions of iron oxide magnetic nanoparticles surrounded by a polymeric layer, which can be further coated with affinity molecules such as antibodies.

From Sahoo SK, Labhasetwar V. Nanotech approaches to drug delivery and imaging. Drug Discov. Today 2003; 8:1112–1120 [35].

Mitoxantrone [45,46], delivered through magnetic nanoparticles, has been shown to be effective in treating squamous cell carcinoma. Organic-coated superparamagnetic iron oxide nanoparticles [47] and ferrofluids (colloidal dispersion of magnetic nanoparticles) [48] have also been proven to be excellent target-specific carriers of anticancer drugs for drug delivery to squamous cell carcinoma cells when compared to the usual systemic administration.

17.7.4 POLYMERIC MICELLES AS DRUG-DELIVERY SYSTEMS

Polymeric micelles (Fig. 17.2) serve as a novel drug-delivery system due to their target specificity and controlled release of hydrophobic anticancer drugs. PEG-based micelles are biocompatible and biodegradable. Therapeutic use of hydroxycamptothecin (HCPT), a promising antitumor agent, is limited by its poor solubility and rapid destruction. PEG-based amphiphilic block copolymer micelle carriers possess significant potential for improving drug solubility and stability. When HCPT was encapsulated in PEG-poly[gamma-benzyl-L-glutamate] (PEG-PBLG) micelles, the antitumor effects of HCPT/PEG-PBLG micelles against oral squamous cell carcinoma in vivo were concluded to be superior to those exerted by HCPT [49]. Polymeric micelles have also been evaluated for photodynamic therapy [50]. The results showed that polymeric micelles were able to encapsulate and solubilize 5,10,15,20-tetrakis(meso-hydroxyphenyl)porphyrin, a photosensitizer, at high loading densities with uniform size distribution. These micelles exhibited fluorescence and photodynamic-therapy-mediated cytotoxicity against head and neck cancer cells in vitro. These studies demonstrate the potential of polymeric micelles as a novel drug-delivery system in oral cancer treatment.

17.8 LIMITATIONS OF NANOPARTICLES AND CONCLUSION

The science and knowledge that the scientific community has today about nanotechnology and its potential versatile applications are based only on the research work done in the laboratories. These research studies are being conducted to understand how matter behaves at the nanoscale level. Factors and conditions governing the behavior of macrosystems do not really apply to the nanosystems. The major limitations and technological hurdles faced by nanotechnology and its applications in the field of drug delivery should be addressed. Scientific community has been trying to completely understand how the human body would react to these nanoparticles and nanosystems, which are acting as drug carriers. Nanoparticles have larger surface area when compared to their volume. Friction and clumping of the nanoparticles into a larger structure is inevitable, which may affect their function as a drug-delivery system. Due to their minute size, these drug carriers can be cleared away from the body by the body's excretory pathways. When these are not excreted, larger nanoparticles can accumulate in vital organs, causing toxicity leading to organ failure. Recent study in mice revealed that tissue distribution of gold nanoparticles is size dependent, with the smallest nanoparticles (15–50 nm) showing the most widespread organ distribution, including blood, liver, lung, spleen, kidney, brain, heart, and stomach [51]. Liposomes have certain drawbacks, such as being captured by the human body's defense system. The drug-loading capacity of liposomes is being tested by researchers and still remains inconclusive. All previous studies resulted in post-treatment accumulation of the nanoparticles in skin and eyes. Gold nanoparticles tend to accumulate in bone joints and organs. Once the nanoparticles are administered into the human body, they should be controlled by an external control, preventing them from causing adverse effects. These drug-delivery technologies are in various stages of research and development. A few of these innovative treatment techniques have made their way into clinical trials. It is expected that these limitations can be overcome and the discoveries to come into practical use for the treatment of oral cancer within the next 5 to 10 years.

REFERENCES

[1] T.G. Burish, D.M. Tope, Psychological techniques for controlling the adverse side effects of cancer chemotherapy: findings from a decade of research, J. Pain Symptom Manage. 7 (1992) 287–301.

[2] C.L. Schwartz, Late effects of treatment in long-term survivors of cancer, Cancer Treat. Rev. 21 (1995) 355–366.

[3] D.C. Tormey, R. Gray, S.G. Taylor, Postoperative chemotherapy and chemohormonal therapy in women with node-positive breast cancer, Natl. Cancer Inst. Monogr. 1 (1986) 75–80.

[4] C. Barton, J. Waxman, Effects of chemotherapy on fertility, Blood Rev. 4 (1990) 187–195.

[5] L.G. Durrant, J.H. Scholefield, Principles of cancer treatment by immunotherapy, Surgery 21 (2003) 277–279.

[6] A. El-Aneed, Current strategies in cancer gene therapy, Eur. J. Pharm. 498 (2004) 1–8.

[7] J. Fayette, J.C. Soria, J.P. Armand, Use of angiogenesis inhibitors in tumour treatment, Eur. J. Cancer 41 (2005) 1109–1116.

[8] J. Van der Zee, Heating the patient: a promising approach? Ann. Oncol. 13 (2002) 1173–1184.

[9] B. Hildebrandt, P. Wust, O. Ahlers, The cellular and molecular basis of hyperthermia, Crit. Rev. Oncol. Hematol. 43 (2002) 33–56.

[10] P. Wust, B. Hildebrandt, G. Sreenivasa, Hyperthermia in combined treatment of cancer, Lancet Oncol. 3 (2002) 487–497.

[11] N.S. Goldstein, Laser therapy for small breast cancers, Am. J. Surg. 187 (2004) 149–150.

[12] J.J. Schuitmaker, P. Baas, H.L.L.M. van Leengoed, et al., Photodynamic therapy: a promising new modality for the treatment of cancer, J. Photochem. Photobiol. B. 34 (1996) 3–12.

[13] J.A. Kim, Targeted therapies for the treatment of cancer, Am. J. Surg. 186 (2003) 264–268.
[14] J.T. DiPiro, R.L. Talbert, G.C. Yee, et al., Cancer treatment and chemotherapy, Pharmacotherapy: A Pathophysiologic Approach, fifth ed., McGraw Hill Publications, New York, 2002, pp. 2175–2222.
[15] L. Feller, J. Lemmer, Oral Squamous Cell Carcinoma: Epidemiology, Clinical Presentation and Treatment, *J. Cancer Ther.* **3**, 2012, 263-268.
[16] K.W. Wee, G.Y. Kang, J. Park, et al., Novel electrical detection of label-free disease marker proteins using piezoresistive self-sensing micro-cantilevers, Biosens. Bioelectron. 20 (2005) 1932–1938.
[17] E.B. Voura, J.K. Jaiswal, H. Mattoussi, et al., Tracking metastatic tumor cell extravasation with quantum dot nanocrystals and fluorescence emission-scanning microscopy, Nat. Med. 10 (2004) 993–998.
[18] T. Neuberger, B. Schöpf, H. Hofmann, et al., Superparamagnetic nanoparticles for biomedical applications: possibilities and limitations of a new drug delivery system, J. Magn. Magn. Mater. 293 (2005) 483–496.
[19] I.H. El-Sayed, X. Huang, M.A. El-Sayed, Surface plasmon resonance scattering and absorption of anti-EGFR antibody conjugated gold nanoparticles in cancer diagnostics: applications in oral cancer, Nano Lett. 5 (2005) 829–834.
[20] J. Wang, D. Mongayt, V.P. Torchilin, Polymeric micelles for delivery of poorly soluble drugs: preparation and anticancer activity in vitro of paclitaxel incorporated into mixed micelles based on poly(ethylene glycol)-lipid conjugate and positively charged lipids, J. Drug Target. 13 (2005) 73–80.
[21] N.B. Dinauer, W. Sabine, K. Carolin, et al., Selective targeting of antibody-conjugated nanoparticles to leukemic cells and primary T-lymphocytes, Biomaterials 26 (2005) 5898–5906.
[22] R. Pandey, Z. Ahmad, S. Sharma, et al., Nano-encapsulation of azole antifungals: potential applications to improve oral drug delivery, Int. J. Pharm. 301 (2005) 268–276.
[23] H.S. Yoo, J.E. Oh, K.H. Lee, et al., Biodegradable nanoparticles containing doxorubicin-PLGA conjugate for sustained release, Pharm. Res. 16 (1999) 1114–1118.
[24] K.K. Jain, Nanotechnology-based drug delivery for cancer, Technol. Cancer Res. Treat. 4 (2005) 407–416.
[25] G.F. Paciotti, L. Myer, D. Weinreich, et al., Colloidal gold: a novel nanoparticle vector for tumour directed drug delivery, Drug Deliv. 11 (2004) 169–183.
[26] G. Miller, Colloid gold nanoparticles deliver cancer-fighting drugs, Pharm. Technol. 7 (2003) 17–19.
[27] V.P. Zharov, E.N. Galitovskaya, C. Johnson, et al., Synergistic enhancement of selective nanophotothermolysis with gold nanoclusters: potential cancer therapy, Lasers Surg. Med. 39 (2005) 219–226.
[28] I.H. El-Sayed, X. Huang, M.A. El-Sayed, Selective laser photo-thermal therapy of epithelial carcinoma using anti-EGFR antibody conjugated gold nanoparticles, Cancer Lett. 239 (1) (2006) 129–135. Epub 2005 Sep 28.
[29] X. Huang, I.H. El-Sayed, W. Qian, M.A. El-Sayed, Cancer cell imaging and photothermal therapy in the near-infrared region by using gold nanorods, J. Am. Chem. Soc. 128 (6) (2006) 2115–2120.

[30] I. El-Sayed, X. Huang, F. Macheret, J.O. Humstoe, R. Kramer, M. El-Sayed, Effect of plasmonic gold nanoparticles on benign and malignant cellular autofluorescence: a novel probe for fluorescence based detection of cancer, Technol. Cancer Res. Treat. 6 (5) (2007) 403–412.

[31] X. Huang, W. Qian, I.H. El-Sayed, M.A. El-Sayed, The potential use of the enhanced nonlinear properties of gold nanospheres in photothermal cancer therapy, Lasers Surg. Med. 39 (9) (2007) 747–753.

[32] B. Kang, M.A. Mackey, M.A. El-Sayed, Nuclear targeting of gold nanoparticles in cancer cells induces DNA damage, causing cytokinesis arrest and apoptosis, J. Am. Chem. Soc. 132 (5) (2010) 1517–1519.

[33] J.F. Hainfeld, F.A. Dilmanian, Z. Zhong, D.N. Slatkin, J.A. Kalef-Ezra, H.M. Smilowitz, Gold nanoparticles enhance the radiation therapy of a murine squamous cell carcinoma, Phys. Med. Biol. 55 (11) (2010) 3045–3059. Epub 2010 May 12.

[34] G. Storm, D.J.A. Crommelin, Liposomes: Quo vadis? Pharm. Sci. Technol. Today 1 (1998) 19–31.

[35] S.K. Sahoo, V. Labhasetwar, Nanotech approaches to drug delivery and imaging, Drug Discov. Today 8 (2003) 1112–1120.

[36] X.Q. Pan, H. Wang, R.J. Lee, Antitumor activity of folate receptor-targeted liposomal doxorubicin in a KB oral carcinoma murine xenograft model, Pharm. Res. 20 (3) (2003) 417–422.

[37] K. Park, J.H. Yang, Y. Choi, C. Lee, S.Y. Kim, Y. Byun, Chemoprevention of 4-NQO-induced oral carcinogenesis by co-administration of all-trans retinoic acid loaded microspheres and celecoxib, J. Control. Release 104 (1) (2005) 167–179.

[38] M.W. Epperly, D. Franicola, X. Zhang, S. Nie, J.S. Greenberger, Effect of EGFR antagonists gefitinib (Iressa) and C225 (Cetuximab) on MnSOD-plasmid liposome transgene radiosensitization of a murine squamous cell carcinoma cell line, In Vivo 20 (6B) (2006) 791–796.

[39] M.W. Epperly, S.Y. Lai, A.J. Kanai, N. Mason, B. Lopresi, T. Dixon, et al., Effectiveness of combined modality radiotherapy of orthotopic human squamous cell carcinomas in Nu/Nu mice using cetuximab, tirapazamine and MnSOD-plasmid liposome gene therapy, In Vivo 24 (1) (2010) 1–8.

[40] M. Young, N. Overlid, K. Konopka, N. Düzgünes, Gene therapy for oral cancer: efficient delivery of a 'suicide gene' to murine oral cancer cells in physiological milieu, J. Calif. Dent. Assoc. 33 (12) (2005) 967–971.

[41] M.W. Epperly, R. Wegner, A.J. Kanai, V. Kagan, E.E. Greenberger, S. Nie, et al., Effects of MnSOD-plasmid liposome gene therapy on antioxidant levels in irradiated murine oral cavity orthotopic tumors, Radiat. Res. 167 (3) (2007) 289–297.

[42] J.H. Xu, C.B. Pan, H.Z. Huang, B. Zhang, J.G. Wang, L.T. Zhang, [Silencing of survivin gene enhances chemosensitivity of human tongue cancer cell line Tca8113 to cisplatin] [Article in Chinese], Zhonghua Kou Qiang Yi Xue Za Zhi 42 (5) (2007) 280–283.

[43] Q.A. Pankhurst, J. Connolly, S.K. Jones, et al., Applications of magnetic nanoparticles in biomedicine, J. Phys. D. Appl. Phys. 36 (2003) R167–R181.

[44] R. Asmatulu, M.A. Zalich, R.O. Claus, et al., Synthesis, characterization and targeting of biodegradable magnetic nanocomposite particles by external magnetic fields, J. Magn. Magn. Mater. 292 (2005) 108–119.

[45] C. Alexiou, R. Jurgons, R.J. Schmid, C. Bergemann, J. Henke, W. Erhardt, et al., Magnetic drug targeting—biodistribution of the magnetic carrier and the chemotherapeutic agent mitoxantrone after locoregional cancer treatment, J. Drug Target 11 (3) (2003) 139–149.

[46] C. Alexiou, R.J. Schmid, R. Jurgons, M. Kremer, G. Wanner, C. Bergemann, et al., Targeting cancer cells: magnetic nanoparticles as drug carriers, Eur. Biophys. J. 35 (5) (2006) 446–450. Epub 2006 Jan 31.

[47] K.J. Landmark, S. Dimaggio, J. Ward, C. Kelly, S. Vogt, S. Hong, et al., Synthesis, characterization, and in vitro testing of superparamagnetic iron oxide nanoparticles targeted using folic Acid-conjugated dendrimers, ACS Nano 2 (4) (2008) 773–783.

[48] C. Alexiou, R. Jurgons, R. Schmid, W. Erhardt, F. Parak, C. Bergemann, et al., [Magnetic drug targeting—a new approach in locoregional tumor therapy with chemotherapeutic agents. Experimental animal studies] [Article in German], HNO 53 (7) (2005) 618–622.

[49] X.Q. Ding, D. Chen, A.X. Wang, S. Li, Y. Chen, J. Wang, Antitumor effects of hydroxycamptothecin-loaded poly[ethylene glycol]-poly [gamma-benzyl-L-glutamate] micelles against oral squamous cell carcinoma, Oncol. Res. 16 (7) (2007) 313–323.

[50] E.M. Cohen, H. Ding, C.W. Kessinger, C. Khemtong, J. Gao, B.D. Sumer, Polymeric micelle nanoparticles for photodynamic treatment of head and neck cancer cells, Otolaryngol. Head Neck Surg. 143 (1) (2010) 109–115.

[51] G. Sonavane, K. Tomoda, K. Makino, Biodistribution of colloidal gold nanoparticles after intravenous administration: effect of particle size, Colloids Surf. B Biointerfaces 66 (2) (2008) 274–280.

CHAPTER

18 Carbon nanotubes: Applications in cancer therapy and drug delivery research

Waqar Ahmed[1], Abdelbary Elhissi[2], Vinod Dhanak[3] and Karthikeyan Subramani[4]

[1]University of Lincoln, Lincoln, United Kingdom [2]Qatar University, Doha, Qatar
[3]University of Liverpool, Liverpool, United Kingdom
[4]Roseman University of Health Sciences, Henderson, NV, United States

CHAPTER OUTLINE

18.1 Introduction 371
18.2 Cellular Uptake of CNTs 373
18.3 CNTs as Carriers for Small and Large Drug Molecules 373
18.3.1 CNTs as Carriers of Small Anticancer Molecules 374
18.3.2 CNTs as Carriers of Immunoactive Molecules, Proteins, and Genetic Materials 378
18.3.3 CNTs as Carriers for Antimicrobial Molecules 381
18.3.4 Photothermal Therapy of Cancer Using CNTs 382
18.4 Carbon Nanotubes for Oral Cancer Therapy 383
18.5 Conclusion 384
References 385

18.1 INTRODUCTION

The aim of developing drug delivery systems is to enhance the therapeutic action or reduce the toxicity of drugs. Liposomes are the prototype nanocarriers that are used for achieving this amin, with evidence of clinically established products such as Doxil (liposomal doxorubicin). As an alternative to liposomes, carbon nanotubes (CNTs) are essentially cylindrical molecules made entirely of carbon atoms and can be used as nanocarriers. CNTs are made from graphene sheets rolled into a seamless cylinder that can be open ended or capped, having a high aspect ratio with diameters as small as 1 nm, and length of several micrometers. CNTs made from a single graphene forms a single-walled nanotubes (SWNT), while several graphen3e sheets make up multiwalled CNTs (MWNT). Ever since

Emerging Nanotechnologies in Dentistry. DOI: http://dx.doi.org/10.1016/B978-0-12-812291-4.00018-2

their discovery in 1991 by Iijima [1], there has been intense interest in these allotropes of carbon due to their unique physical and chemical properties and potential applications in a wide range of fields, from electronic devices and sensors to nanocomposite materials of high strength and low weight. Pristine CNTs are not soluble and it was only after the development of strategies to functionalize these molecules with organic groups and render them soluble, that had opened the way to bio-applications of CNTs. Owing to their high-surface area, they are capable of adsorbing or conjugating with a wide variety of therapeutic molecules. Thus, CNTs can be surface-engineered (i.e., functionalized) in order to enhance their dispersability in aqueous phase or to provide the right functional groups that would bind to the desired therapeutic material or the target tissue to elicit a therapeutic effect. CNTs might help the attached therapeutic molecule to penetrate through the target cell to treat diseases [2–4] and a recent example of CNTs with a variety of functional groups relevant to cancer therapy is shown in Fig. 18.1 [5].

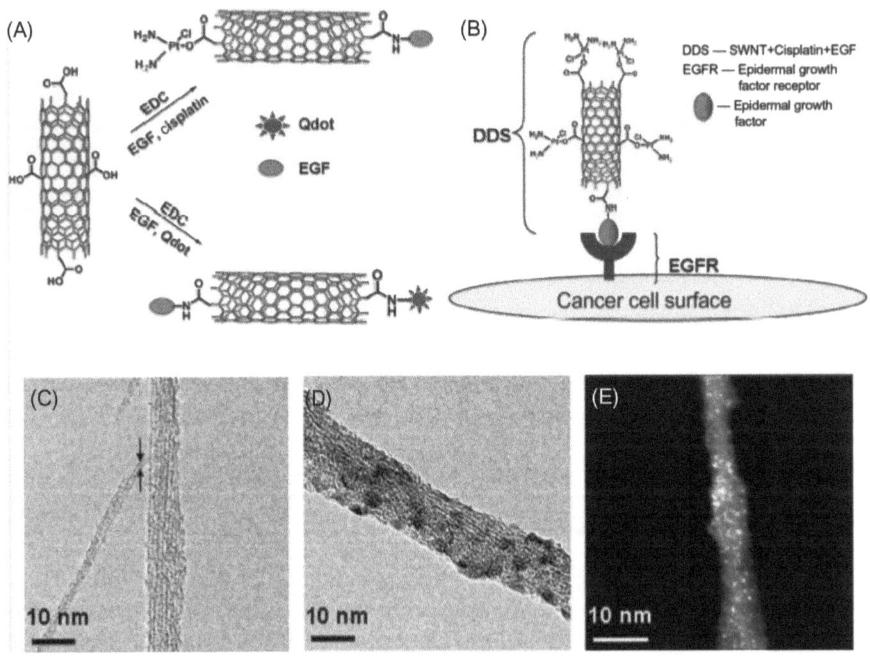

FIGURE 18.1

(A) Schematic representation of functionalization of SWNTs with quantumdots (Qdots), EDC, and Cisplatin; (B) SWNT bioconjugated with Cisplatin and EGF, targeting cancer cell surface receptor EGFR; (C–E) TEM images showing the various functional groups, with Cisplatin shown as bright spots.

From Bhirde, A.A., Patel V., Gavard J., Zhang G., Sousa A.A., Masedunskas A., et al., ACS Nano 2009 (3) 307–316.

Here, we briefly review the therapeutic applications of CNTs with a major focus on cancer and the latest research studies on oral cancer therapy.

18.2 CELLULAR UPTAKE OF CNTs

The cellular uptake of CNTs has been confirmed in many studies but the mechanism of their uptake by cells is still not well understood. Because of their needle-like shapes, CNTs may perforate cellular membrane without causing apparent damage to the cell [2,6–8]. An in vitro CNTs nanoinjector system has been designed by Chen et al. [4] using an atomic force microscope (AFM) tip and functionalized MWNTs attached to a model cargo compound via a disulfide linker [4]. The MWNTs nanoinjector was successful, cell penetration occurred, and the disulfide bond was broken within the cell, resulting in the release of the cargo compound in the cytosol (Fig. 18.2) [4].

The perpendicular positioning of the nanotubes to the cell membranes suggests that uptake of CNTs was similar to that of nanoneedles which diffuse through cell membrane without causing cell death (Fig. 18.3) [3]. In a study conducted by Kam et al. [9], fluoresceinated protein attached to SWNTs-biotin was detected in the endosomes, indicating that uptake of the nanotubes occurred via endocytosis. By contrast, no protein was internalized in the absence of nanotubes. Using epifluorescence and confocal microscopy, functionalized CNTs labeled with a fluorescent agent have shown to penetrate through to the cytoplasm or the nucleus of fibroblast [2]. It has also been reported, in another study, that the uptake mechanism of MWNTs is highly dependent on the length of nanotubes since those shorter than 1 μm were easier to internalize into cells and the process of cellular uptake was reported to be not via endocytosis [10].

18.3 CNTs AS CARRIERS FOR SMALL AND LARGE DRUG MOLECULES

CNTs have been researched as nanocarriers for the delivery of small drug molecules and macromolecules such as genes and proteins, with a special highlight on the delivery of anticancer molecules. The needle-like shapes of CNTs enable them to be functionalized in order to adsorb or covalently bind to a wide range of therapeutic materials, for subsequent internalization into the target cell. The well-established safety of vesicle carriers, particularly liposomes has discouraged many researchers from investigating the potential of CNTs in treatment of many diseases other than cancer; this justifies that most of research involving CNTs for drug delivery applications has been focused on the field of cancer therapy.

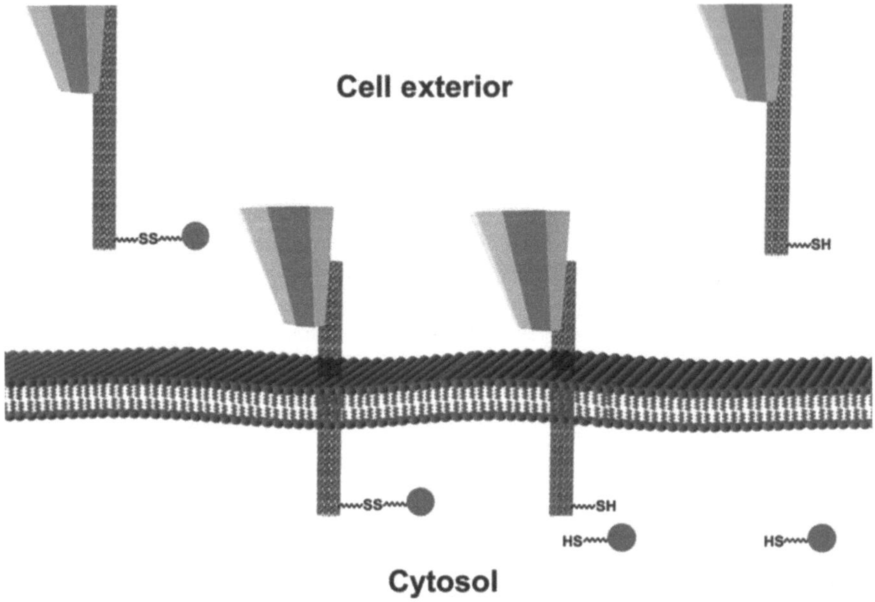

FIGURE 18.2

A schematic diagram showing that an AFM-controlled MWNT-based nanoinjector was able to penetrate into a cell and release the attached cargo compound after the breakage of the disulfide bond. This was followed by successful retraction of the nanoinjector with no apparent cell damage being produced.

From Chen, X., Kis, A., Zettl, A., Bertozzi, C.R. A cell nanoinjector based on carbon nanotubes. Proc. Natl. Acad. Sci. USA 2007 (104) 8218–8222.

18.3.1 CNTs AS CARRIERS OF SMALL ANTICANCER MOLECULES

Cancer cells are able overexpress folic acid (FA) receptors in order to have more nutritional value that is needed for their continuous growth. Therefore, many researchers have designed nanocarriers with engineered surfaces to which FA derivatives can be linked. Moreover, nonspherical nanocarriers such as CNTs might be retained in the lymph nodes for longer periods of time compared to spherical nanocarriers such as liposomes [11]. Thus, CNTs may be used for targeting lymph node cancers as reported by Yang et al. [12,13]. In these studies, magnetic nanoparticles containing the anticancer drug cisplatin were entrapped into FA -functionalized MWNTs. An external magnet was used to drag the nanotubes to the lymph nodes where the drug was shown to be released over several days and the cancer cells to be killed.

Dhar et al. [14] have developed the "longboat delivery system" (Fig. 18.4) [14]. A complex of cisplatin and FA derivative was attached to a functionalized SWNT via a number of amide bonds to make the "longboat" which has been

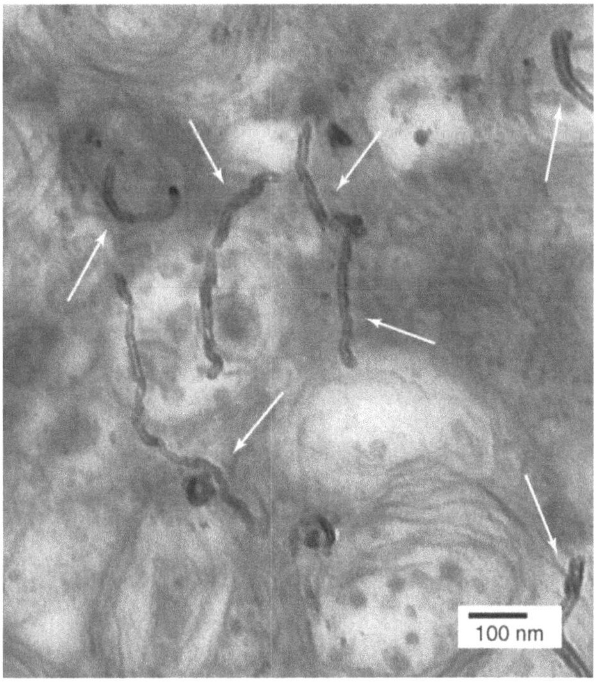

FIGURE 18.3

The perpendicular positioning of MWNTs (pointed at by the white arrows) during internalization into HeLa cells suggests that cellular uptake of CNTs by the cells was similar to that of nanoneedles.

From Bianco, A., Kostarelos, K., Prato, M. Applications of carbon nanotubes in drug delivery. Curr. Opin. Chem. Biol. 2005 (9) 674–679.

reported to penetrate the cancer cells via endocytosis, followed by releasing the drug to target the nuclear DNA. Another platinum anticancer drug, namely carboplatin, can be filled into CNTs; this system was demonstrated to be capable of inhibiting the proliferation of urinary bladder cancer cells in vitro. In another study, the anticancer effect has been shown to be dependent on the method used to entrap the drug in the CNTs, which highlighted the possible effects of preparation conditions on the anticancer activity in CNT formulations [15].

Many anticancer drugs (e.g., paclitaxel) are poorly soluble in water. Taxol is a clinically established formulation of paclitaxel, in which a mixture of Cremophor EL and ethanol is used to solubilize the drug. Unfortunately, Cremophor EL is toxic, which necessitates finding suitable alternative vehicles. Moreover, the circulation time of Taxol in the blood is relatively short. Coating the nanocarriers (e.g., liposomes) with hydrophilic polymers such as polyethylene glycol (PEG) has been used to prolong residence of formulation in the blood

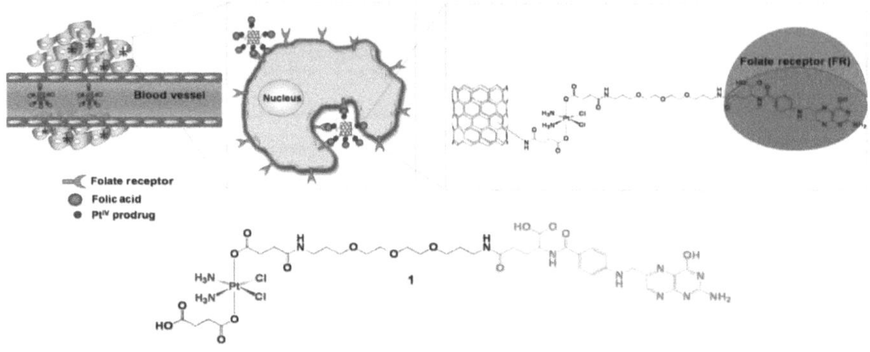

FIGURE 18.4

The "longboat" anticancer system in which the chemotherapeutic agent cisplatin is attached from one end to the FA derivative and from the opposite end to a SWNT via an amide link.

From Dhar, S., Liu, Z., Thomale, J., Dai, H., Lippard, S.J. Targeted single-wall carbon nanotube-mediated Pt(IV) prodrug delivery using folate as a homing device. J. Am. Chem. Soc. 2008 (130) 11467–11476.

circulation through making the carrier system more evasive to uptake by blood macrophages [16,17]. PEGylated paclitaxel formulations have longer circulation time in the blood than Taxol [18]. Functionalized SWNTs were conjugated with paclitaxel through branched PEG chains via a cleavable ester bond. The resultant formulation was more efficient in suppressing tumor growth in vivo than Taxol or paclitaxel-PEG conjugate using a 4T1 breast cancer animal model. The PEGylated nanotubes were able to prolong the circulation and greatly enhance cellular uptake of paclitaxel by the cancerous cells [19]. Similar findings of anticancer activity have been shown when paclitaxel was loaded into PEGylated SWNTs or MWNTs using HeLa cells and MCF-7 cancer cell lines [20]. Multidrug resistance is an obstacle to anticancer therapy since the P-glycoprotein efflux transporter can interfere with the accumulation of anticancer drugs in the target cells, resulting in reduced effectiveness of therapy [21,22]. Using hepatoma cell lines, PEGylated MWNTs may accumulate in multidrug resistant cells as efficient as in nonresistant cells, according to confocal microscopy observations [23]. Interestingly, Liu et al. [24] have reported that although PEGylation of SWNTs can prolong blood circulation time, it may cause an accumulation of the nanotubes in the dermal tissues of mice, suggesting that the degree of coating with PEG should be optimized.

Similar to FA receptors, biotin receptors might be overexpressed on the surfaces of cancer cells. Therefore, biotin-functionalized SWNTs conjugated with the anticancer agent taxoid using a cleavable linker have been designed [25]. The drug was transported via endocytosis, released in the cell, and interacted with microtubules as evaluated by flow cytometry. Thus, stable microtubule–taxoid complexes were formed, causing apoptosis and cell death (Fig. 18.5).

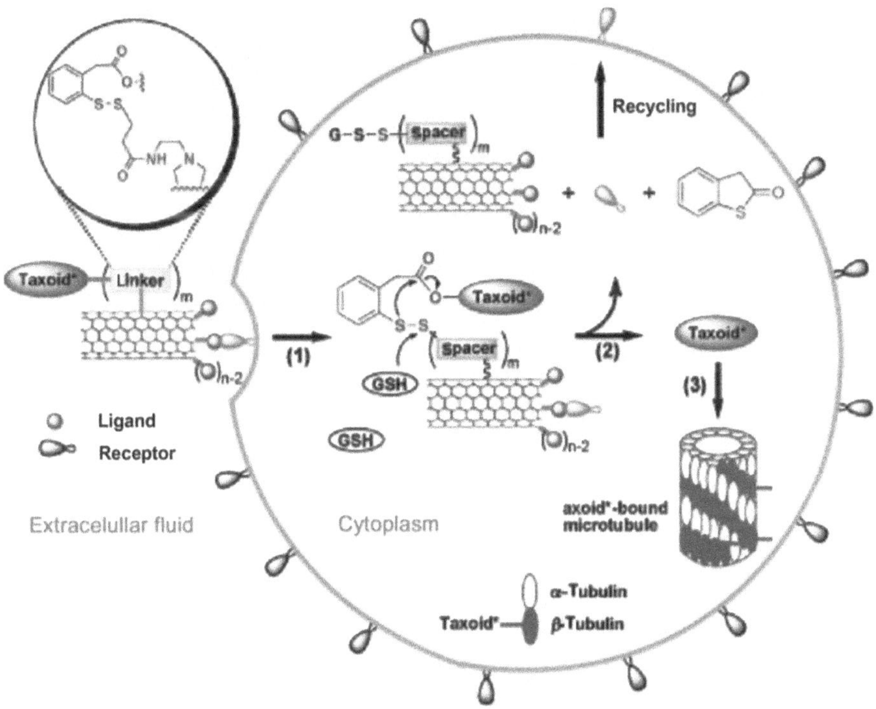

FIGURE 18.5

(1) Internalization of the CNTs carried conjugate into the tumor cell via receptor-mediated endocytosis, (2) taxoid was released by the cleavage of the chemical linker, and (3) the free taxoid molecules were bound to microtubules to form stabilized microtubules, resulting in arrest of cell mitosis and induction of apoptosis.

From Chen, J., Chen, S., Zhao, X., Kuznetsova, L.V., Wong, S.S., Ojima, I. Functionalized single-walled carbon nanotubes as rationally designed vehicles for tumor-targeted drug delivery. J. Am. Chem. Soc. 2008 (130) 16778–16785.

What was applied on paclitaxel was also investigated using other anticancer drugs such as doxorubicin (DOX). Thus, a targeted delivery system of FA-tethered SWNTs-DOX has been designed. In addition, mucoadhesive polymers such as chitosan (CHI) and sodium alginate (ALG) were incorporated to enhance the aqueous dispersability of the nanotubes while FA was employed to improve targeting of the nanotubes (Fig. 18.6). Transmission electron microscopy (TEM) showed that the drug was released after being transported into the tumor cell (HeLa cell) at the low pH of the lysosome but not at the normal physiological pH of 7.4 [26] (Fig. 18.7). PAMAM dendrimers have been attached to FA-treated MWNTs, in order to promote the targeting of the nanotubes. Flow cytometry and confocal microscopy have shown possible targeting of cancer cells that overexpress FA receptors using this formulation [27].

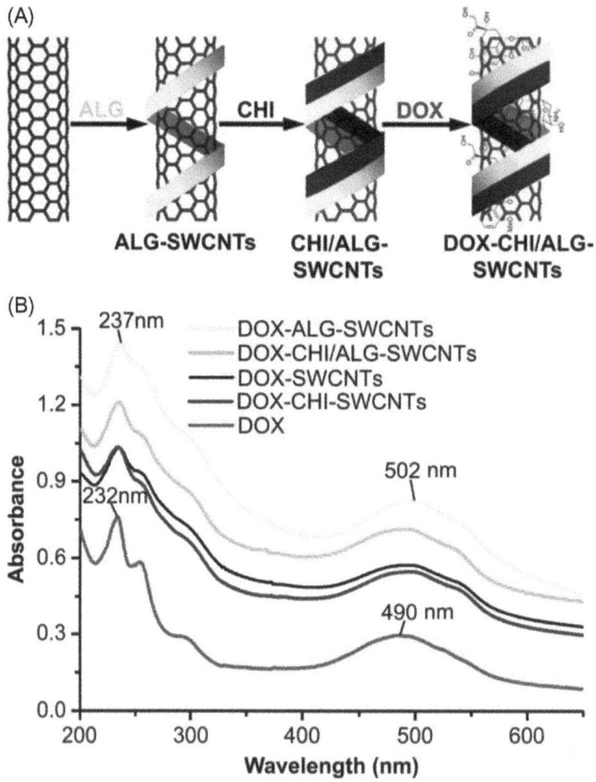

FIGURE 18.6

(A) Preparation of SWNTs-DOX after inclusion of bioadhesive polymers to enhance nanotubes dispersability in aqueous phase. (B) UV absorption spectra of DOX formulations.

From Zhang, X., Meng, L., Lu, Q., Fei, Z., Dyson, P.J. Targeted delivery and controlled release of doxorubicin to cancer cells using modified single wall carbon nanotubes. Biomaterials 2009 (30) 6041–6047.

18.3.2 CNTs AS CARRIERS OF IMMUNOACTIVE MOLECULES, PROTEINS, AND GENETIC MATERIALS

The ability of macromolecules (e.g., genes) to cross the biological barriers and be expressed within target cell is not easy, because of their hydrophilicity and large molecular size. Gene therapy aims to use genes to treat diseased cells by repairing the cause of the disease. Since genetic materials are poorly able to cross the biological membranes, the use of vectors (viral or nonviral) to carry the gene and internalize it into cell is necessary. Nonviral vectors are less efficient than viral vectors [28,29]; however, they are much safer [30,31]. Pantarotto et al. [32] have introduced novel functionalized SWNT-DNA complexes and reported high DNA expressions compared to naked DNA. Functionalized SWNTs have

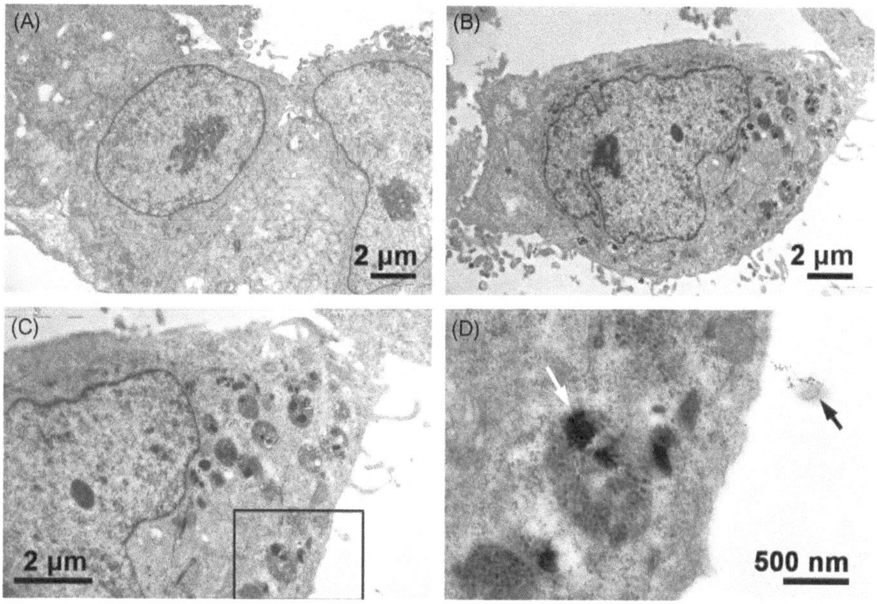

FIGURE 18.7

TEM of (A) HeLa cells before treatment, (B) HeLa cells treated with DOX-FA-CHI-ALG-SWNTs, (C) a magnified image of (B), (D) magnified image of the boxed region in (C). The black arrow points at a SWNT-containing vesicle, and the white arrow points at some aggregated nanotubes inside a lysosome.

From Zhang, X., Meng, L., Lu, Q., Fei, Z., Dyson, P.J. Targeted delivery and controlled release of doxorubicin to cancer cells using modified single wall carbon nanotubes. Biomaterials 2009 (30) 6041–6047.

been suggested as nonviral carriers of macromolecules, and internalization happened via energy-dependent endocytosis [33]. Confocal microscopy and flow cytometry have shown much greater fluorescent activity of protein and DNA when made as conjugates with SWNTs as compared to the naked macromolecules [33], indicating that CNTs are promising nanocarrier vectors for gene and protein. Cai et al. [7] have introduced an approach to gene delivery named "carbon nanotube spearing." Plasmid DNA with a fluorescent protein was immobilized onto nickel-embedded CNTs. The formulation was "speared" into Bal 17 B lymphoma cells using a magnetic field, which produced high transfection in the target cells. In another study, exposure of HeLa cells to DNA attached to SWNTs at 37°C resulted in gene internalization. However, at lower temperature (e.g., 4°C) no internalization occurred (Fig. 18.8), suggesting that gene transport occurs by an energy-dependent endocytosis [34]. In another study, MWNTs grafted with cationic polymers such as polyethylenimine have been shown to be appropriate carriers of pDNA to target cancer cells and gene expression using the nanotubes was higher than that using free pDNA [35].

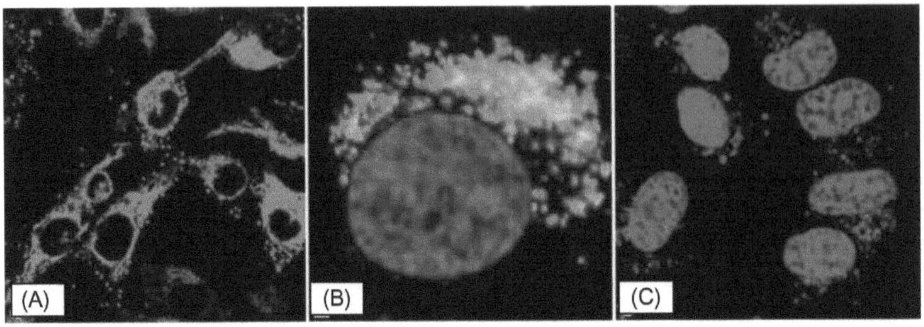

FIGURE 18.8

Confocal microscopy showing the internalization of labeled single strand DNA into HeLa cell using SWNTs. (A) The labeled DNA [green color (grey in print versions)] is surrounding the nucleus (black circles) at 37°C. (B) the nucleus is stained using DRAQ5 [red color (black in print versions)] is surrounded by the labeled DNA [green color (grey in print versions)] after internalization at 37°C. (C) At 4°C, no DNA internalization has occurred.

Adapted from Kam, N.W.S., O'Connell, M., Wisdom, J.A., Dai, H. Carbon nanotubes as multifunctional biological transporters and near-infrared agents for selective cancer cell destruction Proc. Natl. Acad. Sci. USA 2005 (102) 11600–11605.

More recent mechanisms in gene therapy, such as gene silencing, have been introduced using small interfering RNA (siRNA). Gene silencing has been shown to be efficient in the treatment of cancer. In cancer, antitumor immunity might be inhibited by suppressors of cytokine signaling 1 (SOCS1). siRNA can be conjugated to phospholipid-functionalized SWNTs using a cleavable disulfide linker, resulting in efficient gene silencing and death of the targeted cells [36]. In another study, amino-functionalized MWNTs-siRNA complexes can suppress tumor and prolonged survival using a lung tumor animal model [37]. Many types of cancer (e.g., leukemia) can overexpress cyclinA(2) and suppression of this material is expected to prevent tumor growth. Functionalized SWNTs-siRNA has been designed, and internalization into K562 cells can be achieved. This should determine if inhibition of the production of cyclinA(2) can be therapeutically useful in treatment of chronic myelogenous leukemia. It has been found that suppression of cyclinA(2) expression using siRNA-CNTs can promote apoptosis in the targeted tumor [38]. This came consistent with previous findings, for example, functionalized SWNTs conjugated with telomerase reverse transcriptase siRNA can silence the target gene and tumor growth can be inhibited in vitro using murine tumor cell lines and in vivo using a mouse model [39].

Streptavidin is a protein that has anticancer activity [40], however, due to its very large molecular weight (approximately 60,000 Da), its penetration into cells is extremely difficult. Conjugation of streptavidin with SWNTs-biotin may result in internalization of the protein into model cancer cells by adsorption-mediated endocytosis [9]. Transmission electron and confocal microscopy have shown that

MWNTs can act as transporters of the recombinant ricin A chain protein, resulting in high death rates of cancer cells [41].

Immunotherapy may constitute an alternative to gene therapy in the treatment of cancer. Tumor specific monoclonal antibodies, radiometal ion chelates, and fluorescent probe were attached to SWNTs. Targeting the tumor (lymphoma) using a range of techniques has been reported to be possible [42]. MWNTs have been conjugated to tumor lysate protein as an antigen; this specifically increased the antitumor immune responses [43].

Glioma is a brain tumor that is able to evade the host immune system. Consequently, no benefit from conventional chemotherapy has been shown, because glioma cells secrete the immunosuppressive cytokines such as prostaglandins E and TGF-*Beta* and IL-10 [44,45]. Macrophages have a preferential affinity toward CNTs when compared to glioma cells [46]. Using a GL261 murine intracranial glioma cancer model, VanHandel et al. [47] have developed an immunotherapy approach using MWNTs, based on that macrophages prefer to engulf CNTs compared with glioma cells. MWNTs caused an increase in the influx of macrophages into the glioma cells, which was reported to be accompanied by an increase in the levels of IL-10 expression, suggesting that immunomodulation using CNTs can be used to treat cancer.

In another study, angiogenesis targeting antibodies E4G10 were attached to SWNTs via radiometal ion chelates. This delivery system has been reported to reduce the volume of the tumor and prolong survival in animal models [48]. Moreover, oxidized MWNTs can be injected subcutaneously to a hepatocarcinoma-bearing animal to induce an immune response, which has been reported to retard tumor growth [49]. This suggests that CNTs themselves can be surface-engineered to have anticancer activity by inducing an immune response against tumor.

18.3.3 CNTs AS CARRIERS FOR ANTIMICROBIAL MOLECULES

Surface-engineered CNTs may capture pathogenic bacteria in liquid media [50–52]. Thus, CNTs themselves might have antimicrobial activity since microorganisms might be adsorbed onto the engineered surfaces of CNTs. Moreover, using *Escherichia coli* bacteria, it has been reported that the electronic properties of SWNTs may regulate the antimicrobial activity. The antibacterial effect was attributed to CNT-induced oxidation of the intracellular antioxidant glutathione, resulting in increased oxidative stress, causing death of the microbial cells [53].

Functionalized CNTs have been previously demonstrated to be able to act as carriers for antimicrobial agents such as the antifungal amphotericin B. CNTs can attach covalently to amphotericin B and transport it into mammalian cells. This reduced the antifungal toxicity as compared to the toxicity elicited by the free drug (40% of the cells were killed by the CNTs-free formulation compared to no cell death by the CNTs formulation). It has also been reported that the antifungal activity was increased using the CNTs [54].

18.3.4 PHOTOTHERMAL THERAPY OF CANCER USING CNTs

CNTs are capable of absorbing light in the near infrared (NIR) region, resulting in heating the nanotubes [55]. This unique thermal property of CNTs has been exploited to kill cancer cells [34,56–66]. Optical coupling of light with CNTs is predicted to be at highest when the length of the nanotubes is more than half the wavelength of the incident light beam, which has been previously determined by the antenna theory [67]. Engineering the surfaces of MWNTs by creating intentional surface defects might enhance the antenna properties of the nanotubes. These engineered "defects" is known as dopants and can cause scattering in the traveling currents and may also increase the heating of the nanotube. This physicoelectronic characteristic of the engineered MWNTs can thermally destruct the tumor cells by using MWNTs that have good heat conducting properties. The process of involving dopants in the structure of the nanotubes is called doping, and examples of that are boron doping [68] and nitrogen doping (N doping) [57]. N-doped MWNTs have been shown to elicit photo-ablative kill of model kidney cancer cells when NIR light was employed. Moreover, the length of nanotubes has been found to be a major determinant of nanotube ability to transfer heat and kill the tumor with lengths between 700 and 1100 nm being most desirable to kill the tumor (Fig. 18.9) [57].

In a study conducted by Gannon et al. [56], SWNTs were functionalized using Kentera (a polyphenylene ethynylene-based polymer). The incubation of the

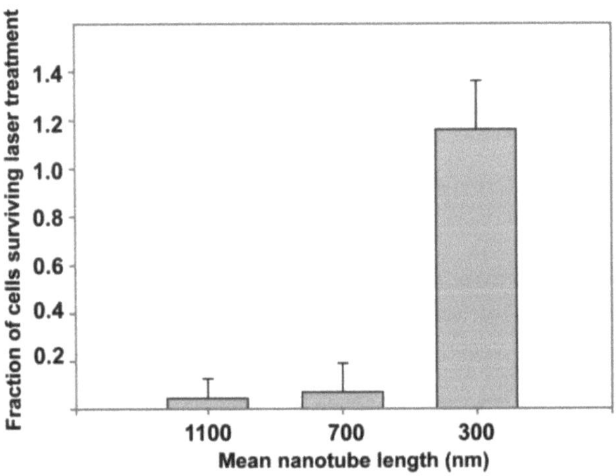

FIGURE 18.9

The relationship between cell survival and CNTs length using photothermal therapy. This has shown that nanotube lengths of 700 and 1100 nm are much more desirable in killing tumor cells compared with the length of 300 nm.

From Torti, S.V., Byrne, F., Whelan, O., Levi, N., Ucer, B., Schmid, M., et al. Thermal ablation therapeutics based on CN(x) multi-walled nanotubes. Int. J. Nanomed. 2007 (2) 707–714.

nanotubes with hepatic tumor cells followed by application of radiofrequency field caused a concentration dependent thermal destruction of the tumor cells which was demonstrated by development of apoptotic cells that caused complete necrosis of the tumor cells. By contrast, tumor cells that were injected with the Kentera alone (without CNTs) were viable after the application of the radiofrequency field. In the same study, it has been reported that in vivo injection of the Kentera functionalized SWNTs was tolerated by rabbits [56]. Unfortunately, the resultant thermal destruction is not selective toward cancer cells and the access to deep tumor areas is poor, thus the inclusion of targeting moieties such as FA on the surfaces of CNTs is needed [34]. Folate-bearing nanotubes having the size of 0.81 nm and a maximum absorbance at 980 nm were used for photothermal therapy of cancer [58]. The tumor cells were exposed to 980 nm laser radiations, resulting in photothermal destruction of cancer cells both in vitro and in vivo.

18.4 CARBON NANOTUBES FOR ORAL CANCER THERAPY

Oral cancer may originate in any of the tissues of the mouth, and may be of varied histologic types:

- Teratoma (true neoplasms composed of multiple tissues foreign to the site from which they originate),
- Adenocarcinoma (derived from a major or minor salivary gland),
- Lymphoma (from tonsillar or other lymphoid tissue), or
- Melanoma (from the pigment producing cells of the oral mucosa).

Oral cancer accounts for approximately 2% of all malignant tumors in the USA and Europe, approximately 30%–40% in the Indian subcontinent, and more than 90% are squamous cell carcinomas, originating in the tissues that line the mouth and lips.

CNT-based drug delivery holds great promise in the treatment of oral cancer. In a recent in vivo study [5], killing of cancer cells using a drug-single wall CNT (SWNT) bioconjugate, was evaluated and it was demonstrated with superior efficacy than the nontargeted bioconjugates. Anticancer drug cisplatin and epidermal growth factor (EGF) were attached to SWNTs to specifically target head and neck squamous carcinoma cells (HNSCC). Initial in vitro imaging studies with HNSCC overexpressing EGF receptors (EGFR) using Quantum dot (Qdot) luminescence and confocal microscopy showed that SWNT-Qdot-EGF bioconjugates internalized rapidly into the cancer cells. Limited uptake occurred for control cells without EGF. Intravital video imaging in vivo showed that SWNT-Qdot-EGF injected into live mice was selectively taken up by HNSCC tumors, but SWNT-Qdot controls with no EGF were cleared from the tumor region in <20 min. The HNSCC cells treated with SWNT-cisplatin-EGF were also killed selectively, while control systems that did not feature EGF-EGFR binding did not influence cell proliferation. Most significantly, regression of tumor growth was rapid in mice treated with targeted SWNT-cisplatin-EGF relative to nontargeted SWNT-cisplatin.

Photodynamic therapy is a type of cancer treatment that uses a drug called a photosensitizer or photosensitizing agent. Photosensitizer is activated by light of a specific wavelength. When photosensitizers are exposed to this specific wavelength, they produce singlet oxygen, which destroys cancer cells. Photothermal cancer therapy is based on generating local hyperthermia by infrared (IR) laser to destroy tumor cells. Single-wall CNTs (SWNTs) with strong optical absorption in the broad visible and near IR offer unique advantages for photothermal cancer therapy. A broad range of wavelengths can be used for the treatment with SWNTs, whereas conventional photothermal therapeutic agent is designed to absorb light only near one selected wavelength. In a recent study [66], SWNTs were injected into the tumor tissue and it was evaluated whether local hyperthermia could destroy tumor cells. Squamous cell carcinoma (SCC) tumor in mice was exposed to 785-nm laser after intratumoral injection of SWNTs with different light and SWNTs dose combinations. The temperatures of the tumor with laser irradiation were monitored. In vivo and ex vivo Raman spectra in different organs were obtained with a rapid Raman system. Tumor responses (tumor volume and mouse survival) were documented daily after treatment up to day 45 to assess the effectiveness of the treatment. The temperature within the tumors increased in a light- and SWNTs-dose dependent manner. The study concluded that SCCs can be eradicated at a moderate light irradiance and fluence (200 mW/cm^2 and 120 J/cm^2). This light dose is also comparable to those used with photodynamic therapy. Tissue Raman spectroscopy measurements revealed that SWNTs remained localized in the tumor even 3 months after injection but was not found in other organs. The result of this study represents a significant step forward toward the goal of advancing SWNTs-based photothermal cancer therapy into clinical applications for treatment of oral cancer and other tumors.

18.5 CONCLUSION

CNTs are potentially promising needle-like carriers of small drug molecules as well as macromolecules such as gene and protein. CNTs can be functionalized so that certain molecules are attached to their surfaces via covalent or noncovalent bonding. The needle-like shape of the CNTs enables them to perforate cellular membranes and transport the carried therapeutic molecules to the cellular components. This process is thought to take place via endocytosis. CNTs have exclusive properties that would make them appropriate in the medical field such as their ability to adsorb pathogenic microorganisms and conduct heat. It is interesting that huge amount of medical research of CNTs has taken place in the area of anticancer therapy at the expense of other areas. From the perspective of the authors of this review, this might be due to the potential toxicity of CNTs as well as the established safety of the vesicle nanocarriers (e.g., liposomes) which was out of the scope of discussion of this paper. CNTs have been introduced to drug delivery research for a limited number of years and therefore extensive amounts of research is expected to be produced in the forthcoming years in order to explore their potential.

REFERENCES

[1] S. Iijima, Nature 354 (1991) 56–58.

[2] D. Pantarotto, J.P. Briand, M. Prato, A. Bianco, Translocation of bioactive peptides across cell membranes by carbon nanotubes, Chem. Commun. (Camb.) 7 (2004) 16–17.

[3] A. Bianco, K. Kostarelos, M. Prato, Applications of carbon nanotubes in drug delivery, Curr. Opin. Chem. Biol. 9 (2005) 674–679.

[4] X. Chen, A. Kis, A. Zettl, C.R. Bertozzi, A cell nanoinjector based on carbon nanotubes, Proc. Natl. Acad. Sci. USA 104 (2007) 8218–8222.

[5] A.A. Bhirde, V. Patel, J. Gavard, G. Zhang, A.A. Sousa, A. Masedunskas, et al., ACS Nano 3 (2009) 307–316.

[6] N.W.S. Kam, Z. Liu, H. Dai, Carbon nanotubes as intracellular transporters for proteins and DNA: an investigation of the uptake mechanism and pathway, Angew. Chem. 44 (2005) 1–6.

[7] D. Cai, J.M. Mataraza, Z.H. Qin, Z. Huang, J. Huang, T.C. Chiles, et al., Highly efficient molecular delivery into mammalian cells using carbon nanotube spearing, Nat. Methods 2 (2005) 449–454.

[8] C. Klumpp, K. Kostarelos, M. Prato, A. Bianco, Functionalized carbon nanotubes as emerging nanovectors for the delivery of therapeutics, Biochim. Biophys. Acta 1758 (2006) 404–412.

[9] N.W.S. Kam, T.C. Jessop, P.A. Wender, H. Dai, Nanotube molecular transporters: internalization of carbon nanotube-protein conjugates into mammalian cells, J. Am. Chem. Soc. 126 (2004) 6850–6851.

[10] V. Raffa, G. Ciofani, S. Nitodas, T. Karachalios, D. D'Alessandro, M. Masini, et al., Can the properties of carbon nanotubes influence their internalization by living cells? Carbon 64 (2008) 1600–1610.

[11] S.T. Reddy, A. Rehor, H.G. Schmoekel, J.A. Hubbell, M.A. Swartz, In vivo targeting of dendritic cells in lymph nodes with poly(propylene sulfide) nanoparticles, J. Control. Release 112 (2006) 26–34.

[12] F. Yang, L. Fu de, J. Long, Q.X. Ni, Magnetic lymphatic targeting drug delivery system using carbon nanotubes, Med. Hypotheses 70 (2008) 765–767.

[13] F. Yang, J. Hu, D. Yang, J. Long, G. Luo, C. Jin, et al., Pilot study of targeting magnetic carbon nanotubes to lymph nodes, Nanomedicine 4 (2009) 317–330.

[14] S. Dhar, Z. Liu, J. Thomale, H. Dai, S.J. Lippard, Targeted single-wall carbon nanotube-mediated Pt(IV) prodrug delivery using folate as a homing device, J. Am. Chem. Soc. 130 (2008) 11467–11476.

[15] S. Hampel, D. Kunze, D. Haase, K. Krämer, M. Rauschenbach, M. Ritschel, et al., Carbon nanotubes filled with a chemotherapeutic agent: a nanocarrier mediates inhibition of tumor cell growth, Nanomedicine 3 (2008) 175–182.

[16] A.L. Klibanov, K. Maruyama, V.P. Torchilin, L. Huang, Amphipathic polyethyleneglycols effectively prolong circulation time of liposomes, FEBS Lett. 268 (1990) 235–237.
[17] T.M. Allen, T. Mehra, C. Hansen, Y.C. Chin, Stealth liposomes: an improved sustained release system for 1-beta-D-arabinofuranosylcytosine, Cancer Res. 52 (1992) 2431–2439.
[18] C. Li, D. Yu, T. Inoue, D.J. Yang, L. Milas, N.R. Hunter, et al., Synthesis and evaluation of water-soluble polyethylene glycol-paclitaxel conjugate as a paclitaxel prodrug, Anticancer Drugs 7 (1996) 642–648.
[19] Z. Liu, K. Chen, C. Davis, S. Sherlock, Q. Cao, X. Chen, et al., Drug delivery with carbon nanotubes for in vivo cancer treatment, Cancer Res. 68 (2008) 6652–6660.
[20] C.L. Lay, H.Q. Liu, H.R. Tan, Y. Liu, Delivery of paclitaxel by physically loading onto poly(ethylene glycol) (PEG)-graft-carbon nanotubes for potent cancer therapeutics, Nanotechnology 21 (2010) 065101.
[21] J.Y. Chan, A.C. Chu, K.P. Fung, Inhibition of *P*-glycoprotein expression and reversal of drug resistance of human hepatoma HepG2 cells by multidrug resistance gene (mdr1) antisense RNA, Life Sci. 67 (2000) 2117–2124.
[22] S.S. Suri, F. Hicham, B. Singh, Nanotechnology-based drug delivery systems, J. Occup. Med. Toxicol. 2 (2007) 2–16.
[23] J. Cheng, M.J. Meziani, Y.-P. Sun, H. Cheng, Poly(ethylene glycol)-conjugated multi-walled carbon nanotubes as an efficient drug carrier for overcoming multidrug resistance, Toxicol. Appl. Pharmacol. 250 (2011) 184–193.
[24] X. Liu, H. Tao, K. Yang, S. Zhang, S.T. Lee, Z. Liu, Optimization of surface chemistry on single-walled carbon nanotubes for in vivo photothermal ablation of tumors, Biomaterials 32 (2011) 144–151.
[25] J. Chen, S. Chen, X. Zhao, L.V. Kuznetsova, S.S. Wong, I. Ojima, Functionalized single-walled carbon nanotubes as rationally designed vehicles for tumor-targeted drug delivery, J. Am. Chem. Soc. 130 (2008) 16778–16785.
[26] X. Zhang, L. Meng, Q. Lu, Z. Fei, P.J. Dyson, Targeted delivery and controlled release of doxorubicin to cancer cells using modified single wall carbon nanotubes, Biomaterials 30 (2009) 6041–6047.
[27] X. Shi, S.H. Wang, M. Shen, M.E. Antwerp, X. Chen, C. Li, et al., Multifunctional dendrimer-modified multiwalled carbon nanotubes: synthesis, characterization, and in vitro cancer cell targeting and imaging, Biomacromolecules 10 (7) (2009) 1744–1750.
[28] E. Fortunati, A. Bout, M.A. Zanta, D. Valerio, M. Scarpa, In vitro and in vivo gene transfer to pulmonary cells mediated by cationic liposomes, Biochim. Biophys. Acta 1306 (1996) 55–62.
[29] T. Rochat, M.A. Morris, Gene therapy for cystic fibrosis by means of aerosol, J. Aerosol. Med. 15 (2002) 229–235.
[30] F.D. Ledley, Non-viral gene therapy, Curr. Opin. Biotechnol. 5 (1994) 626–636.
[31] C. Coutelle, R. Williamson, Liposomes and viruses for gene therapy of cystic fibrosis, J. Aerosol. Med. 9 (1996) 79–88.
[32] D. Pantarotto, R. Singh, D. McCarthy, M. Erhardt, J.P. Briand, M. Prato, et al., Functionalized carbon nanotubes for plasmid DNA gene delivery, Angew. Chem. Int. Ed. Engl. 43 (2004) 5242–5246.
[33] N.W. Kam, Z. Liu, H. Dai, Carbon nanotubes as intracellular transporters for proteins and DNA: an investigation of the uptake mechanism and pathway, Angew. Chem. Int. Ed. Engl. 45 (2006) 577–581.

[34] N.W.S. Kam, M. O'Connell, J.A. Wisdom, H. Dai, Carbon nanotubes as multifunctional biological transporters and near-infrared agents for selective cancer cell destruction, Proc. Natl. Acad. Sci. USA 102 (2005) 11600–11605.

[35] A. Nunes, N. Amsharov, C. Guo, V. Bossche, P. Santhosh, T.K. Karachalios, et al., Hybrid polymer-grafted multiwalled carbon nanotubes for in vitro gene delivery, Small 6 (2010) 2281–2291.

[36] N.W. Kam, Z. Liu, H. Dai, Functionalization of carbon nanotubes via cleavable disulfide bonds for efficient intracellular delivery of siRNA and potent gene silencing, J. Am. Chem. Soc. 127 (2005) 12492–12493.

[37] J.E. Podesta, K.T. Al-Jamal, M.A. Herrero, B. Tian, H. Ali-Boucetta, V. Hegde, et al., Antitumor activity and prolonged survival by carbon-nanotube-mediated therapeutic siRNA silencing in a human lung xenograft model, Small 5 (2009) 1176–1185.

[38] X. Wang, J. Ren, X. Qu, Targeted RNA interference of cyclin A2 mediated by functionalized single-walled carbon nanotubes induces proliferation arrest and apoptosis in chronic myelogenous leukemia K562 cells, ChemMedChem 3 (2008) 940–945.

[39] Z. Zhang, X. Yang, B. Zeng, S. Wang, T. Zhu, R.B. Roden, et al., Delivery of telomerase reverse transcriptase small interfering RNA in complex with positively charged single-walled carbon nanotubes suppresses tumor growth, Clin. Cancer Res. 12 (2006) 4933–4939.

[40] S.L. Hussey, B.R. Peterson, Efficient delivery of streptavidin to mammalian cells: Clathrin-mediated endocytosis regulated by a synthetic ligand, J. Am. Chem. Soc. 124 (2002) 6265–6273.

[41] X. Weng, M. Wang, J. Ge, S. Yu, B. Liu, J. Zhong, et al., Carbon nanotubes as a protein toxin transporter for selective HER2-positive breast cancer cell destruction, Mol. Biosyst. 5 (2009) 1224–1231.

[42] M.R. McDevitt, D. Chattopadyay, B.J. Kappel, J.S. Jaggi, S.R. Schiffman, C. Antczak, et al., Tumor targeting with antibody-functionalized, radiolabeled carbon nanotubes, J. Nucl. Med. 48 (2007) 1180–1189.

[43] J. Meng, J. Meng, J. Duan, H. Kong, L. Li, C. Wang, et al., Carbon nanotubes conjugated to tumor lysate protein enhance the efficacy of an antitumor immunotherapy, Small 4 (2008) 1364–1370.

[44] I.F. Parney, C. Hao, K.C. Petruk, Glioma immunology and immunotherapy, Neurosurgery 46 (2000) 778–791.

[45] L. Yang, K.Y. Ng, K.O. Lillehei, Cell-mediated immunotherapy: a new approach to the treatment of malignant glioma, Cancer Control 10 (2003) 138–147.

[46] B. Kateb, M. Van Handel, L. Zhang, M.J. Bronikowski, H. Manohara, B. Badie, Internalization of MWCNTs by microglia: possible application in immunotherapy of brain tumors, NeuroImage 37 (Suppl. 1) (2007) S9–S17.

[47] M. VanHandel, D. Alizadeh, L. Zhang, B. Kateb, M. Bronikowski, H. Manohara, et al., Selective uptake of multi-walled carbon nanotubes by tumor macrophages in a murine glioma model, J. Neuroimmunol. 208 (2009) 3–9.

[48] A. Ruggiero, C.H. Villa, J.P. Holland, S.R. Sprinkle, C. May, J.S. Lewis, et al., Imaging and treating tumor vasculature with targeted radiolabeled carbon nanotubes, Int. J. Nanomed. 5 (2010) 783–802.

[49] J. Meng, M. Yang, F. Jai, H. Kong, W. Zhang, C. Wang, et al., Subcutaneous injection of water-soluble multi-walled carbon nanotubes in tumor-bearing mice boosts the host immune activity, Nanotechnology 21 (2010) 145104.

[50] L. Gu, T. Elkin, X. Jiang, H. Li, Y. Lin, L. Qu, et al., Single-walled carbon nanotubes displaying multivalent ligands for capturing pathogens, Chem. Commun. (Camb.) 7 (2005) 874–876.
[51] A.S. Brady-Estévez, S. Kang, M.A. Elimelech, Single-walled-carbon-nanotube filter for removal of viral and bacterial pathogens, Small 4 (2008) 481–484.
[52] L.R. Arias, L. Yang, Inactivation of bacterial pathogens by carbon nanotubes in suspensions, Langmuir 25 (2009) 3003–3012.
[53] C. Vecitis, K.R. Zodrow, S. Kang, M. Elimelech, Electronic-structure-dependent bacterial cytotoxicity of single-walled carbon nanotubes, ACS Nano 4 (2010) 5471–5479.
[54] W. Wu, S. Wieckowski, G. Pastorin, M. Benincasa, C. Klumpp, J.P. Briand, et al., Targeted delivery of amphotericin B to cells by using functionalized carbon nanotubes, Angew. Chem. Int. Ed. Engl. 44 (2005) 6358–6362.
[55] M.E. Brennan, J.N. Coleman, A. Drury, B. Lahr, T. Kobayashi, W.J. Blau, Nonlinear photoluminescence from van Hove singularities in multiwalled carbon nanotubes, Opt. Lett. 28 (2003) 266–268.
[56] C.J. Gannon, P. Cherukuri, B.I. Yakobson, L. Cognet, J.S. Kanzius, C. Kittrell, et al., Carbon nanotube-enhanced thermal destruction of cancer cells in a noninvasive radiofrequency field, Cancer 110 (2007) 2654–2665.
[57] S.V. Torti, F. Byrne, O. Whelan, N. Levi, B. Ucer, M. Schmid, et al., Thermal ablation therapeutics based on CN(x) multi-walled nanotubes, Int. J. Nanomed. 2 (2007) 707–714.
[58] F. Zhou, D. Xing, Z. Ou, B. Wu, D.E. Resasco, W.R. Chen, Cancer photothermal therapy in the near-infrared region by using single-walled carbon nanotubes, J. Biomed. Opt. 14 (2009) 021009.
[59] N.H. Levi-Polyachenko, E.J. Merkel, B.T. Jones, D.L. Carroll, J.H. Stewart, Rapid photothermal intracellular drug delivery using multiwalled carbon nanotubes, Mol. Pharm. 6 (2009) 1092–1099.
[60] C.H. Wang, Y.J. Huang, C.W. Chang, W.M. Hsu, C.A. Peng, In vitro photothermal destruction of neuroblastoma cells using carbon nanotubes conjugated with GD2 monoclonal antibody, Nanotechnology 20 (2009) 315101 (7pp).
[61] A. Burke, X. Ding, R. Singh, R.A. Kraft, N. Levi-Polyachenko, M.N. Rylander, et al., Long-term survival following a single treatment of kidney tumors with multiwalled carbon nanotubes and near-infrared radiation, Proc. Natl. Acad. Sci. USA 106 (2009) 12897–12902.
[62] S. Ghosh, S. Dutta, E. Gomes, D. Carroll, R. D'Agostino Jr., J. Olson, et al., Increased heating efficiency and selective thermal ablation of malignant tissue with DNA-encased multiwalled carbon nanotubes, ACS Nano 3 (2009) 2667–2673.
[63] Y. Xiao, X. Gao, O. Taratula, S. Treado, A. Urbas, R.D. Holbrook, et al., Anti-HER2 IgY antibody-functionalized single-walled carbon nanotubes for detection and selective destruction of breast cancer cells, BMC Cancer 9 (2009) 351.
[64] H.K. Moon, S.H. Lee, H.C. Choi, In vivo near-infrared mediated tumor destruction by photothermal effect of carbon nanotubes, ACS Nano 3 (2009) 3707–3713.
[65] A. Burlaka, S. Lukin, S. Prylutska, O. Remeniak, Y. Prylutskyy, M. Shuba, et al., Hyperthermic effect of multi-walled carbon nanotubes stimulated with near infrared irradiation for anticancer therapy: in vitro studies, Exp. Oncol. 32 (2010) 48–50.

[66] N. Huang, H. Wang, J. Zhao, H. Lui, M. Korbelik, H. Zeng, Single-wall carbon nanotubes assisted photothermal cancer therapy: animal study with a murine model of squamous cell carcinoma, Lasers Surg. Med. 42 (2010) 638–648.
[67] J.D. Jackson, Classical Electrodynamics, third ed., John Wiley and Sons, New York, 1999.
[68] J.F. Xu, M. Xiao, R. Czerw, D.L. Caroll, Optical limiting and enhanced optical non-linearity in boron-doped carbon nanotubes, Chem. Phys. Lett. 389 (2004) 247–250.

CHAPTER

Nanodiagnostics in microbiology and dentistry

19

Karthikeyan Subramani[1] and Manjula Mehta[2]

[1]*Roseman University of Health Sciences, Henderson, NV, United States*

[2]*Panjab University, Chandigarh, India*

CHAPTER OUTLINE

19.1 Introduction 391
19.2 Nanomaterials 392
19.2.1 Applications of Nanomaterials 393
19.3 Biomedical Applications of Nanotechnology and Its Limitations 393
19.4 Nanotechnology Applications in Drug-Delivery Systems, Nanodiagnostics, and Various Other Fields 394
19.4.1 Drug-Delivery System 394
19.4.2 Nanodiagnostics and Disease Prevention 395
19.4.3 Disease Prevention 397
19.4.4 Other Applications 399
19.5 Contribution of Microbiology to Nanotechnology 402
19.6 AFM Imaging of Microorganisms 403
19.6.1 Yeast 403
19.6.2 Bacteria 404
19.6.3 AFM Study of the Structure–Function Relationship of the Biofilm-Forming Bacterium *Streptococcus mutans* 407
19.6.4 Viruses 407
19.7 Nanoplasmonic Sensors Detecting Live Viruses 407
19.8 Nanodentistry 410
19.8.1 The Impact of Nanotechnology 411
19.8.2 Nanotechnology in Periodontics 414
19.9 Conclusions 415
References 416

19.1 INTRODUCTION

Nanotechnology is the study, design, creation, synthesis, manipulation, and application of materials, devices, and systems at the nanometer scale (1 m consists of

Emerging Nanotechnologies in Dentistry. DOI: http://dx.doi.org/10.1016/B978-0-12-812291-4.00019-4

1 billion nanometers) [1]. Nanotechnology (sometimes shortened to "nanotech") is the study of manipulating matter on an atomic and molecular scale. Generally, nanotechnology deals with structures sized between 1 and 100 nm in at least one dimension and involves developing materials or devices within that size. Much of the study of nanoscience relates to the phenomenon of self-assembly in which nanoscale building blocks self-assemble to form complex structures. Microbiology relates to nanoscience at a number of levels. Many bacterial entities are nanomachines in nature, including molecular motors like *flagella* and *pili*. Bacteria also form biofilms by the process of self-assembly (e.g., the formation of Curli-film by *Escherichia coli*). The formation of aerial hyphae by bacteria and fungi is also directed by the controlled and ordered assembly of building blocks. In addition, the formation of virus capsids is a classical process of molecular recognition and self-assembly at the nanoscale. Another completely different aspect is the use of nanoordered bacterial assemblies for nanotechnology. Nanoscience impacts several areas of microbiology. It allows for the study and visualization at the molecular assembly level of a process and facilitates identification of molecular recognition and self-assembly motifs and the assessment of these processes. These are relevant to many microbial processes as mentioned above.

19.2 NANOMATERIALS

Nanomaterials have been categorized as those materials which have structured components with at least one dimension less than 100 nm. Materials that have *one dimension in the nanoscale* (and are extended in the other two dimensions) are layers, such as thin films or surface coatings. Some of the features on computer chips come under this category. Materials that are *nanoscale in two dimensions* (and extended in one dimension) include nanowires and nanotubes. Materials that *are nanoscale in three dimensions* are particles, e.g., precipitates, colloids, and quantum dots (tiny particles of semiconductor materials). Nanocrystalline materials, made up of nanometer-sized grains, also fall into this category. The nanomaterial field includes subfields which study or develop materials having unique properties arising from their nanoscale dimensions [2]:

- Interface and colloid science has given rise to many materials which may be useful in nanotechnology such as carbon nanotubes and other fullerenes and various nanoparticles and nanorods. Nanomaterials with fast ion transport are also related to nanoionics and nanoelectronics.
- Progress has been made in using these materials for medical applications.
- Nanoscale materials are sometimes used in solar cells which combat the cost of traditional silicon solar cells.
- Development of applications incorporating semiconductor nanoparticles to be used in the next generation of products, such as display technology, lighting, solar cells, and biological imaging.

19.2.1 APPLICATIONS OF NANOMATERIALS

Most current applications represent evolutionary developments of existing technologies, e.g., the reduction in size of electronics devices.

19.2.1.1 Sunscreens and cosmetics

Nanosized titanium dioxide and zinc oxide are currently used in some sunscreens, as they absorb and reflect ultraviolet (UV) rays and yet are transparent to visible light and so are more appealing to the consumer.

19.2.1.2 Composites

Nanoparticles and nanotubes are used in composites, materials that combine one or more separate components and which are designed to exhibit better overall properties than each of components.

19.2.1.3 Clays

Clays containing naturally occurring nanoparticles have long been important as construction materials and are undergoing continuous improvement. Clay–particle-based composites—containing plastics and nanosized flakes of clay—are also finding applications such as use in car bumpers.

19.2.1.4 Coatings and surfaces

Coatings with thickness controlled at the nano- or atomic scale have been in routine production for some time. Recent applications include the self-cleaning window, which is coated in highly activated titanium dioxide, engineered to be highly hydrophobic (water repellent) and antibacterial, and coatings based on nanoparticulate oxides that catalytically destroy chemical agents.

19.2.1.5 Tougher and harder cutting tools

Cutting tools made of nanocrystalline materials, such as tungsten carbide, tantalum carbide, and titanium carbide, are more wear and erosion resistant, and last longer than their conventional (large-grained) counterparts. They are finding applications in the drills used to bore holes in circuit boards.

19.3 BIOMEDICAL APPLICATIONS OF NANOTECHNOLOGY AND ITS LIMITATIONS

Three applications of nanotechnology in biomedicine [3] are focused on (1) targeted drug delivery, (2) diagnostic techniques, and (3) prostheses and implants. Interest is booming in biomedical applications for use outside the body, such as diagnostic sensors and "lab-on-a-chip" techniques, which are suitable for analyzing blood and other samples, and for inclusion in analytical instruments for R&D on new drugs. For inside the body, many companies are developing nanotechnology applications for anticancer drugs, implanted insulin pumps, and gene therapy. Other researchers are working on prostheses and implants that include nanostructured materials. Nanotechnology applications have not been marketed long enough for claims to be corroborated about risks to human health and environment. Still, small nanoparticles can enter the human body through pores and may accumulate in cells. The health effects of such nanoparticles are unknown. Historical experience with unintended consequences of technologies, such as drug resistance to antibiotics, or the persistence of chemicals, such as DDT in the environment, teaches us to take precautions.

19.4 NANOTECHNOLOGY APPLICATIONS IN DRUG-DELIVERY SYSTEMS, NANODIAGNOSTICS, AND VARIOUS OTHER FIELDS

It is envisaged that nanotechnology should let us make every manufactured product faster, lighter, stronger, smarter, safer, and cleaner. The following are some areas in which nanotechnology can have significant impact in the future.

19.4.1 DRUG-DELIVERY SYSTEM

19.4.1.1 Nanobots and its uses

Nanobots are robots that carry out a very specific function and are ~50–100 nm wide. They can be used very effectively for drug delivery. Normally, drugs work through the entire body before they reach the disease-affected area. Using nanotechnology, the drug can be targeted to a precise location which would make the drug much more effective and reduce the chances of possible side effects. Fig. 19.1 shows

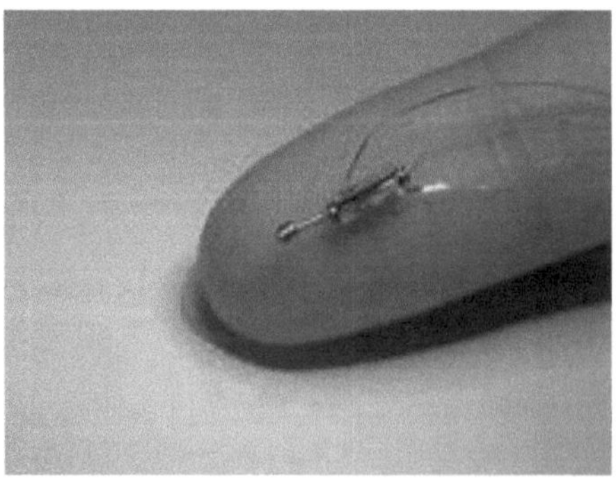

FIGURE 19.1

Device using nanobots for checking blood contents.

Amazing Nanobots (M.J. Perkel, The ups and downs of nanobiotech, The Scientist (2004) 1–8. http://www.the-scientist.com/2004/08/30/14/1, 12 November 2006).

a device that uses nanobots to monitor the sugar level in the blood [4]. Special sensor nanobots can be inserted into the blood under the skin where microchips, coated with human molecules and designed to emit an electrical impulse signal, monitor the sugar level in the blood.

The drug carriers have walls that are just 5–10 atoms thick and the inner drug-filled cell is usually 50–100 nm wide. When they detect signs of the disease, thin wires in their walls emit an electrical pulse which causes the walls to dissolve and the drug to be released. A great advantage of using nanobots for drug delivery is that the amount and time of drug release can be easily controlled by controlling the electrical pulse [5]. Furthermore, the walls dissolve easily and are therefore harmless to the body. Elan Pharmaceuticals has already started using this technology in their drugs Merck's Emend and Wyeth's Rapamune [6]. Nanomedicine could make use of these nanorobots (e.g., computational genes), introduced into the body, to repair or detect damages and infections. Using nanotechnology, the drug can be targeted to a precise location which would make the drug much more effective and reduce the chances of possible side effects. In the future, these nanorobots could actually be programed to repair specific diseased cells, functioning in a similar way to antibodies in our natural healing processes.

19.4.1.2 Use of nanorattles

Tang et al. [7] demonstrated that silica nanorattles (rattle-type nanoparticles consisting of a spherical shell encapsulating a freely moving core particle in solvent) show advantages for in vivo enhancement of therapy efficacy and reducing the systematic toxicity of antitumor drugs. The enhanced tumor inhibition of the docetaxel-loaded silica nanorattles may be attributed to the sustained docetaxel release from the nanorattles in vivo as well as the accumulation of drug-loaded nanorattles in the intratumor due to enhanced permeability and retention effect once intravenously administered (Fig. 19.2).

19.4.2 NANODIAGNOSTICS AND DISEASE PREVENTION

19.4.2.1 Biosensors

Nanodiagnostics utilizes biosensor technology, which is one of the most promising, compact systems consisting of a composite analysis of biological recognition element (DNA, protein, etc.). Detecting an analyte (glucose, antibiotics, etc.) using a transducer element or detector element to quantify the amount of analyte is the working principle of biosensors. The *transducer* or the *detector element* (works in a physicochemical way; optical, piezoelectric, electrochemical, etc.) transforms the signal resulting from the interaction of the analyte with the biological element into another signal (i.e., transducers) that can be more easily measured and quantified.

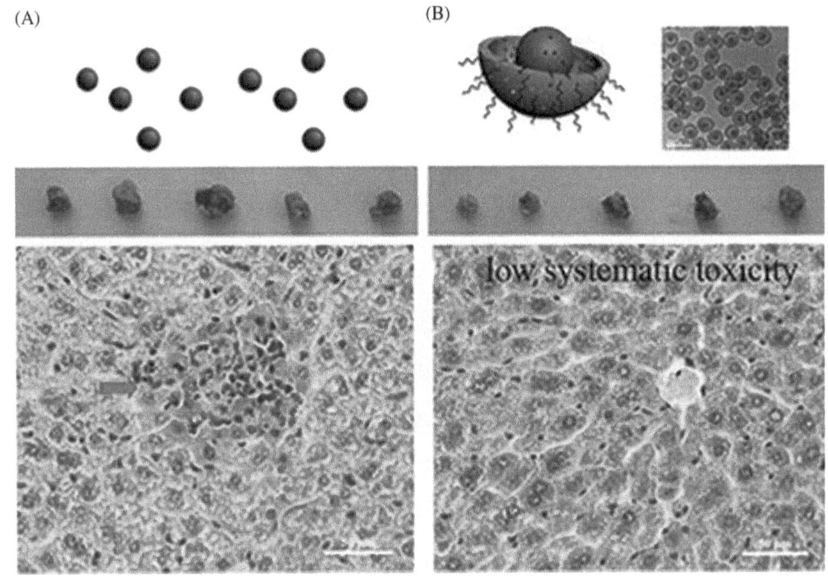

FIGURE 19.2

After encapsulation into the silica nanorattle, the antitumor drug docetaxel had increased therapy efficacy and decreased systematic toxicity for liver cancer therapy: (A) free drug and (B) drug-loaded silica nanorattle.

Tang Group, Chinese Academy of Sciences (L. Tan, D. Chen, H. Liu, F. Tang, A silica nanorattle with a mesoporous shell: an ideal nanoreactor for the preparation of tunable gold cores, Adv. Mater. 22 (2010) 4885–4889).

19.4.2.2 Diagnosis using nanobots

Nanobiotechnology scientists have successfully produced microchips that are coated with biological molecules. The chip is designed to emit an electrical impulse signal when the molecules detect signs of a disease. Special sensor nanobots can be inserted into the blood under the skin where they can check blood contents and warn of any possible diseases. They can also be used to monitor the sugar level in the blood. Advantages of using such nanobots are that they are very cheap to produce and easily portable [5].

19.4.2.3 Quantum dots

Quantum dots are nanomaterials that glow very brightly when illuminated by UV light. They can be coated with a material that makes the dots attach specifically to the molecule they want to track. Quantum dots bind themselves to proteins expressed in cancer cells, thus helping to visualize tumors [8] (Fig. 19.3).

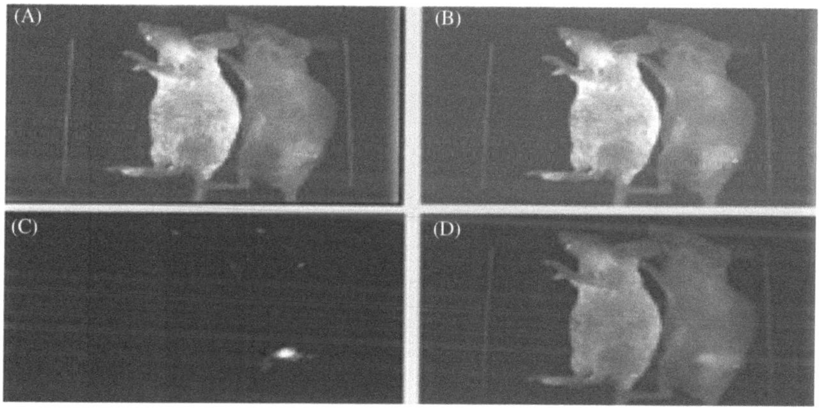

FIGURE 19.3

A light in dark places. Spectral imaging of quantum dots. White spot indicated with white arrow in 19.3D indicate a prostate tumor growing in a live mouse [4].

19.4.2.4 Regenerative medicine

Nanotechnology will also play an important role in the field of regenerative medicine. This area is associated with the use of pluripotent stem cells that can develop into other various types, so that could, at least in theory, replace tissue destroyed by diseases such as diabetes, ischemic heart disease, Alzheimer, Parkinson, spinal cord injuries, muscular dystrophy, retinal degeneration, and many more [9].

19.4.3 DISEASE PREVENTION

19.4.3.1 Cardiovascular interventions

It has been proposed that nanobots can also be used to prevent heart attack (cardiac failure) in the future. Cardiac failure is caused by fat deposits blocking the blood vessels. Nanobots can be directed to remove these fat deposits [5]. Fig. 19.4 shows the artists imagination of nanobots removing the yellow fat deposits on the inner side of blood vessels.

While nano- and microparticle-based imaging of cardiovascular interventions is still in its developing phase, it has already presented the exciting potential to monitor primary interventional procedures for precise therapeutic delivery, enhance the effectiveness of delivered therapeutics, and monitor therapeutic efficiency after interventions performed to treat cardiovascular diseases [10].

19.4.3.2 Nanoparticles and the blood–brain barrier: As treatment opportunity

Nanoparticles can be used as carrier systems to overcome the blood–brain barrier (BBB) and deliver specific medications to regions of the brain that would

FIGURE 19.4

Nanobots preventing heart attacks (artist's view) [5].

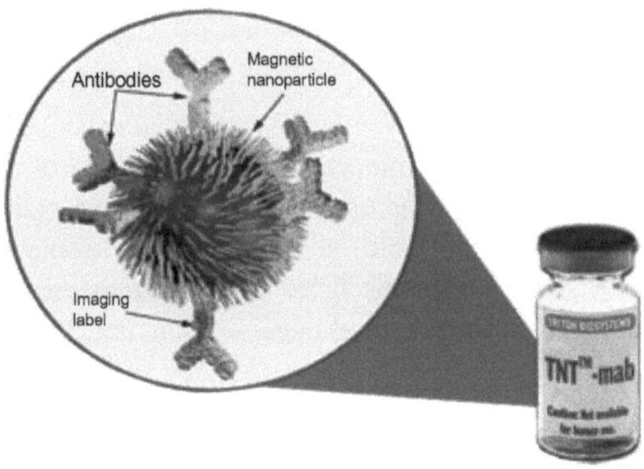

FIGURE 19.5

Cancer Cooker—Triton BioSystems is developing an anticancer therapy using antibody-coated iron nanoparticles [4].

normally be inaccessible. Their surfaces can be coated with certain materials or manufactured such that they can bypass the BBB and transport the drug to the sites where they are needed. One currently used form of therapy, although it does not overcome the BBB, is the so-called hyperthermia therapy, which uses nanoiron particles to treat brain tumors such as glioblastoma [11] (Fig. 19.5).

19.4.3.3 Tissue reconstruction

Nanoparticles can be designed with a structure very similar to that of bone. An ultrasound is performed on existing bone structures and then bone-like nanoparticles are created using the results of the ultrasound [3]. The bone-like nanoparticles are inserted into the body in a paste form [6]. When they arrive at the fractured bone site, they assemble themselves to form an ordered structure which later becomes part of the bone [6].

19.4.3.4 Medical tools

Nanodevices are nanoparticles that are created for the purpose of interacting with cells and tissues and carrying out very specific tasks [3]. The most famous nanodevices are the imaging tools. Oral pills can be taken that contain miniature cameras. These cameras can reach deep parts of the body and provide high-resolution pictures of cells as small as 1 μm in width. (A red blood cell is 7 μm wide [4].) This makes them very useful for diagnosis and also during surgeries. Fig. 19.6 shows such cameras working with other nanoparticles to get rid of a disease.

19.4.4 OTHER APPLICATIONS

19.4.4.1 Treatment of injured nerves

Another key application for nanoparticles is in the treatment of injured nerves. Samuel Stupp and John Kessler at Northwestern University in Chicago have studied on tiny rod-like nanofibers called *amphiphiles*. They are capped with amino acids and are known to spur the growth of neurons and prevent scar tissue formation. Experiments have shown that rat and mice with spinal injuries recovered when treated with these nanofibers [8].

19.4.4.2 Nanocapsules

A nanocapsule consists of shell and a space in which desired substances may be placed. Drug-filled nanocapsules can be covered with antibodies or cell surface

FIGURE 19.6

Miniature cameras inside blood vessels.

Blender Battles (M.J. Perkel, The ups and downs of nanobiotech, The Scientist (2004) 1–8. http://www.the-scientist.com/2004/08/30/14/1, 12 November 2006).

receptors that bind to cancer or various cells and release their biological compound on contact with that specific tissue [12]. Polymeric nanocapsules can now be made in specific sizes and shapes. These can then be functionalized by inserting molecules with a particular property onto the shell of nanocapsules. These molecules cause the release of the contents of the nanocapsule in response to a particular biomolecule which would be the triggering mechanism in a targeted drug-delivery system.

19.4.4.3 Nanotubes

Carbon nanotubes are tubes of graphite sheets with diameter in the nanoscale, including single wall carbon nanotubes and multiwall carbon nanotubes [12]. The ends of some nanotubes are open, the others are closed with full fullerene caps. Carbon nanotubes are named as the "king of nanomaterials." Nanotube drugs have been discovered that kill bacteria. These drugs are more effective than the traditional antibiotics. These nanotubes are about 3 nm in diameter and 6 nm in length. To make them into effective bactericides, they are attached with side chains of the amino acids that make up a tube.

19.4.4.4 Nanosomes

Nanosomes are capable of transporting active ingredients in a specific way and are used as nanoscale vehicles that penetrate into deep layers of skin to deliver vitamin E. Protein and hydrophobic drugs can be encapsulated within these nanosomes. Phospholipid nanosomes may find utility in the enhanced delivery of hydrophobic drugs such as recombinant proteins and nucleic acid as well as hydrophobic anticancer and anti-HIV drugs.

19.4.4.5 Nanowires

Nanowires are the artificial materials that consist of ultrafine wires or linear array of dots. Peptides rich in amino acid histidine have a high affinity for metal ions [13]. Histidines incorporated into nanowires could recruit metals to the surface of wire without attracting the reduced forms. Boron-doped silicon nanowires were used to create highly sensitive, real-time electrically based sensors for biological and chemical species [14].

19.4.4.6 Needle-free, painless vaccinations with nanopatches

Reporting their findings in a recent issue of *Small* ("Nanopatch-targeted skin vaccination against West Nile virus and chikungunya virus in mice"), Kendall et al. [15] describe the nanopatch approach for directly targeting vaccines to thousands of viable skin antigen presenting cells.

19.4.4.7 Nanomagnets remove pathogens from blood

Demonstrating a novel use of nanomagnets, researchers in Switzerland have rapidly and selectively removed heavy metal ions, overdosed steroid drugs, and proteins from human blood. Rapidly accessible surface of the nanomagnets enables

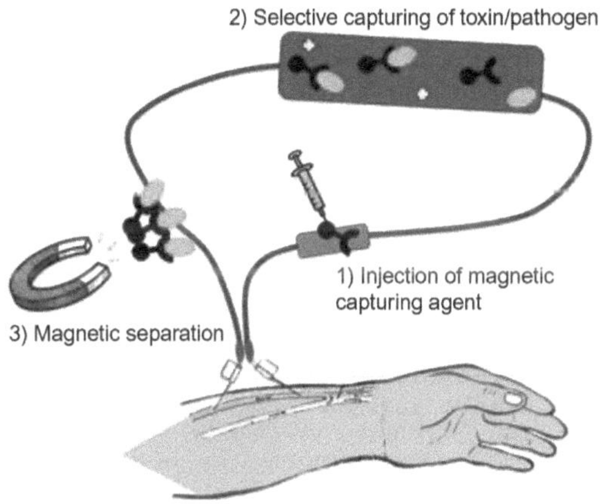

FIGURE 19.7

Schematic depiction of the process of pathogen removal from the blood using nanomagnets [16].

efficient adsorption in contrast to other blood purification techniques available on the market (Fig. 19.7).

19.4.4.8 Nanocrystalline silver

With most applications of nanotechnology in medicine still under development, nanocrystalline silver is already being used as an antimicrobial agent in treatment of wounds [17]. Silver works in a number of ways to disrupt critical functions in a microorganism. It has a high affinity for negatively charged side groups on biological molecules such as sulphydryl, carboxyl, phosphate, and other charged groups distributed throughout microbial cells. Silver attacks multiple sites within the cell to deactivate critical physiological functions such as cell wall synthesis, membrane transport, nucleic acid (RNA and DNA) synthesis and translation, protein folding and function, and electron transport. For certain bacteria, as little as one part per billion of silver may be effective in preventing cell growth [18]. Silver antimicrobial nanotechnology is effective against pathogens associated with biofilms including *E. coli*, *Streptococcus pneumoniae*, *Streptococcus pneumoniae*, *Streptococcus aureus*, and *Aspergillus niger* [19]. Local nanoscale characterization of cellular ultrastructure and functional properties is an important focus for probing cariogenic nature of oral microbes. The development of resistance to antimicrobial silver would be extremely rare because an organism would have to undergo simultaneous mutations in every critical function within a single generation to escape silver's influence. Silver is more efficient than traditional antibiotics because it is extremely active in small quantities, as little as one part per billion of silver may be effective in preventing cell growth.

19.4.4.9 Nanospheres

Nanospheres are hollow nanoscale structures made of polymers. These nanospheres can be loaded with special molecules like anticancer drugs. Injectable nanospheres have important potential applications such as site-specific delivery and medical imaging.

19.5 CONTRIBUTION OF MICROBIOLOGY TO NANOTECHNOLOGY

Considering microbiology in general, nanoscience could be used both to understand and control processes at the molecular assembly level. Just as bacteria were the first models to understand genetics and biochemistry that were later applied to eukaryotic systems, the same is true for the processes of function, recognition, and assembly. Micro (Greek) meaning very small, usually smaller than 1 mm, denotes 10^{-6}, so a micrometer is one millionth of a meter. Nano (Greek) dwarf—goes from small to tiny; the word "nano" means 10^{-9}. So, a nanometer is one billionth of a meter. If it is micro, it is almost nano; however, going down from the micro- to the nanoscale is not a simple matter of size reduction. The transition between the "small" and the "tiny" involves a radical change in the scientific concepts applicable. The laws of physics that govern the macroscopic world of our everyday experience can be readily used to understand microscopic objects. However, they break down completely in the nanoworld because of the radically different length-scale hierarchies and the underlying quantum mechanical behavior of individual atoms, electrons, or photons.

The term "microbiology" generally describes the study of those organisms invisible to the human eye, in particular yeast, bacteria, and viruses. However, these three types of organisms are significantly different from each other. Yeast and bacteria are different cell types (eukaryotic and prokaryotic, respectively), whereas a virus is not strictly a living organism, being an obligate intracellular parasite. The tools to conduct research in microbiology can be separated into two main fields—microscopy, which is required for visualization, and molecular biology, which has been used to characterize (in some cases comprehensively) the genetic and proteomic makeup of these organisms. The initial steps in opening the field of microbiology came with the advent of the first microscopes. Bacteria were first visualized by Antony van Leeuwenhoek, using a simple, self-built microscope, around 1676. The microscopes built by Leeuwenhoek were not compound microscopes, relying instead on a single lens, more like a very powerful magnifying glass. One of his first descriptions of bacteria (referred to as animalcules) was from samples scraped from the teeth of van Leeuwenhoek himself.

Advancements made in light microscopy by the combined work of Ernst Abbe and Carl Zeiss in the 1880s further extended the research of the microbiological world. However, as Abbe himself had described, there is a limit to the resolution

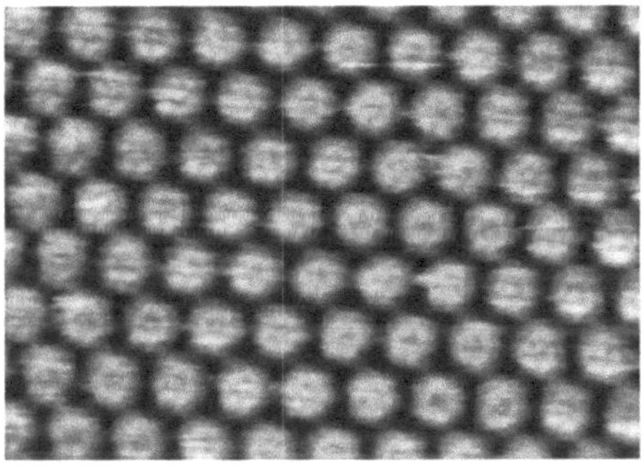

FIGURE 19.8

Hexagonally packed intermediate (HPI) layer from *Deinococcus radiodurans* [20].

of light microscopy, dependent on the wavelength of the illuminating light and the numerical aperture of the lens. In reality, the resolution of light microscopy is limited to half the wavelength of light, or around 250 nm. As such, the study of the structure of viruses had to wait for the advent of the electron microscope in 1931 by Ernst Ruska and the subsequent crystallization of the tobacco mosaic virus (TMV) in 1935 by Wendell Stanley (Fig. 19.8).

19.6 AFM IMAGING OF MICROORGANISMS

More recently, the atomic force microscope has opened a new path for the investigation and manipulation of structures on a very small scale. One of the most often cited advantages of the atomic force microscopy (AFM) in the study of biological structures is the fact that, unlike electron microscopy, high-resolution images can be obtained under physiological conditions. However, there is more to the AFM than just its capacity for high-resolution imaging. The mechanical nature of AFM means that the cantilever, used for imaging, can also be used to measure interaction forces in the piconewton range.

As such, not only can the AFM image the surface of microorganisms at high resolution, under physiological conditions, it can also be used to investigate the binding forces between microorganisms and target surfaces.

19.6.1 YEAST

Saccharomyces cerevisiae, also known as budding yeast, is not only of use in industrial processes from bread making to beer brewing, but it is also a type organism used in the study of eukaryotic cells. The yeast *S. cerevisiae* is surrounded by a cell wall composed of proteins, polysaccharides, and small amounts of chitin. Electron microscopy has shown that this cell wall is a layered structure

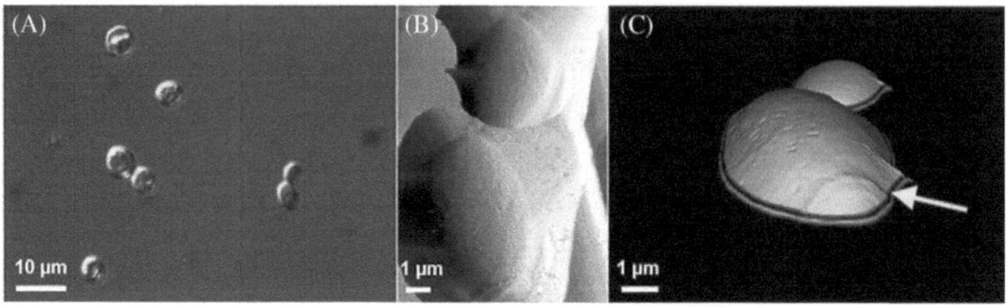

FIGURE 19.9

Imaging of *S. cerevisiae*. Yeast cells were located using DIG microscopy (A) and then imaged in contact mode in fluid with the AFM (B). In (C), a 3D image generated from the height channel is displayed, highlighting the bud scar on the mother cell (white arrow) [20].

ranging up to 300 nm thick. When imaged with AFM, the surface of the cell appears very smooth and is easily deformed, necessitating careful scanning at minimal force (Fig. 19.9). The only apparent surface feature is the bud scar at the opposite end of the mother cell to the newly forming daughter cell. The surface appears smooth as the sugars obscure the membrane.

19.6.2 BACTERIA

As AFM is a surface imaging technique, it has been used to characterize surface structures at high resolution. The most commonly used laboratory bacteria is *E. coli*, a Gram-negative bacteria. Gram-negative bacteria have a plasma membrane, surrounded by a periplasmic space in which there is a rigid but highly porous cell wall of peptidoglycan. This is then surrounded by an outer membrane, from which lipopolysaccharides of varying length extend.

Here, two strains of *E. coli*, DH5a and OP50, have been imaged. The images of DH5a show the classic, rod shape of many Gram-negative bacteria. The cells were scanned in air (Fig. 19.10) and in buffer (Fig. 19.11). When imaged in air, the surface of the bacteria appears highly patterned. In addition, a halo around the bacterium is apparent. These structures likely correspond to pill, which are found at the surface of *E. coli*. In contrast, when imaged in fluid (Fig. 19.11), the surface of the bacterium appears much smoother. In this case, this is due to the fact that the surface structures would be easily displaced by the movement of the tip during scanning, as they are not fixed in place.

The OP50 strain, while also *E. coli*, appears quite different. OP50 was originally isolated as a strain that could be used to feed *Caenorhabditis elegans*. It is a uracil-requiring strain that is more fragile and smaller than other *E. coli* strains. When imaged in air (Fig. 19.12), these bacteria do not exhibit the same structured

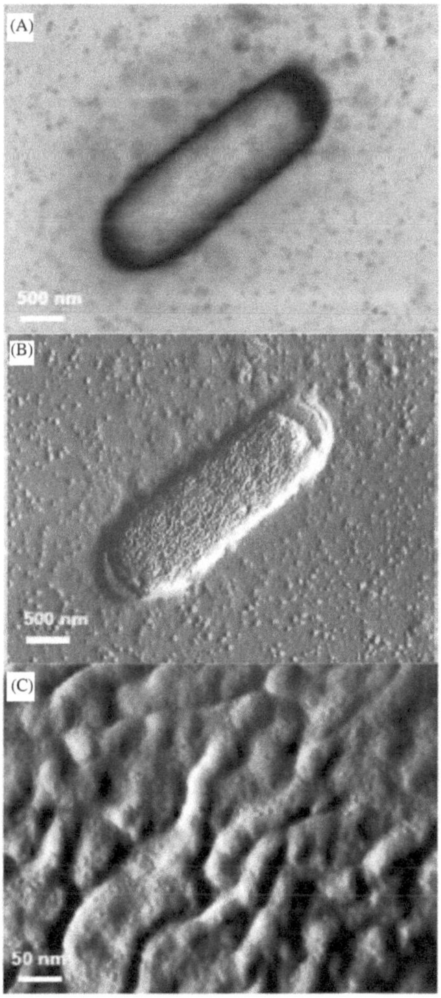

FIGURE 19.10

Intermittent contact mode images of DH5a cells. Overview height (A) and error signal (B) images, and a higher magnification error signal image (C) of the surface of the bacterium [20].

surface as seen for the DH5a. In addition, the cells are more fragile and must be carefully imaged to avoid removing them from the surface. In fluid, the surface is also less structured than that of DH5a, and regions of the cell surface are displaced in the scan direction, likely corresponding to the displacement of the sugars and other flexible structures at the surface of the cell.

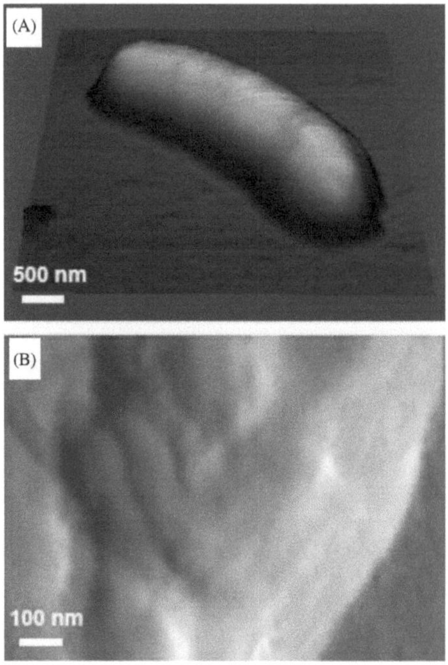

FIGURE 19.11

Intermittent contact mode images of DH5a in fluid. A 3D image generated from topographic data (A) and a higher magnification error signal image (B) are displayed [20].

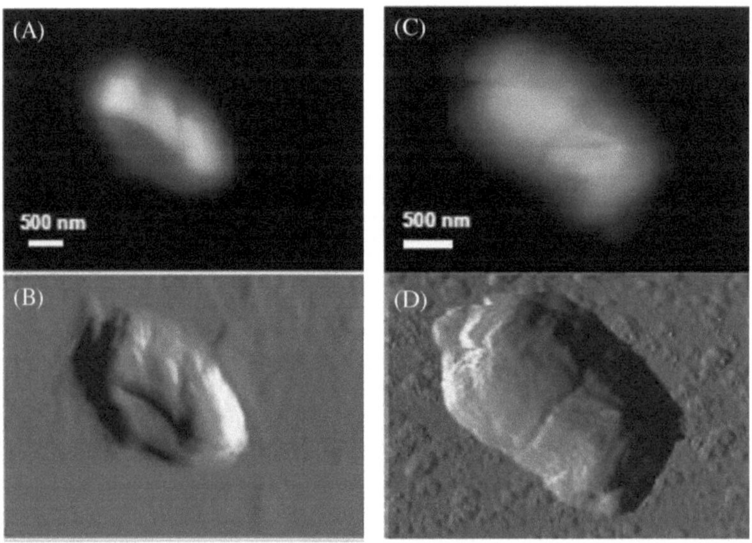

FIGURE 19.12

Images of OP50 bacteria imaged in air (A, topography; B, error signal) and fluid (C, topography; D, error signal) [20].

19.6.3 AFM STUDY OF THE STRUCTURE–FUNCTION RELATIONSHIP OF THE BIOFILM-FORMING BACTERIUM *STREPTOCOCCUS MUTANS*

AFM has garnered much interest in recent years for its ability to probe the structure, function, and cellular nanomechanics inherent to specific biological cells. In particular, AFM has been used to probe the important structure–function relationships of the bacterium *Streptococcus mutans*. *S. mutans* is the primary etiological agent in human dental caries (tooth decay), and it is of medical importance due to the virulence properties of these cells in biofilm initiation and formation, leading to increased tolerance to antibiotics. AFM has been used to characterize the unique surface structures of distinct mutants of *S. mutans*. These mutations are located in specific genes that encode surface proteins, thus using AFM characteristic surface features have been resolved for mutant strains compared to the wild type. Ultimately, characterization of surface morphology has shown distinct differences in the local properties displayed by various *S. mutans* strains on the nanoscale, which is imperative for understanding the collective properties of these cells in biofilm formation [19].

19.6.4 VIRUSES

Viruses are obligate intracellular parasites composed of an outer protein coat surrounding genetic material that consists of either DNA or RNA. This genetic material does not contain all the information required for replication; instead the virus needs to subvert the cellular machinery of the host cell to propagate. Viruses are extremely small, around 20–400 nm. This means that the characterization of viral structure requires high-resolution imaging techniques, such as AFM. Here, TMV has been imaged, the first virus imaged using electron microscopy.

The TMV particles were adsorbed to mica and imaged using intermittent contact mode. These virus particles are known as helical capsids, where the coat protein stacks in a helical pattern around the genetic material. This helical stacking can be seen in the height, phase, and error signal channels (Fig. 19.13). One significant advantage of using a BioAFM, such as the JPK Nanowizard, to image biological samples is that the imaging can be conducted in fluid. As such, virus particles can be imaged on the surface of their target cells in fluid. Here, influenza virus has been imaged, attached to the surface of red blood cells, in fluid, using intermittent contact mode. The virus particles are clearly imaged at the surface of the cells (Fig. 19.14).

19.7 NANOPLASMONIC SENSORS DETECTING LIVE VIRUSES

The recent emergence of H1N1 and H5N1 flu viruses and severe acute respiratory syndrome has highlighted the importance of rapid detection and accurate

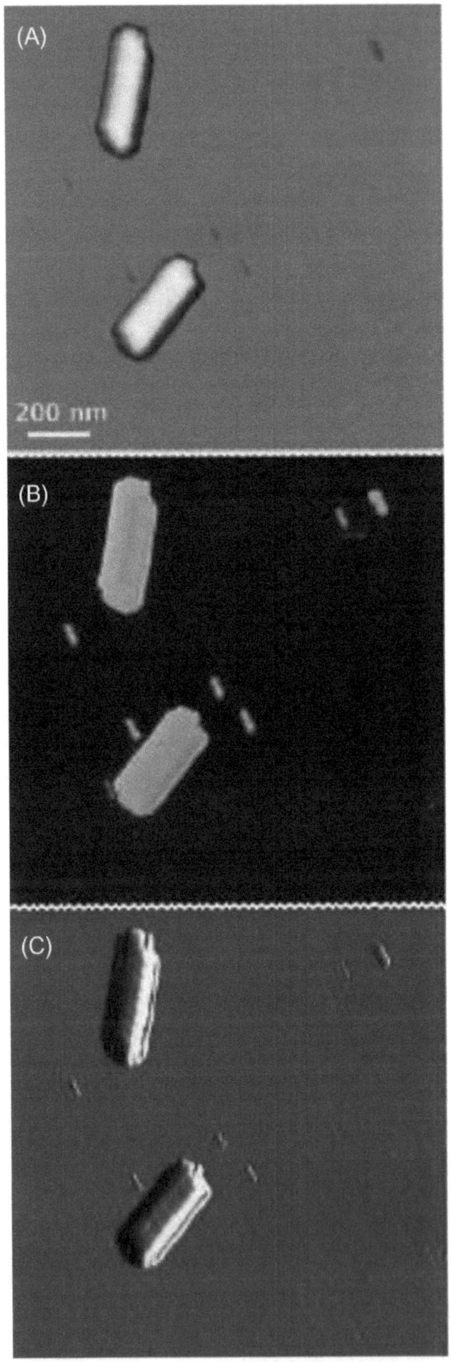

FIGURE 19.13

Intermittent contact mode images of TMV particles. Topography (A), phase (B), and error signal images (C) of the same sample region are presented [20].

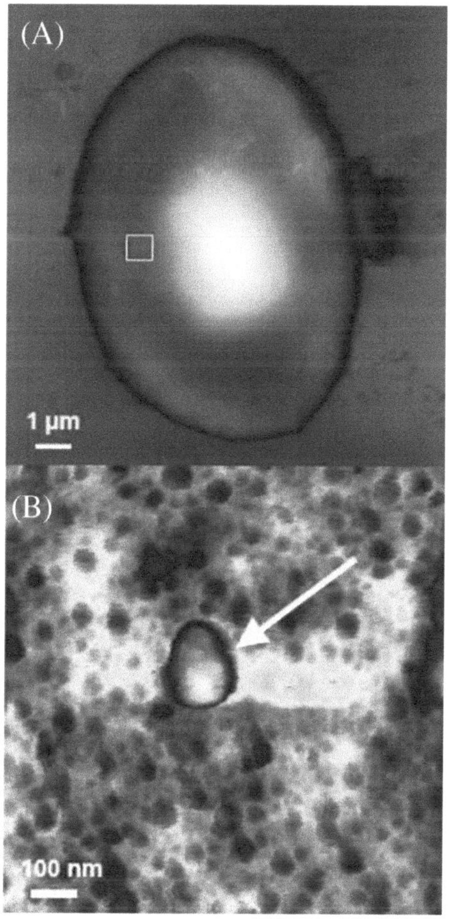

FIGURE 19.14

Influenza virus particles associated with the surface of an erythrocyte. The overview image (A) and higher magnification image (B) of the red blood cell surface show influenza virus particles (white arrow) [20].

diagnosis in health care and preventive medicine. The problem in these areas is that many virus detection platforms have limitations because they are not easily compatible with point-of-care use without the existence of significant infrastructure. Cell culturing is a time-consuming, highly specialized, and labor-intensive process. Therefore, highly sensitive/specific, compact, rapid, and easy to use virus diagnostics are needed to prevent further spread at the onset of a viral epidemic.

Unlike techniques based on external labeling, such resonance shifting operates as a reporter of the molecular binding phenomena in a label-free fashion and enables transduction of the capturing event directly to the far field optical signal.

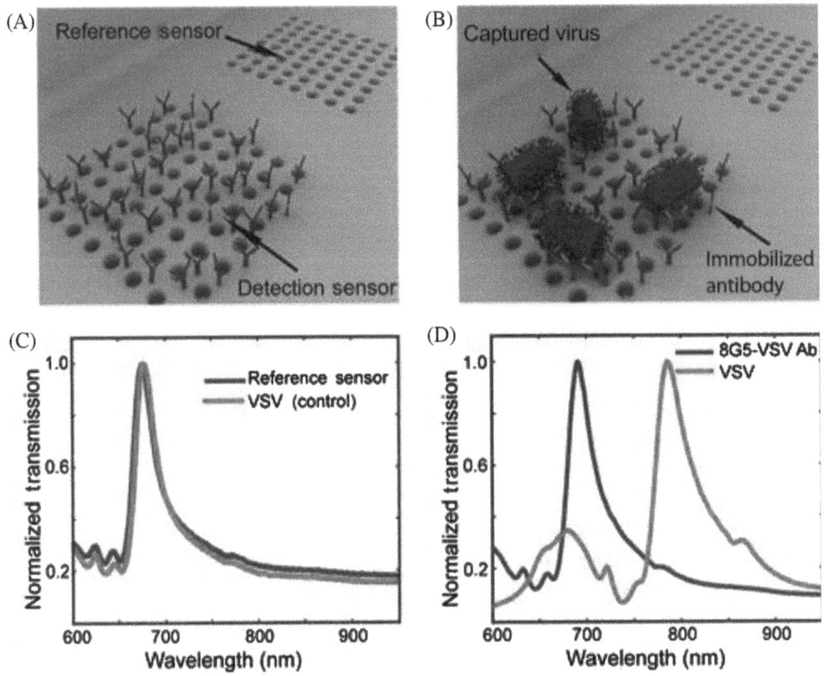

FIGURE 19.15

3D renderings (not drawn to scale) and the experimental measurements illustrate the detection scheme using optofluidic-plasmonic biosensors based on resonance transmissions due to extraordinary light transmission effect. Detection of the vesicular stomatitis virus (VSV) is experimentally observed with a strong red shifting of the plasmonic resonances.

Hatice Altug Research Group, Boston University (A.A. Yanik, M. Huang, O. Kamohara, A. Artar, T.W. Geisbert, J.H. Connor, An optofluidic nanoplasmonic biosensor for direct detection of live viruses from biological media, Nano Lett. 10 (12) (2010) 4962–4969).

Specific detection of viruses in a label-free fashion requires an effective method to distinguish nonspecific binding of the viruses to the plasmonic sensor surface. Selectivity is achieved by surface immobilized highly specific antiviral immunoglobulins showing strong affinity to the viral membrane proteins. Correspondingly, with the use of antibodies, viruses are specifically captured from a sample solution on the surface of our sensor [21]. This is illustrated in Fig. 19.15.

19.8 NANODENTISTRY

There are significant advances in dentistry that illuminated the road for the shift from "macro" to "nano" in dental sciences. It is evident that increases in the versatility of scientific knowledge and the ability to control physical processes at a finer resolution naturally lead to more information and, henceforth, to more questions. The broader our knowledge, the more amazement arises in face of natural wonders [22,23]. The same could certainly be said for the field of dentistry. The historic progress in this area naturally goes hand in hand with many new questions and challenges that provide opportunities for improvement. The progress, admittedly, has been slower than might be considered desirable for those who would wish to put a cutting-edge technology to clinical use. For example, early descriptions of the extraction of teeth with the use of forceps by Hippocrates and Aristotle date back to 500–300 BC, a technique that has remained essentially unchanged till date. Likewise, restorations with amalgam and gold date back to years 700 and 1746, respectively, and are still a part of our clinical setting without much change.

19.8.1 THE IMPACT OF NANOTECHNOLOGY

It is, on the other hand, valid to point out that nanotechnology slowly had made its way from the lab bench to any other technological or medical field. This is hampered not only by slow progress in understanding the basics in controlling the nanoscale phenomena, but by strict regulations in the translational stages too [24–26]. This, however, offers a controlled environment for the timely identification of weaknesses and strengths, which are all critical when it comes to introducing a new material or technology into the clinical setting. As we see, nanotechnologies have favored our understanding of dental tissues at the nanoscale and enabled the design of materials with ultrafine architecture. There is a prospect that probing the structure of dental tissues at ever finer size scales and using the dynamic resolution capabilities of advanced nanotools will give us answers to some of the puzzles that occupy dental researchers.

19.8.1.1 Dynamic view of dental tissues

The significant advances in dental research that revealed the nanoperspective of dental tissues have been explored over the past decade. When viewed at the nanoscale resolution, these structures, otherwise considered as static and stagnant, have fascinating structural dynamism.

Enamel is composed of 92–94 vol% fibrous apatite crystals with ~20 nm in diameter and ultrahigh length-to-width aspect ratio [27]. Precise spatial arrangement of these fibers gives rise to a superstructural organization with symmetry displayed at various size scales, ranging from nano to micro [28]. Morphogenesis of enamel proceeds through interaction between nanospheres or nanofibers of amelogenin, the main protein of the developing enamel matrix, and the growing crystals of apatite [29]. With the advancement of real-time nanoscale visualization techniques [30,31], there is a prospect that a clearer picture of these highly dynamic phenomena that govern the formation of dental and other biological tissues can be visualized. Fostering understanding of the dynamical nature of enamel should thus give rise to finer reparative techniques. For instance, understanding what molecules are able to pass through the biological barrier that enamel offers against the oral environment, as first investigated by Dibdin [32], might offer insights into drug-delivery techniques that can give guidance in reaching more advanced approaches for enamel tissue engineering and regeneration.

The dentin matrix is mainly composed of type I collagen fibrils with associated noncollagenous proteins, forming a three-dimensional organic scaffold that is reinforced by mineral. The mineral is a nanocrystalline hydroxyapatite that is partitioned according to its location with respect to the collagen fibrils into extrafibrillar mineral, which is located in the spaces separating the collagen fibrils [33–35] and intrafibrillar mineral, which is mainly in the gap regions of the fibrils extending between tropocollagen molecules [36–38]. It is noteworthy that the current concepts of restorative dentistry ignore most of the recent findings that elucidate the structure and function of dentin at a nanometer scale, which suggests that the modernity of the technologies currently used may be brought into question.

19.8.1.2 What are we really bonding to?

Hybrid layers, responsible for dentin bonding, form when adhesive comonomers infiltrate demineralized (i.e., acid-etched) dentin collagen fibrils. However, many studies have shown that nanoleakages, which are only around 20–100 nm wide [39,40] compared to 10–20-μm-wide microleakage gaps, occur within the hybrid layer even in the absence of gap formation. Although the spaces are too small to allow for bacterial penetration, they are large enough for enzymes and water to enter and cause degradation of resin/dentin bonds over time. The original interpretation of nanoleakage was that the silver dye used for the leakage studies occupied nanometer-sized spaces around naked collagen fibrils, where resin failed to infiltrate, or where residual water had not been displaced by the adhesive resin [41]. An improved understanding of the nanostructure of collagen fibrils in dentin would lead to reinterpretation of nanoleakage occurrence and may provide insights into the fundamental limitations of adhesives used in the restorative dental materials.

19.8.1.3 "Small is beautiful" of dental science: Small structures, great strength

Nanotechnology has been applied in dentistry in the early 1970s with the beginning of the era of microfills. Microfilled composites comprise silicon dioxide filler particles with less than 100 nm in diameter in conjunction with prepolymerized organic fillers, aggregated by crushing them into larger filler particles. Nowadays, the most commonly used resin composites, i.e., microhybrids and nanofilled composites, comprise filler particles ranging from ~20 to 600 nm. Other dental nanotechnologies rely on the delivery of molecules that facilitate tooth structure remineralization by means of noninvasive dental techniques that forestall caries, the latter being an active area of nanoresearch in dentistry. In agreement with Saunders [42], the most tempting venue for speculation on the next phase of nanorestoration of tooth structure is biomimetics, i.e., mimicking processes that occur in nature [43,44].

In the case of biomaterials, nanoparticulate materials appear to strongly influence the host response at both cellular and tissue levels, which makes nanotechnology particularly attractive for dental implants [45]. Processes such as sol–gel deposition [46], pulsed laser deposition [47], sputtering coating techniques [48,49], ion-beam-assisted deposition [50], and electrophoretic deposition [51] are a few examples of nanotechnology-based approaches that have been used to develop bioceramic thin-film coatings for implant surfaces. The main objective of these newer technologies is to reduce the thickness and the particle size of the coating layer and thereby increase its specific surface area and reactivity, thus improving the interaction with the surrounding living tissue.

19.8.1.4 Biofilm formation and treatment

Nanoscience has also recently promoted emerging concepts in oral microbial ecology, which may soon redefine our understanding of biofilm formation and treatment. Recent analyses with ribosomal RNA-based technologies have revealed the diversity of bacterial populations within dental biofilms and have highlighted their important contributions to oral health and disease [52]. In enamel, the aim appears to be unraveling the ways to mimic nature's own nanotechnological mechanism by which the cooperative interaction between the nanoscale self-assemblies of amelogenin and the uniaxially oriented apatite crystals proceeds [42,53]. Dentin, on the other hand, is linked to far more challenging scenarios. There appears to be a long and tortuous path to make a step from promising results to the actual transition of dental tissue engineering methodology from the lab to the clinical setting. The field of diagnostics of oral diseases is also a subject of rapid evolution. Proteomic analyses by mass spectrometry with their ability to identify proteins at ultralow concentration levels have a chance of drastically improving the diagnostic sensitivity and efficiency [54].

Saliva is now recognized as an excellent diagnostic medium for the detection of malignant tumors that are either within or are remote from the oral cavity [55]. Containing biomarkers for various diseases, the identification of which is currently under investigation, saliva holds great promises for early detection of disease and/or monitoring therapeutic outcomes through a noninvasive approach [56]. Other oral components, such as gingival crevicular fluid, epithelial cells, breath, and dental plaque, also have diagnostic potential [57]. The future of dentistry will thus undoubtedly witness routine and mechanistic restorations ceding place to a more holistic clinical practice where each particular case is analyzed in the context of the organism as a whole.

Finally, in parallel with the strong shift in the field of chemistry away from the traditional reference to strong, chemical bonding effects to the control of weak physicochemical interactions [58] (that has given rise to the prosperous practical framework of self-assembly and soft/wet chemistry [59]), a similar shift away from the mechanically interfering reparative methods toward soft remineralization techniques can present one of the most promising streams in the modern dental science. Despite the seemingly slow development of the dental field, we should keep in mind that scientific fields develop in waves. Computer science has rapidly expanded in the previous two decades or so, whereas the theoretical physics set the quantum mechanical fundaments for its slow subsequent development in only a few decades at the turn of the 20th century. Let us hope that one such big breaking wave is on the horizon for the world of dental science. In our opinion, to surf on that promising wave, learning the art and know-how offered by modern nanotechnologies will be a must.

19.8.2 NANOTECHNOLOGY IN PERIODONTICS

Some of the futuristic applications of nanotechnology in dentistry have been outlined as follows. Freitas has described how medical nanorobots might utilize specific motility mechanisms to crawl or swim through human body tissues with navigational precision, acquire energy, sense and manipulate their surroundings, achieve safe cytopenetration [60]. Functions may be controlled by an onboard nanocomputer executing programed instructions in response to local sensor stimuli. Alternatively, the dentist may issue strategic instructions by transmitting his orders directly to in vivo nanorobots via acoustic signals (e.g., ultrasound).

19.8.2.1 Local anesthesia and hypersensitivity cure

Colloidal suspension containing millions of active analgesic micron-size dental nanorobots will be instilled on the patient's gingiva. After contacting the surface of the crown or mucosa, the ambulating nanorobots will reach the dentin by migrating into the gingival sulcus and passing painlessly through the lamina propria or the 1–3-μm-thick layer of loose tissue at the cemento-dentinal junction. Upon reaching the dentin, nanorobots enter 1–4-μm-diameter dentinal tubule holes and proceed toward the pulp, guided by a combination of chemical gradients, temperature differentials, and positional navigation, all under onboard nanocomputer control. Assuming a ~10-mm total path length from tooth surface to pulp, a very modest nanorobot travel speed of 100 μm/s completes the journey into the pulp chamber in ~100 s. The presence of natural cells that are constantly in motion around and inside the teeth suggests that such journeys should be feasible.

Once installed in the pulp and having established control over nerve impulse traffic, the analgesic dental nanorobots may be commanded by the dentist to shut down all sensitivity in any particular tooth that may require treatment. When the dentist presses the icon for the desired tooth on the handheld controller display, the selected tooth immediately numbs. After the oral procedures are completed, the dentist orders the nanorobots (via the same acoustic data links) to restore all sensation, to relinquish control of nerve traffic, and to egress from the tooth by similar pathways used for ingress, followed by aspiration. Reconstructive dental nanorobots would selectively and precisely occlude selected tubules in minutes, using native biological materials, offering patients a quick and permanent cure.

19.8.2.2 Natural tooth maintenance and repair

The appearance and durability of tooth may be improved by replacing upper enamel layer with covalently bonded artificial materials such as sapphire or diamond, which have 20–100 times the hardness and strength of natural enamel. It can be made more fracture resistant possibly including embedded carbon nanotubes.

Major tooth repair may evolve through several stages of technological development:

- first using genetic engineering,
- tissue engineering, and
- later growing whole new teeth in vitro and installing them.

Ultimately, the nanorobotic manufacture and installation of a biologically autologous whole replacement tooth including both mineral and cellular components (e.g., complete dentition replacement therapy) should become feasible to undertake within the time and economic constraints of an ordinary office visit, using an affordable desktop manufacturing facility in the dentist's office.

19.8.2.3 Nanorobotic dentifrice (dentifrobots)

The futuristic proposals of dentifrobots include subocclusal-dwelling nanorobotic dentifrice delivered by mouthwash or toothpaste could patrol all supragingival and subgingival surfaces at least once a day, metabolizing trapped organic matter into harmless and odorless vapors and performing continuous calculus debridement (invisibly small (1–10 μm) dentifrobots, 103–105 nanodevices/oral cavity, and crawling at 1–10 μm/s).

It would be inexpensive purely mechanical devices which can safely deactivate themselves if swallowed and programed with strict occlusal avoidance protocols.

Properly configured dentifrobots could identify and destroy pathogenic bacteria residing in the plaque and elsewhere, while allowing the ~500 species of harmless oral microflora to flourish in a healthy ecosystem.

Dentifrobots could also provide a continuous barrier to halitosis, since bacterial putrefaction is the central metabolic process involved in oral malodor.

With this kind of daily dental care available from an early age, conventional tooth decay and gum disease will disappear into the annals of medical history.

19.9 CONCLUSIONS

Nanotechnological advances should be viewed in the context of other expected developments relevant to oral health in the coming decades. Nanodentistry faces many significant challenges in bringing its promises to fruition. It is hoped that dental nanorobots in due time will make fast, painless, and precision dentistry a reality. At the same time, continual refinement of traditional methods, development of advanced restorative materials, and new medications and pharmacologic approaches will continue to improve dental care. Nanotechnology is still in its early stages. A few applications discussed in this chapter have already been developed and are already helping patients all over the world. As further research continues in this field, more treatment methodologies will be discovered. Many diseases that do not have a cure today may be cured by nanotechnology in the future. Some of the concerns were also discussed, but with proper care these problems can be avoided. Scientists who are against the use of nanotechnology also agree that advancement in nanotechnology should continue because this field promises great benefits, but testing should be carried out to ensure the safety of the patients. In the near future, nanotechnology will one day become part of our everyday life and will help prevent a lot of dental pathological conditions.

REFERENCES

[1] K. Donaldson, V. Stone, Nanoscience fact versus fiction, ACM (2004) 113. Full Text (ABI/INFORM. ProQuest).

[2] A.J. Clarkson, D.A. Buckingham, A.J. Rogers, A.G. Blackman, C.R. Clark, Nanostructured ceramics in medical devices: applications and prospects, JOM 56 (10) (2004) 38–43. http://dx.doi.org/10.1007/s11837-004-0289-x. PMID 11196953.

[3] A. Silva Gabriel, Introduction to nanotechnology and its applications to medicine, Surg. Neurol. 61.3 (2004) 216–220. http://www.sciencedirect.com/view?cchp87-a2article.pdf (5 November 2006).

[4] M.J. Perkel, The ups and downs of nanobiotech, The Scientist (2004) 1–8. http://www.the-scientist.com/2004/08/30/14/1, 12 November 2006.

[5] H. Roman, Micro and nanotechnology—the next big tiny thing? Mercer Bus. (2005) 1–4. http://findarticles.com/p/articles/mi_qa3697/is_200501/ai_n9521342, 12 November 2006.

[6] R. Adhikari, Nanobiotechnology: will it deliver? Healthcare Purchasing News (2005) 1–3. http://findarticles.com/p/%20articles/mi_m0BPC/is_1_29/ai_n8708476, 12 November 2006.

[7] L. Tan, D. Chen, H. Liu, F. Tang, A silica nanorattle with a mesoporous shell: an ideal nanoreactor for the preparation of tunable gold cores, Adv. Mater. 22 (2010) 4885–4889.

[8] R. Weiss, Nanomedicine's promise is anything but tiny, Washington Post (2005) 1–3. http://www.washingtonpost.com/wp-dyn/articles/A49758-2005Jan30.html, 8 November 2006.

[9] D. Bhowmik, Chiranjib, K.K. Margret Chandira, K.P. Tripathi, Sampath Kumar, Nanomedicine—an overview, Int. J. PharmTech. Res. 2 (4) (2010). 2143–2151 CODEN (USA): IJPRIFISSN: 0974-4304.

[10] X. Yang, Nano- and microparticle-based imaging of cardiovascular interventions: overview, Radiology 243 (2) (2007) 340–347.

[11] M. Simkó, U. Fiedeler, A. Gazsó, M. Nentwich, 2008, Einfluss von Nanopartikeln auf zelluläre Funktionen. NanoTrust-Dossiers No. 007 epub.oeaw.ac.at/ita/nanotrust dossiers/dossier007.pdf.
[12] http://www.aejtudes.club.fr/nano/nanocapsules.htm.
[13] http://www.ntp.com.cn/products.htm.
[14] R. Djalali, Young-fou Cgcnand Hiroshi Matsui, J. Am. Chcm. Soc. (2002).
[15] http://www.nanowerk.com/spotlight/spotid = 17827.php.
[16] I.K. Herrmann, M. Urner, F.M. Koehler, M. Hasler, B. Roth-Z'Graggen, R.N. Grass, et al., Blood purification using functionalized core/shell nanomagnets, Small 6 (13) (2010) 1388–1392.
[17] B. Gibbins, L. Warner, The role of antimicrobial silver nanotechnology, MDDI 8 (2005).
[18] B. Gibbins, The antimicrobial benefits of silver and the relevance of microlattice technology, Ostomy Wound Manage. 49 (6) (2003) 5–6.
[19] Ryan J.M. III, Silver Antimicrobial Nanotech: An Alternative to Antibiotic Use [online] (Longomont, CO: Ionic Fusion Corp; 2005, cited April 15, 2005).
[20] http://jpk.com/jpk-app-microbiology1.download.9c9ea2d47essle9cf9814e975c5fd7df.pdf.
[21] A.A. Yanik, M. Huang, O. Kamohara, A. Artar, T.W. Geisbert, J.H. Connor, An optofluidic nanoplasmonic biosensor for direct detection of live viruses from biological media, Nano Lett. 10 (12) (2010) 4962–4969.
[22] V. Uskoković, On the light doves and learning on mistakes, Axiomathes 19 (2009) 17–50.
[23] V. Uskoković, On science of metaphors and the nature of systemic reasoning, World Futures 65 (2009) 241–269.
[24] C. Petrini, P. Vecchia, International statements and definitions of the precautionary principle, IEEE Tech. Soc. Mag. 1 (2002/2003) 4–7.
[25] G.H. Reynolds, Environmental regulation of nanotechnology: some preliminary observations, Env. Law Rep. 31 (2001) 10681–10688.
[26] A. Mnyusiwalla, A.S. Daar, P.A. Singer, "Mind the gap": science and ethics in nanotechnology, Nanotechnology 14 (2003) R9–R13.
[27] C.E. Smith, Cellular and chemical events during enamel maturation, Crit. Rev. Oral Biol. Med. 9 (1998) 128–161.
[28] J. Kirkham, S.J. Brookes, R.C. Shore, S.R. Wood, D.A. Smith, J. Zhang, et al., Physico-chemical properties of crystal surfaces in matrix–mineral interactions during mammalian biomineralisation, Curr. Opin. Coll. Inter. Sci. 7 (2002) 124–132.
[29] J. Moradian-Oldak, Amelogenins: assembly, processing and control of crystal morphology, Matrix Biol. 20 (2001) 293–305.
[30] H. Zheng, R.K. Smith, Y.W. Jun, C. Kisielowski, U. Dahmen, A.P. Alivisatos, Observation of single colloidal platinum nanocrystal growth trajectories, Science 324 (2009) 1309–1312.
[31] F.A. Scaramuzzo, R. Salvati, B. Paci, A. Generosi, V. Rossi-Albertini, A. Latini, et al., Nanoscale *in situ* morphological study of proteins immobilized on gold thin films, J. Phys. Chem. B 113 (2009) 15895–15899.
[32] G.H. Dibdin, The water in human dental enamel and its diffusional exchange measured by clearance of tritiated water from enamel slabs of varying thickness, Caries Res. 27 (1993) 81–86.

[33] E.P. Katz, S.T. Li, Structure and function of bone collagen fibrils, J. Mol. Biol. 80 (1973) 1–15.

[34] E.P. Katz, E. Wachtel, M. Yamauchi, G.L. Mechanic, The structure of mineralized collagen fibrils, Connect. Tissue Res. 21 (1989) 149–158.

[35] W.J. Landis, Mineral characterization in calcifying tissues: atomic, molecular and macromolecular perspectives, Connect. Tissue Res. 34 (1996) 239–246.

[36] W.J. Landis, M.J. Song, A. Leith, L. McEwen, B.F. Mc-Ewen, Mineral and organic matrix interaction in normally calcifying tendon visualized in three dimensions by high-voltage electron microscopic tomography and graphic image reconstruction, J. Struct. Biol. 110 (1993) 39–54.

[37] M. Balooch, S. Habelitz, J.H. Kinney, S.J. Marshall, G.W. Marshall, Mechanical properties of mineralized collagen fibrils as influenced by demineralization, J. Struct. Biol. 162 (2008) 404–410.

[38] I. Jager, P. Fratzl, Mineralized collagen fibrils: a mechanical model with a staggered arrangement of mineral particles, Biophys. J. 79 (2000) 1737–1746.

[39] H. Sano, M. Yoshiyama, S. Ebisu, M.F. Burrow, T. Takatsu, B. Ciucchi, et al., Comparative SEM and TEM observations of nanoleakage within the hybrid layer, Oper. Dent. 20 (1995) 160–167.

[40] H. Sano, T. Takatsu, B. Ciucchi, J.A. Horner, W.G. Matthews, D.H. Pashley, Nanoleakage: leakage within the hybrid layer, Oper. Dent. 20 (1995) 18–25.

[41] F.R. Tay, M. Hashimoto, D.H. Pashley, M.C. Peters, S.C. Lai, C.K. Yiu, et al., Aging affects two modes of nanoleakage expression in bonded dentin, J. Dent. Res. 82 (2003) 537–541.

[42] S.A. Saunders, Current practicality of nanotechnology in dentistry. Part 1: focus on nanocomposite restoratives and biomimetics, Clinic. Cosm. Invest. Dent. 1 (2009) 47–61.

[43] L.B. Gower, Biomimetic model systems for investigating the amorphous precursor pathway and its role in biomineralization, Chem. Rev. 108 (2008) 4551–4627.

[44] I.C. Gebeshuber, P. Gruber, M. Drack, A gaze into the crystal ball—biomimetics in the year 2059, Proc. IMechE. C J. Mech. Eng. Sci. 223 (2009) 2899–2918.

[45] Coelho, P.G.; Lemons, J.E. Bionanomaterials, in: Proceedings of the Bio Nano Conference and Trade Show, 8–12 May 2005, Anaheim, CA, USA, 2005, pp. 316–319.

[46] W.R. Lacefield, Current status of ceramic coatings for dental implants, Implant Dent. 7 (1998) 315–322.

[47] H. Kim, R.P. Camata, Y.K. Vohra, W.R. Lacefield, Control of phase composition in hydroxyapatite/tetracalcium phosphate biphasic thin coatings for biomedical applications, J. Mater. Sci. Mater. Med. 16 (2005) 961–966.

[48] S. Vercaigne, J.G. Wolke, I. Naert, J.A. Jansen, A mechanical evaluation of TiO_2-gritblasted and Ca–P magnetron sputter coated implants placed into the trabecular bone of the goat: Part 1, Clin. Oral Implants Res. 11 (2000) 305–313.

[49] S. Vercaigne, J.G. Wolke, I. Naert, J.A. Jansen, A histological evaluation of TiO_2-gritblasted and Ca–P magnetron sputter coated implants placed into the trabecular bone of the goat: Part 2, Clin. Oral Implants Res. 11 (2000) 314–324.

[50] C. Qin, R. D'Souza, J.Q. Feng, Dentin matrix protein 1(DMP1): new and important roles for biomineralization and phosphate homeostasis, J. Dent. Res. 86 (2007) 1134–1141.

[51] W.T. Butler, J.C. Brunn, C. Qin, Dentin extracellular matrix (ECM) proteins: comparison to bone ECM and contribution to dynamics of dentinogenesis, Connect. Tissue Res. 44 (2003) 171–178.

[52] S. Filoche, L. Wong, C.H. Sissons, Oral biofils: emerging concepts in microbial ecology, J. Dent. Res. 89 (2010) 8–18.

[53] V. Uskoković, M.K. Kim, W. Li, S. Habelitz, Enzymatic processing of amelogenin during continuous crystallization of apatite, J. Mater. Res. 23 (2008) 3184–3195.

[54] S. Hu, A.J. Loo, D.T. Wong, Human saliva proteome analysis and disease biomarker discovery, Expert. Rev. Proteomics 4 (2007) 531–538.

[55] L.R. Bigler, C.F. Streckfus, W.P. Dubinsky, Salivary biomarkers for the detection of malignant tumors that are remote from the oral cavity, Clin. Lab. Med. 29 (2009) 71–85.

[56] Y.H. Lee, D.T. Wong, Saliva: an emerging biofluid for early detection of diseases, Am. J. Dent. 22 (2009) 241–248.

[57] A.J. Ligtenberg, J.J. de Soet, E.C. Veerman, A.V. Amerongen, Oral diseases: from detection to diagnostics, Ann. N.Y. Acad. Sci. 1098 (2007) 200–203.

[58] V. Uskoković, Theoretical and practical aspects of colloid science and self-assembly phenomena revisited, Rev. Chem. Eng. 23 (2007) 301–372.

[59] M.V. Tirrell, A. Katz, Self-assembly in materials synthesis, MRS Bull. 30 (2005) 700–742.

[60] http://www.foresight.org/Nanomedicine/Nanorobotics.html.

CHAPTER

Neurotoxicity of nanomaterials

20

Feng Xiaoli and Shao Longquan

Southern Medical University, Guangzhou, China

CHAPTER OUTLINE

20.1 Introduction 421

20.2 Possible Dental Nanomaterials 423

20.2.1 Composite Resins and Bonding Systems 423

20.2.2 Root-Filling Materials 423

20.2.3 Bioceramics and Associated Dental Prosthesis 426

20.2.4 Surface Modifications for Dental Implants 426

20.2.5 Target-Delivery and Imaging in Tumor Chemotherapy 426

20.3 Neurotoxicity of Dental Nanomaterials 427

20.3.1 Possible Pathways into the CNS 427

20.3.2 In Vivo Toxicity Studies 429

20.3.3 In Vitro Toxicity Studies 430

20.3.4 Major Toxicity Mechanisms 435

20.4 Future Research Prospects 437

20.5 Conclusion 438

Acknowledgments 438

References 438

20.1 INTRODUCTION

Nanomaterials are defined as materials composed of unbound particles or as particles in an aggregate or agglomerate state with one or more external dimensions with a size ranging from 1 to 100 nm [1]. Such materials possess typical nanostructure-dependent properties (e.g., chemical, biological, optical, mechanical, and magnetic), which distinguish them from their bulk counterparts. Because of their new and unique properties, nanomaterials are becoming ubiquitous in various products, such as sunscreens, cosmetics, medical supplies, clothing, and building materials. The overwhelming increase for nanotechnology-related products has had major impacts on both society and the environment.

Emerging Nanotechnologies in Dentistry. DOI: http://dx.doi.org/10.1016/B978-0-12-812291-4.00020-0

In recent years, nanomaterials have been widely used in the production of dental materials, including light polymerization composite resins [2,3] and bonding systems, coating materials for dental implants [4], bioceramics, endodontic sealers [5], and mouthwashes [6]. However, in addition to yielding significant improvements in clinical treatments, the applications of dental nanomaterials have also created growing concerns regarding their biosecurity. Because the nanomaterials are similar in size to DNA molecules, proteins, viruses, and biological molecules, some of their biological effects may lie in the interaction mechanisms between living things and the environment. When reduced to the nanoscale, many benign materials may exhibit appreciable cellular toxicity. For example, TiO_2, a common substrate material for dental implants, was previously classified as being biologically inert in humans and animals and has been used as a negative control particle in a variety of toxicological studies. Nevertheless, several possible adverse effects of TiO_2 nanoparticles (NPs) on human health have been recently discovered [7,8] Additionally, in vitro data have also demonstrated the cellular toxicity of zinc oxide nanomaterials (nano-ZnO), which have been developed for numerous anti-infection applications [9].

After exposure, some nanomaterials readily travel throughout the body, deposit in target organs and trigger injurious responses. The brain is well protected by the blood–brain barrier (BBB), which can prevent the majority of substances from entering the brain. However, existing research has shown that nanomaterials have relatively easily crossed the BBB into the brain, and the crossing of the BBB by nanomaterials was attributed to their small sizes and high surface activities. Furthermore, these nanomaterials may even translocate into the brain by the olfactory and sensory nerves [10,11]. All of these findings have suggested that the central nervous system (CNS) could be damaged and a range of pathogenic effects may be experienced upon exposure to nanomaterials. Researchers have conducted many in vivo and in vitro studies to explore the interactions between the nanomaterials and biological macromolecules, cells, organs and tissues, and the majority of these studies have found that the effects of the biological toxicities of the nanomaterials may be induced by the mechanisms of oxidative stress, inflammatory reactions, interactions with mitochondria and immune mechanism [12–15]. However, one problem that has arisen is that whether the traditional methods and techniques utilized in the analysis of the toxicities of the nanomaterials are accurate and reliable. Further questions have arisen regarding whether the unique physicochemical properties of the nanomaterials have introduced new mechanisms of injury and whether these new mechanisms will lead to new pathologies. Even if the nanomaterials do not introduce new pathologies, there could be new, novel mechanisms of injury that require special tools, assays, and approaches to assess their toxicities [14]. Therefore, it is still too early to draw explicit conclusions regarding the inherent dangers of nanomaterials, let alone their exact mechanisms of toxicity.

This chapter is mainly intended to provide a detailed introduction of the applications of dental nanomaterials along with their potential neurotoxic effects.

Possible dental nanomaterials and their pathways into the CNS are stated first, and the neurotoxic effects and related mechanisms behind the in vitro and in vivo studies are further discussed. Finally, we will highlight the limitations of the current investigative methods and present some applications of dental nanomaterials in the future.

20.2 POSSIBLE DENTAL NANOMATERIALS

Alongside the industrialization process of nanotechnology, dental nanomaterials have been widely utilized and the opportunities for people to be exposed to nanomaterials have improved greatly. Major applications of nanomaterials in the dental field are described in this section (Table 20.1).

20.2.1 COMPOSITE RESINS AND BONDING SYSTEMS

The matrices of traditional composite resins have generally been comprised of various types of inorganic fillers. Applications of nanosized fillings in the resin matrices have overcome some of the mechanical limitations and have significantly improved their clinical performance. Commonly used nanomaterials include nano-ZnO [2,3], nanosilica [15,16], nanocalcium phosphate and calcium fluoride [nano-$Ca_3(PO_4)_2$ and CaF_2, respectively] [17], and nano-TiO_2 [18]. In addition to composite resins, the utilization of nanomaterials in dental adhesives has also effectively improved their bonding strengths and mechanical properties. For instance, POSS hybrid nanocomposites have polymerized with silicon-based nanomaterials to form a novel type of bonding material that possessed a large mechanical strength and good thermostability [19]. Furthermore, the antibacterial properties of the bonding agents could be greatly improved by the inclusion of nanosized silver and calcium phosphate [20]. Other possible additions have included nanohydroxyapatite (nano-HAp) [21] and nanosilica [22].

20.2.2 ROOT-FILLING MATERIALS

Root-canal-filling materials are supposed to effectively kill the bacteria in the periodical lesions, densely seal the apical zones and promote healing. However, the brittleness of the root canal often increases after treatment due to the large size of the traditional HAp. The mismatch of the elastic modulus between the root dentin and fillings has resulted in percolation. In contrast, nano-HAp represents a unique advantage in this aspect because its structure is similar to natural inorganic bone. Nano-HA was able to induce osteogenesis [23] and further improve the bacteriostatic and antibacterial effects of the root fillings [24]. Considering its good bioactivity, nano-HA was also used as an optimum replacement material in the repair of bone defects [25,26]. For example, Yang [27]

Table 20.1 Summary of the current nanomaterials in the dental field

Major Applications	Nanomaterial Type	Particle Size (nm)	Significant Characteristics	References
Composite resins	Nano-ZnO	125	Better antibacterial and mechanical properties	Kasraei et al. [2]
	Nanosilica	7	Better mechanical properties	Balos et al. [16]
		70		Wang et al. [19]
	Nano-$Ca_3(PO_4)_2$ and CaF_2	112 53	Better stress-bearing capabilities and the inhibition of caries	Xu et al. [17]
	Nano-TiO_2	<20	Improved microhardness and flexural strength	Xia et al. [18]
Dental adhesives	Nano-HAp	20–70	Better bonding strength to dentin	Wagner et al. [21]
	Nanosilver and calcium phosphate	<10 112	Improved antibacterial properties	Melo et al. [20]
	Nanosilica	7	High mechanical strength and good thermostability	Wang et al. [19] Habekost et al. [22]
Root-canal fillings	Nano-HAp	–	Better osteogenesis and improved bacteriostatic and antibacterial effects	Jallot et al. [23] Krisanapiboon et al. [24]
Bone-repair materials	Nano-HAp	100 20	Guiding the regeneration of periodontal and bone tissue	Huber et al. [26]
		3		Qi et al. [25] Yang et al. [27]
	Nano-ZrO_2/HAp composite	70–90/ 500–1000	Guiding bone reconstruction	An et al. [28]

Bioceramics	Nano-ZrO_2/Al_2O_3	–	Better resistance to crack propagation	De Aza et al. [29]
	Nanosilver	10	Increased fracture toughness and Vickers hardness	Uno et al. [30]
Silicone-elastomer material	Nano-Ti-, Zn-, and Ce-oxide	30–40 20 50	Improved mechanical properties	Han et al. [31]
Denture-base materials	Nanosilver	10–20	Better antifungal properties and biocompatibility	Acosta-Torres et al. [32]
Coating materials for dental implants	Nanoporous alumina	20–200	Good cell adhesion and no adverse effect on cell activity	Karlsson et al. [33]
	Nanozirconia/calcium phosphate	360/151	High bioactivity potential and good mechanical stability	Pardun et al. [34]
	Nano-ZnO	10–100	Better antimicrobial and biocompatible properties	Memarzadeh et al. [4
	Nano-Hap	–	Achieving rapid osseointegration	Uezono et al. [35]
Drug delivery	Nanosilica	150	Sustained and controlled release of anticancer drugs (as drug carriers)	Lelcold et al. [37]
	Polymeric NPs (vitamin E TPGS)	300–1000	Controlled release of anticancer drugs (as drug carriers)	Mu and Feng [38]
Tumor imaging	Superparamagnetic iron oxide NPs	82 ± 4.4	Good superparamagnetic and optical properties	Melancon et al. [39]

Abbreviations: TiO_2, *titanium dioxide;* ZrO_2, *zirconium oxide.*

demonstrated that nano-HAp-coated silk scaffolds effectively guided the regeneration of periodontal and bone tissue. Similarly, porous ZrO_2/HAp composite scaffolds were also reported to possess excellent mechanical properties and cellular/tissue compatibilities [28].

20.2.3 BIOCERAMICS AND ASSOCIATED DENTAL PROSTHESIS

Nanostructured bioceramics, which are constructed by a plasma coating or chemical deposition process, generally possess enhanced mechanical properties, such as a better resistance to crack propagation [29] and an increased fracture toughness and Vickers hardness [30]. Additionally, the utilization of nanosized Ti-, Zn-, Ce-oxide has greatly improved the mechanical properties of a maxillofacial silicone elastomer [31]. Nanosized silver may be an effective addition to denture-based materials to improve their antifungal properties and biocompatibility [32]. Thus, the applications of nanomaterials have the potential to effectively improve the comprehensive properties, including the mechanical, chemical and biological properties, of different types of conventional dental materials.

20.2.4 SURFACE MODIFICATIONS FOR DENTAL IMPLANTS

Good osseointegration at the implant–bone interface is essential for the success of dental implantation, but, unfortunately, this still remains a significant challenge. In recent years, a number of studies have reported the increased success rates of implants through the applications of a nanocoating on the surface, a nanoceramic and artificial nanobone materials. For example, nanoporous alumina [33], nanozirconia/calcium phosphate [34], nano-ZnO [4], and nano-HAp [35] have been utilized to increase the surface bioactivities of dental implants to achieve superior osseointegration. The advantages of the nanoscale modifications of dental implant surfaces have been presented in a recent review by Mendonça [36].

20.2.5 TARGET-DELIVERY AND IMAGING IN TUMOR CHEMOTHERAPY

A key problem in the use of chemotherapy for oral malignant tumors is how to improve the local concentrations of the drug while reducing the systemic side effects. To solve this problem, novel NP-based drug delivery strategies have been studied where the NPs are the drug carriers that can transport the anticancer drugs to the tumor sites, which further increases the therapeutic efficacy. For example, Lebold [37] applied mesoporous thin silica films with nanoscale pores as drug carriers that were incorporated with Doxorubicin, a widely used anticancer drug. The mesoporous silica nanomaterials demonstrated a sustained and controlled release of the anticancer drugs. Mu [38] discussed the advantages of manufacturing polymeric NPs (vitamin E TPGS) for the controlled release of paclitaxel and other anticancer drugs. Another application of targeted therapy with novel NPs involves tumor imaging. Superparamagnetic iron oxide with special surface modifications has been utilized to guide the laser ablation of maxillofacial cancer because these manufactured NPs are MR-active and can be selectively heated up for simultaneous imaging [39]. Similarly, liposomal nanocarriers also possess special advantages in their use for tumor radiography and imaging due to their good encapsulation of drugs and gadolinium [40].

Aside from those therapeutic uses, many nanomaterials, such as nano-TiO_2 and nano-ZnO, have been utilized in everyday dental items, including toothpastes and mouthwashes [6,41]. Considering the various applications of dental nanomaterials listed above, we should admit the outstanding contributions of nanomaterials to modern medicine. In the meantime, however, the risks of nanomaterials to human health have also significantly increased accompanied with more exposure opportunities.

20.3 NEUROTOXICITY OF DENTAL NANOMATERIALS

Based on the principles of toxicology, nearly all substances are potentially toxic to humans, and the key lies in the dose and method of exposure. The people who most likely are exposed to dental nanomaterials are the production, research, and development staff, as well as the dental staff and patients. In clinical situations, most of the dental nanomaterials were directly applied in the oral cavity or maxillofacial region, allowing the nanomaterials to easily enter into the bloodstream (or lymph fluid) via absorption through oral mucosa or through the digestive tract after swallowing. In addition, opportunities for exposure to nanomaterials may also occur through the utilization of dental tools containing NPs. Furthermore, many dental prosthetic materials also contain nanometals (e.g., Co, Cr, Au, Ag, Ti), resins (Si), and ceramics (e.g., Zr, Al, Li, Mg, Fe). Once these NPs are absorbed into our bloodstreams, they may distribute to the brain and further induce adverse effects.

20.3.1 POSSIBLE PATHWAYS INTO THE CNS

The brain is well protected by the BBB, which is mainly composed of the cerebrovascular endothelium sealed with tight junctions. Additional structures, such as pericytes, astrocyte end-feet, and a discontinuous basal membrane, are supportive cells to the BBB (Fig. 20.1). In addition to these physical barriers, the BBB also possesses some metabolic barriers to the delivery of therapeutic agents [42]. Another associated structure that serves to prevent potentially harmful substances from entering the brain is known as the blood–cerebral spinal fluid (CSF) barrier [43].

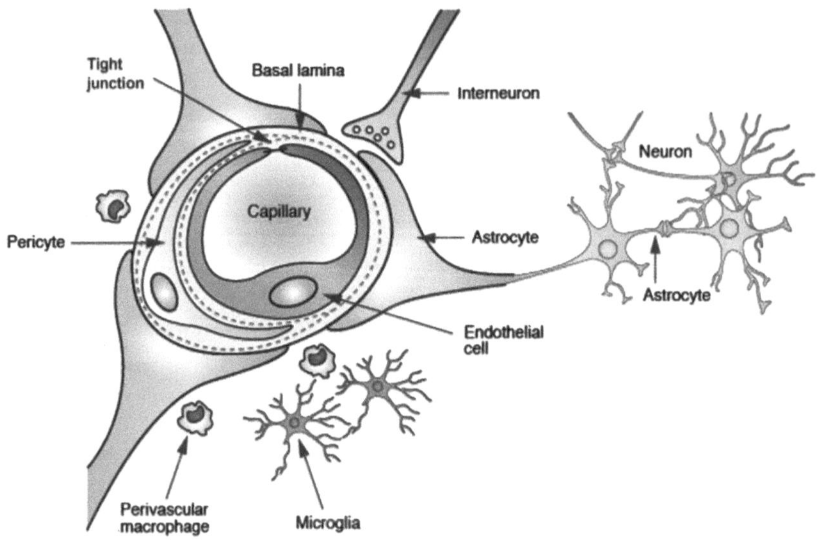

FIGURE 20.1

Schematic of the blood–brain barrier (BBB) and the associated components of the neurovascular unit (NVU).

Reprinted from Adv Drug Deliv Rev, 64(7), Chen Y, Liu L. Modern methods for delivery of drugs across the blood-brain barrier. 640–665. Copyright © 2012, with permission from Elsevier.

Despite these limitations, many studies have shown that nanomaterials, including nanomaterials that have been utilized as drug carriers, can cross the blood–brain (or blood–CSF) barrier into the CNS [44–48]. At present, the majority of the strategies described for the passage of nanomaterials (as drug carriers) through the BBB involve the mechanisms listed as follows (Fig. 20.2): (1) NPs help to enable drugs to penetrate the BBB by opening the tight junctions between the endothelial cells; (2) NPs are transcytosed through the endothelial cell layer; (3) NPs are endocytosed by the endothelial cells and release the drug inside the cell; (4) Coating agents of NPs, such as polysorbates, inhibit the transmembrane efflux systems; (5) NPs may induce local toxic effects on the brain vasculature, leading to a limited increase in the permeability of the neural endothelial cells [47,48]. Additionally, compared with adult brains, fetal brains may be more easily affected by blood-borne substances because the development of the BBBs in the fetal brains is incomplete [49]. Yamashita [50] found that nanosilica and nano-TiO_2 caused complications with the pregnancy when they were injected intravenously into pregnant mice. These NPs accumulated in the placenta, fetal brain and fetal liver. Other toxicity studies have also shown that the nanomaterials could cross the placental barriers in pregnant mice and thereafter cause neurotoxicity effects in the offspring [8,51].

In addition to the systemic pathways, nanomaterials can directly translocate to the brain via nerves. This pathway can be the major route for the CNS delivery of therapeutic agents following intranasal administration. There is a numerous amount

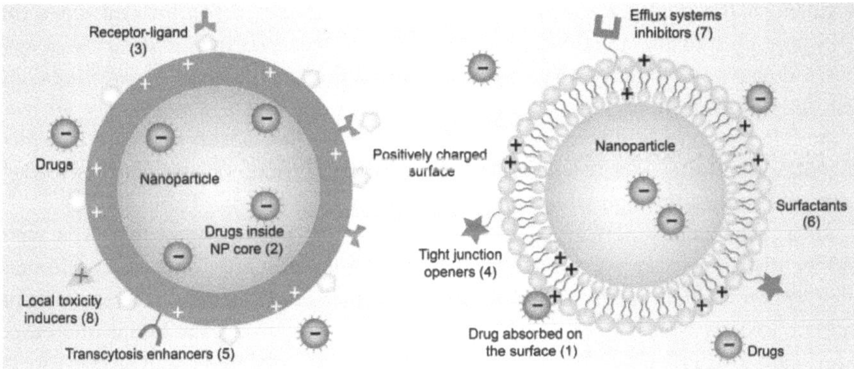

FIGURE 20.2

Schematic representation of multifunctional NPs as drug carriers across the BBB. Drugs can be adsorbed onto the surface of the NP due to the interactions between positive and negative charges (1) and may even be trapped inside its core (2). Different strategies have been applied for the transportation of drugs across the BBB: receptor-ligands for unique recognition and endocytosis (3); tight junction openers for improved intercellular penetration (4); transcytosis enhancers for the promotion of the transport of NP across the membranes (5); surfactants for the enhancement of membrane fluidization (6); efflux system inhibitors for the reduction of drug efflux (7); and local toxicity inducers for the increase in the permeability of the endothelial cells (8).

Redrawn from Expert Opin Drug Deliv. 6(6):553–5650. Copyright © 2009, Informa Healthcare.

of evidence that indicated that inhaled or intranasally instilled NPs can enter into the CNS via the olfactory epithelium and its associated neurons, pass directly into the olfactory lobes of the brain, and then induce significant inflammation-related effects [52–54]. Additionally, an often overlooked but important pathway connecting the nasal passages to the CNS involves the trigeminal nerve, which innervates the respiratory and olfactory epithelium of the nasal passages. Considering that a small portion of the trigeminal nerve also terminates in the olfactory bulbs, cross talk between the trigeminal and olfactory routes of the drug delivery to the brain may occur. Thus, it is often difficult to determine whether intranasally administered drugs reach the olfactory bulb and other rostral brain areas through the trigeminal or olfactory pathways or if both are involved [55].

20.3.2 IN VIVO TOXICITY STUDIES

In clinical situations, nanomaterials are able to enter the body via different routes. For instance, materials utilized in bone repair, periodontal dressing pastes, root canal pastes, and implant surface coatings can be visibly contaminated with blood, thus allowing NPs to rapidly enter the bloodstream in the short term and eventually reach the CNS via transport across the BBB. Furthermore, when materials are being mechanically ground and polished in the clinic or laboratory, particles of the dental restorative materials may spread in the air and enter the CNS via pulmonary inhalation or along the olfactory/trigeminal nerve translocation pathways. Finally, the abrasion and dissolving process of nanofillings and restorative materials and the usage of toothpastes and mouthwashes may all lead to the NP exposure in the oral mucosa, as well as further absorption in the gastrointestinal tract after swallowing. Thus, this part will focus on the main pathways in animal studies through which nanomaterials enter the body (Table 20.2).

In a recent study performed by Vilella [56], the author found that NPs were widely distributed across the regions of the brain 6 h after the intraperitoneal injection on C57Bl6 mice. This finding was further confirmed by the results from another toxicity study [57] in which zinc concentrations in rat brains increased significantly after an intravenous injection of a suspension of ZnO NPs for 14 consecutive days. In contrast, in Zhang's study [58], a significant up-regulation of Ti particles in the cerebral cortex and striatum was reported when mice were intranasally instilled with different types of TiO_2 NPs; this caused clear morphological changes in the neurons and created a significant disturbance in the monoamine neurotransmitter levels. In fact, nanomaterials with the same chemical compositions differed in their toxicological properties according to their shapes, sizes, surface charges, types of coating material and reactivities [59].

Li et al. [60] examined the potential systematic influences of TiO_2 NPs on mice following a 4-week intratracheal instillation. Their results indicated that TiO_2 NPs could transfer through the BBB and thereafter induce an injury to the brain by the activation of oxidative stress responses. In another toxicity study of TiO_2 NPs in mice performed by Ze [61], TiO_2 NPs were translocated and accumulated in the brain through nasal administration, and this led to an oxidative stress and a series of pathological changes, such as an over proliferation of the glial cells and hippocampal cell apoptosis. Similar findings were also reported by Kwon [62], who found that Al NPs exposure modulated the gene and protein expressions of mitogen-activated protein kinases and their activities.

A recent in vivo study [63] investigated the toxic effects of nano-TiO_2, nano-ZnO, and nano-Al_2O_3 in mice through oral exposure (500 mg/kg) for 21 consecutive days. These NPs produced a significant oxidative stress in the brain, as evident from the enhanced levels of reactive oxygen species (ROS) and the altered activities of the antioxidant enzymes. These observations were consistent with another set of research on the toxicity of TiO_2 NPs that was conducted 2 years earlier by Huerta-Garcia [64]. Similar studies of the oxidative stresses that are associated with selective gene expression analyses and immunological biomarkers would further improve our understanding of the mechanisms of neuroinflammation and neurodegeneration associated with NPs [65].

20.3.3 IN VITRO TOXICITY STUDIES

In the brain, the NPs usually contact with three different cell types (Table 20.3) [66]: (1) BBB (specialized endothelium) and/or blood–liquid barrier cells (CSF);

Table 20.2 Neurotoxicity effects of dental nanomaterials under in vivo conditions

Drug Delivery Routes	Nanomaterial Type	Particle Size (nm)	Exposure Model	Neurotoxicity Effects	References
Intravenous injection	Nano-ZnO	30–40	Wistar rats	Increased brain zinc concentrations	Amara et al. [57]
Intraperitoneal injection	G7 NPs	170 ± 18	C57Bl6 mice	Accumulation in brain areas and special cell populations	Vilella et al. [56]
intranasal instillation	Nano-TiO_2	40 50	ICR mice	Morphological changes of neurons Disturbance of monoamine neurotransmitter levels	Zhang et al. [58]
Intratracheal instillation	Nano-TiO_2	3	Kunming mice	Brain injury by inducing oxidative stress responses	Li et al. [60]
Nasal administration	Nano-TiO_2	5–6	ICR mice	Oxidative stress in brain Pathological changes of brain cells; Changes in genes associated with brain toxicity	Ze et al. [61]
Nasal instillation	Al NPs	5–100	SD rats	Alteration of gene and protein expression of MAPK and its activity	Kwon et al. [62]
Oral exposure	Nano-TiO_2 Nano-ZnO Nano-Al_2O_3	<75 <100 45	Swiss albino mice	Oxidative stress in brain	Shrivastava et al. [63]
Intragastric administration	Nano-TiO_2	6–7	ICR mice	Accumulation of ROS in the hippocampus	Hu et al. (2011)

Abbreviations: NPs, *nanoparticles;* G7 NPs, *glycopeptides-modified poly-lactide-co-glycolide NPs;* MAPK, *mitogen-activated protein kinases;* ROS, *reactive oxygen species.*

Table 20.3 In vitro studies of dental nanomaterials with different cell models

Exposure Model	Nanomaterial Type	Particle Size	Cell Type	Main Research Contents	References
The BBB	Nanosilica	50 100 200	hCMEC/D3 cells	Cellular accumulation and transcytosis conditions of NPs in brain	Ye et al. [68]
	Nano-TiO_2	21 nm core	HCECs	Cellular uptake and localization, generation of oxidative stress, and DNA-damaging effects in HCECs	Kenzaoui et al. [70,96]
	PBCA NPs	146 nm	Human BMEC	Cellular uptake and localization	Weiss et al. [69]
	Silica NPs	30 nm 100 nm 400 nm	Rat BMEC + pericytes	Effects of NPs with different size and surface modifications on brain permeability	Hanada et al. (2014)
	Nano-TiO_2	25.2 nm	Rat BECs + glial cells	Effects on the integrity of the BBB and NP-induced inflammatory responses of BECs	Brun et al. [71]
	Magnetic NPs	60 nm	Human BMVEC + AM	Nanocarriers for targeting BDNF across the BBB	Sudheesh et al. (2013)
	Glucose-coated gold NPs	2 nm core	Peripheral vascular endothelial cells + AM	Transfer rate of NPs across the BBB	Gromnicova et al. (2013)
Neuroglial cells	Nano-ZnO	45 nm	Rat astrocytes	Signaling pathways involved in NP-induced apoptosis in primary astrocytes	Wang et al. [19]
	Nano-TiO_2	<25 nm	Human U87 astrocyte	Cytotoxicity of NPs	Lai et al. (2008)
	Si NPs	150–200 nm	Rat microglia cells	Alterations of microglia functions	Choi et al. [15]
	Nano-TiO_2	<50 nm	Human glial cell U373 and Rat glial cell C6	NP-induced oxidative stress and mitochondrial damage in glial cells	Elizabeth et al. (2014)

Neurons	Nano-TiO_2	21 nm	Rat PC12 cells	NP-induced oxidative stress and apoptosis in neurons	Liu et al. [77,83]
	Nano-TiO_2	25 nm	SHSY5Y cells	Cytotoxic and genotoxic effects of NPs on neuronal cells	Valdiglesias et al. [78,79]
	Nano-ZnO	30–50 nm	Rat primary neuronal cells	NP-induced cytotoxicity and DNA damage in neurons	Chiang et al. [80]
	Nano-Ag	50–100 nm	Hippocampal CA1 neurons	Alterations of the action potential of neurons	Liu et al. [83]

Abbreviations: BBB, *blood–brain barrier;* NPs, *nanoparticles;* hCMEC/D3, *human cerebral microvascular endothelial cell line;* HCECs, *human cerebral endothelial cells;* PBCA, *poly(*n*-butylcyanoacrylate);* BECs, *brain endothelial cells;* BMEC/BMVEC, *brain microvascular endothelial cells;* AM, *human astrocytes;* BDNF, *brain-derived neurotropic factor;* C6, *rat brain glial tumor cell line;* U373, *human astrocytoma cell line grade III;* PC12 cells, *rat pheochromocytoma cells;* SHSY5Y cells, *human neuroblastoma SHSY5Y cell line.*

(2) glial cells or neuroglia; (3) two general types of neurons (with (white matter) or without (gray matter) a myelin sheath).

In many situations, the microvascular endothelial cells of the human brain are used as an in vitro BBB model, such as hCMEC/D3 cells [67], human brain microvascular endothelial cells (BMECs) [68], and human cerebral endothelial cells [69]. Rats are another common experimental animal due to their availability on resources and pathological models [70,71]. In addition, the coculture models have also been widely utilized in in vitro BBB models including [72,73]: (1) BMECs + astrocytes and (2) peripheral vascular endothelial cells + astrocytes. Once entering the endothelial cell monolayer, NPs may accumulate along the endo-lysosomal pathway, which affected the normal morphology and function of the BBB itself. For example, Brun [71] observed an accumulation of TiO_2 NPs in the endothelial cells of the brain by using an in vitro cell-based rat BBB model. An intense inflammatory response associated with a modulation of the endothelial cell functioning of the brain was also observed.

Among all of the neuroglial cells, astrocytes and microglia have received the most attention. The astrocyte is thought to induce the barrier phenotype of cerebrovascular endothelial cells during development through the release of soluble factors, such as vascular endothelial growth factor. Recently, the ZnO NP-astrocyte interaction was reported to induce an oxidative stress that could trigger cell apoptosis by activating the c-Jun N-terminal kinase (JNK) signaling pathway in cultured primary astrocytes [74]. A similar finding was observed in the interactions between superparamagnetic iron oxide NPs and astrocytes [75]. As the sentinels of the CNS, microglia are the first cells to respond to a disruption of the brain homeostasis and the entry of foreign particles or infectious agents. Once activated, microglia can generate ROS and reactive nitrogen species (RNS) and even elicit an inflammatory response. In most cases, the macrophage cell line was adopted to evaluate its activation and inflammatory reaction. In a neurotoxicity study of Si NPs performed by Choi [15], even low levels of NPs were capable of increasing ROS and RNS production and inducing cytokine release. These changes had an adverse effect on the microglial function and surrounding neurons.

Neuronal cell lines commonly used for in vitro studies include (1) rat PC12 neuronal cells [76,77], which have been derived from a pheochromocytoma of the rat adrenal medulla; (2) a human SHSY5Y neuroblastoma cell line [78,79], which is perceived as an appropriate cell model for the assessment of neurotoxicity because it possesses many biochemical and functional properties of neurons [80]. Additionally, primary culture cell lines have also been used in the evaluation of the neurotoxicity of NPs, including rat dopaminergic neurons (N27), rat primary neuronal cells [81], and hippocampal CA1 and CA3 neurons [82,83]. It has now been confirmed that some nanomaterials can exploit the endocytotic pathways both to cross the BBB endothelium in vivo and to enter the neurons or glial cells in vitro [84]. For instance, Vilella [56] discovered that there was an uptake of NPs in hippocampal neurons that were prepared from rats at embryonic day 18 or 19. Aside from intracellular accumulation, there was also evidence that different metal-oxide NPs affect the membrane potentials of neurons and increase the neuronal firing rate by changing the responses of the potassium channels [85]. This finding was consistent with a toxicity study of nano-CuO on CA1 hippocampal neurons performed by Xu [86]. Furthermore, this toxic effect may have a physiological impact on animal behavior, which was demonstrated in rats by testing their spatial cognition capabilities [87].

20.3.4 MAJOR TOXICITY MECHANISMS

Because of their small size, NPs can easily reach the brain and are taken up by the brain cells. After interacting with NPs, various mediators released from the microglia and astrocyte will induce inflammation, apoptosis, and oxidative stress in the brain. Considering nervous tissue regeneration is limited, most of the nerve damage is irreversible. From this perspective, it is critical to investigate the mechanisms of nanoneurotoxicity. The possible mechanisms discussed in this chapter are the most commonly proposed in the current literature and include oxidative stress, immune system dysfunction, and autophagy dysfunction.

ROS are capable of inducing membrane damage that affects the structure and function of inner proteins, lipid denaturation, and structural alteration of DNA. DNA oxidation is of the utmost concern because it can lead to mutations and alterations in gene expression. Due to the lack of DNA repair enzymes, mitochondrial DNA appears to be more susceptible to ROS-induced mutations. Protein oxidation may cause the aggregation of insoluble protein, which serves as the molecular basis of some diseases, especially neurodegenerative pathologies [88]. There is an increasing amount of evidence indicating that NPs can cause cytotoxicity due to their large surface area, which is capable of inducing ROS, a key factor that generates cellular stress and disease pathogenesis [89]. However, the impact of NP-induced ROS on the CNS is unclear. Several types of NMs, such as quantum dots and metal-oxide NPs, have been shown to cause the overexpression of intracellular ROS [90]. NPs can provoke oxidative stress through multiple interactions, as reviewed in Fig. 20.3.

Once NPs enter the CNS, they immediately encounter a complex environment of resident microglial immune cells and neurons. As macrophages located in the CNS, the microglial cells are the most likely to respond when NPs enter the brain. Once activated, they are able to change in form and function. Choi et al. [15] observed the microglial uptake of silica-based NPs (Si NPs) using transmission electron microscope (TEM) and fluorescence confocal microscopy. The results demonstrated that low levels of Si NPs are able to alter microglial function by changing the expression of proinflammatory genes and cytokine release. The appropriate response of microglia to various stimuli is thought to be necessary to maintain normal brain function. However, excessively activated or uncontrollable microglia can cause nerve toxicity by inducing proinflammatory factors, such as interleukin-1β, tumor necrosis factor (TNF)-α, and interferon-γ. Gualtieri et al. [91] found that various nanosized organic carbon particles induced cytotoxic and proinflammatory effects on the in vitro

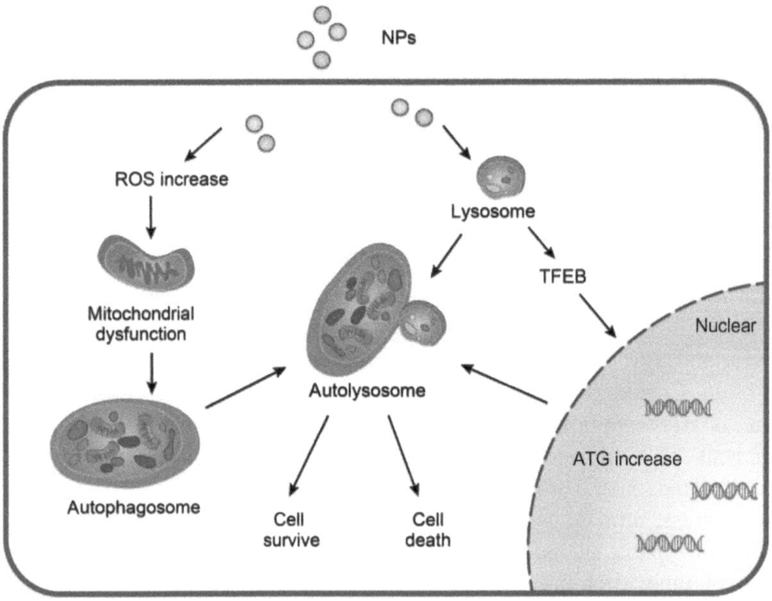

FIGURE 20.3

Overview of the mechanistic steps of NP-induced autophagy. NPs induce autophagy in two ways: ROS-dependent autophagy and lysosome-dependent autophagy. During autophagy, a phagophore is created and it then elongates into a double-membrane autophagosome while sequestering cytoplasmic material. This autophagosome then fuses with a lysosome, resulting in an autolysosome. The enzymes present in the autolysosome lumen eventually degrade the inner membrane and autophagic cargo, thus providing macromolecules that can be transported into the cytosol via permeases. TFeB is a transcription factor that translocates to the nucleus upon activation (similar to dysfunction of the lysosome), where it promotes the transcription of the lysosomal and autophagic genes. Proper stimulation will help cells survive; however, under- and overstimulation of autophagy will lead to cell death.

systems by releasing interleukins 8 and 6. Both TiO_2 NPs and HAp NPs can induce marked inducible nitric oxide synthase expression, resulting in nitric oxide release from the microglia after direct exposure to NPs [92].

Autophagy, a highly regulated cellular process for eliminating long-lived proteins and damaged organelle components via the lysosomal mechanism, has received considerable attention. Increasing evidence shows that autophagy alterations may lie at the root of neurodegenerative diseases, such as alzheimer's disease (AD), parkinson's disease (PD), and huntington's disease (HD) [93,94]. The cell autophagy process can be induced by various forms of stress, such as interactions with various types of NPs. Thus, NPs are considered a novel class of autophagy inducers [95]. A growing number of reports have confirmed that NPs can induce and modulate autophagy in different cells. The autophagy process is induced by uncoated and oleic acid-coated IO NPs, Si NPs, TiO_2 NPs, and polymeric NPs in human brain-derived endothelial cells. This process has been found to be correlated with an increase in DNA strand breaks and defensive mechanisms [96]. In addition, Roy et al. [97] found that ZnO NP exposure induced oxidative stress in macrophages and initiated autophagy and apoptosis simultaneously, suggesting that autophagosomes may be a cellular defense mechanism against oxidative stress. Moreover, autophagy induced by ZnO NPs followed a PI3K/mTOR/Akt signaling cascade. According to the current literature, NPs induce autophagy in two ways: NP-mediated ROS-dependent autophagy and NP-mediated lysosome-dependent autophagy.

20.4 FUTURE RESEARCH PROSPECTS

Nowadays nanotechnology and nanomaterials have shown great application prospects in the field of dentistry. Here we put forward some outlooks on their potential applications in the future: (1) nanomaterials can be utilized in the basic researches of stomatology because of their new and unique properties. For instance, cellular segregation may be realized by combination between target-proteins or molecules with specific antibodies labeled with nanomaterials; nano-DNA probe with high sensitivity and gene vector may be prepared using nanomaterials. (2) Nano-controlled release systems demonstrate unique advantages in local drug administration in root canals and periodontal tissues, such as longer retention, stronger penetrating and higher effective concentrations, etc. (3) Following procedural signal guidance, nanoactive anesthetic agent may show good effects after administration in exposed dentinal tubules, gingival tissue, and gingival sulcus. (4) Quantitative measurements of changes in the body can be acquired by application of nanosensors. (5) Nanomaterials may be used to prepare dental instruments with antibacterial effects. (6) It is also promising to improve therapeutic effects of Chinese herbs with addition of nanomaterials.

On the other hand, alongside the rapid advances in the development of nanotechnology-based materials, it has become imperative to elucidate the toxicity of NPs. Unfortunately, the safety evaluation systems of nanomaterials lag far behind their emerging development and applications. Until now, the understanding of nanomaterial neurotoxicology has been extremely limited. In-depth studies are warranted, particularly when considering the recent emphasis on the use of nanocarriers for drug delivery in the brain [98]. Although researchers have obtained some important information, the risks of NP exposure are not understood sufficiently well to enable the development of a science-based risk assessment. At present, flow cytometry, induced coupled plasma mass spectroscopy, confocal microscopy, the radioactive tracer technique, and transmission electron microscopy in combination with energy-dispersive X-ray spectroscopy are commonly used to study the cellular uptake of NPs [99,100]. However, there is not a single method that is satisfactory in obtaining precise information for all types of nanomaterials. Therefore, better testing and evaluation systems are urgently needed.

20.5 CONCLUSION

Nanomaterials have made major contributions to modern dentistry in various areas, including composite resin and bonding systems, coating materials for dental implants and dental restorations. The wide applications of these dental nanomaterials have created more exposure opportunities to these NPs in both dental staff and patients. Because the CNS may be a potential target organ of nanomaterials, it is essential to determine the neurotoxic effects of NPs. Although the impact of NPs on the CNS has received considerable attention in recent years, the data and findings obtained from the in vivo and in vitro studies are still limited. In the future, more potential applications of dental nanomaterials are supposed to be developed, as well as a better testing and evaluation system. In conclusion, more efforts are required to ensure the wide use of nanomaterials in the field of dentistry.

ACKNOWLEDGMENTS

We acknowledge the Elsevier and Informa Healthcare company for using some drawings taken from their journals.

REFERENCES

[1] E.A.J. Bleeker, W.H. de Jong, R.E. Geertsma, et al., Considerations on the EU definition of a nanomaterial: science to support policy making, Regul. Toxicol. Pharm. 65 (1) (2013) 119–125.

[2] S. Kasraei, L. Sami, S. Hendi, M.Y. Alikhani, L. Rezaei-Soufi, Z. Khamverdi, Antibacterial properties of composite resins incorporating silver and zinc oxide nanoparticles *on Streptococcus mutans* and *Lactobacillus*, Restor. Dent. Endod. 39 (2) (2014) 109–114.

[3] L.N. Niu, M. Fang, K. Jiao, et al., Tetrapod-like zinc oxide whisker enhancement of resin composite, J. Dent. Res. 89 (7) (2010) 746–750.

[4] K. Memarzadeh, A.S. Sharili, J. Huang, S.C. Rawlinson, R.P. Allaker, Nanoparticulate zinc oxide as a coating material for orthopedic and dental implants, J. Biomed. Mater. Res. A 103 (3) (2015) 981–989.

[5] M. Javidi, M. Zarei, N. Naghavi, M. Mortazavi, A.H. Nejat, Zinc oxide nano-particles as sealer in endodontics and its sealing ability, Contemp. Clin. Dent. 5 (1) (2014) 20–24.

[6] E. Frohlich, E. Roblegg, Models for oral uptake of nanoparticles in consumer products, Toxicology. 291 (1–3) (2012) 10–17.
[7] J.A. Keelan, Nanotoxicology: nanoparticles versus the placenta, Nat. Nanotechnol 6 (5) (2011) 263–264.
[8] M. Shimizu, H. Tainaka, T. Oba, K. Mizuo, M. Umezawa, K. Takeda, Maternal exposure to nanoparticulate titanium dioxide during the prenatal period alters gene expression related to brain development in the mouse, Part. Fibre Toxicol. 6 (2009) 20.
[9] S. Lu, R. Duffin, C. Poland, et al., Efficacy of simple short-term in vitro assays for predicting the potential of metal oxide nanoparticles to cause pulmonary inflammation, Environ. Health Perspect. 117 (2) (2009) 241–247.
[10] S.T. Stern, S.E. McNeil, Nanotechnology safety concerns revisited, Toxicol. Sci. 101 (1) (2008) 4–21.
[11] C. Medina, M.J. Santos-Martinez, A. Radomski, O.I. Corrigan, M.W. Radomski, Nanoparticles: pharmacological and toxicological significance, Br. J. Pharmacol. 150 (5) (2007) 552–558.
[12] A. Adamcakova-Dodd, L.V. Stebounova, J.S. Kim, et al., Toxicity assessment of zinc oxide nanoparticles using sub-acute and sub-chronic murine inhalation models, Part. Fibre Toxicol. 11 (2014) 15.
[13] L. Ma, J. Liu, N. Li, et al., Oxidative stress in the brain of mice caused by translocated nanoparticulate TiO_2 delivered to the abdominal cavity, Biomaterials 31 (1) (2010) 99–105.
[14] T.C. Long, N. Saleh, R.D. Tilton, G.V. Lowry, B. Veronesi, Titanium dioxide (P25) produces reactive oxygen species in immortalized brain microglia (BV2): implications for nanoparticle neurotoxicity, Environ. Sci. Technol. 40 (14) (2006) 4346–4352.
[15] J. Choi, Q.D. Zheng, H.E. Katz, T.R. Guilarte, Silica-based nanoparticle uptake and cellular response by primary microglia, Environ. Health Perspect. 118 (5) (2010) 589–595.
[16] S. Balos, B. Pilic, B. Petronijevic, D. Markovic, S. Mirkovic, I. Sarcev, Improving mechanical properties of flowable dental composite resin by adding silica nanoparticles, Vojnosanit. Pregl. 70 (5) (2013) 477–483.
[17] H.H. Xu, M.D. Weir, L. Sun, et al., Strong nanocomposites with Ca, PO(4), and F release for caries inhibition, J. Dent. Res. 89 (1) (2010) 19–28.
[18] Y. Xia, F. Zhang, H. Xie, N. Gu, Nanoparticle-reinforced resin-based dental composites, J. Dent. 36 (6) (2008) 450–455.
[19] W. Wang, X. Sun, L. Huang, et al., Structure–property relationships in hybrid dental nanocomposite resins containing monofunctional and multifunctional polyhedral oligomeric silsesquioxanes, Int. J. Nanomedicine 9 (2014) 841–852.
[20] M.A. Melo, L. Cheng, K. Zhang, M.D. Weir, L.K. Rodrigues, H.H. Xu, Novel dental adhesives containing nanoparticles of silver and amorphous calcium phosphate, J. Biomed. Mater. Res. B Appl. Biomater. 29 (2) (2013) 199–210.
[21] A. Wagner, R. Belli, C. Stotzel, A. Hilpert, F.A. Muller, U. Lohbauer, Biomimetically- and hydrothermally-grown HAp nanoparticles as reinforcing fillers for dental adhesives, J. Adhes. Dent. 15 (5) (2013) 413–422.
[22] L.V. Habekost, G.B. Camacho, G.S. Lima, F.A. Ogliari, G.B. Cubas, R.R. Moraes, Nanoparticle loading level and properties of experimental hybrid resin luting agents, J. Prosthodont. 21 (7) (2012) 540–545.

[23] E. Jallot, J.M. Nedelec, A.S. Grimault, et al., STEM and EDXS characterisation of physico-chemical reactions at the periphery of sol–gel derived Zn-substituted hydroxyapatites during interactions with biological fluids, Colloids Surf. B Biointerfaces 42 (3–4) (2005) 205–210.

[24] A. Krisanapiboon, B. Buranapanitkit, K. Oungbho, Biocompatability of hydroxyapatite composite as a local drug delivery system, J. Orthop. Surg. (Hong Kong) 14 (3) (2006) 315–318.

[25] X. Qi, H. Li, B. Qiao, et al., Development and characterization of an injectable cement of nano calcium-deficient hydroxyapatite/multi(amino acid) copolymer/calcium sulfate hemihydrate for bone repair, Int. J. Nanomedicine 8 (2013) 4441–4452.

[26] F.X. Huber, O. Belyaev, J. Hillmeier, et al., First histological observations on the incorporation of a novel nanocrystalline hydroxyapatite paste OSTIM in human cancellous bone, BMC Musculoskelet. Disord. 7 (2006) 50.

[27] C. Yang, J.S. Lee, U.W. Jung, Y.K. Seo, J.K. Park, S.H. Choi, Periodontal regeneration with nano-hyroxyapatite-coated silk scaffolds in dogs, J. Periodontal. Implant Sci. 43 (6) (2013) 315–322.

[28] S.H. An, T. Matsumoto, H. Miyajima, A. Nakahira, K.H. Kim, S. Imazato, Porous zirconia/hydroxyapatite scaffolds for bone reconstruction, Dent. Mater. 28 (12) (2012) 1221–1231.

[29] A.H. De Aza, J. Chevalier, G. Fantozzi, M. Schehl, R. Torrecillas, Crack growth resistance of alumina, zirconia and zirconia toughened alumina ceramics for joint prostheses, Biomaterials 23 (3) (2002) 937–945.

[30] M. Uno, M. Kurachi, N. Wakamatsu, Y. Doi, Effects of adding silver nanoparticles on the toughening of dental porcelain, J. Prosthet. Dent. 109 (4) (2013) 241–247.

[31] Y. Han, S. Kiat-amnuay, J.M. Powers, Y. Zhao, Effect of nano-oxide concentration on the mechanical properties of a maxillofacial silicone elastomer, J. Prosthet. Dent. 100 (6) (2008) 465–473.

[32] L.S. Acosta-Torres, I. Mendieta, R.E. Nunez-Anita, M. Cajero-Juarez, V.M. Castano, Cytocompatible antifungal acrylic resin containing silver nanoparticles for dentures, Int. J. Nanomedicine 7 (2012) 4777–4786.

[33] M. Karlsson, E. Palsgard, P.R. Wilshaw, L. Di Silvio, Initial in vitro interaction of osteoblasts with nano-porous alumina, Biomaterials 24 (18) (2003) 3039–3046.

[34] K. Pardun, L. Treccani, E. Volkmann, et al., Characterization of wet powder-sprayed zirconia/calcium phosphate coating for dental implants, Clin. Implant Dent. Relat. Res. 17 (1) (2015) 186–198.

[35] M. Uezono, K. Takakuda, M. Kikuchi, S. Suzuki, K. Moriyama, Hydroxyapatite/collagen nanocomposite-coated titanium rod for achieving rapid osseointegration onto bone surface, J. Biomed. Mater. Res. B Appl. Biomater. 101 (6) (2013) 1031–1038.

[36] G. Mendonca, D.B. Mendonca, F.J. Aragao, L.F. Cooper, Advancing dental implant surface technology—from micron- to nanotopography, Biomaterials 29 (28) (2008) 3822–3835.

[37] T. Lebold, C. Jung, J. Michaelis, C. Brauchle, Nanostructured silica materials as drug-delivery systems for doxorubicin: single molecule and cellular studies, Nano Lett. 9 (8) (2009) 2877–2883.

[38] L. Mu, S.S. Feng, A novel controlled release formulation for the anticancer drug paclitaxel (Taxol): PLGA nanoparticles containing vitamin E TPGS, J. Control Release 86 (1) (2003) 33–48.

[39] M.P. Melancon, W. Lu, M. Zhong, et al., Targeted multifunctional gold-based nanoshells for magnetic resonance-guided laser ablation of head and neck cancer, Biomaterials 32 (30) (2011) 7600–7608.

[40] L. Geraets, A.G. Oomen, J.D. Schroeter, V.A. Coleman, F.R. Cassee, Tissue distribution of inhaled micro- and nano-sized cerium oxide particles in rats: results from a 28-day exposure study, Toxicol. Sci. 127 (2) (2012) 463–473.

[41] Y. Chen, L. Liu, Modern methods for delivery of drugs across the blood–brain barrier, Adv. Drug Deliv. Rev. 64 (7) (2012) 640–665.

[42] A. Beduneau, P. Saulnier, J.P. Benoit, Active targeting of brain tumors using nanocarriers, Biomaterials 28 (33) (2007) 4947–4967.

[43] R. Spector, Nature and consequences of mammalian brain and CSF efflux transporters: four decades of progress, J. Neurochem. 112 (1) (2010) 13–23.

[44] Y.Y. Kao, T.J. Cheng, D.M. Yang, C.T. Wang, Y.M. Chiung, P.S. Liu, Demonstration of an olfactory bulb–brain translocation pathway for ZnO nanoparticles in rodent cells in vitro and in vivo, J. Mol. Neurosci. 48 (2) (2012) 464–471.

[45] S.R. Hwang, K. Kim, Nano-enabled delivery systems across the blood–brain barrier, Arch. Pharm. Res. 37 (1) (2014) 24–30.

[46] M. Caraglia, G. De Rosa, G. Salzano, et al., Nanotech revolution for the anti-cancer drug delivery through blood–brain-barrier, Curr. Cancer Drug Targets 12 (3) (2012) 186–196.

[47] E. Barbu, E. Molnar, J. Tsibouklis, D.C. Gorecki, The potential for nanoparticle-based drug delivery to the brain: overcoming the blood–brain barrier, Expert Opin. Drug Deliv. 6 (6) (2009) 553–565.

[48] K.K. Jain, Nanobiotechnology-based strategies for crossing the blood–brain barrier, Nanomedicine (Lond.) 7 (8) (2012) 1225–1233.

[49] R.E. Watson, J.M. Desesso, M.E. Hurtt, G.D. Cappon, Postnatal growth and morphological development of the brain: a species comparison, Birth Defects Res. B Dev. Reprod. Toxicol. 77 (5) (2006) 471–484.

[50] K. Yamashita, Y. Yoshioka, K. Higashisaka, et al., Silica and titanium dioxide nanoparticles cause pregnancy complications in mice, Nat. Nanotechnol. 6 (5) (2011) 321–328.

[51] Y. Okada, K. Tachibana, S. Yanagita, K. Takeda, Prenatal exposure to zinc oxide particles alters monoaminergic neurotransmitter levels in the brain of mouse offspring, J. Toxicol. Sci. 38 (3) (2013) 363–370.

[52] L.E. Yu, L.Y.L. Yung, C.N. Ong, et al., Translocation and effects of gold nanoparticles after inhalation exposure in rats, Nanotoxicology 1 (3) (2007) 235–242.

[53] A.J. De Lorenzo, The Olfactory Neuron and the Blood–Brain Barrier, CIBA Foundation Symposium Series J&A Churchill, London, 1970.

[54] S.V. Dhuria, L.R. Hanson, W.H. Frey 2nd, Intranasal delivery to the central nervous system: mechanisms and experimental considerations, J. Pharm. Sci. 99 (4) (2010) 1654–1673.

[55] G. Oberdorster, E. Oberdorster, J. Oberdorster, Nanotoxicology: an emerging discipline evolving from studies of ultrafine particles, Environ. Health Perspect. 113 (7) (2005) 823–839.

[56] A. Vilella, G. Tosi, A.M. Grabrucker, et al., Insight on the fate of CNS-targeted nanoparticles. Part I: Rab5-dependent cell-specific uptake and distribution, J. Control. Release 174 (2014) 195–201.

[57] S. Amara, I. Ben-Slama, I. Mrad, et al., Acute exposure to zinc oxide nanoparticles does not affect the cognitive capacity and neurotransmitters levels in adult rats, Nanotoxicology 8 (Suppl 1) (2014) 208–215.

[58] L.L. Zhang, R. Bai, B. Li, et al., Rutile TiO_2 particles exert size and surface coating dependent retention and lesions on the murine brain, Toxicol. Lett. 207 (1) (2011) 73–81.

[59] V. Kumar, A. Kumari, P. Guleria, S.K. Yadav, Evaluating the toxicity of selected types of nanochemicals, Rev. Environ. Contam. Toxicol. 215 (2012) 39–121.

[60] Y.F. Li, J.G. Li, J.L. Yin, et al., Systematic influence induced by 3 nm titanium dioxide following intratracheal instillation of mice, J. Nanosci. Nanotechnol. 10 (12) (2010) 8544–8549.

[61] Y. Ze, R. Hu, X. Wang, et al., Neurotoxicity and gene-expressed profle in brain-injured mice caused by exposure to titanium dioxide nanoparticles, J Biomed Mater Res A. 102 (2) (2014) 470–478.

[62] J.T. Kwon, G.B. Seo, Jo, et al., Aluminum nanoparticles induce ERK and p38MAPK activation in rat brain, Toxicol. Res. 29 (3) (2013) 181–185.

[63] R. Shrivastava, S. Raza, A. Yadav, P. Kushwaha, S.J.S. Flora, Effects of sub-acute exposure to TiO_2, ZnO and Al_2O_3 nanoparticles on oxidative stress and histological changes in mouse liver and brain, Drug Chem. Toxicol. 37 (3) (2014) 336–347.

[64] E. Huerta-Garcia, J.A. Perez-Arizti, S.G. Marquez-Ramirez, et al., Titanium dioxide nanoparticles induce strong oxidative stress and mitochondrial damage in glial cells, Free Radic. Biol. Med. 73 (2014) 84–94.

[65] M.P. Thomas, K. Chartrand, A. Reynolds, V. Vitvitsky, R. Banerjee, H.E. Gendelman, Ion channel blockade attenuates aggregated alpha synuclein induction of microglial reactive oxygen species: relevance for the pathogenesis of Parkinson's disease, J. Neurochem. 100 (2) (2007) 503–519.

[66] S. Wohlfart, S. Gelperina, J. Kreuter, Transport of drugs across the blood–brain barrier by nanoparticles, J. Control. Release. 161 (2) (2012) 264–273.

[67] C.M. Powers, A.S. Bale, A.D. Kraft, et al., Developmental neurotoxicity of engineered nanomaterials: identifying research needs to support human health risk assessment, Toxicol. Sci. 134 (2) (2013) 225–242.

[68] D. Ye, M.N. Raghnaill, M. Bramini, et al., Nanoparticle accumulation and transcytosis in brain endothelial cell layers, Nanoscale 5 (22) (2013) 11153–11165.

[69] C.K. Weiss, M.V. Kohnle, K. Landfester, et al., The first step into the brain: uptake of NIO-PBCA nanoparticles by endothelial cells in vitro and in vivo, and direct evidence for their blood–brain barrier permeation, ChemMedChem. 3 (9) (2008) 1395–1403.

[70] B.H. Kenzaoui, C. Chapuis Bernasconi, S. Guney-Ayra, L. Juillerat-Jeanneret, Induction of oxidative stress, lysosome activation and autophagy by nanoparticles in human brain-derived endothelial cells, Biochem. J. 441 (3) (2012) 813–821.

[71] E. Brun, M. Carriere, A. Mabondzo, In vitro evidence of dysregulation of blood–brain barrier function after acute and repeated/long-term exposure to TiO_2 nanoparticles, Biomaterials. 33 (3) (2012) 886–896.

[72] O.O. Ogunshola, In vitro modeling of the blood–brain barrier: simplicity versus complexity, Curr. Pharm. Design 17 (26) (2011) 2755–2761.

[73] D.T. Wiley, P. Webster, A. Gale, M.E. Davis, Transcytosis and brain uptake of transferrin-containing nanoparticles by tuning avidity to transferrin receptor, Proc. Natl. Acad. Sci. USA 110 (21) (2013) 8662–8667.

[74] M.N. Raghnaill, M. Bramini, D. Ye, et al., Paracrine signalling of inflammatory cytokines from an in vitro blood brain barrier model upon exposure to polymeric nanoparticles, Analyst. 139 (5) (2014) 923–930.

[75] M.R. Pickard, S.I. Jenkins, C.J. Koller, D.N. Furness, D.M. Chari, Magnetic nanoparticle labeling of astrocytes derived for neural transplantation, Tissue Eng. Part C Methods 17 (1) (2011) 89–99.

[76] J. Wu, J.A. Sun, Y. Xue, Involvement of JNK and P53 activation in G2/M cell cycle arrest and apoptosis induced by titanium dioxide nanoparticles in neuron cells, Toxicol. Lett. 199 (3) (2010) 269–276.

[77] S.C. Liu, L.J. Xu, T. Zhang, G.G. Ren, Z. Yang, Oxidative stress and apoptosis induced by nanosized titanium dioxide in PC12 cells, Toxicology 267 (1–3) (2010) 172–177.

[78] V. Valdiglesias, C. Costa, G. Kilic, et al., Neuronal cytotoxicity and genotoxicity induced by zinc oxide nanoparticles, Environ. Int. 55 (2013) 92–100.

[79] V. Valdiglesias, C. Costa, V. Sharma, et al., Comparative study on effects of two different types of titanium dioxide nanoparticles on human neuronal cells, Food Chem. Toxicol. 57 (2013) 352–361.

[80] H.M. Chiang, Q. Xia, X. Zou, et al., Nanoscale ZnO induces cytotoxicity and DNA damage in human cell lines and rat primary neuronal cells, J. Nanosci. Nanotechnol. 12 (3) (2012) 2126–2135.

[81] E. Zieminska, A. Stafiej, L. Struzynska, The role of the glutamatergic NMDA receptor in nanosilver-evoked neurotoxicity in primary cultures of cerebellar granule cells, Toxicology 315 (2014) 38–48.

[82] J. Zhao, L. Xu, T. Zhang, G. Ren, Z. Yang, Influences of nanoparticle zinc oxide on acutely isolated rat hippocampal CA3 pyramidal neurons, Neurotoxicology 30 (2) (2009) 220–230.

[83] Z. Liu, G. Ren, T. Zhang, Z. Yang, Action potential changes associated with the inhibitory effects on voltage-gated sodium current of hippocampal CA1 neurons by silver nanoparticles, Toxicology 264 (3) (2009) 179–184.

[84] A.M. Grabrucker, C.C. Garner, T.M. Boeckers, et al., Development of novel Zn^{2+} loaded nanoparticles designed for cell-type targeted drug release in CNS neurons: in vitro evidences, PLoS ONE. 6 (3) (2011) e17851.

[85] S. Krol, Challenges in drug delivery to the brain: nature is against us, J. Control. Release 164 (2) (2012) 145–155.

[86] L.J. Xu, J.X. Zhao, T. Zhang, G.G. Ren, Z. Yang, In vitro study on influence of nano particles of CuO on CA1 pyramidal neurons of rat hippocampus potassium currents, Environ. Toxicol. 24 (3) (2009) 211–217.

[87] L. Chen, Y. Miao, L. Chen, et al., The role of low levels of fullerene C60 nanocrystals on enhanced learning and memory of rats through persistent CaMKII activation, Biomaterials. 35 (34) (2014) 9269–9279.

[88] K. Brieger, S. Schiavone, F.J. Miller, K.H. Krause, Reactive oxygen species: from health to disease, Swiss Med. Wkly 142 (2012) w13659.

[89] A. Nel, T. Xia, L. Madler, N. Li, Toxic potential of materials at the nanolevel, Science 311 (5761) (2006) 622–627.

[90] C. Hanley, A. Thurber, C. Hanna, A. Punnoose, J.H. Zhang, D.G. Wingett, The influences of cell type and ZnO nanoparticle size on immune cell cytotoxicity and cytokine induction, Nanoscale Res. Lett. 4 (12) (2009) 1409–1420.

[91] M. Gualtieri, L. Capasso, A. D'Anna, M. Camatini, Organic nanoparticles from different fuel blends: in vitro toxicity and inflammatory potential, J. Appl. Toxicol. 34 (11) (2014) 1247–1255.

[92] Y. Xue, J. Wu, J. Sun, Four types of inorganic nanoparticles stimulate the inflammatory reaction in brain microglia and damage neurons in vitro, Toxicol. Lett. 214 (2) (2012) 91–98.

[93] R.A. Nixon, The role of autophagy in neurodegenerative disease, Nat. Med. 19 (8) (2013) 983–997.

[94] Z.H. Cheung, N.Y. Ip, Autophagy deregulation in neurodegenerative diseases—recent advances and future perspectives, J. Neurochem. 118 (3) (2011) 317–325.

[95] Y. Zhang, C.G. Yu, G.Y. Huang, C.L. Wang, L.P. Wen, Nano rare-earth oxides induced size-dependent vacuolization: an independent pathway from autophagy, Int. J. Nanomedicine 5 (2010) 601–609.

[96] B.H. Kenzaoui, C.C. Bernasconi, S. Guney-Ayra, L. Juillerat-Jeanneret, Induction of oxidative stress, lysosome activation and autophagy by nanoparticles in human brain-derived endothelial cells, Biochem. J. 441 (2012) 813–821.

[97] R. Roy, S.K. Singh, K.S. Chauhan, M. Das, A. Tripathi, P.D. Dwivedi, Zinc oxide nanoparticles induce apoptosis by enhancement of autophagy via PI3K/Akt/mTOR inhibition, Toxicol. Lett. 227 (1) (2014) 29–40.

[98] N.A. Monteiro-Riviere, A.O. Inman, L.W. Zhang, Limitations and relative utility of screening assays to assess engineered nanoparticle toxicity in a human cell line, Toxicol. Appl. Pharmacol. 234 (2) (2009) 222–235.

[99] H. Kettiger, A. Schipanski, P. Wick, J. Huwyler, Engineered nanomaterial uptake and tissue distribution: from cell to organism, Int. J. Nanomed. 8 (2013) 3255–3269.

[100] R.V. Raspopov, I.V. Gmoshinskii, K.I. Popov, O.V. Krasnoiarova, S.A. Khotimchenko, Methods of nanoparticles control in food and biological objects. Report 1. Use of microscopic and chromatography investigation methods, Vopr. Pitan. 81 (2) (2012) 4–11 [Russian].

Index

Note: Page numbers followed by "*f*" and "*t*" refer to figures and tables, respectively.

A

Abrading
  coefficient of drilling, 277
  with diamond powder, 262
Academisch Centrum Tandheelkunde Amsterdam method (ACTA method), 32–33
Acetate, 51
Acetic acid (CH_3COOH), 127–128
Acetylene, 256–257
Acid
  acid-etched implants, 115
  acidic HF mixtures, 127–128
  etching, 86, 141
  oxidation, 139–141
  treatments, 144*t*
Acrylic resin, 252–253
Acrylic tooth materials, 278, 278*f*, 281*f*
ACTA method. *See* Academisch Centrum Tandheelkunde Amsterdam method (ACTA method)
Adenocarcinoma, 358
Adhesion, MSCs, 89–90, 90*f*
Adhesives
  adhesively bonded composites, 24
  dental, 424*t*
    nanotechnology in, 30
Adper Single Bond 2, 30
Adult bone-marrow-derived mesenchymal stem cells, 211
Adverse effects of chemotherapy, 356–357
AES. *See* Auger electron spectroscopy (AES)
AFM. *See* Atomic force microscopy (AFM)
AFP. *See* Air-formed passive layer (AFP)
Agglomeration, 336–337
Agilent Technologies, Inc, 323
Air-fluid interface, 293–294
Air-formed passive layer (AFP), 105–106
Al NPs. *See* Aluminium nanoparticles (Al NPs)
ALG. *See* Sodium alginate (ALG)
Alginate film, 185–186
Alkaline phosphatase (ALP), 86, 150–151, 221–222
Alkylation agents, 49
All-trans retinoic acid (atRA), 363–365
Allotrope diamond, 260
ALP. *See* Alkaline phosphatase (ALP)
α-aminoisovaleric acid. *See* Valine
Alumina. *See* Aluminum (Al)
Aluminium nanoparticles (Al NPs), 430
Aluminum (Al), 66–67, 106–107. *See also* Titanium alloys (Ti6Al4V)
  oxide, 106–107
  surface modification, 106–109
Amelogenin, 245–246
American Dental Association acceptance guidelines, 35
Amino acids, 233–236
Amino-functionalized MWNTs-siRNA complexes, 380
3-Aminopropyltriethoxysilane (APTES), 106
Ammonium fluoride (NH_4F), 127–128
Amphiphiles, 236, 399–400
Amphiphilic peptides, 236–237, 237*f*
Anesthesia, 414–415
Angiogenesis, 356–357
  inhibitors, 356–357
Animalcules, 402–403
Anisotropy, 289–290
Annealing, 144*t*
Anodic oxidation. *See* Anodization
Anodic Tafel slope, 173
Anodization, 84–85, 105–106, 126
  anodized titanium substrates, 126–127
  titanium nanotubes, 126–127, 126*f*
  titanium surface modifications, 104–114, 142–143, 144*t*
Anti-EGFR antibodyconjugated nanorods, 363
Anti-inflammatory drugs, 127–128
Antibacterial/microbial drugs
  cancer therapy, 382–383
  carbon nanotubes, 382–383
  dental implants, 13
  functionalization, 129–132
  titanium dioxide nanotubes, 129–132
  titanium nanotubes, 13
  titanium surface modifications, 104–114
Antibacterial/microbial nanoparticles, 42–45
  filler phase modification, 43–44
  macromolecules, 45–47
  matrix phase modification, 44–45
Anticancer drug cisplatin, 383–384
Anticancer molecule carriers, 374–377. *See also* Cancer therapy
Anticarcinogenic drug delivery, 361–363
"Antifouling" surfaces, 112
Antifungal toxicity, 381

Antimicrobial agents, 42
nonreleased, 44
released, 43–44
Antimicrobial molecules, CNTs as carriers for, 381
Antineoplastic/cancer drug, 365–366
Apatitic composite scaffolds. *See* Nanoapatitic composite scaffolds
Approaches in nanotechnology, 2–3
APTES. *See* 3-Aminopropyltriethoxysilane (APTES)
Arginine (Arg), 242
Arginine–glycine–aspartic acid sequence (RGD sequence), 112, 130, 242–244
Artificial human teeth, 253
Artificial proteins, 236–237
Astrocyte, 434
Atomic force microscopy (AFM), 8, 32–33, 297–298, 298*f*, 322, 331–333, 403
AFM-controlled MWNT-based nanoinjector, 373, 374*f*
bottom-up approach in nanotechnology, 8–9
cantilever's deflection, 298–299
case studies, 332–333
imaging of microorganisms, 403–407
bacteria, 404–406
structure–function relationship of *S. mutans*, 407
viruses, 407
yeast, 403–404
Atomic level synthesis, 19
atRA. *See* All-trans retinoic acid (atRA)
Auger electron spectroscopy (AES), 343–344, 346–347, 347*f*
Auger process, 346
Autophagy, 436–437

B

Back-scattered electrons (BSE), 327–328
Bacteria, 402–406
Ball mills, 19
Barium (Ba), 66–67
BBB. *See* Blood–brain barrier (BBB)
BCP. *See* Biphasic calcium phosphate (BCP)
BEN. *See* Bias-enhanced nucleation (BEN)
Benzalkonium chloride, 44–45
Berkovich indenter, 320–321
Berkovich tip, 294, 348
β-helices, 242, 243*f*
β-sheet-forming peptides, 246
β-TCP. *See* β-Tricalcium phosphate (β-TCP)
β-tricalcium phosphate (β-TCP), 188, 190–191
Bias-enhanced nucleation (BEN), 263
effect on dental bur, 267–270
BIC. *See* Bone-implant contact (BIC)
BIC ratio. *See* Bone-to-implant-contact ratio (BIC ratio)
Binary alloys, 128
Binder matrix material, 255, 255*f*
Bio-inert implants, 212
Bioactive molecules, 150
BioAFM, 407–409
Bioceramics, 424*t*. *See also* Hydroxyapatite (HA)
and associated dental prosthesis, 426
Biochemical titanium surface modifications, 102*t*, 104–114
Biodegradable self-assembled protein, 236–237
Biofilms, 112–114, 407
formation, 112
and treatment, 413–414
Biological response modifier therapy (BRM therapy), 356–357
Biological therapy, 356–357
Biological tissues, 297
Biomaterials, 138
hydrogel suitability, 111–112
Biomedical applications of nanotechnology, 393–394
Biomimetic(s), 61–63
coating, 109
nanoapatite–collagen fiber scaffold, 215–217, 217*f*, 218*f*
Biomineralization method, 193
Bionanotechnology, 236
advantages and limitations of self-assembling peptide matrix scaffolds, 244
classification of self-assembly, 232–233, 234*f*, 235*f*
mechanism of molecular self-assembly, 232
peptide nanofibers, nanotubes, and nanowires, 236–241
self-assembly of proteins and peptides, 233–236
three-dimensional peptide matrix scaffolds, 242–244
Bioresorbable membranes, 184–185
Biosensors, 395
Biotherapy. *See* Immunotherapy
Biotin receptors, 376
Biotinylation, 240–241
Biphasic calcium phosphate (BCP), 86
Bisphenol A–glycidyl methacrylate (Bis-GMA), 60
Blades, dental burs, 253
Blood
blood–CSF barrier, 427
interaction mechanisms, 86–89
nanomagnets remove pathogens from, 400–401, 401*f*
plasma, 142–143

Blood–brain barrier (BBB), 398–399, 422, 427, 428*f*, 432*t*
nanoparticles and, 397–398
BMECs. *See* Brain microvascular endothelial cells (BMECs)
BMP-2. *See* Bone morphogenetic protein-2 (BMP-2)
BMSCs. *See* Bone-marrow derived mesenchymal stem cells (BMSCs)
Bode plots, 176
Bonding, 412
systems, 423
Bone, 253, 289–290
fractures, 211
grafts, 360
implants, 138
Bone marrow stem cells, 223
nanotopography effects on, 152–153
Bone morphogenetic protein-2 (BMP-2), 106, 109, 129–130, 138–139
Bone morphogenic protein-2, 138–139
Bone regeneration, 88–89, 213, 226
BMP-2, 106
via macroporosity, 212
Bone sialoprotein (BSP), 150
Bone tissue engineering
biomimetic nanoapatite–collagen fiber scaffold, 215–217, 217*f*, 218*f*
bone marrow stem cells, 223
calcium phosphate, 212, 215
cell infiltration into scaffold, 215
collagen fibers, 215, 217*f*
dental implants, 212
dicalcium phosphate anhydrous, 212
human bone marrow mesenchymal stem cells, 223, 224*f*
human umbilical cord stem cell interactions, 220–222, 222*f*, 223*f*
mesenchymal stem cells, 211, 223
nanoapatitic composite scaffolds, 13
development of nanoapatitic and macroporous scaffolds, 213–215
fatigue, 219–220, 220*f*
fracture, 217–219
seeding bone marrow stem cells, 223–226
repeated loadings, 219
umbilical cord stem cell interactions, 220–222, 222*f*, 223*f*
Bone-implant contact (BIC), 131–132
Bone-marrow-derived mesenchymal stem cells (BMSCs), 211
Bone-marrow-derived stromal stem cells. *See* Bone-marrow-derived mesenchymal stem cells (BMSCs)
Bone-repair materials, 424*t*
Bone-to-implant-contact ratio (BIC ratio), 99–100
Borosilicate glass, 273–276, 278–280, 278*f*, 280*f*, 281*f*, 282–283, 283*f*
Borosilicate glass, 278, 278*f*, 281*f*
Bottom-up approach, 2–3, 6–11, 10*f*, 11*f*, 12*f*, 20
nanoparticles synthesis, 19
Bragg's law, 338–340
Brain microvascular endothelial cells (BMECs), 434
Branched polymers, 48
Brittle materials, 19
BRM therapy. *See* Biological response modifier therapy (BRM therapy)
BSE. *See* Back-scattered electrons (BSE)
BSP. *See* Bone sialoprotein (BSP)
Budding yeast (*Saccharomyces cerevisiae*), 404
imaging, 404*f*
Bur performance investigations, 273–284
CVD diamond deposition on dental burs, 273
dental bur machining, 273–274
drilling experiments, 277–278
machining experiments on human teeth, 274–276
performance results, 278–284
performance testing, 276–277, 277*f*
tool preparation, 273

C

$Ca(PO_3)_2$ glass, 335
$Ca_4(PO_4)_2O$. *See* Tetracalcium phosphate (TTCP)
$CaHPO_4$. *See* Dicalcium phosphate anhydrous (DCPA)
Calcium chloride ($CaCl_2$), 292–293
Calcium fluoride nanoparticles-based composites, 28
Calcium phosphates (CaP), 85, 107–109, 212, 215
cements, 212
coatings, 86–87
nanoparticles-based composites, 28
particles, 115
surface modification, 106–109
Cancer
Cooker, 398*f*
photothermal therapy using CNTs, 382–383
Cancer therapy, 356–357
drug-delivery
colloidal gold nanoparticles, 361–363
gold nanoparticles, 361–363
management of oral cancer, 359–360
mechanism, 357–358
nanoparticles limitations, 366–367
nanoparticulate-based, 360–366

Cancer therapy (*Continued*)
oral cancer, 358
polymeric micelles, 366
TNM classification of tumors, 359
photothermal therapy, 382–383
Carbon dioxide laser treatments, 345–346
Carbon nanotubes (CNTs), 125–126, 167, 188, 330, 337–338, 371–373, 400
bottom-up approach in nanotechnology, 8–9
as carriers
for antimicrobial molecules, 381
of immunoactive molecules, proteins, and genetic materials, 378–381
for small and large drug molecules, 373–383
of small anticancer molecules, 374–377
cellular uptake, 373
functionalized, 381
incorporated into nanofibers, 188–190
internalization, 377*f*
for oral cancer therapy, 383–384
photothermal therapy of cancer, 382–383
surface-engineered, 381
Carboxyl group (–COOH), 109–111
Cardiac failure, 397
Cardiovascular interventions, 397
Cariogenic bacteria, 41–42
Carriers
anticancer molecule, 374–377
CNTs as
for antimicrobial molecules, 381
for drug molecules, 373–383
of immunoactive molecules, proteins, and genetic materials, 378–381
of small anticancer molecules, 374–377
drug, 395
Cathodic Tafel slope, 173
Cationic liposomes, 363–365
Cationic polymer disinfectants, 46
Cavities, 252–253
Cell adhesion, 149
Cell behaviors, 138
Cell culture on PLLA/MWCNTs/HA composite nanofibers membranes, 197–199, 198*f*
Cell culturing, 409–410
Cell differentiation, 141, 151
Cell infiltration into scaffold, 215
Cell morphology, 149
Cell nuclei, DNA chemical damage, 357–358
Cell proliferation, 149–150, 194–195, 383–384
Cell sheet technology, 186–187
Cell survival, 190, 382*f*
Cell viability and proliferation, 223–225
Cell-sheet-based technology for GTR, 186–187
Cellular processes, 151
Cellular responses, 143, 144*t*, 150
Cellular uptake, 373
Cellulose acetate membrane, 183–184
Cementum, 289–290
Central nervous system (CNS), 422
possible pathways into, 427–429
Ceramic(s). *See also* Hydroxyapatite (HA)
glass coatings, 166–167
nanoparticle-based drug delivery, 360–361
nanoparticles, 188
Cetuximab, 360, 363–365
Characterization. *See also* Nanocharacterization
conventional characterization techniques, 289–290
inelastic properties
stress–strain diagram, 313–314
yield stress, 315
nanoparticles, 50–51
of PLLA/MWCNTS/HA composite nanofibers, 195–196, 196*f*
Ti–6Al–4V characterization techniques, 169–173
Raman microscopy, 169–173, 171*f*
SEM, 169, 170*f*
time-dependent load response, 317–320, 318*f*
Chemical damage of DNA in cell nuclei, 357–358
Chemical grafting, 86
Chemical treatments, 139–141
Chemical vapor deposition (CVD), 4–5, 101–104, 256–257
bottom-up approach in nanotechnology, 8
diamond deposition on dental burs, 273
of diamond films onto dental burs, 256–272, 256*t*
controlling structure and morphology, 263–272
HFCVD, 259–263
plasma-enhanced CVD, 257–258
top-down approach in nanotechnology, 4–5
Chemisorption, 141
Chemotherapeutic drugs, 356–357
mechanism of action of chemotherapeutic agents, 357–358
chemical damage of DNA in cell nuclei, 357–358
prevention of synthesis of pre-DNA molecule building blocks, 357
Chemotherapy, 356–357, 360
anticancer molecule carriers, 374–377
cancer management plan, 359–360
cancer therapy, 356–357
carbon nanotubes, 125–126, 167, 188, 330, 371–373, 400
DNA building blocks synthesis prevention, 357

Chitosan (CHI), 23, 217–218, 377
lactate, 217–218
Cisplatin, 357–358, 372*f*, 374–375, 376*f*, 383–384
Classification
proteins and peptides self-assembly, 232–233, 234*f*, 235*f*
of tumors, 359
Clays, 393
Clinical applications
clinical experience/investigations, nanoparticle applications, 35
composite resins, 79
CNS. *See* Central nervous system (CNS)
CNT/PCL nanofibers, 189–190
CNTs. *See* Carbon nanotubes (CNTs)
Coated hydroxyapatite on Ti–6Al–4V by electrophoretic deposition, 167
Coating(s), 315–317, 393. *See also* Hydroxyapatite (HA)
biomimetic, 109
CaP, 86–87
collagen, 334
dip, 167
double-layer bioactive glass–ceramic, 168
materials for dental implants, 424*t*
mechanical properties, 315–317
microscale coating of hydroxyapatite/calcium phosphate/alumina, 106–109
SiO_2, 167
SiO_2–CaO, 166–167
sol–gel, 141, 167
TiO_2 nanoparticles, 169
tungsten carbide, 253
Cobalt (Co)–chromium (Cr) alloys, 138
Cobalt etching, 270, 271*f*
Collagen, 23, 191
coating, 334
fibers, 215, 217*f*
membranes, 184
Colloidal gold nanoparticles, 361–363
Colloidal suspension, 414–415
Colonization, 89–90
Commercial nanoindentation devices, 322–323
Compaction, 139, 140*t*, 144*t*
Complex 3D-shaped dental burs, 263–264
Composites, 60, 393
compressive strength, 70–71
diametral tensile strength, 71
flexural modulus, 72–73
flexural strength, 72–73
GTR membranes, 185
based on electrospun CNT/HA nanoparticles, 192–200
cell culture on PLLA/MWCNTs/HA composite nanofiber membranes, 197–199, 198*f*
characterization of PLLA/MWCNTS/HA composite nanofibers, 195–196, 196*f*
electrospun nanofibers, 194–195, 195*f*
fabrication of MWCNTS/HA hybrids, 192–193, 193*f*
fabrication of PLLA/MWCNTs/HA composite nanofibers, 195
in vivo implantation of PLLA/MWCNTs/HA membranes, 199–200, 199*f*
hydrolytic degradation, 68–69
leaching, 68–69
microhardness, 73–74
nanohardness, 74–75
resins, 423, 424*t*
fillers in, 65–66
solubility, 68–69
wear resistance, 75–79
Compressive strength, 70–71
Condensation, 20
Confocal microscopy, 333–335
case studies, 334–335, 334*f*
Conical indenters, 314
Cono-spherical tip, 294, 295*f*
Constant phase element (CPE), 176
Contact mode, 331–332
Contact stiffness, 290–291, 296–297
harmonic, 310–311, 311*f*
monotonic, 310–311
Continuous stiffness measurement (CSM), 310–313
Control issues
diamond films onto dental burs, 263–272
sol–gel synthesis, 21
Conventional characterization techniques, 289–290
Conventional diamond dental bur, 255
Conventional glass ionomer, 29
Conventional liposomes, 363–365
Coprecipitation method, 193
Cord stem cell interactions, 220–222, 222*f*, 223*f*
Corrosion resistance, 138, 142, 165–166
characterization techniques, 169–173
coated hydroxyapatite on Ti–6Al–4V by electrophoretic deposition, 167
corrosion resistance, 165–166
corrosion test with electrochemical techniques, 173–176
double-layer glass–ceramic coatings on Ti–6Al–4V, 168
nanoparticle, 168
nanostructured TiO_2 deposited on, 168–169

Corrosion resistance (*Continued*)
properties of titanium–aluminum–nitride, 344–345
SiO_2 and SiO_2–TiO_2 intermediate coatings, 167
SiO_2–CaO coatings on Ti–6Al–4V alloys, 166–167
Corrosion test with electrochemical techniques, 173–176
electrochemical impedance spectroscopy, 175–176
OCV, 174, 174*f*
Tafel analysis, 174, 174*f*
Cosmetics, 393
Counter ion, 51
Coupling agent, 24
CPC, 212
control, 221–222
CPC-based nanoapatite scaffold, 214*f*
CPC–chitosan–fiber, 219–220, 221*f*
powder, 216–217
CPE. *See* Constant phase element (CPE)
Crack propagation, resistance against, 320–321, 320*f*
Creep, 33–34
Cremophor EL, 375–376
Crosslinking degree, 50
Cryosurgery, 356–357
Crystalline
deposition, 139, 140*t*
glycylglycine bolaamphiphile peptide tubules, 241
CSM. *See* Continuous stiffness measurement (CSM)
CSM Instruments SA, 323
Cube corner tip, 291*f*, 294, 303
Cube-corner indenters, 320–321
Cutting tools, 393
CVD. *See* Chemical vapor deposition (CVD)
Cyclical loading of nanotube, 337–338

D

Data analysis protocol for nanoindentation, 290–292
DC. *See* Direct-current (DC)
DC plasma-enhanced CVD, 258
DCD. *See* Discrete crystalline deposition (DCD)
DCLS, 171–173
DCPA. *See* Dicalcium phosphate anhydrous (DCPA)
Demineralization, 292–293
Denoix, Pierre, 359
Dental adhesives, 424*t*
nanotechnology in, 30
Dental burs, 253–256
chemical vapor deposition of diamond films, 256–272
CVD diamond deposition, 273
with dimensional parameters showing bur head, 254*f*
geometry of typical tungsten carbide bur, 254*f*
machining, 273–276, 274*f*
effect of negative BEN, 267–270
Dental composite resin, 51
Dental fillers. *See* Filler(s)
Dental implants, 83
titanium oxide blasted and acid-etched, 115
Dental implants, 83, 212
blood interaction mechanisms, 87–89
coating materials for, 424*t*
confocal microscopy, 333–335
interaction mechanisms
blood, 86–89
hematopoietic stem cells, 86
MSCs, 86, 89–91
platelets, 86–88
interferometry, 335–336
measurement of composition of nanostructures, 342–347
measurement of mechanical properties of nanostructures, 347–349
measurement of nanostructure internal geometries
focused ion beams, 338
mercury porosimetry, 340–342
transmission electron microscope, 336–338
X-ray diffraction, 338–340
measurement of topology of nanostructures
atomic force microscope, 331–333
field emission scanning electron microscope, 327–329
scanning probe microscopy, 330–333
scanning tunneling microscope, 330–331
nanoscale surface modifications, 86–87
impact of nanotechnology, 86
MSCs, 89–91
nanoscale surface modifications, 86–87
surface dental implants interactions with blood, 87–89
surfaces interaction, 89–91
tissue integration, 91–93
nanotopography, 138–141
osseointegration, 83–84, 86–89
surface modifications, 86–87
Dental materials
nanoindentation, 289–290, 291*f*, 309–310
characterization of inelastic properties, 313–315

characterization of time-dependent load response, 317–320
data analysis protocol for, 290–292
devices for, 322–323
hydration, 292–294
indenter tips, 294–295
information from load–displacement curves, 309–313
load function and data analysis, 295–297
microstructural influence, 297–302
properties in nonhomogeneous bodies, 315–317
resistance against crack propagation, 320–321
sample preparation, 292
scratch tests for evaluation of friction and wear resistance, 322
nanoparticles, 23–30
calcium fluoride nanoparticles-based composites, 28
calcium phosphate nanoparticles-based composites, 28
clinical experience and investigations, 35
mechanical properties, 33–34
nanocomposites containing oxide nanoparticles, 23–27
nanotechnology in dental adhesives, 30
optical properties, 30–32
properties, 30–34
silica nanoclusters in Filtek Supreme, 25*f*
silsequioxane-based composites, 27–28
wear properties, 32–33
Dental nanomaterials, 423–427. *See also* Neurotoxicity of dental nanomaterials
bioceramics and associated dental prosthesis, 426
composite resins and bonding systems, 423
root-filling materials, 423–426
surface modifications for dental implants, 426
target-delivery and imaging in tumor chemotherapy, 426–427
Dental science, 412–413
Dental tissues, 411–412
Dentifrobots, 415
Dentin, 41–42, 289–290
matrix, 412
Dentistry, 73, 244–246, 314, 322–323. *See also* Operative dentistry nanotechnology
Denture-base materials, 424*t*
Deposition. *See also* Chemical vapor deposition (CVD); Electrophoretic deposition (EPD); Titanium dioxide (TiO_2)—deposition
calcium phosphate, 141
electrolytic deposition, 144*t*
electrophoretic deposition, 167
ion beam deposition, 139, 140*t*, 144*t*
nanoparticle synthesis, 19–23
physical vapor deposition, 4–5, 114
pulsed laser deposition, 144*t*
titanium surface modifications, 104–114, 142–143, 144*t*
Depth-sensing indentation, 290–291
Detector element, 396
Development
macroporous scaffolds, 213–215
nanoapatitic scaffolds, 213–215
Devices for nanoindentation, 322–323
Dexamethasone, 131–132
DH5a cells, 404–405, 405*f*, 406*f*
Diagnostic sensors, 394
Diagnostics, microbiology and dentistry, 13
Diametral tensile strength, 71
Diamond
diamond-coated dental bur, 282
diamond-coated WC–Co dental burs, 270
nucleation process, 262–263
Diamond deposition, substrate preparation effects on, 270–272, 272*f*
Diamond films
chemical vapor deposition, 256–272
controlling structure and morphology, 263–272
HFCVD, 259–263
plasma-enhanced CVD, 257–258
Dicalcium phosphate anhydrous (DCPA), 212
Differentiation, MSCs, 90–91
Digital Instruments Dimension 5000 SPM, 32–33
Dimethacrylated POSS, 27–28
Dip coating, 167
Dip-pen nanolithography (DPN), 8–9
Dipping, 141–142, 144*t*
Direct-current (DC), 257–258
Discrete crystalline deposition (DCD), 115
Disease prevention
cardiovascular interventions, 397
medical tools, 399
nanoparticles and BBB, 397–398
tissue reconstruction, 398–399
DMEM. *See* Dulbecco's modified Eagle's medium (DMEM)
DNA
building blocks, 357
chemical damage in cell nuclei, 357–358
prevention of synthesis of pre-DNA molecule building blocks, 357
Double-layer bioactive glass–ceramic coatings, 168
Double-layer glass–ceramic coatings, 168
Doxil, 371–373

Doxorubicin (DOX), 363–365, 377
DPN. *See* Dip-pen nanolithography (DPN)
Drilling
borosilicate glass, 279
experiments, 273–274, 274*f*
tools, 277
Drug carriers, 395
Drug delivery, 424*t*
Drug molecules, CNTs as carriers for, 373–383
Drug-delivery
cancer therapy, 356–357
colloidal gold nanoparticles, 361–363
gold nanoparticles, 361–363
management of oral cancer, 359–360
mechanism of action of chemotherapeutic agents, 357–358
nanoparticulate-based drug delivery in cancer treatment, 360–366
oral cancer, 358
TNM classification of tumors, 359
limitations of nanoparticles, 366–367
liposomes, 363–365
magnetic nanoparticles, 365–366
nanobots and uses, 394–395
nanorattles, 395
polymeric micelles, 366
Drug-filled nanocapsules, 400
DUH. *See* Dynamic Ultra Micro Hardness Testers (DUH)
Dulbecco's modified Eagle's medium (DMEM), 219–220
Dynamic elastic modulus, 310–311
Dynamic mode, 331–332
Dynamic self-assembly, 232–233, 235*f*
Dynamic SIMS, 345
Dynamic Ultra Micro Hardness Testers (DUH), 323
Dynamic view of dental tissues, 411–412

E

E4G10, 381
ECM. *See* Extracellular matrix (ECM)
Ectopic bone formation, 106, 109
EDS. *See* Energy dispersive spectroscopy (EDS)
EDX. *See* Energy dispersive X-ray spectroscopy (EDX)
EGF. *See* Epithelial growth factor (EGF)
EGF receptor (EGFR), 361–363, 383–384
EIS. *See* Electrochemical impedance spectroscopy (EIS)
Elan Pharmaceuticals, 395
Elastic modulus. *See* Young's modulus
Elastic-plastic
indentation, 310
materials, 308
response, 312
Electric motor, 256
Electrical equivalent circuits, 175*f*, 176
Electrochemical impedance spectroscopy (EIS), 173–176
Electrochemical polymerization, 142
Electrochemical techniques, corrosion test with, 173–176
electrochemical impedance spectroscopy, 175–176
OCV, 174, 174*f*
Tafel analysis, 174, 174*f*
Electrode, 168–169
preparation of Ti–6Al–4V electrode
Electrolytic deposition, 144*t*
Electromagnetic
energy, 343–344
radiation, 342–343
Electron(s), 327–328, 342–344
beam lithography, 5
confinement, 7, 8*f*
microscope, 322
Electronics industry, 4, 4*f*
Electrophoretic deposition (EPD), 167
coated hydroxyapatite on Ti–6Al–4V by, 167
Electrospinning, 187–188
use for preparation of nanocomposites, 187–191
carbon nanotubes incorporated into nanofibers, 188–190
organic–inorganic composite nanofibers, 190–191
Electrospun CNT/HA nanoparticles incorporated composite nanofibers
cell culture on PLLA/MWCNTs/HA composite nanofibers membranes, 197–199, 198*f*
characterization of PLLA/MWCNTS/HA composite nanofibers, 195–196, 196*f*
electrospun nanofibers with different fiber arrangements, 194–195, 195*f*
fabrication of MWCNTS/HA hybrids, 192–193, 193*f*
fabrication of PLLA/MWCNTs/HA composite nanofibers, 195
GTR membranes based on, 192–200
in vivo implantation of PLLA/MWCNTs/HA membranes, 199–200, 199*f*
Electrospun nanofibers with different fiber arrangements, 194–195, 195*f*
Elongation, 71, 139*t*
Embryonic stem cells, 211
Emission current, 269, 269*f*

Enamel, 41–42, 289–290, 315, 320–321, 412–413
Enameling technique, 168
Endothelial progenitor cells (EPCs), 151–152
  nanotopography effects on, 151–152
Energy dispersive spectroscopy (EDS), 66–67
  evaluations, 66–67
  nanofilled composites submitted to, 67*t*
Energy dispersive X-ray spectroscopy (EDX), 328–329, 342–343, 346
  case study, 343
Enterococcus faecalis (*E. faecalis*), 47
EPCs. *See* Endothelial progenitor cells (EPCs)
EPD. *See* Electrophoretic deposition (EPD)
Epidermal growth factor. *See* Epithelial growth factor (EGF)
Epithelial cells, 182
Epithelial growth factor (EGF), 361–363, 383–384
ePTFE membranes. *See* expanded polytetrafluoroethylene membranes (ePTFE membranes)
Erythroleukoplakia, 358
Erythroplakia, 358
Escherichia coli (*E. coli*), 47
  strains
    DH5a, 404–405, 405*f*, 406*f*
    OP50, 404–405, 406*f*, 407
Esthet-X
  compressive strength, 70–71, 71*f*
  diametral tensile strength, 72*f*
  filler concentration in percentage by weight, 67, 68*f*, 68*t*
  flexural strength/modulus, 72–73, 73*f*
  microhardness, 74, 74*f*
  morphological comparisons, 66, 66*f*
  nanohardness, 74–75, 75*f*
  solubility, 67–69, 70*f*
  water sorption, 67–69, 70*f*
Etching process, 270
Ethicon, 213, 218
expanded polytetrafluoroethylene membranes (ePTFE membranes), 183–184
Extracellular matrix (ECM), 151, 184
Extrafibrillar mineral, 412

F

FA receptors. *See* Folic acid receptors (FA receptors)
Fabrication
  of MWCNTS/HA hybrids, 192–193, 193*f*
  of PLLA/MWCNTs/HA composite nanofibers, 195
(1 1 1) Faceted diamond crystals, 271, 271*f*
F-actin, 142–143
FAK activity. *See* Focal adhesion kinase activity (FAK activity)
Fast fracture of nanoapatite scaffold, 217–219
Fatigue
  of nanoapatite scaffold, 219–220, 220*f*
  processes, 308–309
FBS. *See* Fetal bovine serum (FBS)
FC. *See* Fibrillar collagen (FC)
FDA. *See* US Food and Drug Administration (FDA)
Ferrofluids, 365–366
FESEMs. *See* Field emission scanning electron microscopes (FESEMs)
Fetal bovine serum (FBS), 219–220
Feynman, Dr. Richard Phillips, 1, 8–9, 10*f*
Fiber arrangements, electrospun nanofibers with, 194–195, 195*f*
Fibrillar collagen (FC), 105–106
Fibroblast growth factor, 138–139
Fibroblast growth factors, 138–139
Fibronectin (FN), 88–89, 104–105, 142–143, 334
FIBs. *See* Focused ion beams (FIBs)
Field emission cathode, 328
Field emission scanning electron microscopes (FESEMs), 327–329
  case studies, 328–329, 329*f*
Filament
  carburization process, 262
  characteristics, 261–262, 262*t*
Filler weight content (wt%), 67
  of nanofilled composites, 68*t*
Filler(s), 20
  in composite resins, 65–66
  modification, 44
  phase modification, 43–44
    nonreleased antibacterial agents, 44
    released antibacterial agents, 43–44
  wear resistance, 75–79
Fillipodia, 152–153
Filtek Supreme, 33–34, 66
  compressive strength, 70–71, 71*f*
  diametral tensile strength, 72*f*
  filler concentration in percentage by weight, 67, 68*f*, 68*t*
  flexural strength/modulus, 72–73, 73*f*
  microhardness, 73–74
  morphological comparisons, 66, 66*f*
  nanohardness, 74–75, 75*f*
  solubility, 67–69, 70*f*
  tensile strength, 71
  wear resistance, 75–79
Filtek Supreme Plus

Filtek Supreme Plus (*Continued*)
nanoclusters, 25–26
system, 25–26
Filtek Supreme translucent shades (FST), 31*f*
Filtek Supreme XT, 66, 66*f*
Fitting functions, 316
Flank wear, 282–283, 283*f*, 284*f*
Flat-end tip, 293*f*, 294
Flexural modulus, 72–73, 73*f*
Flexural strength, 72–73
Flu viruses, 409–410
Fluid-specimen interface, 293–294
Fluoride, 43–44, 115
Fluoride-releasing filler systems, 43–44
FN. *See* Fibronectin (FN)
Focal adhesion kinase activity (FAK activity), 151–153
Focused ion beams (FIBs), 338
case studies, 338, 339*f*
milling, 299–301
Folic acid receptors (FA receptors), 374
Foods and drinks act chemically on enamel, 315
Fourier-transform infrared spectroscopy (FTIR spectroscopy), 349–350
Fracture, 308–309
of brittle materials, 320–321
of nanoapatite scaffold, 217–219
toughness, 308–309, 321
Friction, scratch tests for evaluation of, 322
FriosBoneShields, 183–184
FST. *See* Filtek Supreme translucent shades (FST)
FTIR spectroscopy. *See* Fourier-transform infrared spectroscopy (FTIR spectroscopy)
Functionalization
antibacterial drugs, 129–132
antiinflammatory drugs, 129–132
osteogenic growth factors, 129–132
of TiO_2 NT, 129–132
Functionalized SWNTs-siRNA, 380
Futuristic applications, 14, 414

G

GBR. *See* Guide bone regeneration (GBR)
GECs. *See* Gingival epithelial cells (GECs)
Gefitinib, 363–365
Gel, 20
Gene silencing, 380
Gene therapy, 356–357, 363–365, 378–379
Genetic materials, CNTs as carriers of, 378–381
Gentamicin-loaded NTs (NT-G), 132
Geometries measurements, 336–342
Gingival epithelial cells (GECs), 185–186
GL261 murine intracranial glioma cancer model, 381
Glass, $Ca(PO_3)_2$ glass, 335
Glass ionomer systems
nanoparticles in, 28–29
Glass penetrometer, 341
Glass–ceramics, 166
Glioma, 381
Gloss retention of composites, 30–32, 34*f*
Glossectomy, 360
Gold
nanoparticles, 239–240, 366–367
for anticarcinogenic drug delivery, 361–363
principle of selective nanophotothermolysis of cancer cells, 362*f*
nanorods, 363
Gore-Tex, 183–184
Grafting PLA onto HA nanoparticles (HA–PLA), 191
Grain boundary diffusion, 345
Gram-negative bacteria, 46, 404
Gram-positive bacteria, 46
Grandio
clinical applications, 80*f*
compressive strength, 70–71, 71*f*
filler concentration in percentage by weight, 67, 68*f*, 68*t*
flexural strength/modulus, 72–73, 73*f*
microhardness, 73–74
morphological comparisons, 66, 66*f*
nanohardness, 74–75, 75*f*
Graphene sheets, 371–373
Grid-blasting
anodization, 86
technique, 109
Griffith theory, 19
Grinding, 19, 101
Growth factor(s)
epithelial growth factor, 361–363, 383–384
fibroblast growth factor, 138–139
MSCs, 89–90
osteogenic growth factors, 129–132
vascular endothelial growth factor, 142
Growth mechanisms, HFCVD, 259–261, 260*f*
Growth morphology, titanium nanotubes, 127
Growth periods, sol–gel synthesis, 21
GTR membranes. *See* Guided tissue regeneration membranes (GTR membranes)
Guide bone regeneration (GBR), 185
Guided tissue regeneration membranes (GTR membranes), 182
based on electrospun CNT/HA nanoparticles, 192–200
cell-sheet-based technology for, 186–187

electrospinning, 187–191
carbon nanotubes incorporated into nanofibers, 188–190
electrospinning, 187–188
organic–inorganic composite nanofibers, 190–191
layer-designed membranes, 185–186
periodontal defects and, 183–187
cell-sheet-based technology, 186–187
layer-designed membranes, 185–186
studies using bioresorbable membranes, 184–185
studies using nonresorbable membranes, 183–184

H

H1N1 flu viruses, 409–410
H5N1 flu viruses, 409–410
HA. *See* Hydroxyapatite (HA)
HA-βTCP. *See* Hydroxy apatite-β tricalcium phosphate (HA-βTCP)
Hand pieces, dental burs, 255–256
Hank's balanced salt solution (HBSS), 292–293
HA–PLA. *See* Grafting PLA onto HA nanoparticles (HA–PLA)
Hardness
composite resins, 73–74
and elastic modulus, 309–310
Hardness and elastic modulus ratio (H/E ratio), 321
Harmonic contact stiffness, 310–311, 311*f*
hBMSC. *See* human bone marrow mesenchymal stem cells (hBMSC)
HBSS. *See* Hank's balanced salt solution (HBSS)
HCPT. *See* Hydroxycamptothecin (HCPT)
H/E ratio. *See* Hardness and elastic modulus ratio (H/E ratio)
Head and neck squamous carcinoma cells (HNSCC), 383–384
Heads, dental burs, 253, 254*f*
HeLa cells, 378–379, 379*f*
Helical dipolar peptides, 239–240, 240*f*
Hematopoietic stem cells, 86
Hertz' formulae, 310
Hexafluoropropane (HFP), 191
Hexamethyldisilazane (HMDS), 342
Hexyl-poly(vinyl-*N*-pyridinium) surface coating, 46
HF. *See* Hydrofluoric acid (HF)
HFCVD. *See* Hot filament CVD (HFCVD)
HFP. *See* Hexafluoropropane (HFP)
High molecular weight polyacrylic acid, 23
High-energy Bremsstrahlung X-rays, 344
High-pressure, high-temperature (HPHT), 256–257
Higher-frequency process, 258
Histidines, 400
Historical review, 60–61
HMDS. *See* Hexamethyldisilazane (HMDS)
hMSCs. *See* Human mesenchymal stem cells (hMSC's)
HNSCC. *See* Head and neck squamous carcinoma cells (HNSCC)
Homogeneous material, 317
Homogenous passive polyacrylic acid films, 142
"Honeycomb" morphology, 346–347
Hook's law, 331–332
Hot filament CVD (HFCVD), 259–263, 259*t*
diamond nucleation process, 262–263
filament characteristics, 261–262
growth mechanisms, 259–261
HPHT. *See* High-pressure, high-temperature (HPHT)
HPI layer from *Deinococcus radiodurans*, 403, 403*f*
HPV. *See* Human Papilloma Virus (HPV)
hUCMSC. *See* Human umbilical cord mesenchymal stem cell (hUCMSC)
Human
bone, 215
enamel, 301–302, 301*f*
teeth, 278, 279*f*
machining experiments on, 274–276, 275*f*, 276*f*
human bone marrow mesenchymal stem cells (hBMSC), 223, 224*f*, 225*f*
Human mesenchymal stem cells (hMSCs), 142, 152–153
Human Papilloma Virus (HPV), 358
Human SHSY5Y neuroblastoma cell line, 434–435
Human umbilical cord mesenchymal stem cell (hUCMSC), 213–215
Human umbilical cord stem cell interactions, 220–222, 222*f*, 223*f*
Hybrid fillers, 30–32
Hybrid layers, 412–413
Hydration, 292–294
Hydrocarbons, 259–260
Hydrofluoric acid (HF), 127–128, 139–141
Hydrogel, 236–237
on titanium surface, 111–112
Hydrogen (H_2), 263–264
Hydrogen peroxide (H_2O_2), 139–141
Hydrolysis, 20
Hydrolytic degradation, 67–69
Hydrophobic drugs, 400

Hydroxy apatite-β tricalcium phosphate (HA-βTCP), 129–130
Hydroxyapatite (HA), 84–85, 106–107, 138–139, 166–167, 185, 188, 334
  surface modification, 106–109
Hydroxycamptothecin (HCPT), 366
Hydroxycarbonate apatite, 166–167
Hydroxyl group (–OH), 109–111
Hypersensitivity cure, 414–415
Hyperthermia, 356–357, 398–399
Hysitron, Inc., 323

I

Immune system, 356–357
Immunoactive molecules, CNTs as carriers of, 378–381
Immunoliposomes, 363–365
Immunotherapy, 381
Implants. *See* Dental implants
In vivo implantation of PLLA/MWCNTs/HA membranes, 199–200, 199*f*
In vivo toxicity tests, 53
Incorporation effects, 52–53
Indentation
  size effect, 317
  testing, 290–291
  toughness, 321
  work and information from *P–H* curves, 311–312
Indenter
  calibration, 312–313
  tips, 294–295
Induction period, 21
Inelastic properties, characterization of
  stress–strain diagram, 313–314
  yield stress, 315
Influenza virus particles, 407–409, 409*f*
Infrared laser (IR laser), 384
Injection molding, 252–253
Injured nerve treatment, 399
Inorganic materials, 138–139
  Ti surface modification, 139–141
Instrumented indentation. *See* Nanoindentation
Integrins, 151
Interaction mechanisms
  blood, 86–89
  hematopoietic stem cells, 86
  MSCs, 86, 89–91
  platelets, 86–88
Interferometers, 335
Interferometry, 335–336
  case studies, 335–336, 336*f*
Interferon-γ, 435–436
Interleukin-1β, 435–436
Intermetallic titanium compounds, 128
Intermittent mode, 331–332
Intermolecular self-assembly, 233
Internal geometries measurements, 336–342
Interproximal restorative procedures, 79
Intracellular parasites, 402–403, 407
Intrafibrillar mineral, 412
Intramolecular self-assembly, 233
Iodide, 51
Ion
  beam lithography, 5
  implantation method, 114
  ionic implantation, 84–86
  sputtering, 167
IR laser. *See* Infrared laser (IR laser)
Iron oxide nanoparticles, 365–366
Irregular macrofiller, SEM of, 65*f*
Irreversible processes, 315
Isterix, 138–139

K

Kentera, 382–383
KFE8 peptide complex, 242
  inner and outer β-helices, 243*f*
  molecular model, 242*f*
"King of nanomaterials". *See* Carbon nanotubes (CNTs)
Klebsiella pneumoniae (*K. pneumoniae*), 114
Knoop indenter, 73, 74*f*

L

"Lab-on-a-chip" techniques, 394
Label-free fashion, 410–411
Laryngectomy, 360
Laser ablation, 115
Laser micromachining, 115, 335
Laser spot deflection, 331–332
Laser therapy, 356–357
Lateral force microscopy (LFM), 11*f*
Layer-designed membranes for GTR, 185–186
Leaching, 43, 45, 68–69, 187
Leukoplakia, 358
LFM. *See* Lateral force microscopy (LFM)
Lifecore surface, 335
Light microscopy, 348, 403
Liposomal doxorubicin. *See* Doxil
Liposomes, 371–374
  different nanotechnology-based drug-delivery systems, 364*f*
  in oral cancer treatment, 363–365
Liquid helium cooling, 6–7
Lithography, 84–85

bottom-up approach in nanotechnology, 8, 10*f*, 11*f*
methods, 8
top-down approach in nanotechnology, 4–5
Load displacement curve, 293–294, 293*f*
Load function and data analysis, 295–297
Load response, dental material nanoindentation, 317–320
Load-bearing function of mineralized tissues, 297
Load–displacement curves, information from
hardness and elastic modulus, 309–310
harmonic contact stiffness, 310–311
indenter calibration, 312–313
work of indentation and information from *P–H* curves, 311–312
Load–displacement data, 299–301
Loadings, nanoapatite scaffold, 217–219
Local anesthesia, 414–415
Longboat anticancer system, 374–375, 376*f*
Longboat delivery system, 374–375
Lymphoma, 358, 378–379, 381
3M ESPE Filtek Supreme Plus, 25, 27*f*
3M ESPE Ketac Nano, 28–29
3M ESPE Scotch Bond SE adhesive, 30
3M ESPE Single Bond Plus, 30

M

Machining performance, 273–276, 278–284
Macrofiller, SEM of irregular, 65, 65*f*
Macromolecules
antibacterial/microbial nanoparticles, 45–47
Macrophages, 381
Macropores, 213
of CPC scaffold, cell infiltration into, 215, 216*f*
Macroporous scaffolds, development of, 213–215
Macroscopic scales, proteins and peptides, 232–233
Magnetic nanoparticles in oral cancer treatment, 365–366
Magnetron sputtering, 344–345
Maintenance and repair, periodontics, 415
Management plans, cancers, 359–360
Mandibulectomy, 359
Manufacturers, nanoindentation commercial devices, 323
Mastication, 32–33
Material degradation, 335
Matrix phase modification, 44–45
nonreleased antibacterial agents, 45
released antibacterial agents, 44–45
Matrix scaffolds, 242–244
Matrix-bound fluorides, 44–45
Maxillectomy, 359
Mean contact pressure, 310, 313–314, 314*f*
Measurements, nanostructures
composition, 342–347
internal geometries, 336–342
mechanical properties, 347–349
topology, 327–336
Mechanical attrition synthesis, 19–20
Mechanical behavior, 69–79
compressive strength, 70–71
diametral tensile strength, 71
flexural modulus, 72–73
flexural strength, 72–73
microhardness, 73–74
nanohardness, 74–75
wear resistance, 75–79
Mechanical characterization
dental implants, 13
mineralized tissues, 13, 289–290
oral cavity, 13, 289–290
Mechanical properties, 33–34, 307–308
coatings, 315–317
nanoindentation techniques
coatings, 315–317
multiphase microstructure, 317
properties in nonhomogeneous bodies, 315–317
surface layers, 315–317
Mechanical stability, 183
Mechanism, molecular self-assembly, 232
Medical tools, 399
Melanoma, 358, 383
Mercaptohexadecanoic acid, 8–9, 10*f*
Mercury porosimetry, 340–342
case studies, 341–342, 342*f*
Mesenchymal stem cells (MSCs), 86, 211, 223
adhesion, 89–90, 90*f*
differentiation, 90–91
interaction mechanisms, 89–91
migration, 89–90, 90*f*
origin of, 89
proliferation, 89–90, 90*f*
Mesoscopic scales, self-assembly of proteins and peptides, 232–233
messenger RNA, 106
Metabolic activity, 221
Metal alkoxides ($M(OR)_n$), 20
Metal implants, 83–84
Metallic salts (M_mX_n), 20
Metallic surfaces, 166
Metalloporphyrins, 241
Metals, 112–114, 141
Methacrylate nanocomposites, silsesquioxane-based nanocomposites, 27–28

Methacrylate nanocomposites, silsesquioxane-based nanocomposites (*Continued*)
3-methacryloxypropyltrimethoxysilane (MPTS), 24
12-methacryloyloxydodecylpyridinium bromide (MDPB), 44–45
Methane (CH_4), 263–264
Micro Materials Ltd, 323
Micro-fabrication, 8
Microbiology, 392–393
contribution to nanotechnology, 402–403
Microcantilevers, 290–291, 299–301, 300*f*
Microfills, 24
composites, 12–13, 413
Microhardness, 73–74
Microhybrids, 30–32
Microorganisms, 403–407
Micropillars, 290–291
Microscale coating of hydroxyapatite/calcium phosphate/alumina, 106–109
Microscale to nanoscale, titanium surface modifications, 13
Microscopy, 402–403. *See also* Atomic force microscopy (AFM); Scanning electron microscopy (SEM)
confocal microscopy, 333–335
lateral force microscopy, 11*f*
light microscopy, 348, 403
Raman microscopy, 169–173, 171*f*
Microscratch testing, 347–348
Microsphere analysis, focused ion beams, 338
Microstructural influence, mineralized tissue nanoindentation, 297–302
Microvascular endothelial cells of human brain, 434
Microwave plasma-enhanced CVD, 257–258
Migration, MSCs, 89–90, 90*f*
Milling process, 19–20
Millipore filter, 183–184
Mineralization, 242–244
Mineralized tissues, 289–290
hydration, 292–294
mechanical properties, 289–290
microstructural influence, 297–302
sample preparation, 292
Miniature cameras inside blood vessels, 399, 399*f*
Miniature compressed air, 256
Minneapolis (MN), 223–225
Mitoxantrone, 365–366
MN. *See* Minneapolis (MN)
Modification. *See also* Surface modification; Titanium surface modification techniques
filler phase modification, 43–44
titanium surfaces, 13
Molecular imaging, 363
Molecular level synthesis, 19
Molecular self-assembly, 232–233
Molecular switch peptides, 239–240
Molecular weight, 51
Molecule carriers, 374–377
Molybdenum, 259
Monochromatic aluminum Kα X-rays, 344
Monomethacrylate functionalized POSSs, 27–28
Monomethacrylated POSS, 27–28
"Monotonic" contact stiffness, 310–311
Morphology
controlling structure and, 263–272
effects of substrate preparation on diamond deposition, 270–272
effects of temperature, 264–267
effect of negative BEN on dental bur, 267–270
MSCs. *See* Mesenchymal stem cells (MSCs)
Multiphase microstructure, 317
Multiwalled carbon nanotubes (MWCNT), 188–189, 338, 371–373, 381–382
AFM-controlled MWNT-based nanoinjector, 373, 374*f*
amino-functionalized MWNTs-siRNA complexes, 380
bottom-up approach in nanotechnology, 7–8, 9*f*
fabrication MWCNTS/HA hybrids of, 192–193, 193*f*
perpendicular positioning, 375*f*
MWCNT. *See* Multiwalled carbon nanotubes (MWCNT)

N

N-alkylated poly(ethyleneimine), 47
N-alkylation, 49, 52
"Nano" wine glass, 5, 6*f*
Nanoapatitic composite scaffolds
biomimetic nanoapatite–collagen fiber scaffold, 215–217, 217*f*, 218*f*
bone marrow stem cells, 223
calcium phosphate, 212, 215
cell infiltration into scaffold, 215
collagen fibers, 215, 217*f*
dental implants, 212
development of nanoapatitic and macroporous scaffolds, 213–215
dicalcium phosphate anhydrous, 212
fast fracture of nanoapatite scaffold, 217–219
fatigue of nanoapatite scaffold, 219–220, 220*f*
fracture of nanoapatite scaffold, 217–219
hUCMSC, 220–222, 222*f*, 223*f*

human bone marrow mesenchymal stem cells, 223, 224*f*
mesenchymal stem cells, 211, 223
repeated loadings, 219
seeding bone marrow stem cells on nanoapatite scaffolds, 223–226
Nanobased indentation system, 348–349
Nanobiotechnology-based approaches, 360–361
Nanobots, 394–395, 394*f*, 397
diagnosis using, 396
preventing heart-attacks, 398*f*
Nanocapsules, 399–400
Nanocarbonated HA/collagen/PLGA (nCHAC/PLGA), 185–186
Nanocharacterization techniques
composition of nanostructures measurement, 342–347
confocal microscopy, 333–335
interferometry, 335–336
mechanical properties of nanostructures measurement, 347–349
nanostructure internal geometries measurement
focused ion beams, 338
mercury porosimetry, 340–342
transmission electron microscope, 336–338
X-ray diffraction, 338–340
topology of nanostructures measurement
atomic force microscope, 331–333
field emission scanning electron microscope, 327–329
scanning probe microscopy, 330–333
scanning tunneling microscope, 330–331
Nanoclusters, 25, 30–34, 63*f*
Nanocomposites, 12–13, 35
clinical experience and investigations, 35
electrospinning use for preparation of, 187–191
carbon nanotubes incorporated into nanofibers, 188–190
electrospinning, 187–188
organic–inorganic composite nanofibers, 190–191
mechanical properties, 33–34
optical properties, 30–32
oxide nanoparticles, 23–27
properties, 30–34
Nanocrystalline silver, 401
Nanodentistry, 11–13, 410–415
Nanodevices, 399
Nanodiagnostics, 395–397
biosensors, 395
diagnosis using nanobots, 396
quantum dots, 396
regenerative medicine, 397
Nanoelectromechanical systems, 125–126
Nanofabrication, 8
Nanofibers, 63
carbon nanotubes incorporating into, 188–190
electrospun, 194–195
organic-inorganic composite, 190–191
peptide, 236–241
synthetic, 187
Nanofill composites, 24–26, 27*f*
clinical experience and investigations, 35
commercial, 21–22
gloss retention, 30–32
resin materials, 60–61
Nanofillers, 14, 30, 53–54, 60–61, 63, 74–75, 188–189
Nanofluorapatite, 29
Nano-HA, 423–426
Nano-HAp. *See* Nanohydroxyapatite (Nano-HAp)
Nanohardness, 74–75, 348–349
Nanohybrid composites, 24, 26–27
Nanohydroxyapatite (Nano-HAp), 28–29, 423–426
Nanoindentation, 289–290, 291*f*, 309–310
characterization of inelastic properties, 313–315
characterization of time-dependent load response, 317–320
data analysis protocol for, 290–292
devices for, 322–323
information from load–displacement curves, 309–313
of oral mineralized tissues
hydration, 292–294
indenter tips, 294–295
load function and data analysis, 295–297
microstructural influence, 297–302
sample preparation, 292
properties in nonhomogeneous bodies, 315–317
resistance against crack propagation, 320–321
scratch tests for evaluation of friction and wear resistance, 322
Nanomagnets removing pathogens from blood, 400–401, 401*f*
Nanomanufacturing
bottom-up approach, 6–11, 10*f*, 11*f*, 12*f*
nanotechnology to, 3–11
top-down approach, 4–5, 4*f*, 5*f*, 6*f*
Nanomaterials, 18, 392–393
applications, 393, 421–422, 428–429, 435
in dental field, 424*t*
Nanomechanical characterization
basic data analysis protocol for nanoindentation, 290–292
indenter tips, 294–295
mineralized tissues, 289–290

Nanomechanical characterization (*Continued*)
nanoindentation of oral mineralized tissues, 292–302
oral cavity, 289–290
Nanomedicine, 53–54, 395
Nanomers, 24–26, 30–32
Nanometer-scale, 115, 125–126, 152–153
Nanoparticles (NPs), 18, 47–53, 337, 421–422, 428, 435
and BBB, 397–398
dental materials using, 23–30
calcium fluoride nanoparticles-based composites, 28
calcium phosphate nanoparticles-based composites, 28
clinical experience and investigations, 35
mechanical properties, 33–34
nanocomposites containing oxide nanoparticles, 23–27
nanotechnology in dental adhesives, 30
optical properties, 30–32
properties, 30–34
silica nanoclusters in Filtek Supreme, 25*f*
silsequioxane-based composites, 27–28
wear properties, 32–33
in glass ionomer systems, 28–29
polyethylenimine nanoparticles, 48–53
synthesis of, 19–23
mechanical attrition, 19–20
polymer-templated nanoparticles, 23
silsesquioxane nanoparticles, 22–23
sol–gel synthesis, 20–22
Ti surface modification, 139–141
Nanoparticulate-based drug delivery, 360–366
Nanopatches, 400
Nanopillars, 144*t*
Nanoplasmonic sensors detecting live viruses, 407–410
Nanorattles, 395
Nanorobotic dentifrice. *See* Dentifrobots
Nanorobots, 414–415
Nanorods, 363
Nanoscales
cantilevers tool, 360–361
coating of hydroxyapatite/calcium phosphate/alumina, 106–109
deposition of calcium phosphate, 141
self-assembly, 233
surface manufacturing techniques, 115
in three dimensions, 392–393
in two dimensions, 392–393
Nanoscience, 413
Nanoscratch testing, 347–348
Nanosized silver, 426
Nanosomes, 400
Nanospheres, 402
Nanostructured bioceramics, 426
Nanostructured TiO_2 deposited on Ti–6Al–4V, 168–169
nanoparticle, 168
preparation of Ti–6Al–4V electrode, 168–169
TiO_2 nanoparticles coating, 169
Nanostructures
composition measurement
AES, 346–347
EDS, 342–343
SIMS, 345–346
XPS, 343–345
mechanical properties measurement
nanohardness test, 348–349
nanoscratch testing, 347–348
nanostructure internal geometries measurement, 336–342
topology measurement, 327–336
"Nanotech". *See* Nanotechnology
Nanotechnology, 1–2, 18, 60–61, 84–85, 355, 392–393
AFM imaging of microorganisms, 403–407
applications
disease prevention, 397–399
in drug-delivery system, 394–395
nanocapsules, 399–400
nanocrystalline silver, 401
nanodiagnostics, 395–397
nanomagnets remove pathogens from blood, 400–401, 401*f*
nanosomes, 400
nanospheres, 402
nanotubes, 400
nanowires, 400
needle-free, painless vaccinations with nanopatches, 400
treatment of injured nerves, 399
approaches, 2–3
biomedical applications and limitations, 393–394
in dental adhesives, 30
impact, 411–414
biofilm formation and treatment, 413–414
bonding, 412
on dental implants, 86
dynamic view of dental tissues, 411–412
MSCs, 89–91
nanoscale surface modifications, 86–87
"small is beautiful" of dental science, 412–413
surface dental implants interactions with blood, 87–89
surfaces interaction, 89–91

tissue integration, 91–93, 92*f*
microbiologic contribution, 402–403
multidisciplinary nature, 3*f*
nanodentistry, 11–13, 410–415
to nanomanufacturing, 3–11
nanoplasmonic sensors detecting live viruses, 407–410
in periodontics, 414–415
dentifrobots, 415
local anesthesia and hypersensitivity cure, 414–415
natural tooth maintenance and repair, 415
NanoTite surface, 115
Nanotopography
bone marrow stem cells, 152–153
dental implants, 138
EPCs, 151–152
osteoblast responses, 143–151
bioactive molecules, 150
cell adhesion, 149
cell morphology, 149
cell proliferation, 149–150
cellular responses, 143*f*
osseointegration, 150–151
tissue responses, 139*t*
and protein absorption, 142–143
stem cell responses, 151–153
surface, 149–151
Ti surface modification, 138–142
bone marrow stem cells, 152–153
EPCs, 151–152
with inorganic materials/nanoparticles, 139–141
osteoblast responses, 143–151
polymers, 141–142
and protein absorption, 142–143
stem cell responses, 151–153
Nanotubes (NT), 125–127, 236–241, 400
alignment, 338
carbon, 188–190, 383–384
titanium, 126–127
Nanowires, 236–241, 400
Nano-ZnO. *See* Zinc oxide nanomaterials (Nano-ZnO)
Nano-ZrO_2. *See* Zirconium dioxide nanoparticles (Nano-ZrO_2)
Natural bone, 213
Natural human teeth, 253, 282
Natural tooth maintenance and repair, 415
nCHAC/PLGA. *See* Nanocarbonated HA/collagen/PLGA (nCHAC/PLGA)
N_{DEAD}. *See* Number of dead cells (N_{DEAD})
Near-infrared region (NIR), 363, 382
Neck
cancers, 360
of dental implants, 83, 115
Needle-free, painless vaccinations with nanopatches, 400
Negative BEN effect on dental bur, 267–270
Nerve treatments, 399
Neuroglial cells, 432*t*
Neuronal cell lines, 434–435
Neurons, 432*t*
Neurotoxicity of dental nanomaterials, 427–437. *See also* Dental nanomaterials
future research prospects, 437–438
possible pathways into CNS, 427–429
toxicity mechanisms, 435–437
in vitro toxicity studies, 430–435
in vivo toxicity studies, 429–430, 431*t*
with different cell models, 432*t*
Niobium, 139*t*
NIR. *See* Near-infrared region (NIR)
Nitrate, 51
N-methylation effect, 48–52
Nominal load, 319
Nondimensional weight function, 316
"Nonfouling" surfaces, 112
Nonhematopoietic cells expressing markers, 89
Nonhomogeneous bodies
multiphase microstructure, 317
properties in, 315–317
surface layers and coatings, 315–317
Nonhomogeneous materials, 308
Nonmonochromatic magnesium X-rays, 344
Nonreleased antibacterial agents, 44–45
Nonresorbable membranes, 183–184
Nonwoven nanofibrous membranes, 194
NPs. *See* Nanoparticles (NPs)
NT. *See* Nanotubes (NT)
NT-G. *See* Gentamicin-loaded NTs (NT-G)
Nucleation
density, 267–269, 268*f*
of diamond, 263–264
process, 261
sol–gel synthesis, 21
Number of dead cells (N_{DEAD}), 223–225
Nyquist plots, 175–176, 175*f*

O

Obligate intracellular parasites, 407
OC gene expression. *See* Osteocalcin (OC) gene expression
OCT. *See* Octenidine dihydrochloride (OCT)
OCT/PLGA-TNTs, 132
Octenidine dihydrochloride (OCT), 132
OCV. *See* Open-circuit voltage (OCV)

Oliver–Pharr method, 348
One dimension in nanoscale, 392–393
OP50 strain, 404–405, 406*f*, 407
Open-circuit voltage (OCV), 173–174, 174*f*
Operative dentistry nanotechnology, 60–61. *See also* Nanotechnology
  biomimetics, 61–63
  clinical application, 79–80
  filler weight content, 67
  fillers in composite resins, 65–66
  mechanical behavior, 69–79
  nanotechnology in CAD/CAM, 63–64
  SEM and EDS evaluations, 66–67
  water sorption, 67–69
Optical properties, 30–32
Optically active reagents, 363
Optofluidic-plasmonic biosensors, 410–411, 410*f*
Oral cancer therapy, 358
  CNTs for, 383–384
  liposomes in oral cancer treatment, 363–365
  magnetic nanoparticles in oral cancer treatment, 365–366
  management, 359–360
  oral cancerous lesions, 358
Oral cavity, 289–290
  biofilm, 54
  mineralized tissues, 289–290
Oral components, 413–414
Oral mineralized tissues, 289–290
  nanoindentation of, 292–302
Oral pathogens, 47
Order of diffraction, 338–340
Organic nanoscale SAMs, 109–111
Organic nanoscale self-assembled monolayers, 104, 109–111
Organic treatments, 332–333
Organic-coated superparamagnetic iron oxide nanoparticles, 365–366
Organic–inorganic composite nanofibers, 190–191
Origin of MSCs, 89
Osseointegration, 83, 91–92, 138, 150–151
OsseoSpeed implant surface, 115
Osteoblast responses, 127, 143–151
Osteocalcin (OC) gene expression, 199–200, 221–222
Osteoclast cells, 86–87
Osteoconductive coating, 106
Osteogenic growth factor carriers, 13, 129–130
Osteoinductive biomolecular cues, 104–106
Oxide nanoparticles
  nanocomposites, 23–27
  sol–gel synthesis, 22
Oxide-layer-coating approach, 176

P

Paclitaxel, 375–377
Painless vaccinations, 400
PAMAM dendrimers, 377
PAN. *See* Polyacrylonitrile (PAN)
Parasites, 402–403, 407
Particles
  fracture, 19
  effect of particle size, 51
  sol–gel synthesis, 20
Pathogen removal from blood, 401, 401*f*
Pathogenic microorganisms, 42
PCD. *See* Polycrystalline diamond (PCD)
PCL nanofibers. *See* Polycaprolactone nanofibers (PCL nanofibers)
PDL. *See* Periodontal ligament (PDL)
PDL-derived cells (PDLCs), 184, 197, 198*f*, 199
PEG. *See* Poly(ethylene glycol) (PEG)
PEG-poly[gamma-benzyl-L-glutamate] (PEGPBLG), 366
PEGylated paclitaxel formulations, 375–376
PEI. *See* Polyethyleneimine (PEI)
Penetration, nanoindentation, 290–291, 294–295, 309–310
Penicillin, 131–132
Penicillin/streptomycin (PS), 219–220
Peptides, 400
  matrix scaffolds
    advantages and limitations of self-assembling, 244
    three-dimensional, 242–244
  nanofibers, 236–241
  self-assembly of peptides and peptides, 233–236
Percentage of live cells (P_{LIVE}), 223–225
Performance investigations, dental burs, 273–284
Peri-implant bone formation, 131–132
Peri-implantitis, 112, 113*f*, 131–132
Periodontal defects and GTR, 183–187
  cell-sheet-based technology for GTR, 186–187
  layer-designed membranes for GTR, 185–186
  studies using bioresorbable membranes, 184–185
  studies using nonresorbable membranes, 183–184
Periodontal ligament (PDL), 182
Periodontal structures, 182
Periodontal therapy, 182
Periodontics
  local anesthesia and hypersensitivity cure, 414–415
  nanorobotic dentifrice, 415
  nanotechnology in, 414–415

natural tooth maintenance and repair, 415
Periodontium, 182
Peripheral blood stem cell transplantation, 356–357
Peroxidation, 139–141
PGE2. *See* Prostaglandin E2 (PGE2)
P–H curves, indentation work and information from, 311–312
Phase angle, 310–311
Phosphate ($-PO_4H_2$), 109–111
Phospholipid nanosomes, 400
Photodynamic therapy, 356–357, 384
Photofunctionalization, 115
Photolithography, 141
top-down approach in nanotechnology, 4–5
use, 141
Photomasks, 111–112
Photopolymerization, 111–112
Photosensitizer, 356–357
Photothermal therapy
of cancer using CNTs, 382–383
relationship between cell survival and CNTs length using, 382*f*
Physical data, hot filament chemical vapor deposition, 261, 262*t*
Physical titanium surface modifications, 114
Physical vapor deposition (PVD), 102*t*, 114
Physicochemical titanium surface modifications, 101–104, 102*t*
PI3K/mTOR/Akt signaling cascade, 436–437
PLA. *See* Polylactic acid (PLA)
Plaque, 41–44, 112, 253
Plasma blood interactions, 88–89
Plasma etching, 4–5
Plasma spray technique, 106–109
Plasma-enhanced CVD, 257–258
DC, 258
microwave, 258
RF, 258
Plasma-sprayed HA coated dental implants, 86–87
Plasmid DNA, 378–379
Plastic deformations, 308, 310, 317, 319–320
Plasticity index, 311–312
Platelet-rich plasma (PRP), 88–89
Platelets, 86–88
PLD technique. *See* Pulsed laser deposition technique (PLD technique)
Pleiotropic manufacturing effect, 115
PLGA. *See* Poly(lactic-co-glycolic acid) (PLGA)
P_{LIVE}. *See* Percentage of live cells (P_{LIVE})
PLLA. *See* Poly (L-lactic acid) (PLLA)
Pointed indenters, 313, 318–319
Polarization resistance, 173
Polishing, 102*t*
Poly (L-lactic acid) (PLLA), 185–186, 192
PLLA/MWCNT/HA composite nanofibers membranes, 192
cell culture on, 197–199, 198*f*
characterization of, 195–196, 196*f*
fabrication of, 195
in vivo implantation of PLLA/MWCNTs/HA membranes, 199–200, 199*f*
Poly (L-lactide). *See* Poly (L-lactic acid) (PLLA)
Poly methyl methacrylate, 292
Poly(ethylene glycol) (PEG), 111–112, 242–244, 363–365, 375–376
Poly(L-lactic acid), 138–139
Poly(lactic-co-glycolic acid) (PLGA), 132, 138–139, 184
Poly(methyl methacrylate)-grafted nanoclays, 23
Polyacrylonitrile (PAN), 189
Polycaprolactone nanofibers (PCL nanofibers), 187
Polycationic
antimicrobials, 45
disinfectants, 46
Polycrystalline diamond (PCD), 256–257, 280*f*
Polyethyleneimine (PEI), 46–47, 46*f*
nanoparticles, 48–53
characterization, 50–51
incorporation of, 52–53
synthesis, 48–50
Polyglactin fibers, 213–215, 218–220
Polyhedral oligomeric silsesquioxane (POSS), 22
Polylactic acid (PLA), 184–185
Polymer chains, 22, 111–112
Polymer grafting, 191
Polymeric antimicrobials, 45
Polymeric micelles, 366
Polymeric nanocapsules, 400
Polymers, 138–139
bone replacement implants, 141–142
dental implants, 141–142
polymer-templated nanoparticles, 23
Ti surface modification, 141–142
Porcelain, 252–253
Porous ePTFE membrane, 183–184
Porous materials, 340–341
Porous ZrO_2/HAp composite scaffolds, 423–426
Porphyrins, 241
POSS. *See* Polyhedral oligomeric silsesquioxane (POSS)
Potentiodynamic polarization, 173
Pre-DNA molecule building blocks, prevention of synthesis of, 357
Preparation
diamond deposition, 273
titanium alloy (Ti6Al4V) electrodes, 267
Primary electrons, 327–328

Primary stability, 83–84
Pristine CNTs, 371–373
Proliferation, MSCs, 89–90, 90*f*
Properties
  dental materials, 30–34
  nanofill composites, 24–26
  nanohybrid composites, 26–27
  nanoparticle applications, dental materials, 30–34
Prophylactic treatments, 359
Prostaglandin E2 (PGE2), 150
Protein(s), 400
  CNTs as carriers of, 378–381
  conformational diseases, 242–244
  nanotopography protein absorption, 142–143
  self-assembly, 233–236
Protons, 342–343
PRP. *See* Platelet-rich plasma (PRP)
PS. *See* Penicillin/streptomycin (PS)
Pseudomonas aeruginosa, 52–53
Pulsed laser deposition technique (PLD technique), 107–109, 167
Pure Ti, 139*t*
Pure titanium, 99–100, 148*t*
PVD. *See* Physical vapor deposition (PVD)

Q

Qdot luminescence. *See* Quantum dot luminescence (Qdot luminescence)
QPEI. *See* Quaternary ammonium PEI (QPEI)
Quantitative real-time reverse transcription polymerase chain reaction (qRT-PCR), 221–222
Quantum corrals, 7, 8*f*
Quantum dot luminescence (Qdot luminescence), 383–384
Quantum dots, 360–361, 396, 397*f*
Quasi-static test mode, 290–291
Quaternary ammonium PEI (QPEI), 46–47
  particles, 48–52, 49*f*

R

Radiation therapy, 356–357, 359–360
Radical neck dissection, 360
Radio frequency plasma treatments, 84–85
Radio-frequency (RF), 257–258
  plasma-enhanced CVD, 258
Raman microscopy, 169–173, 171*f*
Raman microspectroscopy, 173
Raman spectra, 266, 266*f*
Raman spectroscopy, 267, 349–350
Ras homolog gene family member A (RhoA), 194–195
Rattle-type nanoparticles, 395
Reactive nitrogen species (RNS), 434
Reactive oxygen species (ROS), 430, 435
Real-time reverse transcription polymerase chain reaction (RT-PCR), 221–222
Recombinant human bone morphogenetic protein-2 (rhBMP-2), 130
Regenerative biology, self-assembly in, 244–246
Regenerative medicine, 244–245, 397. *See also* Tissue
Relaxation, 308
Released antibacterial agents, 43–45
Repair, periodontics, 415
Repeated loadings, 219
Resin composites. *See* Composites
Resin matrices, 423
Resin-modified glass ionomers (RMGIs), 28–29
Resin–tissue interfaces, 292
Resistance against crack propagation, 320–321, 320*f*
Respiratory syndrome, 409–410
Restorative composite
  antibacterial/microbial nanoparticles, 42–45
  filler phase modification, 43–44
  macromolecules, 45–47
  materials, 41–42
  matrix phase modification, 44–45
Restorative dentistry, 341–342
RF. *See* Radio-frequency (RF)
RGD sequence. *See* Arginine–glycine–aspartic acid sequence (RGD sequence)
RGD–cysteine peptide sequence (RGDC peptide sequence), 106
rhBMP-2. *See* Recombinant human bone morphogenetic protein-2 (rhBMP-2)
RhoA. *See* Ras homolog gene family member A (RhoA)
RMGIs. *See* Resin-modified glass ionomers (RMGIs)
RNA, 357–358
RNS. *See* Reactive nitrogen species (RNS)
Robots, 14, 395, 415
Room temperature (RT), 170–171, 172*f*
Root-canal fillings, 424*t*
Root-filling materials, 423–426
ROS. *See* Reactive oxygen species (ROS)
Roughness, 86, 149–150, 336
RT. *See* Room temperature (RT)
RT-PCR. *See* Real-time reverse transcription polymerase chain reaction (RT-PCR)

S

Saccharomyces cerevisiae.. *See* Budding yeast (*Saccharomyces cerevisiae*)

Saliva, 413–414
SAM. *See* Scanning Auger microscopy (SAM)
SAMP. *See* Self-assembled monolayer of phosphonate molecules (SAMP)
Sample preparation, 292
SAMs. *See* Self-assembled monolayers (SAMs)
Sandblasting and acid-etching method (SLA method), 141
SBF. *See* Simulated body fluid (SBF)
Scaffolds. *See also* Nanoapatitic composite scaffolds
  development of, 212
  macroporous scaffolds, 213–215
  three-dimensional peptide matrix scaffolds, 242–244
Scanning Auger microscopy (SAM), 346
Scanning electron microscopy (SEM), 52, 169, 170*f*, 213, 237, 238*f*, 264, 266*f*, 268*f*, 270*f*, 282*f*, 327–328, 342–343
  of cured nanomer, 29, 29*f*
  evaluations, 66–67
  of hUCMSC attachment, 222*f*
  of irregular macrofiller, 65*f*
  nanoclusters, 63*f*
  wear resistance, 75–79
Scanning force microscopy (SFM). *See* Atomic force microscopy (AFM)
Scanning probe microscopy (SPM), 330–333
  atomic force microscope, 331–333
  scanning tunneling microscope, 330–331
Scanning tunneling microscopy (STM), 6–7, 330–331
SCC. *See* Squamous cell carcinoma (SCC)
Schottky in-lens thermal FESEM electron gun, 328
Scratch tests, 322
Secondary electrons, 327–328
Secondary ion mass spectroscopy (SIMS), 345–346
  case studies, 345–346
Seeding
  bone marrow stem cells on nanoapatite scaffolds, 223–226
  with diamond powder, 262
Selective neck dissection, 359
Self-assembled monolayer of phosphonate molecules (SAMP), 109–111
Self-assembled monolayers (SAMs), 104, 233
  organic nanoscale, 109–111
Self-assembly, 231–232
  bionanotechnology, 236
  classification, 232–233, 234*f*, 235*f*
  mechanism, 232
  nanotubes, 236–241, 239*f*, 240*f*, 241*f*
  nanowires, 236–241
  peptide matrix scaffolds, 244
  peptide nanofibers, 236–241, 239*f*, 241*f*
  proteins and peptides, 233–236
  regenerative biology and dentistry, 244–246
  three-dimensional peptide matrix scaffolds, 242–244
Self-cleaning window, 393
"Self-polishing" effect, 35
SEM. *See* Scanning electron microscopy (SEM)
Serine, 242–244
Shank, dental burs, 253
Shimadzu Corporation, 323
Si NPs. *See* Silica-based NPs (Si NPs)
Silane molecule, 106
Silanization technique, 106–107
Silica, 22
Silica nanorattle, 395, 396*f*
Silica-based filler particles, 42
Silica-based NPs (Si NPs), 435–436
Silicates (Si), 66–67
Silicon
  atom manipulation, 23
  features size evolution in silicon chips, 4*f*
  substrate, 112
Silicon dioxide (SiO_2), 60–61, 166
  coatings on titanium and Ti–6Al–4V alloy, 167
Silicone-elastomer material, 424*t*
Silicones (R_2SiO), 22
Silsequioxane-based composites, 27–28
Silsesquioxane, 22
Silsesquioxane nanoparticles, 22–23
Silver, 43, 402
  nanoparticles, 43
  silver-based antimicrobials, 114
Silver-coated titanium surface (TiAg), 114
SIMS. *See* Secondary ion mass spectroscopy (SIMS)
Simulated body fluid (SBF), 105–106, 126–127, 191
Single-walled carbon nanotubes (SWCNTs), 188–189
Single-walled nanotubes (SWNT), 125–126, 371–373, 372*f*, 384
  bottom-up approach in nanotechnology, 7–8, 10*f*
  functionalized, 378–379, 382–383
  HeLa cells, 379*f*, 380*f*
  SWNT-Qdot-EGF bioconjugates, 383–384
  SWNTs-DOX preparation, 378*f*
Sintering, 255
SiO_2–CaO, 166
  coatings on Ti–6Al–4V alloys, 166–167
SiO_2–TiO_2, 166

SiO_2–TiO_2 (*Continued*)
intermediate coatings on titanium and Ti–6Al–4V alloy, 167
siRNA. *See* Small interfering RNA (siRNA)
SLA method. *See* Sandblasting and acid-etching method (SLA method)
Slender, 255–256
Small anticancer molecules, CNTs as carriers of, 374–377
Small interfering RNA (siRNA), 380
SOCS1. *See* Suppressors of cytokine signaling 1 (SOCS1)
Sodium alginate (ALG), 377
Sodium fluoride (NaF), 127–128
Sodium hydroxide (NaOH), 129–130
treatments, 139–141
Sol, 20
Sol–gel
coating method, 141, 167
nanoparticle synthesis, 20–22
process, 167
SiO_2–CaO coatings, 166–167
surface modification of Ti implants, 139–141, 140*t*, 144*t*
synthesis, 20–22
technology, 166–167
Solubility, 68–69, 70*f*
Soluble antimicrobial, 44–45
Solvent-casting technique, 132
Sonication, 144*t*
Special sensor nanobots, 395–396
Spectroscopy
Auger electron spectroscopy, 343–344, 346–347
electrochemical impedance spectroscopy, 173, 175–176, 175*f*
energy dispersive spectroscopy, 66–67, 342–343
Fourier-transform infrared spectroscopy, 350
Raman spectroscopy, 267, 349–350
secondary ion mass spectroscopy, 345–346
X-ray photoelectron spectroscopy, 343–345
Speed, dental burs, 255–256
Spherical indenter, 312, 312*f*
SPM. *See* Scanning probe microscopy (SPM)
Squamous cell carcinoma (SCC), 358, 384
Stainless steel, 138, 254–255
dental burs, 253–255
Staphylococcus aureus (*S. aureus*), 52–53, 114
Staphylococcus epidermidis(*S. epidermidis*), 52–53, 114
Static self-assembly, 232–233, 234*f*
Static SIMS, 345
Steel wears, 254–255
Stem cell delivery, nanoapatitic composite scaffolds for
biomimetic nanoapatite–collagen fiber scaffold, 215–217, 217*f*, 218*f*
bone marrow stem cells, 223
calcium phosphate, 212, 215
cell infiltration into scaffold, 215
collagen fibers, 215, 217*f*
dental implants, 212
development, 213–215
dicalcium phosphate anhydrous, 212
fast fracture of nanoapatite scaffold, 217–219
fatigue of nanoapatite scaffold, 219–220, 220*f*
fracture of nanoapatite scaffold, 217–219
human bone marrow mesenchymal stem cells, 223, 224*f*
human umbilical cord stem cell interactions, 220–222, 222*f*, 223*f*
mesenchymal stem cells, 211, 223
repeated loadings, 219
seeding bone marrow stem cells on nanoapatite scaffolds, 223–226
Stem cell responses, nanotopography, 151–153
nanotopography effects on EPCs, 151–152
Stents, 5, 6*f*
Sterilization process, 254–255
STM. *See* Scanning tunneling microscopy (STM)
Strain, nanoindentation, 295–296, 303, 308, 313–314
Strength, nanoindentation, 308
Streptavidin, 380–381
Streptococcus mutans (*S. mutans*), 41–42, 44, 51–52, 407
structure–function relationship, 407
Streptomycin, 131–132
Stress, nanoindentation, 308
Stress intensity factor, 308–309, 320–321
Stress–strain
curves, 308, 312
diagram, 313–314
Strontium fluoride (SrF_2), 43–44
Structure–activity relationship analysis, 47
Structure–function relationship of *S. mutans*, 407
Substrate preparation effects, 270–272, 272*f*
Substrate temperature, 264, 265*f*, 267*f*
Sulfonic acid ($-SO_3H$), 109–111
Sunscreens, 393
Suppressors of cytokine signaling 1 (SOCS1), 380
Supramolecular
micelle formation, 233
self-assembly, 237
Surface chemical analysis, 51
Surface coating, 144*t*
Surface engineering

bur performance investigations, 273–284
CVD of diamond films onto dental burs, 256–272
dental burs, 253–256
tooth materials, 251–253
Surface interactions
impacts of nanotechnology, 86–89
mesenchymal stem cells, 86, 89–91
Surface layers, 315–317
Surface modification. *See also* Titanium surface modification techniques
for dental implants, 426
methods, 138–139
Surface roughness, 149–150, 335
Surface tension, 340–341
Surface-binding peptides, 237, 239*f*
Surfaces, nanomaterial microbiology and nanodiagnostics, 393
Surfactant peptides, 240–241
Surfactant-like peptides, 237, 239*f*
Surgical excision, 356–358
Surgical flaps, 360
Surgical treatments, 356–357
Survival rates, cancer management, 360
SWCNTs. *See* Single-walled carbon nanotubes (SWCNTs)
SWNT. *See* Single-walled nanotubes (SWNT)
Synthesis of nanoparticles, 19–23
mechanical attrition, 19–20
polymer-templated nanoparticles, 23
silsesquioxane nanoparticles, 22–23
sol–gel synthesis, 20–22
Synthetic and naturally derived GTR barriers, 183
Synthetic nanofibers, 187

T

Tafel analysis, 173–174, 174*f*
Taniguchi, Norio, 60–61
Tantalum, 259, 261
Tapping mode, 331–332
Target-delivery and imaging in tumor chemotherapy, 426–427
Targeted cancer therapy, 356–357
Taxol, 375–376
TCPS. *See* Tissue culture polystyrene (TCPS)
Teeth, 320–321
TEM. *See* Transmission electron microscopy (TEM)
Temperature, diamond films onto dental burs, 264–267, 264*f*, 265*f*, 267*f*
Tensile strength, 71, 138, 139*t*
Teratoma, 358
Tests
corrosion test, 173–176
dental burs, 276–277
electrochemical tests, 173–176
scratch tests, 322, 347–348
Tetracalcium phosphate (TTCP), 212
5,10,15,20-Tetrakis(meso-hydroxyphenyl) porphyrin, 366
TGA. *See* Thermogravimetric analysis (TGA)
TGF-β. *See* Transforming growth factor-beta (TGF-β)
Thermal therapy. *See* Hyperthermia
Thermodynamic near-equilibrium, 259–260
Thermogravimetric analysis (TGA), 67, 68*f*
Thermogravimetry, 67, 68*f*, 68*t*
Thermotherapy. *See* Hyperthermia
Three-dimensional peptide matrix scaffolds, 242–244
Three-electron process, 346
Ti6Al4V. *See* Titanium alloys (Ti6Al4V)
Time of Flight-SIMS (ToF-SIMS), 345–346
Time-dependent load response, characterization of, 317–320, 318*f*
TiN matrix. *See* Titanium nitride matrix (TiN matrix)
TiOblast implants, 115
Tirapazamine, 363–365
Tissue
integration, 91–93, 92*f*
reconstruction, 398–399
Tissue culture polystyrene (TCPS), 215
Titania nanoparticles, 29
Titania NTs (TNTs), 132
Titanium (Ti), 99–100, 109, 138
dental implants, 252–253
SiO_2 and SiO_2–TiO_2 intermediate coatings on, 167
surface modification
with inorganic materials/nanoparticles, 139–141
nanotopography, 138–142
polymers, 141–142
Titanium alloys (Ti6Al4V), 13, 138, 139*t*, 142, 165–166, 242–244
characterization techniques, 169–173
Raman microscopy, 169–173, 171*f*
SEM, 169, 170*f*
corrosion resistance, 165–166
corrosion test with electrochemical techniques, 173–176
double-layer glass–ceramic coatings, 168
by electrophoretic deposition, coated hydroxyapatite, 167
nanoparticle, 168

Titanium alloys (Ti6Al4V) (*Continued*)
nanostructured TiO_2 deposition, 168–169
SiO_2 and SiO_2–TiO_2 intermediate coatings, 167
SiO_2–CaO coatings, 166–167
Ti–6Al–7Nb, 138, 139*t*
Ti–13Nb–13Zr, 139*t*
Titanium dioxide (TiO_2), 126, 138–139
antibacterial drugs, 129–132
antiinflammatory drugs, 129–132
deposition
characterization techniques, 169–173
nanoparticle, 168
Raman microscopy, 169–173, 171*f*
SEM, 169, 170*f*
functionalization, 129–132
nanoparticles, 166
nanoparticles coating, 169
nanotube arrays, 86
NPs, 430
NT for implant fabrication, 127–128, 128*f*
osteogenic growth factors, 129–132
Raman spectra, 171–173
recent advancements, 132
Titanium nanotubes, 126–127
anodization experiments, 126*f*
antibacterial drugs, 129–132
antiinflammatory drugs, 129–132
functionalization, 129–132
osteogenic growth factors, 129–132
recent advancements, 132
TiO_2 NT for implant fabrication, 127–128
Titanium nitride matrix (TiN matrix), 344–345
Titanium oxide, 112–114
blasted implants, 115
layers, 100–101
Titanium surface modification techniques, 99–114
biochemical modification, 104–114
antibacterial titanium surfaces, 112–114
hydrogels on titanium surface, 111–112
microscale and nanoscale coating of hydroxyapatite/calcium phosphate/alumina, 106–109
organic nanoscale SAMs, 109–111
osteoinductive biomolecular cues, 104–106
DCD, 115
laser ablation, 115
mechanical modification, 101, 102*t*
photofunctionalization, 115
physical modification, 114
physicochemical modification, 101–104
titanium oxide blasted and acid-etched implants, 115
TMV. *See* Tobacco mosaic virus (TMV)
TNF. *See* Tumor necrosis factor (TNF)
TNM classification of tumors, 359, 359*t*
TNTs. *See* Titania NTs (TNTs)
Tobacco mosaic virus (TMV), 403, 407–409, 408*f*
ToF-SIMS. *See* Time of Flight-SIMS (ToF-SIMS)
Tooth, 253
materials, 251–253, 252*f*
"Top-down" approach, 2–5
demonstration of three-dimensional nanostructure fabrication, 5*f*
examples of stents coated with diamond-like carbon, 6*f*
features size evolution in silicon chips, 4*f*
nanoparticles synthesis, 19
process sequence in electronics industry, 5*f*
"Top-down" manufacturing process, 19
Topology, 327–336. *See also* Nanotopography
Toxicity, 42
Traction force, 294
Transducer, 396
Transforming growth factor-beta (TGF-β), 89–90
Transmembrane integrins, 152–153
Transmission electron microscopy (TEM), 25, 328–329, 336–338, 377
case studies, 337–338, 337*f*
Triethylene glycol dimethacrylate, 42
TTCP. *See* Tetracalcium phosphate (TTCP)
Tumor necrosis factor (TNF), 361–363
TNF-α, 150, 435–436
Tumor(s)
imaging, 424*t*
target-delivery and imaging in tumor chemotherapy, 426–427
TNM classification of, 359, 359*t*
Tungsten, 259, 261
Tungsten carbide, 255
coating, 253
dental burs, 253–255, 254*f*, 273
2D island nanotopography, 152–153

U

UHV. *See* Ultrahigh vacuum (UHV)
Ulcer, 358
Ultimate tensile strength, 138, 139*t*
Ultra Micro-Indentation System (UMIS), 348–349
Ultrahigh vacuum (UHV), 330–331
Ultrasonic measurements, 299–301, 303
Ultraviolet (UV), 393
light, 115
rays, 393
treatment of dental implant surfaces, 115
Umis (IBIS), 323
UMIS. *See* Ultra Micro-Indentation System (UMIS)

Uncoated WC–Co dental bur, 279, 280*f*
Universal rheologic models, 318
Urethane dimethacrylate, 42
US Food and Drug Administration (FDA), 212
US National Nanotechnology Initiative, 18
UV. *See* Ultraviolet (UV)

V

V6D amino acid sequence (VVVVVVD), 239–240
V6D peptide
  nanovesicle and nanotube, 239–240, 240*f*
  quick-freeze/deep-etch transmission electron micrographs, 239–240, 241*f*
Vaccinations, 400
Valine, 239–240
Valve metal. *See* Titanium (Ti)
Vanadium (V), 99–100, 112–114, 138, 139*t*. *See also* Titanium alloys (Ti6Al4V)
Vancomycin-modified titanium surface, 114
Vascular endothelial growth factor (VEGF), 142
Venus, 70*f*, 74
Vickers indenter, 320–321
Videomicroscope, 322–323
Vinculin, 152–153
Virus capsids, 392–393
Viruses, 407
Viscoelastic
  effects, 294, 296–297, 318
  materials, 295–296, 296*f*, 308, 309*f*
  processes, 320
Viscoelasticity, nanoindentation, 289–290, 296–297, 308, 317–318
Vitamin E TPGS, 426–427
Vitronectin, 88–89, 104–105, 142–143
Vroman effect, 88–89

W

Washburn equation, 340–341
Water absorption, 67–68
Water solubility (WSL), 69
Water sorption (WS), 67–69
Water-cooling device, 282
Water-soluble pyridinium-type polycations, 46
Wavelength dispersive X-ray spectroscopy (WDS), 343
WC–Co dental burs, 271
WDS. *See* Wavelength dispersive X-ray spectroscopy (WDS)
Wear
  properties, 32–33
  resistance, 75–79, 322
Weight functions, 316
Wet etching, 4–5
WS. *See* Water sorption (WS)
WSL. *See* Water solubility (WSL)
Wst-1 assay, 221

X

X-ray diffraction (XRD), 338–340
  case studies, 340, 340*f*
X-ray photoelectron spectroscopy (XPS), 343–345
  case study, 344–345
X-rays, 342–343
  diffractometer, 328–329
  lithography, 5
  photoemission spectrometer, 328–329

Y

Yeast, 403–404
Yield strength, 308, 313
Yield stress, 315
Young modulus (*E*), 292
Young's modulus, 308, 316–318, 316*f*, 320
Ytterbium trifluoride (YbF_3), 43–44

Z

Zinc oxide (ZnO), 112–114
Zinc oxide nanomaterials (Nano-ZnO), 421–422
Zircon, 138, 139*t*
Zirconia nanoparticles, 21–22
Zirconium dioxide nanoparticles (Nano-ZrO_2), 43
Zirconium/silica-based filler particles, 42

Kolibri 360 GmbH
Ferdinand-Jühlke-Straße 7
99095 Erfurt, Deutschland
produktsicherheit@kolibri360.de